SEVEN PLANTS TO SAVE THE WORLD

SEVEN PLANTS TO SAVE THE WORLD

The Rise of the Weeds and an End to the Diseases of Civilization

Karl Elliot-Gough

AEON

First published in 2025 by
Aeon Books

British Library Cataloguing in Publication Data

A C.I.P. for this book is available from the British Library

ISBN-13: 978-1-80152-162-8

Typeset by Medlar Publishing Solutions Pvt Ltd, India

www.aeonbooks.co.uk

Please note:

Every effort has been made to ensure that the information in the book is accurate. The information in this book may not be applicable in each individual case so it is advised that professional medical advice is obtained for specific health matters and before changing any medication or dosage. The author accepts no legal responsibility for any personal injury or other damage or loss arising from the use of the information in this book. Errors may have crept in, due diligence and repeated fact-checking is still fallible. If you find any errors please contact the author or publisher so any further editions can be corrected. The author is a caveman after all, what do you expect!

Thanks:

Cover watercolour by the wonderful Kari McGowan (www.karimcgowan.com), edit and polish by the amazing Matthew Seal, whilst Paul Kellett's assistance was valued enough to get this individual mention. And, of course, everyone I spoke with, visited, emailed, who offered advice, encouragement, or just a cup of herbal tea, it was all greatly appreciated, cheers guys. Last but far from least, Oliver Rathbone and all at Aeon Books for resonating with the plants and putting this book before you now, respects … Positive Harmonious Vibrations.

Dedicated to our beautiful companion Bella, without whose walks this book would never be … we all thank you for everything … Namaste … x

CONTENTS

INTRODUCTION xiii

Human Plants xiv
The diseases of civilization xv
Oikonomia xvi
Disconnection xvii

NETTLE 1

Humble pie 1
It's a family affair 2
Nettle history, ethnobotany and folklore 5
"It's got more chi, man!" 9
A whole heap of benefits 11
Side effects 12
Tea terrorists and vultures 13
Equipoise (ek-wuh-poiz) 15
"Who's the daddy?" 18
A bit more ChAT 19
Phenomenal phenols 21
The first epidemiological study 21
Nettles for textiles 24
What's wrong with cotton? 28
The nettle industrial revolution 33
Get foraging 38
Recipes 39
Nettle neglect 41

DANDELION 43

The best things in life are free 43
Etymologicum 45
What's in a name? 46
Chronological cappers 49
When folklore was the law 51
Dawn of the lawn 52
The Big C 53
Duelling with the diuretics 57
Warning! 58
Nutrients 59
Phreaky phytonutrients 60
The inconvenient truth 61
More than an opinion 63
Taraxagum 65
Dandelions are proper cool 68
Foraging/Harvesting 70
Culinary dandelions 70
Final word 72

YARROW 75

Armageddon 75
Them Greeks, they knew a thing or two 78
History and names 78
Ethnobotanical 81
Traditional 83
"It's all about the phytonutrients, darling!" 85
Axis of evil leading the way (again) 86
Here's more countries' opinions 88
OTC cash registers on fire 91
Aedes albopictus 92
Radiation 95
Divination 99
"Need ya aura cleansing or chakras aligned, Mister?" 101
Anything else? 102
It can't all be good, can it? 104
Foraging, growing and culinary curios 105
Summation 108

PLANTAIN 109

Canine crisis 109
The people will not revolt. They will not look up from their screens long enough to notice … 111
Species specific 113
Names 115
First Nation peoples loved them 116
History 117
Traditional uses 120
Firm scientific credentials 121
Forbidden knowledge 127
Psyllium (*silly om*) 130
Farmers need to become the guardians of the soil 131

CANNABIS 133

Human Plant numero uno? 133
Mary Jane, what's in the name? 134
"Pssst! … you wanna buy sum smoke?" 135
Hamp or hemp 136
Cannabis history 137
BC 138
AD 142
Into the twentieth and twenty-first centuries 148
Building a better world—Hempcrete 153
Fabric 157
Paper production 160
The decorticator 162
Plastic, bioplastic and oil 164
Hempire 168
The seed—health 169
CBD 173
Some science 176
Side effects 180
Smacked out or stoned? 181
Wrapping it up 184

HAWTHORN 187

Barbs of wisdom 187
Hawthorn (*Crataegus* sp.) 188
Botanical 188
Folk and ethnographic 190
Bubbling up from the depths of time 192
Jesus Christ and the soul of London 194
Vibrations, Green Men, fairies and pixilated superstitions 197
Traditional uses 201

Heart disease 203
Global science 209
Fishy science with an industrial whiff 213
Eating 214
Pleaching and a smeuse 215
Enclosure 218
It's not 'just' the land that got stolen/reappropriated 223

BLACKBERRY 227

Botanically and plantality 227
Challenging etymology 229
So little folklore, so much Devil's piss 230
History, ethnobotanic and traditional uses 232
Any problems? 236
The biggest threat to the future of all Gaia's organisms 237
Last scientific studies to digest 240
Can't get away from those polyphenols 245
The B17 story 247
Foraging and Shinrin-Yoku 250
Blackberry cultivation 254
Yum! 255

SOLUTIONAL SUMMATION 259
Systemic reformation 261
1. Health system 261

An end to the diseases of civilization 261
Medicine 263
2. Economic system 265
What is money? 265
Crypto or Green Currency? 270
Show us the money 271

A Green-backed economy 276
3. Industrial system 276
Building 277
Fabrics 277
Transport 278
Petroleum 280
4. Agricultural system 283

The great agricultural myth 283
PermaLife 1:0 288
A nation is only as healthy as the soil it stands upon 289
Permaculture 296
Ten acres with planning permission 298

How to get the land … the peasants are revolting 301
Forest charter, allotments and PCCs 305
Land prices and subsidies 308
Pharmas to farmers 311
5. Cultural system 312
Obscenity law 312
The Co-operator 314
Wetiko and follow the money 315
Socialism? 317
Take a bow, the Seven Plants 318

APPENDIX 321
1. Nutritional profiles 321
2. Pesticide sprayed since 1950 326
3. Pollution 328
4. GDPs vs corporation revenues 329
5. Humans 332
6. Historical pandemics 333

BIBLIOGRAPHY 337

INDEX 371

INTRODUCTION

OPEN YOUR EYES

To shop, and school, to work and play,
The busy people pass all day;
They hurry, hurry, to and fro,
And hardly notice as they go
The wayside flowers, known so well,
Whose names so few of them can tell.

O, if these people understood
What's to be found by field and wood;
What fairy secrets are made plain
By any footpath, road, or lane—
They'd go with open eyes and look.

These words were written in the 1920s by Cicely Mary Barker, a time when most literature, pictures and stories revolved around the natural world. Barker gave us a great mnemonic by way of her wonderful *Flower Fairies* books, suggesting there was already a noticeable disconnection from nature occurring in her time. It is even more true a century later, as now is the time when more and more people need to 'go with open eyes and look'.

This book is about what happened when I had my eyes opened by being introduced to the plants that are literally 'hiding in plain sight'. As I looked deeper I was shown that many of the plants shared a very special and intricate history, not just for centuries or millennia, but entwined within our evolution as an animal on this miraculous planet. From the book's research it became evident that if our species is to survive for many more generations we need to heal the disconnection with nature we have been forced to endure as a consequence of our cultural programming.

For seven years I've been immersed in the world of seven plants, during which time they have conveyed to me a somewhat radical reformation of our cultural systems. There is no claim that these seven plants are the best plants or the only plants; they literally presented themselves and chose me, and what they showed me well and truly blew my mind. From what at first glance appeared unbelievable potentials, were corroborated and expanded upon after scientific research and observation. It is the plants that hold the key to a transformation of the health, agricultural, food and economic systems; also the construction industry, transportation, land reform, a reconnection to the land; and most importantly they offer a solution, an alternative to a cultural system that we all know is unsustainable, environmentally unfriendly, unethical, immoral and redundant—here's an answer offered up by Gaia to show us how to switch to a sustainable, respectful and abundant future for all.

Before we delve deeper into what these seven plants can offer us, it'll be good to begin with a brief expansion of a few terms, which in some sense are the fundamental philosophical foundations of this book—

Human Plants

A Human Plant is a plant humans have literally evolved on, a plant eaten by our pre-human ancestors—meaning that our body and digestive system were designed around this food input—or more pertinently the nutrients within the food items. These plants are not called Human because we control or own them, but because it is the consumption of these very plants that made us human.

Often people will mention how they prefer to eat the food local to where they live as this is best for them. This is true on so many levels including the bacterial level, as we'll see later, but it is no coincidence that these Human Plants find themselves globally wherever humans found themselves, thus they are always local plants.

There's no hard and fast definition of what a Human Plant is, especially as I've made the term up! Human Plants are among the ones we most commonly come across, noticed or not. In among the grass of the garden, the paving or park, there is often more than just Timothy ryegrass no. 5. If you stop and have a good look there will be dandelions, clover, plantain, yarrow, daisies and dock with nettles, hawthorn and brambles in abundance around disturbed ground and the edges of gardens. It's as if these Human Plants are in some way following us, requiring little or no effort to encourage them, so abundant as they are, yet usually unnoticed, especially as their role in our evolution has gone mainly unreported.

* * *

What this book is attempting to show is that these seven plants—nettle, dandelion, yarrow, plantain, cannabis, hawthorn and blackberry—contain unique phytonutrients (*phyto* is Greek for plant), whose biological benefits have only just begun to be understood by Western medicine. Ominously most of these phytonutrients, aka phytochemicals, are absent from our plate, while many recent scientific studies clearly point out the potential of these phytonutrients as being extremely healthy additions to the diet. This book is the first epidemiological study of these seven plants that also presents their history and herbal folklore, corroborated in most instances by recent scientific studies. *Materia Medicas* (books on healing materials) have been written for millennia, as humans have long wanted to document the associations between plants, health and illness. This field of interest—the observation and collection of information about the plants around where you live—is the most fundamental and important aspect of our chance of survival. *Seven Plants* shows what these plants can achieve, with the surprising possibility of a systemic reformation, where civilization becomes harmonious and sustainable with ultimately everyone thriving, as opposed to hardly surviving.

The diseases of civilization

Sixteen years of researching plants, health and disease has resulted in me becoming a herbalist, writing two books and establishing a food supplement business (The Culinary Caveman). Both the books share a central thesis, *The Diseases of Civilization*, while the herbal powders act as a preventative against these very diseases. It sounds a little paradoxical, this concept of civilization causing our death, be that from cancer, heart disease, iatrogeny (medical intervention, notably prescribed pharmaceuticals, the third biggest killer of civilized people), diabetes, depression, dementia, obesity, neurological dysfunctions, all manner of degenerative diseases and—the most recent mental health disease of civilization—the mobile phone and social media.

This shouldn't be happening if we were truly at the pinnacle of what is considered civilized, with heightened scientific enquiry and superior health observation and facilities. Instead each year more people begin to take harmful drugs for the rest of their life. If we are destined to die from our lifestyle, I for one am not so happy with this. There are many of us, I'm sure, who would rather have lived and died a little closer to that which nature prescribed. Today we live in the most polluted, toxic-filled environment *ever*, saturated in poison, from the air full of smog and radiation, the water full of toxins or the barren soil providing food with no sustenance and a dose of deadly agricultural chemicals to boot.

* * *

The word 'disease' comes from the Old French *desaise*, meaning discomfort, inconvenience, distress—although it could just as easily have come from *ease*, which is to help, assist; and while *to be at ease* is to be tranquil, at peace and in physical comfort, the addition of the *dis* prefix reverses this. The word civilization is formed of *civil*—a citizen/civilian and *ization*—the act, process or result of making or doing, so *civilization* means little more than being a person in a city. Today, 1 in 3 of us will die of heart disease, 1 in 3 from cancer and 1 in 3 from pharmaceuticals or obesity, diabetes, depression or dementia.

According to the *Oxford English Dictionary* (OED), the word 'health' derives from the Old High German for whole—thus not an absence of illness, as we see it today, but from wholeness, togetherness, all parts present and functioning correctly, also known as homeostasis. Civilization is killing us with diseases almost absent to our hunter-gatherer ancestors. These diseases are a side effect of the unsustainability of our civilization.

Whichever disease gets us from our polluted lifestyle there are drugs to alleviate some symptoms, with health now dominated by pills that have no curative effect. Instead of leading natural lives where our food acts as a potent prevention, we are faced with a diabolical predicament where our food is an antagonist, preventing us from ever attaining real health. Which brings to mind Hippocrates, hailed as the father of Western medicine, who 2,500 years ago stated that food is our medicine. Today it is sadly far more true to say that food is our poison—we lead lives full of processed carbohydrates, fats, sugar, harmful chemicals and devoid of nutrition.

This book is written from a holistic, herbalist perspective, while the greater part of the scientific corroboration was undertaken from a reductionist perspective, thus joining the two disciplines instead of continuing the diverging dualism so ever-present in our culture. This is the middle ground where the truth dwells. Etymology also plays a key role in this book as words and their meanings tell a story of their own, not always meaning what we assume. So, continuing down this etymological route we next come to—

Oikonomia

Oikonomia is from ancient Greek and is the root of the word economics. Having been somewhat corrupted recently, *oikonomia* is more sustainably defined as 'good household management', a term as relevant on the micro-level (running an individual house or small business) as the macro-level (global corporations and countries). Today we have to ask: whose house is being managed well? Economics is dominated by one item, money, forever decried as the Devil, something to sell one's soul for, as well as becoming attached to a whole diaspora of pollution, wars, greed and short-term profit. It is time we all take a long, hard look at this misguided economic world we live in, as economics is responsible for most, if not all of our present problems.

In our current economic system, if these plants do indeed hold preventative and curative abilities, if the sale of them as medicine is not equal to the lost revenues generated by pharmaceuticals, it is seen as an economic loss, with less revenue to add to the gross domestic product (GDP) of countries—it'll never happen, even if the plants work better. Also in our current system, profits aren't good enough: there has to be more, bigger profits. Being so myopic as only registering growth and only seeing profit has sorely proven to be terrible household management for the vast majority.

Economics run on principles closer to *oikonomia* wouldn't mean changes that make life harder or more austere—as if it could get much worse for so many already. The solution is a viable, sustainable alternative that is radically different and positive. By continuing with our current unsustainable cultural system it is impossible to become sustainable. The only way it can become sustainable is by stopping and starting something new. If we don't, we're screwed. This book proposes new agricultural, food, health, lifestyle, economic, industrial and transport systems, along with land reform as well as uncovering some interesting truths about money, because in a world dominated by money it would be foolish to ignore it.

Disconnection

Cut off from our roots, which once united us to the whole planet, there can only be one possible future path with no roots. Death.

Our civilization has made us progressively more disconnected from Mother Earth, soon to enter a dystopian sci-fi realm where we will no longer be from, or understand our place within nature. This will be a disastrous dislocation, causing as big a quake in the consciousness as the change from hunter-gatherer to agriculturalist. We are faced with a crossroads requiring a truly intelligent choice, which is exactly what the Latin route of the word intelligence means—'to choose between'. With this choice between artificial intelligence or natural intelligence being made for us, what with algorithms, systems and apps, true Intelligence has been side-stepped.

This book is a scientific epidemiological research tool, a manual for our survival by completing the cycle back to Earth wisdom. By looking at seven plants, hopefully our position on this planet can be re-established. As this reconnection triggers the soul, it reawakens our sense of purpose, our sense of unity. This book shows seven plants, each a monograph in itself with the folklore and traditional uses outlined, then the scientific evidence shown, seven times, as well as mention of a vast range of industrial applications. All this is then condensed into a Solutional Summation with agriculture, land reform and economic and financial systems exposed and explored, and simple solutions offered, with lists and interesting stuff sneaking into the appendices.

Our book (as *you* are reading it and investing *your* time) presents a sustainable future with a revolution of the systems supporting our current cultural incarnation, a revolution not, as erroneously assumed, involving placards, harsh words, police batons and smashed bank windows—this revolution will take no more than a change of mind, making it possible to simply say 'enough is enough'.

> Our Grandparents of old, they are saying, 'Listen to Her, all, to the Earth our Mother, to what She is saying.' People, listen all.
>
> (Mohawk saying)

NETTLE

(Urtica dioica)

Humble pie

Is what I'm constantly being forced to eat when remembering back to how I was when fresh out of town. That town, Brighton, is now a city, but many Brightonians of my age and older were reluctant to consent to this change. Which turns out to be a common trait among us humans, this reluctance towards change. It is a trait we need to drop because usually the procrastination and worries are unfounded, and when whatever change is actually initiated and completed it's usually for the better, whether it's building an extension, changing your diet or overhauling health, food or economic systems. This incorporates the bigger picture about change this book wants to paint, with new systems to enable a new culture to emerge—a more sustainable, environmental, equal and peaceful culture. There is abundance beyond our limited vision; we just need to learn how to refocus on what has always been there.

On a personal level regarding change, my increasing family needed a change in our residential location while giving the children a fuller taste of nature. Every weekend we would drive somewhere in the countryside, until, deciding to make it permanent, we moved 18 miles out of town, to be surrounded by trees and plants instead of buildings and cars. Opening the door to be greeted by grass and a weeping willow was indeed the required change from the pavement and junkies.

Wanting to get active outdoors just after moving in, a path was cleared in the garden and up through the sloping bank of weeds to create a bendy walkway, continuing up the slope and into the woods. The task was set about with much gusto and enthusiasm, wanting to eradicate those barbaric weeds, mainly nettles, whose only mission in life it seemed to me at the time being their potential to sting the kids. Like a Conquistador assured that God was on my side, I went in for the slaughter, and out they came. I emitted a wry chuckle whenever a length of root with several vertical shoots attached was pulled out, imagining this to be the nettle hunter's bullseye. The battle lasted all afternoon and, at the end, with a few stings and plenty of sweat, the bank was clear. Mowing a windy path up into the woods (also rather cheekily extending this path to the woods to create a mountain bike track that went through the woods, down some cool hillocks, skirting a lake and back through the woods to the mowed path, which really teed off the local gamekeeper, but that's another story!). The mini nettle-apocalypse was complete.

Since this time, twenty years ago, I have learnt so much about plants and the incredible world they inhabit. I've also learned loads of etymology, such as the definition of that word 'apocalypse', which has nothing to do with death and destruction, but is far more passive, meaning 'to lift the veil (of truth)', 'to uncover'. The assumed apocalypse inflicted on the nettles has of course turned full circle and become nothing short of apocalyptical for me, revealing what lies behind this veil of truth embedded in our twenty-first- century, culturally indoctrinated lenses. The future role of nettles is staggering, and it would be true to say it not only blew my mind, it made me drop loads of crumbs from my humble pie* all very unattractively into my beard.

By the end of this chapter, I want to have shared with you all there is to know, or at least as near as it's possible to do so in one chapter, about nettles. I'm so excited about this insignificant plant at the edge of the garden or field. It has a real gift and we've allowed it to be absent from our lives for the past 70 and more years. This nettle neglect could quite possibly have been extremely costly, and hopefully it is not too late to make the necessary changes.

It's a family affair

Among the nettle family, *Urticaceae*, there are 54 genera and 4,918 species, ranging from herbs, shrubs, small trees and a few vines. The *Urtica* genus has 282 species (most are remarkably similar) with only three found in Britain.

It's also quite reassuring to know that however painful or inconvenient we may find stinging nettles to be in Europe and America, we should be grateful that an absent-minded brush past, or stumble into, will only result in a temporary rash and an itch, not the mind-numbing torture inflicted by some of nettle's close relatives from around the world. Take, for example, the innocent-looking *Urtica ferox from* New Zealand, responsible for several fatalities among horses and other domesticated animals. A single sting from this bad boy is unbelievably painful, with the toxin tryfydin accumulating after multiple stings and potentially causing paralysis. Apparently five stings are sufficient to kill a guinea-pig (shame on those scientists carrying out such experiments on poor fluffy!); in one instance a hunter died within 5 hours from multiple stings, another was close to death and no doubt others suffered anaphylactic shock. Or there is *Urtica urentissima*, found in Java, Indonesia, which is said to have a sting on a par with having a branding iron pushed into you, causing immense distress for up to

---

*Technically a lie as I'm a vegetarian and humble pie would originally have been made from the innards and umbles of a deer; as those upstairs ate the venison, the poorest living below stairs would have eaten the umbles, which got corrupted in time to humble.

a year, with reported fatalities as well. Perhaps even worse is the Australian gympie-gympie (*Dendrocnide moroides*, a close relative in the *Urticaceae* family), charmingly also known as The Suicide Plant. This delivers excruciating and debilitating pain from the neurotoxin moroidin, which often kills dogs and horses, and drives humans to insanity as the pain subsides devilishly slowly, sometimes continuing for a whole year. The pain is said to feel like being burned with acid, electrocuted or squashed by giant hands. Perhaps it's an urban myth, but there's a story of a man who once foolishly used some gympie-gympie to wipe his bum with in the bush, and so intense was the pain the only remedy and relief was to shoot himself in the head!

* * *

That last paragraph is perhaps not the best introduction to the notion that nettles may be capable of saving the world, although it clearly illustrates the inherent power in this family of plants. The following *Urtica* members are more akin to the common stinging nettle found in Britain, *Urtica dioica*, in appearance, sting, biological constituents and nutrients—

Australia has the *Urtica incise*, utilised for food by the Aboriginals, as is *Urtica massaica* in much of central Africa from Congo to Tanzania; this is also used as a medicine for many ailments from stomach ache to diarrhoea, venereal disease and malaria. Gorillas eat loads of this one as well, being one of their best sources of protein as they are mainly vegan, apart from a bit of entomophagy (insect eating); being undeniably powerful, gorillas are a great advertisement for the benefits and potential of nettles—there should be a nettle drink called 'Gorilla Juice'.

The East Asian countries of China, Japan and Taiwan have utilised *Urtica thunbergiana*, while the south-eastern United States and Mexico have *Urtica chamaedryoides*. Peru has *Urtica lalibertadensis*, and *Urtica aestuans* is spread from Puerto Rico to Central America, India and Indonesia. This is not to mention the forty or so other similar species from the *Urtica* genus.

Finally, there is the River Nettle (*Laportea peduncularis*), a closely related member of the *Urticaceae* family with very similar leaves and sting, although the flowers are considerably larger. It is eaten extensively from Tanzania down to Mozambique, KwaZulu Natal in South Africa and Eswatini (Swaziland); it is even more nutritious than *Urtica dioica*.

* * *

These *Urtica* species grow across the whole world and undoubtedly, if soils were similarly mineralised, their potential nutrient profile would be similar. They have also been mentioned to show that the whole world has been eating and utilising members of the nettle family, not so much today but definitely in the recent past (from before 1950). That's essentially the whole ethos of the book, the rather obvious statement that we, as the species *Homo sapiens*, have always consumed and utilised nettles (just as with the other plants in the book). A consummate Human Plant.

* * *

There are three species in Britain with the potential to sting. Not a sting as found on a scorpion or jellyfish, it's a minuscule silica hypodermic needle, called a trichome or spicule, which lodges into the skin, breaking at the tip upon the slightest contact and injecting the irritating contents. The biological weaponry possessed by *Urtica dioica* is a compound of histamine, serotonin, 5-hydroxytryptamine, formic acid, oxalic acid and acetylcholine, a chemical that plays an unexpectedly important role, as revealed later. The reason they sting is because they are so nutrient rich—which is all that food is, a convenient delivery system for the nutrients essential for life. But it is also exactly what modern food is not—an inconvenient delivery system for anti-nutrients, all those fats, processed carbs and sugar. *Urtica* got so annoyed by being munched on all the time that the only course of action was to punish those pesky herbivores with a sharp sting to the tongue. It worked a little too well, because now most of us sensitive humans have forgotten what we should be doing with this plant.

This *Urtica dioica* is the common nettle seen everywhere (if you're reading this in Europe and North America), and this has been shown to retain its sting for at least two centuries, as proven when Carl Linnaeus' collection of dried specimens was moved from the Linnean Society in London to their new home at Woburn Abbey, Buckinghamshire. It was Linnaeus who gave the names to and put into a structured order all the plants known at his time, the mid-eighteenth century. All the specimens were recatalogued with new photographs, and when the nettle was being mounted it duly stung the photographer on the arm.

Urtica dioica is a perennial (meaning it lives for more than two years) with an extensive root system, or more botanically correct it's termed a rhizome. This is an underground, horizontally creeping plant stem capable of producing the shoot and root systems of a new plant. These rhizomes seek out nutrients from the soil voraciously, with a preferred habitat being rich in nitrogen and phosphorus, best represented by soils disturbed by humans, our own and domestic animals' excrement and more recently the nitrogen fertilisers (most modern fertiliser is a mix of nitrogen, phosphorus and potassium, NPK).

Another stinging nettle found in Britain is *Urtica urens*, also called the Small Nettle. Being smaller than its close cousin *Urtica dioica*, it's also only an annual (lives for one year), is quite rare and is very hard to distinguish a fully grown one from a young *Urtica dioica*. The third and final stinging nettle is *Urtica pilulifera*, which may or may not have been introduced by the Romans, as it's common in Italy. It can still be found in very small pockets around Britain, mainly in the east—the Romans unsuccessfully attempted to introduce their own nettle from Italy into a culture and cuisine already eating nettles.

There is another nettle, looking near-identical to the normal stinging nettle, with leaves a little spear-shaped, although this one, the Fen or Stingless Nettle (*Urtica dioica* ssp. *galeopsifolia*), is stingless, being still hairy but lacking the stinging trichomes. Confusingly, I've come across a stingless variety on numerous occasions when out walking, with deep green, large, flat and soft leaves (not spear-shaped), often in the locality of an old farm or large country house. This has often got me thinking of how useful it would be to be growing a large patch of stingless nettles with a bigger, juicier leaf. This preference of softer and less stingy plants, after a few generations of deliberate selection, would have resulted in plants with these traits, but because it wasn't documented we can never be sure. It's as if nettles never formed a significant part of the diet, which is untrue as they were a staple of the diet, so a stingless nettle would have been a boon, with the seeds of these large stingless varieties bartered at the regular markets and seed-swap festivals of days gone by. Since those days, these stingless nettles and their stinging cousins have crossbred to produce all manner of nettles with varying amounts of stinginess and soft leaves, as found today.

It's also worth a quick mention that nettles are the preferred food for what many consider Britain's most beautiful butterflies, the Peacock, Red Admiral, Painted Lady, as well as the Comma, Small Tortoiseshell and Gatekeeper among others; in addition there are many moths, including the Mother of Pearl, Silver Y and Small Magpie. Up to 30 insects, including the Nettle Weevil and many birds benefit from their seeds such as bullfinches. Flies have a distaste of them, so a dried bunch in the larder keeps flies away.

Nettle history, ethnobotany and folklore

Nettles, even if forgotten about today, played a prominent role in history, often associated with bravery, health and strength, providing not just food but also medicine and clothing. The following is a brief synopsis of nettle's appearance in history and folklore. The word folklore has been deliberately corrupted to give it a quaint, sort of pretend air, similar to the words 'complementary' and 'alternative', as if they are not quite the real thing, a pale replica, but a good effort. That is what words do: they are quite literally spells with their spellings, where even the sound of them can put you in a trance, or the entrance to another world. Of which plants have been denigrated to the lowest levels of life barely living, sentient, unintelligent, underlined by the fact that a human who becomes unresponsive is referred to as a 'vegetable'. Yet vegetables sustain us.

In the third century BC, Greek physicians prescribed nettle juice externally for snakebites and scorpion stings, and internally for plant poisons such as hemlock and henbane. Dioscorides, a Roman physician in the army, wrote the amazing *De Materia Medica* between AD 50 and 70, covering 600 plants and 1,000 medicines made from them, with nettles being put forward as a purgative and detoxicant. *De Materia Medica* was used as a reference book for over 1,500 years until replaced by the encyclopaedic herbals in and after the Renaissance, which although newer and fashionable, with lovely illustrations, were questionably better informed. Roman soldiers posted in the far north east of their British empire were reported to flog themselves with nettles to alleviate rheumatism and sore limbs brought on by the wet, cold British weather.

Milarepa (1052–1135), one of Tibet's best-known yogis and poets, changed from a murderously vengeful dark magician to one of Tibet's most respected and humane heroes. His transformation was achieved after 12 years' training in the cold, high mountains, by meditating and living on soup made from nettles and other herbs, which is why he is always portrayed with a green tinge. In Nepal and Sikkim a nettle soup, *shisinu*, is still very popular.

Monks throughout Europe were apparently forbidden from eating nettle seeds in the Middle Ages. This could have been because they are rich in a chemical called beta-sitosterol, which acts like a hormone with aphrodisiacal effects, as well as increasing sperm production. Randy monks is a worrying vision … Is it wrong to say that?

There is the fascinating twelfth-century saint (there is dispute she was never properly canonised, but by all accounts she is more than deserving of such a title), Hildegard of Bingen, a Benedictine Abbess, considered in her native Germany to be the founder of scientific natural study. A veritable intellectual, she composed at least 70 works of music, invented an alphabet and wrote medical literature, of which *Physica* is perhaps her most memorable, documenting over 500 herbs, plants, animals, precious metals and stones, along with 2,000 remedies.

The word *Physica* in this form has nothing to do with physics. It means pharmacology, the study of the effects of a drug on the human organism, which would have been a tincture or wine back then made from herbs and plants, not the toxic pharmacology of today dominated by denatured man-made pills. *Physic* is a word originating from the first Western medical school in southern Italy in the city of Salerno, founded in the ninth century. Salerno had a famous library containing the writings of Hippocrates, Dioscorides, Galen and various Arabic luminaries, and many herbal doctors, leading to people flocking there to learn or be healed.

Hildegard didn't visit Salerno, but she had access to copies of some of the books, as well as documenting the knowledge of the plants that grew in monastic *Physic* gardens. Hildegard had this to say about nettles:

> The nettle is very warm in its type. It is not good eaten raw, because of its harshness. But when it sprouts fresh from the earth, it is useful cooked as food for people, because it purges and cleans the stomach and removes mucus from it.

The sixteenth-century English herbalist John Gerard (1545–1612) authored his impressive *Herball* in 1597 (with 1,484 illustrations) and used nettles as an antidote to poisons. Nicholas Culpeper (1616–1654), famous for his book *The English Physician*, later known as *The Complete Herbal*, mentions that "the decoction of leaves in wine is singularly good to provoke women's courses". Culpeper also stated that it was Mars that held dominion over nettles, and as Mars is hot and dry and winter is cold and wet, they complement each other and make a great spring-clean, or as he prosaically said, "eaten in the spring, consumeth the phlegmatic superfluities in the body of man, that the coldness and moistness of winter hath left behind". Culpeper also mentions nettles as being useful for the lungs and wheezing, as an expectorant, for the removal of polyps, parasitic worms in children, as well as a wash for gangrene, fistulas, scabs, aches and pains, gout and sciatica.

His near-contemporary, Samuel Pepys, the OG blogger, wrote in his diary on 25 February 1661: "I enjoyed very much a nettle porridge." A porridge in those days would have been more akin to a broth, probably made from grains such as oats and barley, with some vegetables and even bits of meat added. John Josselyn's *New-Englands Rarities Discovered*, from 1672, mentions nettles as a favourite of domestic animals, particularly goats and chickens, as well as being an excellent supplement to hay for cattle and horses. More than just a homophone, John Evelyn's 1699 book *Acetaria: A Discourse of Sallets* notes nettles as being hot, dry, solvent and a blood purifier.

Pioneer in a sexist world (even less known than Hildegard), Elizabeth Blackwell (1707–1758) was possibly Europe's first professional female botanical illustrator.

Her superb work in *A Curious Herbal* (1737–39), with 500 illustrations of medicinal herbs, most from the collection at the Chelsea Physic Garden, recommends the seeds of nettles for coughs, shortness of breath and obstructions of the lungs, making them a popular therapy for the deadly and fairly common consumption (now known as pulmonary tuberculosis).

As rare as it is for you to be reading a chapter devoted to Nettles, there is another to be recommended from the Danish publication *Folk-Liv 1942*, in a chapter called 'The Nettle as a Culture Plant' by Margrethe Hald. Hugely informative, especially from an ethnographic perspective, she teaches us about nettle-cloth, which remained popular and sought after as under-garments until the early 1800s, where those produced in France, Holland and England commanded a high asking price, until the turn of the nineteenth century. At that time nettle use fell to all-time lows, with only small pockets of continuance, such as among the Russian tribes, the Bashkirs, Koibales and Tartars (until Stalin killed most of them) and through parts of Eastern Europe (Hungary and Lithuania). Mrs Hald had samples of ancient Voldtofte linen (late Bronze Age c. 800–500 BC) analysed to discover they weren't flax but nettle, possibly a common misidentification. Danish national costumes were made from nettle until the 1820s and probably a little later as well. Hald also mentions the health benefits for rheumatism, encouraging flagellation/urtication with nettle stems; she notes that juice of nettle leaves is good for the lungs, and the roots can be employed for many symptoms from lungs to hair, TB, eczema, liver disease and viper bites (all of which have recent scientific corroboration, apart from the viper bites, but I did just see an Adder in the garden, which is a bit of a worry!).

Other nineteenth- and early twentieth-century texts from Denmark and Sweden confirm the fact that nettles were eaten abundantly and sold throughout Sweden as food, while in Denmark nettles have to be included among the nine Maunday Thursday's cabbage (broth) to keep people safe from sorcery for the year. There are also plenty of accounts from Sweden and Denmark throughout the eighteenth century of the benefits of feeding nettle leaf and seed to horses, pigs and poultry, with hemp seeds particularly recommended.

There is no question we ate nettles. It might be a laugh or a curiosity to try nettle soup today, but it has to be understood that nettles once formed a crucial component of the diet as recently as just before World War II (and during rationing). It's important to appreciate that these plants being perceived today as having no use is merely indicative of the cultural amnesia we are suffering from, as not so long ago nettles (and all other foraged plants) contributed a significant proportion to the daily life and diet. Perhaps those English villages still surviving with names such as Nettlebed, Netttlecombe, Nettleton, Nettleham and Nettlestead indicate the everyday usefulness of this plant and shows that nettles were once worthy enough to honour the name of a settlement.

* * *

Combining all the North American First Nation People's (the indigenous original inhabitants) ethnobotanical uses of nettles that can be sourced, both materially and medicinally, it can be seen there was extensive usage across the entire continent. The Cherokee used nettles for upset stomach, fever and for making bowstrings, while the Cree used them as a gynaecological aid after childbirth. The Hesquiat rubbed nettles on a sore stomach, and made a poultice of steamed roots and stems to be applied to swollen, sore, arthritic legs, ankles and joints, as well as using nettles for twine, ropes and herring nets. The Iroquois cooked nettles as a vegetable and added them to dried snake's blood for a 'witching medicine'. The Kwakiutl rubbed the plant on the skin for chest pains, while rubbing the juice of the plant into the scalp to prevent hair loss, and used the plant as a drink for overdue pregnant women, and also as a venereal aid and for locomotor ataxia (trouble controlling bodily movements).

The Lakota took an infusion of nettle roots for stomach pains and for making cordage, while the Makah rubbed the leaves on a fishing line to imbue it with magical properties as well as to dye it green. The Makah also applied the leaves to the bodies of whale hunters to make them strong, as a stimulant in the morning, as well as made an infusion of the roots for expectant mothers and use the plant to make basketry, cordage and food. The Nitinaht ate the young shoots as a general tonic, flagellated the body with nettles for arthritis and rheumatism, and used them in a ritual by married couples for affection and faithfulness; they rubbed the leaves onto hands to remove any human smell when fishing and hunting as well as for cordage, fish and duck nets. The Okanagan-Colville used them similarly to the Nitinaht, as well as for a tea drunk during 'sweathousing' and as a body wash, also as a wash to combat witchcraft, hexes and jinxes. The Palute used nettles for flagellation and an infusion of roots for hives and itches; the Shuswap used them for bathing, drinking and rheumatism; the Thompson also flagellated for rheumatism/arthritis, ate the plant tops for skin diseases and as a general tonic, a poultice for paralysed limbs and as a source for making twine.

To avoid repetition, pretty much every use mentioned above was also shared by the Apache, Bella Coola, Cahulla, Carrier, Chehalis, Chippewa, Costanoan, Cowichan, Cowlitz, Dakota, Diegueno, Gitksan, Halsla and Hanaksiala, Hoh, Inuktitut, Kawalisu, Klallam, Klamath, Luiseno, Lummi, Mahuna, Menominee, Miwok, Montana, Ojibwa, Omaha, Oweekeno, Pawnee, Pomo, Ponca, Potawatomi, Quilleute, Saanich, Samish, Sanpoil, Shoshoni, Sioux, Skagit, Skokomish, Snohomish, Squaxin, Swinomish, Tanaina, Tolowa, Tsimshian, Ute, Winnebago, Yuki, Zuni and many more.

American First Nation women believed nettles strengthened the foetus and eased delivery, helped to stop uterine bleeding after delivery and increased milk production, a function that gives it the verb of being a galactagogue—a great word, sounds like a 1950s B-Movie, 'Invasion of the Galactagogues'—an ability attributed to the workings of the beta-sitosterol.

* * *

There is also a strong link between nettles and magic (as attested to by the First Nation Peoples above), although sadly most of the ancient oral traditions and stories have been lost, obliterated throughout time by invaders, Christians, religious zealots and the like, so the extent of the loss is impossible to know. A different type of Christian, the Danish author Hans Christian Andersen (1805–1875), managed to preserve some of these stories, mixing them with his vivid imagination, giving us *Wild Swans*, in which a young Princess has to make shirts out of nettles to break an evil spell that has turned her 11 brothers into swans, as nettles at the time had a popular tradition of protecting the wearer from sorcery.

In Scandinavian mythology, nettles are identified with Thor, with it being the practice to throw nettles onto a fire during a thunderstorm to protect the home from lightning, a practice also prevalent across Britain, as was carrying some nettles around upon the person to bestow courage, as well as drive fear away. An old charm to help with fever of unknown origin was to grasp the nettle plant and pull it out of the earth while saying the name of the patient and their parents; in Yorkshire they were used to exorcise the Devil.

Considering Britain's (maybe the world's) oldest surviving nettle clothing has come from Dartmoor, it's fitting that a local name for the nettle down that way is Hokey Pokey, as in the nursery rhyme, "you do the hokey pokey and you turn around"; it's also known as the sting-nittle there, where folklore says the crushed-up root was used as a stypic to stop nosebleeds. In north Dartmoor around the Moretonhampstead area an infusion of the leaves was a remedy for rheumatics, while just a few miles north, in Okehampton, a handful of nettle leaves was placed under the pillow of a sick person, if the nettles remained green the prognosis was positive, not so good if the leaves changed colour. Interestingly, both of these towns have 'hamp' within their names, a source of confusion, as discussed in the Cannabis chapter.

Just six miles south east of Moretonhampstead, around Bovey Tracey, children whipped each other's legs on 3 May. The rationale has sadly been lost, although it does bear a close resemblance to an old Sussex custom of 29 May, known as Oak Apple Day, where if a child was not displaying an oak leaf or apple (not a fruit, but a wasp's gall) they would be susceptible to a whipping on the legs from some nettles. In my hometown of Brighton this day is called 'Pinch-bum-day', being the anniversary of the return from exile in 1660 of Charles II, who passed through Brighton on his secretive exit to France, as well as being his birthday. If you did not 'sport their oak', by wearing an oak apple or leaf, you would be leaving yourself open to a pinch on the bum.

"It's got more chi, man!"

Sometimes when giving a talk or a foraging walk, I would offhandedly mention that it's best to pick a leaf not too long after it emerges from the ground or branch, as it has more energy then, whether that's a hawthorn leaf, dandelion or nettle. Although every time I say this, there is a truth nebbish on my shoulder berating me for having no evidence for this statement, it's just a convincing story about nature, vibes and chi (Qi). Fortunately, there is some excellent research work from Virginia State University (Rutto et al. 2013) addressing this, as they had the wherewithal to analyse nettles in the spring with fresh shoots and in autumn just before flowering. The difference in vitamins and mineral content depending on the season was significant for a few nutrients, such as there being double the vitamin A precursor available in spring as opposed to autumn. There was also, incredibly, nearly treble the calcium, with improvements in sodium and iron as well, although vitamin C levels halved. So next time plants are claimed to be more vibrant, alive and better for you in spring, it's true, they are indeed, except for vitamin C. Rutto et al. concluded by recommending nettles as a highly nutritive and functional food that should become a part of the food chain as soon as possible.

Analysis from Lithuania (Paulauskiene et al. 2021) on leaves picked every month from April to September from the same place pretty much corroborated the above, also showing that antioxidant activity was much higher in May thanks to greater chlorophyll, phenols and carotenoids at this time, with levels of calcium, magnesium and boron also diminishing throughout the year.

Another nutrient analysis, this one conducted in Spain (Guil-Guerrero et al. 2003), looked at the essential fatty acids (EFAs) in nettle seeds, stems and roots, as well as young and mature leaves. All showed a good spectrum of fatty acids, with mature leaves having a remarkable Omega3 content of 40% of its saponifiable oils; young leaves had 30% of this well-known anti-inflammatory oil, as well as an excellent content of carotenoids, luteins and xanthophylls.

There's no publicly available data from any human trials available on the seeds or roots' efficacy with certain diseases. We have no idea of the extent of any industrial knowledge about herbs and wild plants, as research paid for by corporations is their research and they are free to do as they please with it. As one can imagine, if the results on nettle roots confirmed the belief that they act as an anti-carcinogen (recent tests have shown very good response to prostate, colorectal and breast cancer cells), how does this equate into their business model? If their drugs are currently being sold for as much as £15,000/year while a viable alternative will take an hour or so to collect, and then left hung up to dry, who's being hung up to dry at the moment? What if their research showed that suffering from arthritis is eased considerably within just 6 weeks of a daily infusion of nettle tea? It's clearly not in the pharmaceutical corporation's business interests to release the results, especially not before they can patent an active component and turn it into a drug, then a news story, while promoting it with an extensive marketing campaign as a major breakthrough and discovery in medical and health care. They can make it sound even better than nature Herself!

* * *

Considering their abundance, there is surprisingly little available information about the nutrient profile of nettles. The commonest go-to reference for the nutritional profile of many plants is from nutritionvalue.org, who get their figures from the USDA's (United States Department of Agriculture) National Nutrient Database—a great resource with quite a broad range of plants on the site and definitely worth a browse for an in-depth analysis of the food on your plate. They had blanched nettles, from the Northern Plains Indians territories, with figures indicating that nettles have a good range of nutrients, although nothing superlative, except their vitamin K at about 500mcg (300% RDA) and calcium at 441mg (30% RDA) per 100g; there is also a good range of other minerals, such as iron, manganese, magnesium, potassium, zinc, sulphur and the vitamins A, B1, B2, B3, B6 and C, and some Omega3 and 6. Nettles, as one would expect from a Human Plant, have an excellent range of all nutrients in ratios best absorbable by humans, with all the co-factors, such as calcium, which requires magnesium and potassium to make it usable. All are found in fantastic ratios, getting the calcium to where it's most needed—the bones, teeth and nerves. The nutritional figures put nettles in a similar ball-park as the popular healthy green plants, such as kale, chard and spinach, a good potential health provider, although nothing exceptional so far.

That's because our current understanding of health and nutrients has focused on the 50 essential nutrients—minerals, vitamins, water and a few amino acids—and this is where most knowledge ends, with these 50. What has been neglected are the tiny, under-researched phytonutrients, with about 25,000 identified so far. They include groups of polyphenols, flavonoids and carotenoids, and loads of acids such as butyric, caffeic, chlorogenic, coumaric and formic; then there's acetylcholine, chlorophyll, histamine, quercetin and many more we will meet in this and the following chapters, with even longer lists in the Appendix. Most of these phytonutrients appear to be good for us and do something beneficial, with the endless chains and connections of these phytonutrients being quite possibly the real source of disease prevention and cure. Being so new to science these phytonutrients are rarely mentioned, let alone looked for in any nutritional analysis of plants. With nettles, as we are about to find out, these phytonutrients are found in abundance and some exclusively. It will be these phytonutrients that will in the near future be known as the real health nutrients.

A whole heap of benefits

The highest concentration of many nutrients is found in the freshly cut leaves of nettles, as is the case with the flavonoids (a polyphenol), a group of nutrients we will be seeing a lot more of throughout this book. Another phytonutrient, beta-sitosterol, which has been mentioned briefly already in aiding the production of milk and semen, has also been shown to reduce prostate inflammation. It's also worth noting it's never going to be just one chemical enacting specific biological activity, as there will be many chemicals all playing a role, known as co-factors, interacting and making it all work as it should.

Another phytonutrient found abundantly in nettles, caffeoylquinic acid, has been shown to have excellent anti-influenza, anti-inflammatory, astringent and pain-relieving abilities. Again, it's not just the caffeoylquinic acid having this anti-inflammatory effect because this job is also attributed to the superb omega-oil ratio found in nettles. Both would be working together in this anti-inflammatory role, making nettles ideal for osteoarthritis and rheumatoid arthritis. Nettles have so many of the good things, all in the best ratios to benefit humans. Coincidence or Gaia-driven?

The nutrients found in nettles also have a detoxifying, astringent effect on blood and skin conditions, as well as stimulating the digestive glands across the whole system, from the stomach to pancreas, intestines and liver, improving metabolism and lowering blood glucose levels. Discomfort from urinary tract infections can be eased, as well as helping in the treatment and prevention of kidney stones, because nettles are a mild diuretic, increasing urine flow and decreasing the residual amount kept on board. Nettle's efficacy with arthritic conditions has already been mentioned, and is further corroborated by studies from the Universities of Frankfurt and Munich, proving nettles to be an effective pain reliever. In traditional German medicine, nettles have a long and ancient list of uses, from hair loss to menopause, from osteoporosis, anaemia and fatigue to allergies.

Being one of the earliest plants to make an appearance in early spring, nettles are the perfect spring tonic, helping to clear the biological systems from all the fats and salts of preserved foods, a necessity until relatively recently if one was to survive throughout the winter in a northerly location. Nettles stimulate the immune system and help clear metabolic waste, which was essential in days before central heating, refrigerators and supermarkets, although still essential today, as everyone should have a spring-clean with nettles and other plants that present themselves in the spring. From the Dublin National Folklore Collection comes the saying "Three doses of nettles in the month of April will prevent any disease for the rest of the year."

* * *

Here are some of the medical terms, verbs and words, in alphabetical order, that nettles and its phytonutrients could be said to be or to act as: an alterative, analgesic, anodyne, anthelmintic, anti-allergenic, anti-arthritic, antibacterial, anti-carcinogen, antifungal, anti-histamine, antihyperglycemic, anti-inflammatory, anti-influenza, antilithic, anti-microbial, antioxidant, antiplatelet aggregation, anti-rheumatic, anti-spasmodic, anti-ulcer, astringent, decongestant, detoxifier, diuretic, expectorant, febrifuge, galactagogue, haemostatic, herpetic, hypoglycaemic, hypotensive, immunomodulator, immunostimulant, insecticide, kidney depurative/nephritic, nutritive, pectoral, relaxant, restorative, rubefacient, stimulant, styptic, a tonic and tons more.

Which all means nettles benefit these ailments and biological bits: acne, allergies, anaemia, arthritis, asthma, bladder infections, bleeding piles, blood purifier and builder, bronchitis, burns and scalds, bursitis, cancer, chicken pox, circulation, coeliac disease, congestion, coronary vascular disease, coughs, dandruff, debility, diabetes, diarrhoea and dysentery, eczema, enlarged prostate, excessive menstruation, external bleeding, fatigue, fever and cold, gingivitis, goitre and scrofula, gout, gravel from bladder, haemorrhoids, hair, hair loss, hay fever, heart, hives, increases in lactation and urine output, insect bites, internal bleeding, joints and muscles, kidneys, kidney stones, laryngitis, liver, loss of muscular power, lowering blood sugar, lung congestion, lupus, menopause, metabolic disorders, MS, neuralgia, neurological disorders, nose bleeds, obesity, paralysis, parasites, pelvic decongestant, PMS, prenatal/eases labour, prostate, raising blood pressure, restoring colour to skin, rheumatism, rhinitis, scalp conditioner, sciatica, seasonal allergies, scurvy, skin complaints, stones, tendonitis, tuberculosis, ulcer, urinary tract infection and weight loss.

Wow! The stinging nettle does all that. I trust you are suitably impressed and not just thinking, "Yeah! Whatever!" All these benefits coming from a plant seen by most as nothing more than an irritant, with the substantial benefits substantiated by the scientific studies later in this chapter and some of the books in the Bibliography.

It is also definitely worth pointing out that so far there have been no serious side effects reported from stinging nettles. When prepared as a drug, usually as a capsule, a low incidence of mild gastrointestinal upset was reported (only from 300mg capsules, never from fresh leaves or roots), or from numbness for up to 24 hours after direct stinging (urtication via flagellation, which was and still should be prescribed for arthritis). As a double bonus, nettle has also been shown to actually benefit many medications such as Diclofenac, Ibuprofen and Naproxen, as well as lessening the incidences of side effects. Boom!

Side effects

Let's mention now that little folded-up piece of paper that comes in the prescription's box, the one that's never read. If it is, often the word 'death' is on the list, and you shrug your shoulders and think "that's unlikely, isn't it?" Yes it is, most unlikely—for some drugs there's up to a 0.1% chance of it happening, sometimes as small as up to 0.01% (a one in ten thousand). Which is not such small odds when millions of people are taking them; it actually means it is a certainty for quite a few. It's true, because the people that make the drugs know all about it, and they are legally obliged to tell you, so they do, and it's on that little folded-up piece of paper that always gets scrunched up and blocks the pill tray.

This is because prescribed pharmaceuticals most definitely DO have side effects, being the third biggest individual killer of Western humans, with around 750,000 deaths every year and rising. Just as an example, from those drugs just mentioned that nettles benefited, let's start with a very brief selection from what are 136 side effects reported from Diclofenac (this goes under many trade names, Diclofenac being the name of the drug in them). So, up to 24% of users suffer from nausea; up to 13% with constipation; 1 to 10% suffered either abnormal renal function, anaemia, bleeding, bronchitis, diarrhoea, dizziness, duodenal ulcers, dyspepsia, headaches, oedema, tinnitus, urinary tract infection or vomiting; 0.1% to 1% reported bronchospasm; and fewer than 0.1% (and more than 0.01%) suffered exfoliative dermatitis, hallucinations, hepatitis, hyperglycemia, jaundice, meningitis, myocardial infarction, pancreatitis, pneumonia or renal failure; while fewer than 0.01% got struck down with bullous eruptions, depression, hepatic necrosis, loss of hair, leukopenia (loss of white blood cells) or psychotic disorder—none of which you want to get when you're just trying to calm your arthritis!

Naproxen comes with 291 chances in the side effects lottery. Here's just a few—15% complained of headaches; 14% complained of dyspepsia, nausea or constipation; 10% of influenza-like syndrome; 1 to 10% found it caused back pain, confusion, diarrhoea, dizziness, drowsiness, infection, skin eruptions or visual disturbances; fewer than 0.1% suffered from myalgia; fewer than 0.01% had convulsions or a stroke, myocardial infarction, renal papillary necrosis or toxic epidermal necrolysis …

I'm sure you get the idea. There's an awful lot of side effects, some really serious. Considering that the numbers of people given these drugs is sensational, with a minimum of 30 million prescribed daily users for this class of drug—the nonsteroidal anti-inflammatory drugs or NSAIDs, prescribed for anything from arthritis to ankylosing spondylitis, with Diclofenac commanding 30% of this market and Naproxen with just under 10%—there would have been a staggering 2.16 million reports of Diclofenac causing nausea, up to 900,000 suffering from anaemia and vomiting, as well as there being up to 9,000 myocardial infarctions (aka a heart attack) and up to 900 patients had their livers fail. Naproxen saw 450,000 patients get headaches, up to 300,000 had dizziness and were confused (I was prescribed this once, I took one and one only and can confirm the mental confusion and nausea) or had skin eruptions, and up to 300 had either a heart attack, stroke or toxic epidermal necrolysis (when your skin rots off with a possible death from sepsis).

Tea terrorists and vultures

Can you imagine the press coverage, hate mail and fuss that would be made if I was to give 30 million people a cup of nettle tea, three times a day and if after a year the incidences (side effects) mentioned above came back as being reported from those drinking it? Do you seriously think I would be allowed to continue making a vast profit from my nettle tea service? Even if I paid my taxes (although of course I wouldn't, because I'd locate my business somewhere nice to go with the accountant, such as the Bahamas or

Panama) I still strongly think I'd not only be asked to stop, I'd be arrested as well, facing some time at Her Majesty's Pleasure for sure.

Diclofenac is never going to cure arthritis, never ever. Nettles, on the other hand, appear to have this potential, so why isn't this advice available? Irrespective of the health benefits or the savings to the NHS and so much more, the only consideration for the NSAIDs is the £20 billion in global sales that might be put at jeopardy, as this is the potential loss to the economy in revenue and numbers that contribute towards GDPs (Gross Domestic Products). Sales have quadrupled in under a decade for this sector of healthcare and are set to rise to over £30bn by 2027, with a surge in new customers from MS, myalgia, arthritis, respiratory ailments and a thousand other complaints.

This loss of £20bn is always going to be a bad thing within the current economic system unless it can be replaced by something else to sell. For now the pharmaceutical corporations employ lobbyists to heckle the politicians about how allowing these 'dangerous wild plants' will lever out their 'clean safe drugs', causing economic and health ruin, arguments backed up by their own pseudo-scientific research. Imagine how much louder the voice is for a drug that suppresses a symptom of a disease that has £20bn worth of sales to protect, over that of the nettle, with its inaudible curative murmur and perhaps £30m worth of sales globally. Who's getting heard?

* * *

As an aside, to underline how far and wide the mess and utter devastation caused by pharmaceutical drugs is spreading, this tale carries an ominous message. To the Zoroastrians (the oldest religion still practised, probably closely related to the first patriarchal religion) fire and earth are sacred, so the dead cannot touch them. Instead they are exposed to the air, and since at least the 3rd century AD the dead have been put on top of a circular raised mound, lyrically called a Tower of Silence. Here the body is left to the elements and stripped by vultures. The Iranians no longer carry out this practice, although the Parsi in India still do, or at least they used to until a few years ago. This is because the vulture population of the nine species found in India has plummeted. Worst hit was the White-Rumped Vulture, which dramatically reduced from around 80 million in the late 1980s to just several thousand in 2005 (99.7% reduction). The Indian Vulture fared little better with a 97.4% reduction, and other species coped slightly better, with some areas still suffering roughly a 20% year-on-year decrease. With so few vultures remaining and with some cadavers taking six months to be stripped, this ritual has been sadly discontinued in India now.

It wasn't just the Zorastrian Parsi having trouble getting their dead stripped quickly enough. There were also all the cows, sacred to the Hindus, and of the 500 million or so alive in India, only about 4% end their days being consumed by humans, while the vast majority of the rest were taken to carcass depositories where they were eaten by vultures. No longer of course, with many carcasses now left rotting in fields, contaminating watercourses, while the rat and dog populations have risen to fill the void a little; the downside is that they still carry any pathogens carried by the dead cows, such as anthrax, rabies, even plague, whereas vultures miraculously acted as a dead-end for pathogens. Tragically for the vultures, irrespective of their biological magic in terminating diseases, they were ill-equipped to cope with a drug widely used in the veterinary field, given for inflammation in livestock and arthritis in humans. This drug is the one and only Diclofenac from a couple of pages back, with it being fatal for a vulture to feed on an animal who has been given some and with 10% of cows being treated with it, or rather were. It has been banned in India for veterinary purposes as of 2006 (although still used well over a decade later as stocks remained, perhaps still today), with Diclofenac being held solely responsible for this tragedy.

That's totally insane, isn't it? Particularly, as we just saw, nettles make Diclofenac more effective and lessen its side effects. Nettles by themselves could even see greater efficacy without all the contraindications, as they are not deliberately assisting the Diclofenac, more so getting on with their job, which is supplying the ingredients for the human immune, nervous and digestive systems to work more efficiently.

Diclofenac and the whole array of pharmaceuticals given to livestock, from growth hormones to antibiotics, definitely doesn't make the meat taste better or have any nutritional benefit. Quite the opposite in fact, as drug-fed animals become increasingly toxic, with these chemical residues quite happily passing from host to host when consumed. Which is why it's so dangerous for humans to be at the top of the food chain, particularly in the current bio-accumulating chemical world we inhabit, especially one of fat-soluble toxins in an ever-expanding waistline world. Also it encourages drug-resistant pathogens to evolve and mutate, which will soon make all our antibiotics redundant. Nettle-fed livestock, on the other-hand, require less drugs, tastes better and is more nutritious.

* * *

It's also worth considering that every class of drug, for every ailment, carries warnings of side effects on those folded-up bits of paper. It wasn't as if I randomly chose the only drugs with side effects; they all do. Diclofenac was chosen because nettles have a proven efficacy with an ailment Diclofenac suppresses—arthritis—so it was a pertinent example. Many other drugs, such as those for cancer and mental health, have a greater and an increased risk of side effects, although not much trumps a heart attack or a liver eating itself! Consider also, only a fraction of people complain directly to the GP, as opposed to the postman, shopkeeper or neighbour about a side effect, as well as the fact that some folk are unaware their extra suffering is even as a consequence of a side effect. This flood of pharmaceuticals is causing planetary-wide toxic shock, which might be mitigated if they were creating a healthier public, which clearly is not the case, as illness appears to be having a bonanza.

Equipoise (ek-wuh-poiz)

Let us put in some balance as it may not all be completely perfect with nettles. For those advocating or even prescribing nettles, the delivery usually comes as a capsule or a pill, with in some cases a standardised amount of certain active constituents as these are recognised as having a beneficial biological effect. These constituents are isolated and added, guaranteeing a definite amount of a perceived beneficial compound or chemical. This all negates from the very beauty of nettles, in that while they obviously have the isolated bit, they also have everything else as well, all acting in perfect synergy within the nettle; the fantastic balance appears to maximise biological assimilation and utilisation in humans. Which makes the scientific approach of reductionism highly questionable, especially as we have already seen that health is defined as being whole.

Some nettle capsules are just the root, some just the leaf, some everything dried and powdered from the root, stem and leaf. The mere fact that some of the ingredients have been isolated and seen as healthy with a need to get a guaranteed amount of them is both interesting and encouraging—such as standardising the glucoside scopoletin, a coumarin at 15–30ppm (parts per million); or 3mg of silicic acid; or making sure the formulation contains 0.8% vegetable sterols. There are so many of these phytonutrients when it comes to nettles, who on earth deemed these few the best? Especially as nettles have some phytonutrients unique to them, it's disrespectful and not necessary to standardise (which means process and change, rather like fortification), as this disrupts the natural balance nettles possess with regards passing on to us the good bits we need for optimum health.

These pills and capsules usually come in at a size of about 250–300mg, which equates to just 3 grams or so of fresh nettles being dried and put in a capsule. I do recommend these supplement-type products, having seen their efficacy for myself, although the question has to be asked: how much better would they be if nettles were consumed fresh and frequently in the diet? Everyone I have suggested to take nettles, for whatever ailment they were complaining of, particularly those who actually bothered to follow a fortnight regime, usually of some capsules, saw their symptoms rapidly disappear. No one has actually gone and picked enough for a bowl of soup every day for a fortnight to see if this works, apart from me, but I'm totally biased and not ill. Although most people agree nettles are amazing, they do nothing about eating any!

* * *

Anyway, this section was meant to be pointing out some reported problems with nettle consumption, to get some balance, instead of constantly singing their praises. Let's start with the main consideration heard repeatedly about not eating too many of them, or when in flower, because of crystals that form on the plant when it is about to go into flower or seeding. These cystoliths are made of calcium carbonate, which is good for bones, although these are biologically inactive apart from being a risk factor in kidney stone formation. The problem is easily avoided by only eating the top few shoots from young plants.

There is also some oxalic acid in nettles (it could well be a part of the pain sensation of the sting as well), and it is true that too much of this, too regularly, will be a definite risk factor in developing kidney stones and gout-type symptoms. However, blanching destroys this acid, so unless loads of raw nettles are repeatedly eaten there will be no problem. Oxalic acid is far more prevalent in spinach, and there are no advisories about consuming spinach.

The main precaution is regarding the consumption of nettles when pregnant or breastfeeding, with the concern that nettles might stimulate uterine contractions, causing a miscarriage. Although this goes against traditional knowledge, it's best to veer on the side of caution and not to consume nettles in too great a quantity during pregnancy. The precaution to avoid them when breastfeeding is completely unfounded and appears nonsense, considering that nettle is a well-known galactagogue, that is, increases milk production in livestock and presumably therefore humans also, as suggested from traditional uses, with the nutritional spectrum of the nettle bestowing benefits on the milk produced.

For those already on a pharmaceutical regime, there are also some provisos, such as for diabetics as nettles can lower blood sugar levels, and medication plus nettles might lower it too much. Similarly, nettles can lower blood pressure, so if on medication to lower blood pressure, nettles might make it too low—which beggars the question, why not just eat the nettles? Also nettles are a mild diuretic, so may interfere with drugs taken for the kidneys.

They also decrease lithium excretion, so if on pills containing lithium nettles might cause a dangerous build-up. Nettles are also mildly soporific, so if taking sedatives you might become a little too chilled out. Their rich vitamin K content (a blood coagulant) may also interfere with blood-thinning drugs such as warfarin. Finally, nettles caused a few people to suffer from a slight gastrointestinal disturbance, such as a bit of farting or bloating with occasionally diarrhoea, which is probably a result of the cleansing effect (known as the detox or toxic flush), all of which probably subsides after a few days and doesn't come remotely close to those side effects mentioned from the pharmaceuticals above, which invariably results in another prescription to counter the side effects.

A big concern about nettles and other plants suddenly becoming fashionable as super-plants to cure everything, is that the diet of the herbalist is first and foremost a preventative one, and the curative aspect is a more grey area. I'd still stand by plants before the toxic drugs that more and more doctors themselves are having to admit are simply not working—not least the 2013 book *The Drugs Don't Work*, not by a hippy activist, or a reference to the great 1997 Verve song but none other than the ex-Chief Medical Officer for England, Professor Dame Sally Davies.

Not in any shape or form is Western medicine being totally dismissed here. There is most definitely a place for it in the strategies that can be applied to healing and curing. However, drug use should only be temporary, usually during the initial stage of suppression of symptoms, while either the body is left to heal itself (it's great at doing this, having done it for aeons) or use nutrition and herbalism to bolster bodily systems and even enact a cure. The problem today is that there is no middle ground, which there has to be if we are to progress from this scary world of side effects and contraindications.

If such a small amount, those 275mg capsules equalling just 3 grams of fresh nettles, is considered beneficial to consume by all the companies now selling them, for anything from prostate issues to tiredness and arthritis, what are the benefits from eating a big bowl full of the plant itself?

Just eat the nettles, think about yourself first, then others can follow. Nettles sting you, that's about the only downside. They certainly could never have caused all those problems from vulture genocide to heart attacks and kidney failure, all from something that only ever offers relief. It is mind-blowing—as Spock would have said, "It's illogical, Captain."

"Who's the daddy?"

And there it was, embedded in a very dry scientific Iranian research paper. I had read it through once already, and it was clear that the chickens had benefited from both a 1.5% and a 2.5% addition of nettle to their diet, yet it felt as if something had been missed. Reading it through again was when this sentence smacked me in the face like a sock full of snooker balls! "An acetylcholine synthesising enzyme, choline acetyltransferase, was found and it appears that nettle is the only plant to have this enzyme" (Navid et al. 2011).

I'm sure you want to know why this deserved such a gory analogy, but I'm not going to tell you, though I will explain about acetylcholine and why something that facilitates its biological assimilation might be not just beneficial for humans, it could also be seriously game-changing for many dementia and neurological-type diseases, such as Parkinson's and Alzheimer's. That's potentially massive. I had to take a walk and try to let it sink in before piecing together the implications of that sentence.

It means that without this enzyme, the acetylcholine has fewer enzymes that will assist it in it being biologically assimilated, and it appears there's only one other, which is always referred to as if it is the one and only, acetylcholinesterase. This unique enzyme found in nettles, choline acetyltransferase (ChAT) would be inconsequential if we had little use for this acetylcholine; it wouldn't matter what and where the enzyme was, if we didn't need it. Well, we most definitely need acetylcholine, and to be neglecting the one plant that provides the only source of an enzyme to make the acetylcholine do its job may have been a huge mistake with regards our health (the human body makes its own enzymes, a process easily hampered by man-made chemicals, so feeding the body the right ingredients is the best start). Think of those animals surviving on just one food source, the koalas on eucalyptus leaves, or giant pandas and bamboo shoots, leaves and stems.

We feel for these animals, how precarious their life must be that their survival depends on just one plant—a little bit silly, we think, without realising we may also have some plants that exclusively we rely on, only we've forgotten. How silly that would be to forget the Human Plants.

* * *

So, this acetylcholine (ACh) is what is known as a neurotransmitter. It was actually the first one identified in 1921 by Otto Loewi, after it had been first identified by Henry Dale in 1915, both men winning the Nobel Prize in Physiology or Medicine for this research in 1936. This neurotransmitter is a chemical released by nerve cells to send signals to other cells; it is the messenger that allows the motor neurons of the nervous system to activate muscles. In confusing biological speak, this is what happens:

> When a motor neuron generates an action potential, it travels rapidly along the nerve until it reaches the neuromuscular junction, where it initiates an electrochemical process that causes acetylcholine to be released into the space between the presynaptic terminal and the muscle fiber. The acetylcholine molecules then bind to nicotinic ion-channel receptors on the muscle cell membrane, causing the ion channels to open.

> Calcium ions then flow into the muscle cell, initiating a sequence of steps that finally produce muscle contraction. (https://en.wikipedia.org/wiki/Acetylcholine)

Many types of poison work on activating or inhibiting these receptors and/or the enzymes acetylcholinesterase and choline acetyltransferase (ChAT—the one uniquely found in nettles). For instance Sarin, VX nerve gas and the organophosphate and carbamate pesticides all inhibit these ACh enzymes—as is the uncontrollable spasms associated with an ever-increasing list of neurological diseases, notably Parkinson's. Botox acts by suppressing the release of ACh, as ACh inhibition causes paralysis, whereas the venom from a black-widow spider has the reverse effect, causing a flood of ACh; with this wastage of ACh the muscles begin to contract, and if the ACh reserves run out, paralysis occurs, followed by death as all muscles stop working, including the heart and lungs.

This acetylcholine chemical also has a critical role in memory and memory retrieval, bringing it even further into the dementia fold, as this partly defines Alzheimer's. It makes it even more imperative that regardless of when science and its economic subservience catches up, this information could potentially save you from dementia and arthritis (and other ailments) when you get older—all from regular consumption of fresh or dried nettles.

Acetylcholine therefore is beyond important, it is absolutely vital for any chance of a normal life, from not just muscular movement but also mentally, concerning arousal, attention, memory and motivation; even depression is linked to a deficit of ACh. It's obviously of utmost importance to maintain the health of the pathways of this ACh, particularly as neurological disorders and depression are on the increase.

It's definitely worth considering that the food-sourced availability of the enzyme for this ACh to function properly has been destroyed by modern processed food, which is eaten by most people. And of course hardly anyone is regularly consuming nettles, which leads to the conclusion that there has been a significant interference in the pathways of this most important neurotransmitter, acetylcholine. It's not my place here in this book to announce a major medical breakthrough, nor am I in a position to ask for a multi-million pound research grant to undertake a scientific experiment to validate the assertion that nettles are essential, where a deficiency in the diet has opened the door to many forms of what are classed as neurological and dementia-type ailments. It might still be the truth, though, and 'You read it here first—Nettles are the daddy!'

A bit more ChAT

So who is to say that there has not been an environmental pollutant that has either directly interfered with the manufacture of the transferase enzymes, or has caused a genetic mutation that has damaged the natural production of the enzymes or acetylcholine (this is made from foods high in choline, such as peanuts, eggs, brassicas and meat—most foods have some, although organic ones come with fewer detrimental man-made chemicals); or the transference, conversion and assimilation of them; or that of another essential co-factor? It's all a complicated, interconnected web, although what is making a lot of noise in my head at the moment (apart from the ringing caused by the snooker balls!) is this fact that nettles, upon investigation, contained compounds and chemicals found in no other plants researched and that pesticides are made to target this acetylcholine and its enzymes—a most important point that needs to be remembered and will come up again later in the book.

We didn't need to isolate the compounds, reduce the plant to its constituent bits, give every minuscule microscopic part a name and a mission, role and purpose. Humans might never be able to see via our scientific analysis and methodology what role these plants might play in our health. This lack of ascertaining scientific knowledge most definitely does not mean that the plant doesn't bestow unique health potential. Our ancestors didn't worry about the fact they didn't know the names of all the bits within the nettle that might make them feel better. It just did; they knew it had potential, they had seen its power numerous times before and stuck with it. It could have been an innate knowledge, which could be called our 'humanness', which has been all but wiped clean by our current cultural existence.

* * *

Leaving this section with a little recap on choline acetyltransferase (ChAT). This ChAT breaks down the acetylcholine (ACh), which is essential for all proper motor movement, nerve and thought processes. Movement and thought, they are quite important, eh? Is it therefore any surprise that with the total exclusion of one of these essential enzymes, and the destruction of the other by modern food production, there is an increase in diseases displaying those very symptoms their deficiency would cause?

Further research revealed that this ChAT was first mentioned in 1973 as being active in nettles in the *Journal of Biochemistry* (Barlow & Dixon, vol. 132 and later in 1981, vol. 194); these authors also showed there were minuscule quantities available from peas, spinach and green algae, with the rather spooky correlation that the nettle ChAT activity was equal to that found in housefly heads, a superlatively rich non-plant source, making the mind boggle with images of harvesting bowls full of fly heads! The report also mentioned another key role of acetylcholine, as being a biochemical necessity in the earliest photosynthetic organisms, meaning the beginnings of life on Earth would not have happened without acetylcholine. With enzymatic activity of the ChAT at a maximum at human body temperature, this highlights if not verifies its candidacy as a Human Plant.

Detrimentally messing with something that we know is vital in the transference of nerve signals doesn't sound like a good thing to do. We might not know why nettles bestow the health potential they do, but what we do know is that it would be good for them to re-enter the food chain. If half of the claims about them can be substantiated by hard, unbiased scientific facts (as we will see later, it's more like 90% corroboration), that's all going to help with the marketing, because the scientific facts for processed food have never been positive, unless the producers have paid for it themselves (which of course they do). The fact that nettles have unique properties and could be of value for neurological ailments is yet another exciting avenue nettles can take us down.

"EAT MORE NETTLES". This needs to be shouted from the rooftops! We are koalas, and nettles are our eucalyptus. We need nettles.

Phenomenal phenols

This wasn't all that was revealed in the Iranian study on broiler chickens. It also mentioned that among the plethora of phytonutrients, nettles had other unique chemicals apart from the choline acetyltransferase enzyme; there were eight lignans, some of which were previously unknown and ... "there's more" ... they also found five new monoterpenoid components, which probably have antibacterial activity ... Kerching!

Lignans are polyphenols found in plants, and although their role is not very well understood, they do appear to have a preventative function in regards cardiovascular disease. The problem with looking for information on these lignans is flaxseed (linseeds). Because flaxseeds are by far the best source of lignans, all research on lignans is done using linseed, definitely not nettles, even if nettles have unique lignans. The lignans from flaxseed have been shown to have anti-carcinogenic abilities, a benefit probably transferred across much of the lignan group. Polyphenols are being touted for many roles at the moment, with lignans being just another group among loads of polyphenols. The important part of the life of a lignan, as with many compounds, is that it takes place in the gut, being dependent on the gut bacteria to break down substances so that others can be made.

The phenolic compounds of nettle, such as carvacrol and thymol, have demonstrated considerable antimicrobial, antiseptic, antioxidant and antifungal activity, as well as lowering low-density lipoprotein (LDL, the one with the bad rep) and cholesterol levels and improving the hair. Nettles have a minimum of 18 phenolic compounds found lurking within their fibre—another nutrient most humans are deficient in.

Studies from Turkey (Ege University, Otles & Yalcin 2011) looking at the phenolic compounds present in all parts of the nettle plant found that the leaves had 300% more than other parts (root and stalk). There was a tremendous and varied amount of these phenolic compounds from nettles collected all around the Mediterranean, all having only a slightly different nutrient and chemical composition, with agreement on fresh leaves being the best way to go. We are going to see a lot more from these polyphenols as we continue through the book, as they are definitely not confined to just nettles.

The first epidemiological study

Here's just a small amount of information from some of the research carried out on nettles to date.

Owing to the very worrying emergence of drug-resistant pathogens and the ever-present threat of a global pandemic, the hunt is on for natural antimicrobials and antivirals. Leading this hunt into a wide array of investigations and disease arenas are the Iranians, with this little gem from the Islamic Azad University, in Rasht (Salehzadeh et al. 2014). Using isolated methanolic extracts from *Urtica dioica* and *Sambucus ebulus* (a dwarf elder) they tested their effectiveness against the really dangerous Methicillin Resistant *Straphyloccus aureus* (MRSA). Their conclusion was that both extracts possessed antibacterial potency against MRSA and should be used as antiseptics and microbial agents in medicine. The antibacterial properties of *Urtica dioica* were further investigated by the University of Ahvaz, Iran (Motamedi et al. 2014), being found effective against the common strains of *Bacillus cereus, Staphylococcus aureus, Staphylococcus epidermidis* and *E. coli*, all *in vitro* (in a test-tube—*in vivo* is within a body). The extract was unsuccessful against *Salmonella typhi*.

Further research from the University of Ahvaz, Iran (Behzadi et al. 2015) showed the antioxidant abilities of nettles, which also decreased the risk factors of cardiovascular complications, such as raising HDL levels for diabetic patients (a good thing), all from an alcohol extract. Other studies on diabetes saw the Research institute of Medicinal Plants, Karaj, Iran (Kianbakht et al. 2013) conclude that nettles safely improved glycaemic control in Type 2 diabetic patients who required insulin therapy. Diabetes was also looked at at the Babol University of Medical Sciences, Iran (Qujea et al. 2013), using both an ethanolic extract and an aqueous one of nettle leaf, with positive results for protection and stimulation, the release and regeneration of beta cells and the pancreas as a whole.

More Iranian studies, reported in the *Asian Pacific Journal of Cancer Prevention* (Fattahi et al. 2013), showed that an aqueous extract of nettle leaf had anti-proliferative activity on human breast cancer cell lines, as well as aiding apoptosis (cancer cell death and removal, also known as cell suicide).

A dichloromethane extract was obtained from nettles at the Tabriz University of Medical Sciences, Iran (Mohammadi et al. 2016) that significantly inhibited breast cancer cell proliferation while causing no damage to healthy human cells, as well as causing the cancer cells to undergo the desired condition of apoptosis. Further studies by the same team showed the dichloromethane extract to significantly destroy colorectal cancer cells. Dead.

An Iranian study on rats and livers demonstrated the benefit from nettle extract (the rats would have fared even better without the hepatectomy), as reported in the *Journal of Toxicological Health*, 2015 by Oguz et al., with significant regeneration and protection offered to the liver.

The biggest killer of humans is heart disease, and a significant risk factor in this is atherosclerosis (thickening and hardening of artery walls). Nettles showed protection from this thanks to research at Shiraz University, Iran (Namazi et al. 2018), where an ethanolic extract also prevented the establishment of lesions, with positive effects on serum lipid profile. This adds to an impressive gamut of diseases covered already by nettles and Iranian researchers.

* * *

An epidemiological study from the University of Campania, Italy (Esposito et al. 2019) looked at global scientific research focused on cancer, finding plenty of evidence for anticancer activities from breast, cervical, epidermoid, colon, gastric, lung and prostate cancers. The authors' conclusions were that nettles should be a nutraceutical food for the prevention and treatment of these cancers, without the side effects of current treatments. More Italian cancer studies from the same university (D'Abrosca et al. 2019), this time on the third most common cancer, non-small cell lung cancer, discovered nettle extract to be both anti-proliferative as well as causing cell death (apoptosis).

This anti-proliferative cancer effect demonstrated by Mohammadi above had already been established by a German study in 2000 (Konrad et al.) at the Marburg Philipps-University, using a methanolic extract of the roots of *Urtica dioica*, although this study showed the effectiveness against human prostate cancer cell lines as opposed to the breast.

Research in the *International Journal of Oncology* (Jiang et al. 2011) explored the anti-cancer effects of a dietary supplement and showed it to be effective against human prostate cancer cells—the ingredient list of their supplement looks great, from medicinal mushrooms, turmeric, green tea, sarsaparilla root, dandelion, nettle and broccoli with some vitamins and minerals added.

A study (Hodroj et al. 2020) on something as simple as nettle tea, the leaves steeped in water, found there to be anti-cancer activity against the most common form of leukaemia, acute myeloid leukaemia, the nettle demonstrating both anti-proliferation and apoptosis action.

In work at the Annaba University in Algeria (Siouda et al. 2015), rats with mercury poisoning were given boiled nettle leaves. The results were quite astonishing. In those rats given the nettles there was a marked improvement in poison elimination and increased levels of glutathione (a protective antioxidant which cleans the body) and sperm levels.

The University of Maryland, USA (Fan et al. 2020) concluded that nettles eaten in the diet protected against diet-related obesity, positively affecting lipid accumulation and glucose metabolism. From the University of California (Johnson et al. 2012) the anti-inflammatory abilities of various methods of extraction were noted, seconding the opinion of Tabriz University that a dichloromethane extract yielded the best results, as well as concluding that nettles were probably best used in conjunction with other known anti-inflammatory plants such as *Tripterygium wilfordii* (Thunder God vine) and *Zingiber officinale* (ginger).

Specific to the recent Covid-19 confusion, fascinating research on mice at the Institute of Anti-viral Research, Utah, USA (Kumaki et al. 2011) looked at the inhibitory effects of a tiny lectin found within *Urtica dioica* on the SARS-CoV strain—SARS stands for Severe Acute Respiratory Syndrome. This lectin, agglutinin, showed powerful inhibitory effects on the mice tested, notably that the lungs did not increase in weight, a symptom of the deaths attributable to Covid-19 with blood clots settling on the lungs. It was extremely positive news for those looking for the best prevention against Covid-19 (a SARS virus)—not that there was any research done with nettles in the recent scramble for a vaccine or preventation, nor for that matter any involving yarrow or blackcurrants, both very effective unpatentable antivirals.

* * *

A disease revisited in the yarrow chapter, leishmaniasis, was also shown to be benefited by nettles at the University of Medical Sciences, Tehran, Iran (Badirzadeh et al. 2020). The researchers concluded that nettles were an effective and harmless herbal compound for killing parasites without toxicity to the host, proclaiming them to be the perfect therapeutic herbal drug.

Also sharing another of yarrow's attributes (for which it is legendary), wound healing, Greek researchers (Kasouni et al. 2021) discovered that wounds healed quicker, from 13 to 9 days with a nettle extract ointment, identifying also anti-inflammatory and antimicrobial properties.

A study from the Universities of Exeter and Plymouth (Randall et al. 2007) looked at how effective nettle sting was in treating severe and chronic knee pain. By directly applying the nettle to the knee and causing the stinging sensation, knee pain was reduced with the sting being acceptable for the patients, with little possibility of any other side effects. There was no follow-up to what is a unique human trial with nettles and one of the only UK studies.

A massive potential that nettles (and the other plants) hold is for something known as phytoremediation, which is when the plants act as hoovers, sucking up contaminants, heavy metals and pollution from the soil. With an ever-increasingly polluted planet this is going to be an absolute necessity in the very near future. One of the only studies on this phytoremediation with regards nettles was reported in *PLOS One* (Viktorova et al. 2016). This study demonstrated that nettles cultivated in soils contaminated with heavy metals (zinc, lead and cadmium) and polychlorinated biphenyls (PCBs) managed to remove 33% of the PCBs and 8% of the heavy metals in just four months. The authors also made the valuable observation that it was the bacteria, specifically endophytic bacteria, that accumulate around terminal root nodules that were acting as the detoxifying mechanism, removing the toxins and protecting the plant. The researchers were unhappy with the results, expecting better, but as we will see, if they had combined them with any of the following six plants the results would undoubtedly have been significantly improved.

A Turkish study (Yildirim et al. 2013) highlighted that the rich mineral, vitamin and the phenolic content of nettles gave them serious potential as an alternative source for food and within the pharmacology and medicine sectors—not that you'd know it from the lack of activity on this front from these sectors.

Albeit only a small selection of the positive research, these samples demonstrate how nettle has preventative capabilities in regards inflammatory diseases, diabetes, heart disease and cancer, as well as possessing antibacterial and antifungal properties, with their neglect possibly being causative in the rise in neurological conditions. It is surely a good start, and considering we've only looked at the food and medicinal potentials, take a look at what else nettles can do—

Nettles for textiles

The oldest confirmed nettle textile found in Britain is from Whitehorse Hill on Dartmoor, Devon, dated to 1690–1620 BC. It is an extraordinarily rare find considering there is little chance of any textile, or any other organic matter, surviving once buried for a century let alone thirty-seven centuries. The fabric even had an amazing leather applique, showing how sophisticated ancient clothing was. It's reasonable to assume such garments would have been worn for some time before then as well, particularly as nettle textiles have to be a contender as one of the first plants utilised for their fibrous material to make yarn and weave with. Although an older nettle fabric has yet to be found and confirmed, a carbonised ball of yarn, either flax or nettle, has recently been found in France and dated to 3900 BC.

The famous Copper Age find in 1991 in the Otztal Alps of the murdered Otzi the Iceman, some 5,300 years ago, had no textiles made from nettles on him, but his knife had a nettle cordage to attach it to a belt. It also looks as if Otzi's arrows used nettle twine to secure the arrow to the shaft, making it most probable nettles were also put to many other uses. The fact that nettles live all around us, loving our excrement, means they would have found us and would have always been readily available; we would have never needed to go looking for them.

The next oldest British nettle textile is from a remarkably well-preserved Bronze Age community at Must Farm, Cambridgeshire, dating from about 1000 BC, where the inhabitants utilised a range of materials for textiles, including flax and nettles. A fire destroyed the settlement, catastrophic for them although a goldmine for archaeologists, with much of the contents preserved under the collapsed roofs, providing a very rare and tantalising view back three millennia.

Further evidence of the ancient use of nettles for textiles comes from Denmark a little later at c. 800 BC, with the textiles found at Voldtofte reattributed as being made from nettles, not flax. Analysis showed they had also been imported from another part of Europe even though nettles were and are abundant in Denmark, which suggests the status of this fabric as being significant. This was a surprise for Danish archaeology as nettles weren't even considered as a source for textiles until the finds from this site were unearthed and later reattributed. Flax and hemp were assumed to be the main contenders for early plant fibres for clothing, so when the fibres of other previous excavations are re-examined—even under high magnification the fibres are very similar, but differentiating markers have been found—nettles will show their true worth. This misidentification of ancient textiles appears to be rather common, as shown from a Finnish study published in *Studies in Conservation* (Suomela et al. 2018), where a number of museum textiles were re-examined and many had to be reclassified as being made from nettle and not the perceived cotton, flax or hemp as had been originally attributed.

Seriously extending the textile timeframe, an archaeological site in Georgia unearthed a 30,000-year-old piece of dyed flax in 2009, smashing all the records for the antiquity of a woven fabric. It will be interesting to see why it's definitely flax and not nettle, perhaps needing a second opinion, particularly as Bergfjord et al. in the *Science* journal of June 2010 thought it had been incorrectly identified. Nettles have been neglected from much conjecture of the past, no doubt because of the low esteem they are held in today, with focus being on hemp and flax as both are coincidentally still utilised today. There have been plenty of bone needles found in archaeological excavations, many dating to 10,000 years ago, with not just a sharp point, also a hole to put the thread through, so something was being sewn then. Amazingly predating this by a stunning 40,000 years, a needle with a hole has been unearthed in a cave in Siberia from the famous Denisova Cave, the home of *Homo denisova* who was evidently also sewing 50,000 years ago.

* * *

A little etymological break now for the word Nettle, as deriving from the Middle English *Netle*, Anglo-Saxon and Dutch *Netel*—which itself is a word derived from *Noedl*, a needle, which either refers to the sting or more likely that it was this plant supplying the thread, owing to the root of both words being from the Proto Indo-European language group (in effect the mother tongue), meaning 'to bind or tie'. This would give nettles an illustrious past, where it was once the preeminent source of fibre to make clothing and nets, themselves named after nettles. The word 'textile' comes from the Latin adjective *textilis*, meaning 'woven', with nettles being among the first fibrous plants to be woven, so the words nettle and textile would long ago have been synonymous with one another. The botanical name *Urtica dioica* is derived from *Urtica*, being Latin for sting or burn, and *dioica* is Latin for two houses, a reference to the fact the plant is either male or female, growing in groups/houses of the same sex.

* * *

Up until the sixteenth century, flax, hemp and nettle would have been the main plants used for clothing in Europe, alongside wool and leather. As Europeans extended their global reach with ships, cotton soon became accessible and fashionable, with its benefit of being easier and faster to turn to yarn. By the end of the eighteenth century, nettle, hemp and flax clothing had all but disappeared, as cotton coming from the enslaved farms in America and the other imperialist tendencies in Asia saw cotton supply increase hugely, with an associated drop in prices, massively undercutting all other fabrics.

Thomas Green, in his beautifully illustrated *Universal Herbal* from 1820, noted that the fibre from nettles made a stronger twine for fishing nets than flax, and was not so harsh as the ones of hemp. Nettle clothing has taken a back seat for centuries now, although its previous use and importance has been documented by ethnographers across the world, with only a small remaining foothold in some fairly remote parts of Poland, Scandinavia, India, Nepal and Peru still carrying on the ancient traditions. Ethnographically, from the early twentieth century, it can be seen that the last surviving tribes of Europe, such as the Eastern Ugric tribes of Siberia and the Sámi of Finland utilised nettles for their traditional clothing, as indeed did the Japanese for their kimonos (often mixed with hemp).

There is today a slow resurgence, with nettle clothing being available from some niche fashion houses in Europe. The Nettle Fibre Company makes blends of nettle and hemp into usable yarns for a multitude of finishes and uses. Such makers recognise that nettle will never be able to compete with cotton on price, but as cotton prices rise owing to consumer demands for ethically sourced cotton, hemp and nettles will become more competitive—although it's questionable whether competition should ever be based on price as opposed to quality, coupled with the fact it is a rarer, higher-quality commodity, which should make it more appreciated and command a higher price as of right.

* * *

The last time nettles were planned to be used commercially on a large scale was to make the uniforms for German soldiers in WWI, as cotton supplies had run very low following blockades by the British, preventing cotton from entering Germany from either America or India. Whether any were actually made is uncertain as there are no extant uniforms to confirm this, and it may just have been German propaganda against the blockades. Whether made or not, the Germans do provide us with some interesting statistics, which can be of assistance in understanding the potential viability of nettles today. In 1915, 1,300 tons of nettles were foraged in Germany, increasing to 2,700 tons in 1916, while the quantity of nettles growing wild in Germany was estimated at 60,000 tons (probably a huge underestimate). However, wild ones do not guarantee the best stems for processing, so the Germans set about cultivating them, a task that proved significantly harder than imagined. In 1917, Germany put 70,000 hectares to nettle cultivation, with it being estimated that a million hectares would provide the equivalent of 18% of their cotton requirement (while at war with all the uniforms etc.), albeit nettle clothing is hardier and lasts longer than cotton.

A good friend just gave me a lovely shirt made from nettles, and he also had a nettle waistcoat and a jacket from Yunnan Province, China. The shirt I can now wear to a talk provides a perfect example of the quality of clothing already being made from nettles. Although it wouldn't have been made from *Urtica dioica*, it would have been either ramie (*Boehmeria nivea*), also known as China Grass, or *Girardinia diversifolia*. Both species are in the *Urticaceae* family, looking exactly like giant nettles with bigger and thicker stems and big leaves; they are used for food and the stems for textiles, grown for these very purposes from north India across to China. The Rai people of Nepal have utilised this giant nettle for centuries and today still make a small supplementary income from this plant.

The potential of using nettles in the textile industry is enormous, and there are companies popping up all over Europe offering nettles as a commercial textile. When processed well, nettles easily match the best cotton can offer, with nettles providing a fibre from one inch to two and a half inches in length, comparable with the finest Egyptian cotton coming in at just under two inches. Nettles also have the versatility to produce a tough cloth, soft cloth, sailcloth, tablecloth, sack cloth, hanging cloths, bed linen, cordage, curtains, baskets, trays, and other household accessories, as well as being considered superior for making velvet from. Aided by the outstanding ability of nettle fabric to not only be able to absorb water, it can be partially waterproof at the same time. Magic!

* * *

Which is all well and good, but you might still be a bit confused about how a nettle becomes a string of yarn and then a shirt or any other fabric. So now might be a good time to explain how it has been done since at least 1700 BC in Devon and probably much longer elsewhere.

Firstly, in around late June to July, when the nettles are a good height, about a metre and a half, they are cut at the base so as to not damage the root, with picking possible up until October. Then with a tough cloth or gloves, you rub the stem clean of stalks, which also flattens out any remaining stinging trichomes.

Next is the retting stage, which allows microbial enzymes to rot the stems, turning the fibrous outer layer from a solid casing into individual strands; it is these strands that will be used. The commonest two types of retting are dew retting and water retting. Dew retting simply involves spreading the stalks on the ground and leaving them exposed to the elements, when after about three weeks of daily turning, depending on the weather, the outer casing turns from a solid green to light brown colour. The retting process removes the pectin and lignin content, eaten by the microbial enzymes, although careful observation is required as nettles can rot and ferment. Water retting is the other method, done in a bath of water, or a stream or pond for a day, then drain the water if in a bath and soak for a further two or three days.

The marvellous Allan Brown from nettlesfortextiles.org.uk clearly demonstrates that green processing is possible with no retting, using just fresh stalks and a serrated finger-sized flint tool, the likes of which have been found extensively throughout European Neolithic sites. Allan's film, *The Nettle Dress*, was easily my favourite film of 2023. Go and seek.

These outer fibres are all that is required for textile production. Called the bast fibre, this accounts for less than 15% of the dried weight of the nettle plant. All the rest of the stem, mainly the woody inner pulp, is known as the shiv or hurd, being predominantly cellulose. It has many industrial uses, from insulation to organic bricks, animal bedding, plastic and paper manufacture. When grown in significant amounts the rest of the plant can be utilised to establish alternate revenue steams, such as the leaves being used as a food ingredient, colourant, supplement and natural drug (nutraceutical), with the roots also having many uses in the food and medicinal sectors.

The process to transform the external nettle fibres to a yarn is definitely best to watch as it makes for a rather complicated paragraph, so if interested to see how it is done, head on over to nettlesfortextiles.org.uk to see how nettles are processed from Peru to India and Europe. It'll keep you fascinated for hours, if not days, even a lifetime! It's also a really good introduction as to how hemp would have been traditionally processed. This cottage-industry level is still the predominant technological level for textiles made from nettles and hemp on the market today, especially those originating from Nepal and China. It is a time-consuming process, but, as we will see later, there is no reason for change or to be obsessed with upscaling, economies of scale or continuing with desecrating monoculture.

Once a yarn, the fibre can then be dyed, with nettles readily absorbing and retaining wonderful colours, which would have increased its desirability and tradable value in the past. A vast number of natural dyes are available, such as blues from woad, alkanet and berries; yellows from weld, tansy and dyer's greenweed, or nettle roots boiled with alum; reds from madder, dyer's woodruff and cochineal; orange from urine and onion skins. Urine, fermented fruits and vinegar are also required to 'fix' the colours. Greens are surprisingly tricky, requiring combinations of dyes, but nettle leaves provide a green permanent dye still used in parts of rural Russia today.

So, a whole section that didn't have a negative yarn spun into it about politics and pollutants. That's all about to end because the main player in the clothing world, apart from oil, plastics and polyester awfulness, is of course cotton. Nettle clothing will be taking a percentage away from the cotton traders, who will fiercely protect their interests, so here's a brief lowdown on cotton—

What's wrong with cotton?

> Without firing a gun, without drawing a sword, should they make war on us, we could bring the whole world to our feet ... What would happen if no cotton was furnished for three years? ... England would topple headlong and carry the whole civilized

> world with her save the South. No, you dare not to make war on cotton. No power on the earth dares to make war upon it. Cotton is king.
>
> (James Hammond, Senator of South Carolina, 1858)

Cotton truly was that important and influential. Cotton was pivotal in Britain becoming a superpower in the nineteenth century, amassing a Commonwealth spanning half the world. In 1794–96 British cotton goods accounted for 16% of all exports, rising to 42% in less than a decade. The way the British seized this global dominance in cotton was perhaps a brilliant business strategy, although the citizens of India would not see it this way; it was a plan that held no morals or ethics, but it did help to put Britain at the top of the industrial pile.

India in the seventeenth century was by far the world's largest cotton grower and producer, with Britain's East India Company importing much of the harvest to Europe. Not satisfied, the EIC wanted more, intent on destroying the Indian cotton industry and moving the manufacturing to Britain, to the new textile mills of Lancashire and Derbyshire in the early eighteenth century. By 1839 the cotton mills had at least 200,000 children in their 'employ' in awful working conditions. Karl Marx saw this many times, and it influenced his writings. With heavy taxation and import and export quotas and Acts of Parliament, it actually became cheaper to buy imported British cotton goods in India than Indian ones. The labour-intensive cotton industry in India was switched to a mechanised industry in Britain, diverting much of India's wealth to itself, as well as opening up a market in India for British goods. This resulted in India becoming a grower of cotton only, seeing many people starve, while Britain created a global monopoly for cotton goods and many British millionaires. As with sugar, the very beginnings of the cotton trade when dealing with the white man were soaked in the blood of impoverished locals and barbaric international slavery.

* * *

So what if it might have had an unsavoury introduction, that's historical, what's wrong with cotton today?

In many respects nothing is wrong with cotton *per se*, it's the industry itself that has so much wrong with it. The biggest problem has to be the devastation caused by the crazy amount of pesticides sprayed on the crop. Before WWII all cotton was grown chemical-free (as were most crops), with pests dealt with by natural predators, such as birds and spiders or hand-picked, to protect in most instances infestation from the boll weevil. Farms were small, and millions survived and lived happily this way.

Today nearly £2 billion is spent on pesticides globally for cotton every year, with knock-on costs far more than numbers on paper. Cotton is cultivated on about 2.5% of the world's total farmed land, although it uses 16% of the pesticides (50% in Uzbekistan, India and Pakistan). These pesticides use some of the deadliest chemicals in their formulations, some of which can kill if just a single undiluted drop comes into contact with the skin, such as aldicarb and parathion. Both of these are classified by the World Health Organization as extremely dangerous and carry the highest possible warning of WHO1a. Not only do these kill thousands of people every year, they do it by disrupting the enzymes for acetylcholine, therefore impairing the nervous system, disabling the synapses, detrimentally affecting movement and all muscles including the lungs as well as cerebral functionality—coincidentally or not, these are symptoms associated with an array of ailments on the rise, and all correlating with the introduction of these pesticides into the food chain.

Remember the tragedy in Bhopal, India? It was the world's worst industrial accident, when in 1984 a factory manufacturing aldicarb blew up, killing 20,000 instantly, injuring at least 120,000. It is still causing severe inherited genetic problems three generation later. Aldicarb (an N-methyl carbamate), made by Bayer and finally banned in 2010, can still be found being used extensively around the world and is making a comeback under a new name, AgLogic 15G. Considering it has been shown to have catastrophic effects in doses as low as one part in a billion, acting as a powerful immune suppressant, we'd surely agree we'd rather not see it make that comeback.

Not only is our food becoming progressively demineralised from virtually barren soils, the immune system is having the compounded negativity of not having the minerals to maintain homeostasis (healthy balance) nor to deal with the detrimental agricultural chemicals bio-accumulating, making it severely unhealthy at any level of exposure, let alone constant throughout a lifetime. The mere fact that humans are even managing to continue surviving faced with the blatant affront to humanity and food that is modern agriculture, says as much about our superb resilience (which is why humans have survived through many ice ages and meteor impacts) as it does the severe lack of intelligence of those responsible for the travesty of nearly destroying the human genome and planet in just 70 years with these chemicals.

* * *

This all might seem like a million miles away from affecting us, but it does affect us, up close and personal. So pernicious are these chemicals, they can still be found on clothing, surviving harvesting, processing and manufacture, to still be entwined without the fibre, even after you've washed them. And from the harrowing, sickening and distressing accounts of farmers and their families from around the world, who have lost children, parents and friends from a misplaced drop or two getting into drinking water, it can be seen that cheap cotton to us comes at a heavy price to someone else.

The whole chemical world of agriculture has to stop. There is no question that it will eventually stop, as the health costs will start to outweigh the benefits—benefits it appears are only very vociferously argued for by the agrochemical giants themselves. In the real world, far greater yields have been ascertained from man-made, chemical-free, organic smallholdings and further exceeded by biodiverse and permacultural smallholdings. There is no need or justification for deadly chemicals to be poisoning everything on the planet, mutating genes and chromosomes, causing cancers and respiratory diseases and many other maladies to boot.

In 1935, before chemicals were extensively used, 30% of crops worldwide were destroyed by crop diseases, insects and weeds. By 2000, after millions of tons of deadly agricultural chemicals have made a global toxic soup of the entire planet, at least 37% of crops worldwide are destroyed by diseases, insects and weeds. That's the reality—more crops are today destroyed with chemical application than before it, as well as adversely affecting the health of the soil, water, air and all living organisms, depleting the countryside of the majority of small mammals, birds, insects, plant diversity and fungi. It's been a floral and faunal genocide, known as ecocide, and we have been witnessing it since Rachel Carson's book *Silent Spring* in 1962, and before that Albert Howard and Eve Balfour in 1946. The only gains made have been in the profits and pockets of the agricultural and chemical corporations. It is complete insanity to be spraying tons of the deadliest chemicals ever invented by science directly onto our crops. For a comparison, permacultural plots lose between 3 and 10% to pests, disease and weeds.

There is 1kg of pure undiluted pesticide used on average for every hectare of land under cotton cultivation, a figure that rises to a stupendous 90kg/ha in Uzbekistan, where in one region, Karakalpakstan, 85% of the population suffer poor health directly attributable to these pesticides. Also for every 1kg of finished cotton garments up to 29,000 litres of water is used, and with about 29 million tons of cotton goods produced every year, that's more than a splash in the ocean.

* * *

It is only getting worse and more strung out for cotton, because 23 million tons of cottonseed are fed direct to livestock annually, with all the fat-soluble chemicals happily hitching a ride, ending up in the meat, milk and eggs eventually consumed by humans. These chemicals all bio-accumulate within us, although there have been no studies on what this chemical soup is doing to us. On the other hand, plenty of research shows we have these chemicals currently residing within us and that individually they are carcinogenic and genome-mutating. What is known is that this bio-accumulation of pesticides leads to chronic, life-long diseases suffered by 50 million farmers worldwide.

Still the end is not in sight. Making the problems even worse, as cotton is not a human food, there is little restriction on the amount of pesticide that can be applied, even if the cottonseed does do a full circle by ending in our stomachs via the seed being fed to chickens etc. There's an even more direct route, and that's through peanuts because the poison-soaked soil is not exclusively used for cotton growing. To improve the soil, peanuts are grown intermittently as a nitrogen fixator for the soil. Peanuts are of course a food item, definitely consumed by humans. This low-growing, shrubby plant has seen a significant rise in allergies to it, coincidentally after this agricultural method became common practice. There might be a correlation between pesticides and the way they cling to the proteins of the host plant, becoming biologically intertwined. These are then violently rejected by the human organism as this alien protein shouldn't be allowed in; the body blocks it, manifesting as any number of allergic reactions, with death by anaphylaxis being the worst (acetylcholine is involved in this also). Peanuts kill about 100 people this way in the UK every year, a similar number to those killed by insects. Airborne pathogenic, bacterial human-invaders get the blame from the industry for these deadly anaphylactic peanut deaths … really? Are these new also? Considering that imported peanuts are already soaked in antifungal agents to prevent mould growing on these naturally oily legumes during transportation, it's unlikely. It is a little peculiar that there is little to no mention of any peanut allergy among those who ate it historically and still do in some quantity today, such as the Chinese and Indonesians … Just saying.

There is already a precedent involving allergic reactions to pesticides in food ingredients. That is regarding wheat, with a newly recognised gluten intolerance, which has come about because the proteins of the wheat, the gluten, get invaded by the proteins from the pesticides attaching themselves, as they were designed to do. Unaffected by processing, this toxic-laced gluten when presented to the human body as bread, pasta or a million other things gets accepted at first as the body recognises it, but is then rejected as it enters the digestive system, causing myriad problems as the toxic-laced proteins wreak biological havoc. Suddenly this new gluten-intolerance has become more prevalent and fashionable than a moccasin-wearing Hoxton vegan. I wonder why? Blaming the pesticides will always see you in court, so it's all just allegedly.

* * *

Well, after all that I hope you can see why an alternative to cotton might be necessary. There are plenty of good cotton manufacturers, and clothes made from organic cotton, natural dyes and fair wages. These all need to be fully encouraged with that good old much-hyped-up consumer power. The £4 t-shirts from Primark are the problem—making it as cheap as possible is never going to achieve a better product. Consumers demand cheapness, apparently, and if the t-shirt fades or loses its shape after two washes, it's no bother—just go and buy a new one and feel that rush of excitement that comes from consumer spending. I've still got hemp t-shirts over 20 years old that have kept their shape and colour; they were a little more expensive, but are cheaper over time, winning hands down for personal economic sense—which is not actually what modern economic sense is all about. It's not personal, it's business and it's all about making as much money as possible.

It does require some mental understanding and readjustment to want to and actually transform our consumer habits. But if the consumer demanded an item that lasted for years, which was less environmentally destructive and was prepared to pay a real and fair price for it, so be it. Manufacturers will have to give in, otherwise they will go out of business. More importantly, it will encourage the re-emergence of quality and artisans, creating a whole new workforce, led by craftspeople and artists, which when combined with some of the impressive latest technological inventions across all industrial sectors would offer a future that begins to look aesthetic, environmental and sustainable.

* * *

Pesticides are a very depressing scenario with cotton, but this is not the case with nettles, whose biggest demand is a rich nitrogen soil. Nettles have very few natural predators/pests, and therefore can be grown without the use of pesticides, with phytochemicals within the nettle acting as a natural insecticide; and of course nothing 'needs' a deadly man-made poison to grow! The water table, rivers and seas have all been depleted and desecrated, while nettles are nothing like as thirsty as cotton plants and can end the cycle of poison and run-off and soil depletion, with their roots preventing soil erosion as well as cleaning them of contaminants via phytoremediation.

All global cotton farms should undergo a year or two of soil improvement, then begin again. All large farms should be divided back to smaller ones, grow only organic cotton with other crops in rotation, as before monoculture, only this time with the advantages of a whole world interconnected via the internet and fairer profit distribution. Farmers can share growing and pest avoidance techniques and strategies, a little like the good initial intentions of the Codex Alimentarius when it was set up in 1963, before it was completely hijacked and distorted to allow the monopoly of the market to be retained by the international mega-corporations.

The nettle industrial revolution

So far in this chapter, we've already seen nettles as a source of food, medicine and clothing. This all gives it the sheen of viability, although the fact it isn't being done on an appreciable level tarnishes this potential far too readily. One of the major reasons it isn't being done is because the current economic system has several fail-safe mechanisms that protect the status quo and make it very hard for any new entrants, especially those not wearing the same tie, let alone those attired in a nettle shirt!

Currently, it would be true to say there isn't a nettle industry, which gives it the positive spin of there being 100% potential still available. There are people employed in cultivating nettles, selling the dried leaves and roots to herbal wholesalers and some making yarn from Poland, Germany, Nepal and China. Few people rely on nettles for their regular income, with no tax returns for nettle sellers in England, Ecuador or Estonia or elsewhere, as employment via nettles is certainly thin on the ground at the moment. The best start is to have more nettles grown, as the supply is imperative.

Slow inroads are being made, as nettles are beginning to appear in many products these days. Some sweets are coloured by nettles, with chlorophyll listed as a food colourant, E140, indicating that someone, somewhere is processing nettles to a relatively large extent and is supplying these green nettle dyes to not only the food industry, possibly also the plastic, clothing, printing and paint industries as well. The whole global market for colourants is somewhere in the region of £20 billion per year, so nettles can certainly capture some of this, particularly with their superlative environmental credentials.

* * *

There has long been available from the heath store herbal extracts containing nettles, from lozenges to capsules, pills, tonics and tinctures, with an established association of nettles and health with some standardised constituent elements, which is good for the sake of medical and industrial recognition, but bad for the problems mentioned earlier about refuting holism when nature Herself and the definition of health is when whole and in balance.

Nettles are out there, but a reason they are not taking off could be because of the general disdain and lack of enthusiasm or knowledge about them. When the word gets out of the real benefits experienced from just a few days' continuous consumption in any formulation, social media will be full of people curing themselves. Nettles will become the new miracle plant that was hiding in plain sight. Then the pharmaceutical giants will be fighting to get to the patents, with the food industry making nutritious, revitalising, nettle processed foods (stealing the Gorilla Juice idea, no doubt) and key celebrities being seen in nettle trousers and hats. As with so much, the chance of success can be a fickle thing, dictated by fashion and trendsetters not necessarily relating to quality or a meritocracy.

An easy launch-pad for the soon-to-be-burgeoning nettle industry is to piggyback off the success of hemp, as the machinery to process hemp requires only a slight alteration to make it useable for nettles as well, as both plants share many qualities and uses. As we'll see in the Cannabis chapter, a multi-billion-pound industry has rapidly emerged, achieved in the main by many small-scale growers and producers co-operating.

The best way to make nettles a viable plant is to use every part, utilising them in as many industries as possible and creating multiple revenue streams. So the textile industry could use the bast (outer fibrous bits) and the animal bedding and paper industries use the shiv (inner pith), while the food, toiletries and medicine industries make use of the leaves, seeds and roots.

Making money from nettles will cease to seem like a strange proposition when plenty of people are actually making money from the growing and utilisation of nettles. Then the weird becomes normal, and it will even begin to seem weird that there was a period from 1950 to 2020 when we completely neglected the nettle.

* * *

A good venture to raise nettle's profile would be to use nettles as fertiliser, a well-known and ancient practice. To encourage farmers to use them, give hard-hitting marketing talks, with impressive PowerPoint presentations showing lush fields of crops all fertilised by nettles, with loads of graphs all pointing up and upwards for yields, profits and sustainability. Trials will have to be conducted demonstrating the efficiency of growing plants with nettle fertiliser, as well as growing lots of nettles while making a big fuss about the minimal water and pesticides required. These nettle fertilisers can be rolled out from an agricultural level and sold by the ton, or as small containers and sprays for the domestic garden and home.

It's not news to farmers that nettles make a fantastic fertiliser, as there are still many farms that make a liquid feed from nettles, with comfrey being a popular addition and a little yarrow as an activator. After just two weeks covered in water, the nettle soup takes on a most revolting smell and can be sprayed on fields and allotments. Nettle's rich nutrient content perfectly breaks down to a nutrient-rich fertiliser that makes these nutrients easily available to the growing plants it's fed to. Different dilutions can be used for different plant and crop solutions, with the added bonus that it also acts as a great natural insecticide. It's also a wonderful first-aid for houseplants—if a liquid nettle fertiliser will not reinvigorate a sickly houseplant, nothing will. Curiously or conspiratorially, nettle manure has been banned in France since this legislation was passed in Loi n° 2006–11 du 5 juillet 2006 d'orientation agricole NOR: AGRX0500091L. With a penalty of €75,000 there is a ban on nettle use and even a prohibition on disclosure of information, so to even suggest its use is finable, as is therefore this book in France. It's a common and popular feed in France, called *Purin d'ortie*, with different diluted strengths being used for centuries for its feeding, weedkilling and insect-repelling abilities. Banned from 2006 to 2011. Why weren't the French farmers dumping loads of nettle manure on the steps of the French Parliament in protest? This surely is a result of lobbying pressure from the £150 billion a year fertiliser industry, an industry responsible, via phosphates, for spreading cadmium, a known carcinogen and a cause of bone and joint diseases, extensively and globally. With an average of half of all applied fertiliser ending in waterways, the pollution becomes global and pan-species, and it certainly isn't only cadmium found in fertilisers, there's also, among many others, chromium, lead, mercury, nickel and vanadium.

* * *

It's not just the land nettles work wonders on, they also do a great job with the hair and scalp. The seeds have shown particular success in this area, with tonics and lotions being a well-known treatment for millennia among herbalists, utilising the leaves and roots as well. This is a massive industry to tap into, one that with a little marketing can easily explode, helped by simply highlighting the inherent dangers in modern shampoos and the so-called beauty products. Market research will reveal the consumer interest in clean, environmental alternatives to the incredibly harmful products on the market at the moment, especially when the dangers are more widely understood. The hair products sector is worth in global revenues annually somewhere in the region of £65 billion, so there has to be a potential for safe and natural alternatives, particularly when the results achievable from nettles (along with rosemary and others) are promoted.

The chemicals being allowed to be added to bathroom products are incomprehensible and often unpronounceable. These so-called beauty products are definitely not health products, and a shampoo with cancer risks doesn't sound very beautiful! If the public knew of the diseases and ailments the ingredients have been linked to, they might be more careful about what is being rubbed into the scalp or soaked-in in the bath. Then if the opportunity arose to buy a safe, sustainable and environmental product for only a little more money—wouldn't that be a desirable product? Those cheap shampoos are cheap because they are full of cheap, dangerous chemicals. The product may be emblazoned with pictures of flowers and harmonious slogans, implying a natural product, when all it will have is less than 0.5% of anything vaguely from a natural source, which has been thoroughly denatured by the time you squeeze a palm-full out in the shower.

Nettles offer a fantastic and viable alternative to many bathroom and cosmetic products, from shampoos, conditioners, scalp and dandruff treatments to soaps and lotions—already a brand in the UK has jumped on the nettle seed and hair connection, albeit encased in a slurry of man-made chemicals.

* * *

Paper manufacture is also possible from nettles, as it is from a plethora of plants from grasses to trees, most commonly from conifers, although hemp, jute, sugar cane, kenaf, bamboo, cotton and even seaweed are all alternatives. With nettles, both the bast fibre and shivs can be used for paper production as it's all mashed, beaten and blended to a pulp. This pulp is then spread across frames submerged in water, resulting in a strong fibrous paper. Naturally light green, it can be readily bleached and dyed, with plenty of safe, environmental options available to achieve this. It is also possible to upscale and use nettles as an ingredient in more modern processes, where the paper produced can be a mix of several other plant materials, such as bamboo for a water-resistant paper, toilet papers or recycled papers for wrapping paper. The possibilities are enormous, and papermaking is a definite market that has yet to be exploited by nettles. It certainly is a possible revenue stream as soon as more nettles begin to be cultivated and processed, with any waste becoming earmarked for other industrial sectors' use.

This exploitation of wastes has made a new and huge revenue stream among many industries, such as the waste stalks of corn being turned into sweeteners, starch, vitamins, protein, paper, plastic, ethyl alcohol, biofuel and loads more—nettles have the same potential for pretty much the same things.

Even if not used for anything and just left in the field as a groundcover, nettles act as a nutrient accumulator while also sequestering carbon, helping in the ongoing battle to harness greenhouse emissions and rein in climate change, as well as removing toxic chemicals from the soil via phytoremediation, making spillages and former industrial polluted sites safe. Nettle's preference for soils rich in nitrogen and phosphorus can also be put to use, making them ideal plants to break down human and animal waste, as done by the reed beds already in use for soak-aways from households to industrial facilities. For those nettles deemed unsuitable for any other uses, such as those used for phytoremediation, or the waste from any of the industrial uses, electricity can be produced from incineration in biomass generators.

Elaborated upon in the Cannabis chapter as it's already being done within that sector, nettles can be used in the building industry, from producing a type of insulation to a type of concrete 'nettlecrete', mixed-density fibre boards for walls, floors and even furniture, as well as textiles for curtains and soft furnishing. When there are guarantees of availability and it's more competitively priced, there is no reason why nettles can't begin to be incorporated as one of the raw materials used in the building industry.

Nettles even make the meat of pigs and cows taste better and of course impart nutrients, which are transferred to the human digestive and immune system when eaten. This is an important consideration when weighing up whether Diclofenac or nettles should be given to an animal soon to be slaughtered anyway. Diclofenac drops vultures like a stone, while nettles make the gorilla the king of the jungle. The healthier the animals become fed on nettles the less drugs are required, significantly reducing livestock pollution of soils and waterways, as well as lessening the side effects the livestock suffer; importantly it means less chemicals transferred up the food chain and more nutrients for those who eat meat.

Nettles have long been known as a valuable addition to fodder, dried out and added to hay. Dry nettles retain a great quantity of minerals and protein, and are not only a great feed, they have curative potential from being so rich in nutrients, as the immune, nervous and digestive systems are all given a good feed, as indicated in the previous positive scientific studies. In Sweden and Russia it is still common to utilise the galactagogue benefits of nettle, increasing the amount and improving the nutrient profile of milk in cows, sheep and humans ... Why did it stop elsewhere?

Chickens also benefit from nettles, as shown in the Iranian study, exposing nettles' unique role in making sure we have good transmission of thoughts and movements. When dried, crushed and added to chicken feed, egg production is increased, the health of the chicken is raised, drug use reduced and the nutrient profile of the egg is improved, as is the colour—nettle's high content of lutein, carotene and zeaxanthin all act as a yellow pigment with the added bonus of being super-healthy phytonutrients. All this is achievable by giving the chickens less than 1% of their total feed as nettles. The seeds are good for fattening up fowls, and in Egypt and Holland horse-dealers mix nettle seeds with oats and other food to improve the sheen of the coat, making the animals more sellable. Tinctures from the pressed seeds were used to restore human libido, sexual energy and general vitality throughout Europe in the nineteenth century and before.

Perhaps the person who knew the most and devoted more time to nettles than anyone else was Prof. and Dr Gustuv Bredemann (1880–1960). He was a German botanist and researcher, who from the 1920s to the 1950s grew lots of nettles. During this time he cultivated and bred a nettle where the fibre content had been raised from 5% to 17%, with the intention of using them for fabrics. Bredemann's work was forgotten about until fairly recently, but fortunately varieties of his fibre nettles are still dotted around several academic institutions in Germany.

* * *

From the conclusions of all those scientific studies and research mentioned earlier, the oft-repeated recommendation was that nettles should be grown commercially whether for human or animal food, medicines and tonics, toiletries, clothing and fabrics, biofuel or fertiliser. These studies represent a growing body with scientific backing corroborating the traditional and ancient uses of this most noble and proud plant. It wouldn't take much to elevate nettles back to a common everyday food. When one of the big soup companies puts a celebrity as the face of their new timeless caveman nettle soup, then people will want to know about the benefits, and when they hear the truth, everyone will want to be eating nettles.

There's not going to be a marketing campaign any time soon coming from the nettle growers of Poland or England. It is what happens when a group of growers club together and, depending on their size and the potential market to capture, their marketing campaign will range from internet adverts, a door-to-door leaflet drop, to national newspapers and magazine and television ads, and it may even involve very dubious scientific tests to give it a valuable scientific endorsement. Think of any of the past 'superfoods' this century, and behind most of them would be a conglomerate of growers paying for marketing and advertising. Avocados shared the limelight with fast food and fast cars at recent American Football Superbowl finals (2015–20), costing over £3m each time, paid for by Avocados from Mexico (a non-profit 'charity', which obviously seeks profits for its clients) for a 30-second ad. Demand is now so high for avocados that the price is beyond the poorest from Peru and other countries where they're grown; it was a similar scenario with corn and quinoa once they became fashionable in New York, London and Paris.

The current rise in the numbers of vegans, bolstered by loads of celebrities choosing this path, could be a great help in spreading the word about the power of plants—

come on Stevie Wonder, Al Gore, Jennifer Lopez, Joaquin Phoenix, Michelle Pfeiffer, Ted Danson, Capleton, RZA, Ariana Grande, Miley Cyrus, Mike Tyson, Woody Harrelson, Demi Moore, Brad Pitt, Pamela Anderson, Russell Brand and loads of others. There needs to be a really cool advert made for veganism, getting as many of these newly proclaimed vegans as possible to appear in it, even if they have since fallen off the fashion-fuelled vegan train.

The more vegans there are, the more restrictions there are on which materials to use for clothing, such as wool and of course leather—this should mean there is a growing demand for non-wool textiles, especially if the celebrity vegans make people aware of the alternative fabrics. Vegans as a demographic tend to be more environmentally astute (not always), and they will more readily understand and accept nettles, hemp and flax as being the quality materials they are. Also, all three plants can potentially share any technological breakthroughs—diversity is always going to be a safer model, as opposed to putting all your nettle-fed, super-yellow and uber-healthy eggs in one nettle-spun basket.

Get foraging

> Tender-handed, stroke a nettle,
> And it stings you for your pains.
> Grasp it like a man of mettle,
> And it soft as silk remains.
>
> (A. Hill, 1685–1750)

True that. The first consideration when foraging for nettles is to have some gloves. You won't need them when you know what you're doing, but if you need to read how to pick a nettle, you'll need gloves for now, possibly a pair of secateurs and of course a bag to put them in.

Second consideration is where to get them? Best place is back of a friend's or your own unsprayed garden, or public land—the last is harder to know if sprayed, although an abundance of weeds in the locality is a good sign it hasn't been—or any field away from a main road to limit petroleum pollutants. Use your common sense and show respect to other people, plants and animals, and don't trespass or climb over fences—unless lost or running away from a bull!

When a nettle patch is found, your next consideration is what to pick and when. Best is to pinch out the tip, the top two or three layers of leaf growth. This causes no harm to the plant, as it will still happily continue in the vegetative phase, with the bonus of two tops replacing the one, which can both be pinched out again in a month, from whence four tops will emerge, allowing the plant to produce more seeds or flowers. The ideal height is about 18 inches, and always before the plant has begun to set flowers or seed. We want culinary nettles, so throughout spring is the best time; larger leaves can also be picked at this time as well, and thereafter is the possibility of picking second- or even third-growth nettles before they flower, as late as November.

We need to briefly rectify the paucity of mention about the seeds. To begin with it's a bit tricky initially identifying which is a male or female plant. The male has tiny flowers, which look a bit like they might be seeds, although as each sex tends to grow in clusters it is possible to see a whole group of females. When in seed the female plants will have bracts full of seeds hanging downwards; wait till these turn brownish and then harvest, the easiest way being to cut the heads and hang to dry, and then tease the seeds out into a bowl or bag. Seeds from many plants are an extremely rich and concentrated source of nutrition, and nettle seeds are no exception, with the seeds being rich in both serotonin and acetylcholine, acting as an effective pick-me-up, revitalising the adrenals (consider nettle seeds if you are a long-time sufferer of hypothyroidism), skin and hair. Miles Irving's superb forager.org.uk site has some great recipes and links using the seeds (notably the nettle seed and dandelion bars). Lightly roasted, they are absolutely divine, sprinkled or crushed on whatever takes your fancy.

The benefits of picking some seed for yourself come with many positive side effects. Depending on where you have to go to get them, there's the fresh air, exercise, and if you go with a friend, some companionship as well. Also going to pick a wild plant opens a connection with Nature, one that has become very nearly severed. Apparently spending time in the woods or nature can bolster your immune system by a staggering 50%, which if it were a pill with only a 25% increase in immunity would be seen and sold as a miracle drug. It also brings humans together, shows us our common unity and reminds us of a time when we were all free and we literally went out and picked what we needed. The actual act of foraging is incalculably more rewarding and educational than just reading about it (which is of course the perfect first step). So, get a carrier bag full and let's see what is possible—

Recipes

When you've successfully managed to forage some lovely deep green nettle tops and leaves, you'll hopefully be itching (not literally, one hopes) to use these as the main ingredient in some food items. The stinging trichome is made from water-soluble silica, so blanching quickly with a stir removes all worries. Everyone has different tastes and every chef offers a different version. Here's what I make with nettles; feel free to change and add whatever, it's just to show how easy it is to incorporate something that actually tastes good as well as being superlatively healthy and of course timeless into your diet—

This **soup** gets pretty much the thumbs-up from everyone, with my kids moaning there's never thirds as everyone has seconds of it!

Take 2 largish sweet potatoes, 2 largish normal potatoes, large carrot, large onion, 6 cloves of garlic (or a smallish bulb), a small colander of freshly picked nettle tops, a teaspoon of cumin seeds, good pinch pink salt and pepper, 1 litre water. Chop up the veg and steam the potatoes and carrot for 5 minutes, then fry the potatoes and onion and carrot for 3 minutes, add the garlic and fry for 2 minutes more, then add the water and boil. Add salt, pepper, freshly ground cumin seeds, nettles and any seasonal fresh herbs such as parsley, wild garlic, oregano etc. Turn off heat, leave to stew for 5 minutes, blend thoroughly and serve.

Pesto is amazingly versatile, as a base for other sauces and dishes, especially good with tomato sauces, dolloped on a pizza, added to a stir-fry, eaten neat on bread, crackers or anything you can imagine. The pesto coming out of our own kitchen changes with the seasons, just as with the hummus coming up next, the biggest seasonal change is that from March to May wild garlic (ramsons) are used liberally.

The standard pesto recipe is—Half a colander of nettle tops, 4 halves of sun-dried tomato in oil, medium white onion, a handful of whatever fresh herbs can be found from basil, parsley, oregano, sorrel, jack-by-the-hedge, dandelion, cleavers, dead-nettles, plantain, ramsons, hawthorn, beech, birch and hazel leaves, 3 tablespoons of ground seeds (sunflower, hemp, linseed and pumpkin, for instance) and nuts, 4 garlic cloves (not if adding ramsons), juice of a lemon, extra-virgin olive oil, pink salt, freshly ground black pepper and a glug of balsamic vinegar.

Chop and lightly fry the onions for 4 minutes, add garlic, add washed nettles, letting them blanch in the little water left from washing for 2 minutes, add the rest of the herbs, seeds, salt, pepper and balsamic. Blend, adding olive oil to desired consistency (around 4 tablespoons). This will last in the fridge for at least 2 weeks if it's not eaten straight away!

Hummus is about a third of a cup of extra virgin olive oil, 3 cloves of garlic or more to taste, a spoonful of tahini and a tin of chickpeas. Add to this some roasted vegetables (such as red pepper and aubergine) and seasonal herbs from the above. Supermarket hummus is rarely good for you: check out those oils being used such as sunflower, rapeseed and palm—every time you eat something with palm oil in you are effectively encouraging the spilling of orangutan blood—it has to be extra-virgin olive oil every time for proper hummus.

Switching to just these three (soup, pesto and hummus) more regularly could improve your diet, especially if you were feeling it was becoming dominated by processed foods, as it is for most people.

For **nettle tea**, what I do, about 11am, is put a berry fruit tea bag in my big mug, then go and find what can be picked as quick as possible. So in will go a few from the following: nettle top, a sprig of rosemary, oregano, plantain leaf, dandelion leaf, clover, mint, meadowsweet, star anise, cloves and a little honey. I'll then top this up two or three times throughout the day, maybe adding another nettle and mint top and a bit more honey. By the last mouthful, after stewing all day … aahh! I love that mouthful.

* * *

The Albanians make a layered bread, called *borek*, where nettles with other herbs and rice are filled between layers of dough, while the Greeks make a similar dish called *hortopita*. In Georgia a dish of boiled nettles seasoned with walnuts is still fairly common, while the Romanians make a soup from fermented wheat bran, vegetables and young nettle leaves. In Scotland a vegetarian version of haggis uses nettles, leeks, broccoli, cabbage and rice, all boiled in a muslin bag. In Ireland, nettle cream soup is still a favourite (as it once was everywhere), and in the Basque region of Spain they are put in omelettes. I always put a handful of blanched leaves in a nut roast.

In Europe and North America a cordial was once commonly made, with nettle leaves steeped in a sugar solution; citric acid was added as a preservative. Modern versions are still available. Nettles along with other healthy and tasty herbs were added to beers, wines and tonics for much of history and no doubt deep into prehistory.

Northumberland's Blagdon Estate makes a cheese with nettles in, as does varieties of Dutch cow and goat milk Goudas, while Cornish Yarg is covered and cured in a nettle skin. Rather surprisingly, boiling nettles in salt will give a juice capable of curdling milk, a superb healthy and vegetarian alternative to rennet in cheese. The same decoction can be rubbed liberally into leaky seams in wooden tubs and boats to fix them.

* * *

There is no reason why a considerable amount of fresh soups can't be rolled out across supermarkets on a seasonal basis, which for nettles can be a long season from late March/early April right through to October. Nipping out the tips, as explained before, means that as little as a sixteenth of an acre can provide a considerable amount of nettles. When farms can start to understand the benefits of downsizing, adopting permacultural philosophies and diversifying, as well as making a range of end products, such as sauces, pesto, hummus, pies, soups and yarns, farmers will enjoy a much wider revenue stream. As it is farmers have been bled dry by the supermarkets in producing food as cheaply as possible. It is time to turn the tables and start to compete on quality and variety, then the farmers will see a rise in income, as the consumer will be happy to pay for real quality, as well as enabling far more local people to be employed in a range of activities and production.

And, to finish with, the World Nettle Eating Championship, established in 1986, takes place at The Bottle Inn in Marshwood, Dorset, drawing a crowd of thousands these days. It sees competitors served two-foot lengths of nettles, with the leaves stripped and eaten. The current record stands at 80 feet in an hour!

Nettle neglect

Most people look at nettles as a useless weed that needs to be eradicated—hopefully no one who has read this chapter does. Up until about 70 years ago it would not be an exaggeration to state that everyone ate nettles regularly, along with a variety of other preferred foraged foods. When you eat nettles, you are eating the same nutrients enjoyed by herbalists, health practitioners, midwifes, witches, Romans, pagans, Danes, cavemen, Neanderthals and everyone else, all benefiting from the healthy nutrient profile we evolved on.

It is time to recap on all that has been mentioned, with so many facts to digest during it. We can start with nettles being a possible answer to neurological ailments, as the past 70 years of nettle neglect has also quite possibly opened the door to an imbalance in our acetylcholine pathways, responsible for our movement and thoughts, because only nettles have one of the two known enzymes that control this acetylcholine, with the other one destroyed in the food processing industry.

Nettles were also shown to be an analgesic, anti-allergenic, antibacterial, anti-carcinogenic, antifungal, anti-histamine, anti-inflammatory, anti-influenza, antimicrobial, antioxidant, anti-rheumatic, astringent, decongestant, detoxifier, diuretic, expectorant, galactagogue, haemostatic, immune-modulator, immunostimulant, insecticide, nutritive and more.

Nettles are also good for the skin, hair, digestion, bones, teeth; they make the best clothing, twine, baskets, nets, ropes and fertiliser; they feed our most beautiful butterflies, make animals healthier and taste better, can insulate the house, make the electricity to power it, help reduce carbon dioxide and assist with climate change, clean the soil, as well as keeping you well and replace some drugs for humans and animals.

Next up is a plant given even less thought than the nettle and quite possibly has even more potential …

DANDELION

(Taraxacum officinale)

The best things in life are free

I'm not alone in admiring the dandelion. Flickering across my imagination are timeless scenes of children blowing the puffball seeds, laughing and making wishes as the air fills with little fairies taking flight. Such scenes fill me with warmth. Fickle gardeners bemoaning their taproots make me chuckle at their ignorance, while chickens and rabbits just devour the plants anyway.

This chapter will hopefully open your heart to the humble dandelion. They've been knocking at our door for generations now, and we ignorantly won't let them in. Their tenacity and toughness has to be admired, having never given up. Now is the time to shine a light on them, as we discover yet another super-plant hiding in plain sight.

* * *

Finally, the time was right to write, having waited patiently for this moment, which is nothing compared to the serene Buddha patience of this golden beauty. She was in no rush, it is our species that seems to be in a relentless rush, scurrying here and there like caged rats, being told what to do and when to do it, with little purpose or understanding of what we are doing on this planet. This constant toing and froing has been a massive diversion, always busy, never stopping to actually observe where it is we are going. Being shepherded to a life that is not our true destiny. Education, employment, consuming, sickness and death—not really worth the rush, is it? Especially when and if you stop, it can be seen that the best things in life are free.

It might cost £100 million for a landscape by J.M.W. Turner, but all he was doing was trying to capture a moment of natural beauty he was witnessing with his own eyes. This same natural beauty is accessible to us every day if we just hit the pause button and have a look around. Even the concrete jungle has beauty by the bucket load, with many sunsets providing a vista more than comparable to the best Turner could muster up. The lines and juxtapositions of buildings and trees, skies and hills, colours and textures hold their own priceless aesthetic, when the only expense is the time to stop and the appreciation of what can often be awe-inspiringly dramatic. We are not only blessed to be living on such a varied and beautiful planet, we are also fortunate enough to comprehend and appreciate this beauty.

The blue sky, clouds, birds, seeds and plants all growing from urban decay, the dandelions and trees in blossom, the scampering squirrel, all hold more warmth, companionship and love than we can comprehend in our blinkered world. If you can stop and admire what is around you, this is a priceless gift. Saint Attenborough has been tirelessly showing us this all our lives, from *Life on Earth* to *Blue*, *Green* and *Frozen Planet*, clearly flaunting the wonderment we reside among. Isn't it about time we stopped rushing to a degenerative grave and start to open our hearts and souls to what has been given to us for no charge?

There is no financial cost for this appreciation, but it does have a payback in an increase in happiness, as a pleasure is felt and mental health gets a credit. Real health, homeostasis, is a combination of physical health, gut health and mental health. To collect some wild nettles or dandelions will boost appreciation and satisfaction levels of yourself and the plant world; as well as companionship if you go with someone, because humans are social creatures and we yearn to share experiences and time with others—it's not just guinea-pigs who die of loneliness when solitary. With that reported 50% boost to the immune system from being in the countryside, regular visits would clearly be a bonus.

By eating the collected plants we are directly feeding the brain and the other parts of the health equation, as dandelions are nutritive, superlatively so in many instances. The nutrients unequivocally feed all of the human systems beneficially, be that the skeletal, muscular, nervous, circulatory, digestive, urinary, reproductive, lymphatic, endocrine or respiratory.

For no cost, real health is possible. The only requirement is a little bit of real knowledge. Although the 'powers that be' would make it an offence if it became public knowledge, not just the information, also the picking of anything for free. For your safety and welfare will be the claim, and this is happening today, such as with Directive 2004/24/EC, discussed in the Plantain chapter, where EU membership is restricted to countries until their herbal lore is done away with. The corporations want to cleanse all ethnic knowledge of the plant world, in very much the same way as conquistadors and colonialists with their trademarks of arrogance and white supremacy.

* * *

We've just seen that nettles have the potential to play a significant role in this reformation of our corrupt cultural systems and destructive lifestyles. Dandelions have perhaps even more to offer, because they are beyond amazing and incredibly they are also free, with no strings attached, with only those life-affirming strings back to Mother Nature, slowly weaving humans back into the tapestry of life on this planet.

To set the scene, here's a quick perusal of the etymology, history and botany of dandelions, for what I hope you will find is interesting, or, as a good friend just said—"It's not another sacred herb, is it? It's a bleeding dandelion this time, mate, you're off your ferking head!"

Etymologicum

Not long after deciding to start studying herbalism, reading through a herbal book I drifted off into a daydream, imagining I was strolling in the countryside, sun shining, marigolds mingling among a sea of dandelions waving in the breeze, as I crossed an open field, bordered by hemp agrimony, fat hen, melilot and the wafts of citrus from some lemon balm poking out of the ditch along with some borage and the similar-looking comfrey, with the light catching their silvery, downy leaves. Standing firm in a row of some hawthorn bordering the field, I smile and wink at a cluster of white dead-nettles, camouflaging themselves as nettles—clever move.

Clearly I was starting to become besotted with flowers, judging by these daydreams. Although what was strange was that all these plants had a very different look if written as their botanical names. It would now read—

"… strolling in the countryside, sun shining, *Calendula officinalis* mingling among a sea of *Taraxacum officinale* waving in the breeze, as I crossed an open field, bordered by *Eupatorium cannabinum*, *Chenopodium album*, *Melitotus officinalis* and the wafts of citrus from some *Melissa officinalis* poking out of the ditch along with some *Borago officinalis* and the similar-looking *Symphytum officinale*, with the light catching their silvery, downy leaves. Standing firm in a row of some *Crataegus monogyna* bordering the field, I smile and wink at a cluster of *Lamium album*, camouflaging themselves as *Urtica dioica*—clever move."

The first impression when written like this is that it just lacks any flavour or poetry. I knew that the *album* word indicated the flower is probably white, but what was all this *officinale* and *officinalis* business? What makes some official and the others bogus? There must be something important about bestowing such a grandiose name on these plants. For starters, the meanings of the names had a pretty good ring to them. 'Official' in my heavy two-volume edition, massive in size, tiny in eye-aching font, *The New Shorter Oxford English Dictionary* says,

> performing some office or service, serving the needs or purpose of a vital organ; with 'officinal' being a herb, drug, etc.: regularly sold or used as a medicine. Of a medicinal substance: kept as a standard preparation by apothecaries and pharmacists.

Wow! Says it all really.

All this talk of apothecaries conjures up another vivid image, crossing times and flying way back to the ancient Greeks, to a herbalist's store, terracotta floor-vases all full of sun-dried herbs, the aroma heavy and layered. Scales and weights. Fields of herbs in flower, spread around the Mediterranean and Europe and across the Silk Road, all to be dried and stored in these vases. How exciting. I wanted a slice of this action. This name, *officinale*, provided just that; it was a wormhole back to a time when it was these herbs being served as foods and medicines. The actual naming of them is down to Carl Linnaeus (1707–1778), who invented the binomial system of nomenclature (that is, he gave them two names), giving those with an established and documented medicinal or culinary use the *officinale*, or *officinalis* name, to about 60 plants in total.

This *officinale* suffix was not a secret code that unearthed the holy and sacred union of 23 herbs that held the key to longevity, health and full mental and physical potential. Oh well, that wasn't expected anyway, but it did open a new avenue for some botanical detective work. All these new paths leading to wellness and mindfulness, some of which had yet to be retrodden for generations. Many paths, all strewn with the very plants we're talking about.

What's in a name?

An awful lot. Here's a very important point about language and cultural knowledge—the erosion of languages around the world also erodes our knowledge of the plant kingdom. Going through those couple of sentences again from the beginning of the last section, this time instead of using the common or botanical names, we use the regional/ vernacular names. Here's how it sounds now—

"... strolling in the countryside, sun was shining, Gold Bloom mingling among a sea of Piss the Beds waving in the breeze, as I crossed an open field, bordered by Ague Weed, Melde, Yellow Sweet Clover and the wafts of citrus from some Cure All poking out of the ditch with some Star Flower Oil and the similar-looking Boneset, with the light catching their silvery downy leaves. Standing firm in a row, some Bread-and-Cheese bordering the field, I smile and wink at a cluster of Adam-and-Eve-in-the-bower, camouflaging themselves as Devil's Plaything—clever move."

Common words now lost would have been *ague*, meaning fever (a range of ailments from flu to malaria), while *melde* in Old English would have indicated the plant was suitable for food or fabric. These three same paragraphs written in effectively three different languages show that the common names aren't particularly descriptive; the botanical names offer a little more information, but you'd need to know Latin and Greek to understand it otherwise it's just a foreign language; while the local vernacular names provide a much better description of what the plant is, where it might be found and the best uses of it. Regrettably and irretrievably, most of these regional names have already been lost to us.

This is what has happened around the world as smaller cultures are absorbed into larger ones, or the locals are simply exterminated. When the Spanish or English language got enforced on the tribal people and they were forced to be resettled into reservations, as their languages disappeared the local descriptive names for the flora and fauna also disappeared, which equates to all the local knowledge, gathered forever, lost for all time.

* * *

Folk and regional names remain in a hugely diminished quantity for the dandelion (with more lost, yet still more than the local nettle names surviving, which is literally a handful); still with us are bitterwort, blowball, cankerwort, clock flower, crow-parsnip, doonheadclock, fortune-teller, Irish daisy, lion's tooth, piss the bed, priest's crown, puffball, swine snort, telltime, wild endive and yellow gowan.

These names apply to the species abundant throughout the British Isles, *Taraxacum officinale*, as well as being found as far as New Zealand, Australia, India, southern Africa, all the states of America and most provinces of Canada, across Europe in Poland, France, Spain, Belgium, Holland and Greece (with some botanists claiming Greece to be their original homeland). If it's not *Taraxacum officinale* it's going to be a close cousin, both visually and nutritionally, with a single flower per stem—unlike the similar-looking flower of the hawkbit (*Leontodon autumnalis*), from the same daisy family with a very different leaf shape and multiple flower heads per stem.

It appears that the whole field of naming and differentiating these species and understanding their global movement, is, to say the least, an educated, convoluted mess. Some authorities say there are as many as 1,000 *Taraxacum* species in Europe, while others go for nearer 60. With crossbreeding and polyploidy common (creating chromosomes), the task is never going to be an easy one. Interestingly, there is a micro-botanical fraternity that looks at just dandelions, the taraxacologists, who differentiate 235 British dandelion species from *Taraxacum aberrans* to *Taraxacum xiphoideum*. Within a 500m circumference of my cottage, there are dandelions with small leaves, big leaves, thin leaves, fat leaves, deeply serrated leaves, nearly smooth oval leaves, reddish veins, tall stems and small stems, all definitely dandelions, although exactly which species or subspecies is best left to the taraxacologists. All I know is, I'm always envious seeing loads of massive leafy specimens around someone else's house, as mine are in the main fairly small in comparison. A patch of dandelions with big leaves when out walking is manna from heaven—thank goodness for (unscented and biodegradable) pooh bags!

* * *

The botanical name, often erroneously called its Latin name, was known as New Latin when Linnaeus was using it, not names ascribed by the ancient Romans. *Taraxacum officinale* has an interesting etymology and corroborates the lofty position of dandelions as a symbolic representation of the sun, lion and the Goddess. *Taraxacum* may well have originated from the Arabic for wild endive, *tarakhshaqun*, a word that when used appears to denote the root, or a dried preparation made from the root. The Greek origin of *taraxacum* is made of two words, *taraxos*, meaning to stir up, disorder and *akos*, meaning remedy. Which makes sense and clearly alludes to the health potential of the plant, especially when combined with its official designation as a healing plant via *officinale*.

As an almost irrelevant aside, in Chinese, the name translates as 'flower growing in public spaces by the riverside'. Which is just nice.

* * *

The name dandelion is not as it may sound, a *dandy* lion. The name is said to be from the French for the teeth of the lion, *dent de lion*, although it's mighty similar to the Spanish *diente de leon*, or the Portuguese *dente-de-leao* or Italian *dente di leone*, the Latin *Dens leonis*, or the Greek, as originally assigned by Linnaeus, *Leontodon*. Which is fair enough, as some edges of the leaf can loosely be seen as being a profile of a set of serrated teeth … sometimes, but this doesn't cut the mustard. Why are the French, Spanish, Italians and ancient Greeks even comparing the leaf of a dandelion to an animal that no longer lives in Europe? And Linnaeus named the hawkbit *Leontodon*, absolutely not because the smooth oval leaves resembled a silhouette of lion's teeth! Perhaps it was more a reference to the flower head, the mane of flowers. Although this lion thing goes quite a bit deeper than this, deep into history and through the back of prehistory.

The dandelion is actually a plant with a completely forgotten illustrious past. Dandelions are symbolic of not just a lion, as the lion itself is also emblematic and symbolic of the sun, and of course the astrological sign, Leo. The lion represents all that is strong and majestic, the lion rules over the land, is invulnerable, and will protect you from all evil and give you the strength and purpose in your stride to succeed. It's a stereotypical image of strength.

The flower of the dandelion, picture it in your head, is a fantastic representation of the sun, being a better pictographic representation of the sun than a sunflower, for instance—which is a better example of the Golden Ratio, the Fibonacci sequence, Pi and all that. The flowering head of a dandelion is actually composed of many individual flowers, called discs and/or rays. The flower of the dandelion opens with the sunrise, moves itself to face the sun as it moves across the sky and closes just before sunset. It is a little sun on earth. I trust you've got a good mental fix on what a dandelion flower looks like, as it keeps on getting deeper, because we need to consider a little further—why a lion?

A lion is the symbolic representation of the Egyptian Goddess of fire, war, fertility and healing, Sekhmet (in Lower Egypt at the Nile Delta end the equivalent was Goddess Bastet), a major player in the pantheon of ancient deities, who coincidentally is always depicted with a sun disc, a ray, being also the daughter of Ra, the Sun God. Lion and Sun together. Then there is of course the Sphinx and Leo, when the celestial constellation Leo housed the sun on the spring equinox, with the Sphinx orientated towards Leo on the horizon. Leo is the sign of fire and of course the sun, an astrological sign from 20 July to 20 August, the month of the sun's strongest influence in the year (in the northern hemisphere). This sun and lion symbol was also a key to belief in Mesopotamia over 5,000 years ago and continued with the Mongols during the era of the great Khans in the thirteenth century, with Ethiopia and the RasTafari also having the sun and lion as central themes entwined in their theological cosmology, and the duo of sun and lion being clearly depicted on the Iranian flag pre-1979. This is the lion the dandelions in our garden represent. The lion is the sun, as the sun is a lion and a dandelion is both the sun and lion. It's quite simple really, the dandelion is *awe*some.

That word awe, like so many words, derives from Mother Nature, from Aurora to auspicious and euphoric to Eucharist—the original was not a cracker and wine or has anything to do with Jesus Christ, it was an hallucinogenic brew allowing the Mother to be consulted. Each time the word was said it was done so with veneration for the Goddess as Aur and Eu is the word or sound for the Goddess—the word veneration itself comes from the Roman Goddess of love and beauty, Venari. The Greeks and Romans continued the tradition, and we still do today by living in Europe, named after the Goddess Europa, who was reattributed by the Greeks as the Mother of King Minos of Crete who slept with Zeus, although she was originally the Cretan Moon Goddess.

We are as detached today from the real meaning of words, numbers and symbols as we are from Nature, bumbling along in a bubble. Words are amazing things, as mentioned before; they put you in a spell, and to change anything you need to know how to really spell it, because words have been taken away from the public domain in many respects (which is exactly why understanding Latin and books were kept away from the public back in the day), being distorted and corrupted, often twisting the true meaning to avoid any suspicion of something being hidden behind the veil of truth. This apocalypse is coming, that lifting of the veil of truth, exposing the lies of our culture, but it's not coming from words because words are meaningless in a world full of lies. It's coming from the plants.

Chronological cappers

Starting way back in time, to Greek mythology and the age of the Gods, in celebration of his great feat of slaughtering the Minotaur, Theseus, with the planet's culinary delights at his beck and call, ate a dandelion salad, or with a slight variation in John Evelyn's *Acetoria, grande salad* (1699) Theseus's wife feeds him a dandelion salad for 30 days before he kills the Minotaur. Thirty is the number of days in a lunar cycle—well, it's actually 29.53 days—with 30 a fairly common number of stones at many megalithic stone circles.

Historically, the dandelion is St Brigid's flower. Brigid is one of Ireland's earliest recorded saints (actually shanghaied by the Christians as She is Brigit, the Goddess) and along with St Patrick is one of Ireland's most famous saints. She was born around AD 455 in County Louth, and died on 1 February AD 524, making this her feast day thereafter, conveniently the same day as the Celtic festival of Imbolc. The first green shoots of the dandelion can emerge at this time, with the dandelion also being one of the first noticeable flowers of the year (and one of the last also as late as December), acting as a good barometer for the soon-to-follow spring and abundance of new plant growth.

The first documentation of dandelions comes from the 7th century AD Tang Dynasty's *materia medica*, the *Yaoxing Lun*, suggesting their use for digestive disorders and tumours. Then the Iranian physician, Avicenna (AD 980–1037), who continued the work of Aristotle, Socrates, Hippocrates and Galen, while adding his own important theories on mental health and self-esteem, as well as arguing that understanding the root of disease is the essential first step if a cure is to be found—his *Canon of Medicine* was a standard reference book in medical studies till the seventeenth century across Europe—mentions that dandelions regulate menstruation and other female ailments, with further writings on *tarakhshaqun*.

The Welsh used dandelion to treat jaundice (mixed with cornflower and parsley) in the thirteenth century, as mentioned in a herbal written by the physicians of Myddfai in Wales. The famous dandelion and burdock beverage also hails back to the thirteenth century, with both of the roots being fermented, and a little honey added, making a refreshing, slightly boozy, healthy mead. It was apparently enjoyed by St Thomas Aquinas (he of the saying "To one who has faith, no explanation is necessary. To one without faith, no explanation is possible"), getting a mention in 1265 with his original recipe idea for these fermented roots. Dandelions, it would be safe to say by this time, could be found growing abundantly in the physic gardens of the various monasteries dotted around Britain and Europe.

Culpeper tells us in the mid-seventeenth century:

> It is under the dominion of Jupiter. It is of an opening and cleansing quality, and therefore very effectual for the obstructions of the liver, gall and spleen, and the

> diseases that arise from them; it opens the passages of the urine both in young and old; powerfully cleanses impostumes (pus) and inward ulcers in the urinary passages, and by its drying and temperate quality doth afterwards heal them.

He is surprisingly spot-on, even down to the autophagic qualities (elimination of dead cells) of dandelion, as after taking down the baddies, it can also eliminate them with the added bonus of feeding all the right parts of the host's organism from its nutritional excellence.

* * *

Among the First Nation Peoples, the list of ethnobotanical uses for the dandelion is rather extensive. The Aleut used a poultice of steamed or wilted leaves for stomach aches, sore throats and indolent ulcers. The Algonquin ate the greens to purify the blood and used the leaves as a poultice and for plasters for cuts, sores and skin ailments. The Apache drank the flowers to make them stronger, being known for their enthusiasm in scouring the countryside in search of them, for days if need be. The Bella Coola used an infusion of the roots for stomach pain, while the Cahulla ate them as a vegetable, as did all the peoples mentioned here and more.

The Cherokee chewed the root for toothache, used an infusion of the roots for the blood, with an infusion of the plant to calm nerves and as a general medicine. The Delaware used them as a laxative tonic, while the Hesquiat used the hollow stems for whistles to communicate incognito. The Hoh used dandelion as a general medicine, and the Iroquois took an infusion of all parts for back pain, anaemia, bad blood, constipation, pain, dark puffy eyes, kidney problems, water retention, dropsy, a laxative, swollen lungs, tooth decay, swollen testicles, to make wine from, as an anti-witch medicine, a love potion, and with the roots roasted and ground for a hot beverage.

The Kiowa used a decoction of the young leaves for menstrual cramps and ate them as greens. The Malecite did the same, as did the Menominee who ate a speciality of the leaves with maple sap vinegar. The Meswaki used infusions as a pulmonary aid, with the root taken for chest pain and as a vegetable, a speciality with pork. The Micmac ate them as a food, as did the Mohegan, who also used them for a spring tonic and wine. The Navajo made a poultice from dandelion leaves for swellings and to speed the delivery of a baby, while the Ojibwa used the roots as blood medicine, for heartburn and also enjoyed them with pork and maple sap vinegar, as did the Potawatomi.

The Okanagan-Colville ate them as greens, as did the Papago, also using an infusion of the leaves for menstrual cramps. The Pillager-Ojibwa used dried leaves for a tea for heartburn, and the Quilleute as a general medicine. The Rappahannock drank an infusion of the roots as a blood tonic and for dyspepsia/indigestion, while the Tewa dressed bone fractures with them and ate as a fresh green, as did the Ute.

All of which makes the general consensus that dandelion was introduced to America via seeds on the *Mayflower*—taken by the early pilgrims, being an indispensable food and medicine in England, spreading like wildfire to cover the whole new country—a little hard to believe.

The *Mayflower* sailed in 1620, and so good were these dandelions that the indigenous American Peoples adopted them from the white man because they recognised them as being so healthy, or so the story goes. It was the displaced locals who were indebted to the white man for the dandelion after the white man has just been responsible for up to 50 million people dying because Christopher Columbus got so terribly lost. I'm not so sure the indigenous tribes would be too interested in accepting the ways of the white man, as the white man's ways didn't appear to be very accepting of any other peoples' ways.

Of course dandelion seeds would have been taken by people undertaking the daunting experience of not just moving house, moving continent to America on the *Mayflower* etc. (or Australia a century and a half later), they had no idea what flora and fauna awaited them, so the most important ones from home would have been the ones taken. Dandelions were already such an important food and medicinal plant it would have been folly to not take them. So European dandelions would have been taken, which completely ignores the fact that there was already several species of very similar *Taraxacum* growing in America. These would have been the dandelions being utilised by the indigenous tribes, such as *Taraxacum laurentainum*, *Taraxacum palustre* and possibly even *Taraxacum officinale*, but unsurprisingly none of the experts can agree with one another about this, not helped by all the plants looking nearly identical. Perhaps it was considered by writers and chroniclers at the time as insensitive to suggest that the savage dark natives lauded similar plants as the pure white angelic Christians, being easier instead to suggest that the Satan-worshipping, cannibalistic natives adopted the use of the dandelion (and others) from the white invaders' clearly greater intellect. It's a similar path of egotism and ethnocentrism we saw with nettles, where some authorities are adamant the Romans introduced them to Britain.

When folklore was the law

Today, even for most of those living in the countryside, dandelions, apart from being a lawn irritant, go all but unnoticed. Although, from the frequency of dandelions appearing in folklore and traditions, it is very evident they held a far more central role in the past.

Dandelions woven into a wedding bouquet are supposed to bring good luck and continued love for the newly married couple. As is their appearance in dreams, seen as a good omen by representing a happy union. They are often associated as a symbol representing happiness, faithfulness, desire and love, as they do in the Language of Flowers, as well as being a little flirtatious. The Language of Flowers is a secret language, full of romance and subterfuge, brought back from Turkey in the eighteenth century and used as a way to carry out relations covertly; it could well have a considerable lost antiquity. There are over 650 meanings, stretching from a red rose indicating love and respect to a variegated tulip meaning beautiful eyes; a hydrangea could mean frigidity or a yellow lily suggest walking on air, with an ivy for marriage or grass for submission.

There are many versions of what blowing the fluffy seed ball of the dandelion can be interpreted as meaning, depending on how many times blown and the remaining seed heads still attached etc. Some of the romantic ones entail such variations as blowing on the seed head three times and if any seeds remain that person's lover is thinking of them still and not forgotten; or that the seeds once blown off will travel on the wind, delivering affectionate thoughts and good luck to loved ones. Imagine WWI, with the savage dislocation and barbaric slaughter of a whole generation of young men, which must have been a time ripe for such stories in an unimaginably sad period, with every village and town losing young men to bloodshed, horror and PTSD. Blowing the seeds and dispersing them all has been interpreted as love being held on to or love lost, or bestowing a wish if all can be scattered with a single puff; or even the number of seeds remaining will be the number of children a girl will have later in life, or more scarily they represented the number of years left to live. This could have come from the ancient belief that dandelions were seen as a clock, no doubt owing to the regularity of the dandelion flower to open an hour after sunrise and close at sunset, providing the many local variations on names such as Shepherd's Clock, a time gauge for the day and also the year. The flower head also rotates with the sun, an act known as heliotropism. The seed head can be used as a barometer to predict the weather, as in fine weather the seed ball will extend fully, but as rain or showers approach the seed ball will close up tight, only opening again when the chance of rain has passed.

Just as the reflection of a buttercup under the chin indicates the depth of appreciation of butter, a dandelion put under a baby's chin predicts gold, with the clearer the reflection the richer the child will become, or the sweeter and kinder for those not besotted with wealth. The tallest stem found by a child in early spring was supposed to be an indication of how much the child will have grown by the end of the summer.

The roots, drunk as a tea, are said to increase a person's psychic abilities, and if this tea is left steaming next to the bed at night the spirits will be called, while flowers picked on St John's Eve (23 June) are considered as a useful witch repellant. Folklore also attests to the many curative and health capabilities of the dandelion, as well as them being eaten everywhere.

Stories and songs arise in folklore because it is information worthy of remembrance, knowledge needing to be passed down the generations, with the songs and stories an aid to remember and continue the wisdom. All the stories and songs were part of a greater story and picture, culminating in a cultural tapestry of health and wellbeing of Nature and human's place within it. Today we have no cultural stories, only Hollywood movies, merchandise and memes. With folklore and traditional uses standing up to the scrutiny of modern science, proving the traditions and old stories are not fantastical, where often the truth has gotten lost and distorted in translation through time.

We have been programmed to believe that all the new stuff today is the best ever and we are so much better informed than in the dumb old days. It's not necessarily true. Dandelions and their position with human health are very real, and it is a rather strange circumstance of fashion that may well have begun the vilification of this particular Human Plant as well as helping to distance us from Nature.

Dawn of the lawn

War was declared on the dandelion, culminating in the selling of herbicides for domestic use to kill the tap-rooters and the broadleaves, while leaving the grass alive and giving us an unhealthy dose of chemicals, being shown by science to be contributing to people's cancers and respiratory failures. It's the Devil's manicured garden of doom, masquerading as a nice green lawn.

Before the lawn we know of today, the garden from the Middle Ages may have had an area of shorter grasses, with other areas of wild flowers and a selection of preferred herbs. It was not until the Jacobean period of the seventeenth century that a closely cut lawn began to appear. The year 1661 could very well mark the moment where our more modern disconnect with Nature took place, with the first lawn conceived in this year, being incorporated into the first gardens of note at Chateau de Vaux-le-Vicomte (34 miles south east of Paris). So impressive were they that the owner Nicholas Fouquet was arrested and imprisoned for life after a lavish play for allegedly wasting public funds. He may have been innocent of this charge but three villages were levelled in the construction of his garden. Louis XIV promptly hired Fouquet's architect and landscaper and got them to replicate these gardens, only better, at Versailles in 1662.

Lancelot 'Capability' Brown is responsible for hundreds of public and private parks, incorporating lawns, rolling gardens, water features and flower gardens in his designs from the mid-eighteenth century in Britain. From then on it trickled down, first for the aristocrats until by the 1930s the lawn takeover had become almost complete and we all believed the garden lawn was a good thing to aspire to.

If our culture continues for much longer on its current trajectory, within a few years this disconnect from nature will be complete and this is where the realm of sci-fi, governed by computers, trans-humanism and artificial intelligence takes over. Then we will be truly screwed with little hope of turning back or indeed being 'saved'. All thanks to Capability Brown and that little aristocratic thought of having so much land and wealth that this much, the lawn, can be made into something as totally fickle and useless as short grass, a green desert, whose clear-cut lines allowed the neat and tidy monocultural system to appear beautiful and proper, like a massive lawn.

* * *

Interestingly, the etymology of the word 'garden' is the same as for 'paradise', *pariaida za* referring to the Zoroastrian gardens and orchards of Iran. A garden can definitely be a paradise, a tranquil retreat and reconnection. Originally in a garden the dandelions and other salad leaves such as fat hen, clover and shepherd's purse were encouraged and valued for both their food and health benefits; now with the advent of the lawn they were seen as weeds and would have been hand-picked; then, to save us precious time to do less boring things instead, we allowed the deadliest chemicals ever invented by science to enter our private home space and do our little bit in our back gardens, normalising the ecocide that was under way in the countryside thanks to monoculture. Drummed into our consciousness from constant adverts about these weedkillers they have become normalised, a normal part of our normal modern life—although of course to many there is no greater insult than to be called 'normal'.

The Big C

The traditional wisdom of an old man in Ontario, Canada has possibly been the biggest kick-up the backside cancer research has possibly ever had. After several rounds of chemotherapy a few years ago, this particular gentleman was thoroughly unimpressed with what the world of Western medical oncology was doing for him, so he decided to take matters into his own hands by preparing a daily tea from dandelion roots. Upon his next check-up, the oncologist was so impressed with the 'miracle' numbers that the Dandelion Root Project at the University of Windsor in Ontario came about. The oncologist immediately set about preparing laboratory trials to establish how it was that dandelion root extract (DRE) dealt with various cancers. The initial tea was no more than roots seeped in water, so the actions initiated came about from a tiny quantity of a possible anti-carcinogen ingested. With more concentrated DRE, the research team at Windsor have already shown its efficacy in controlling and killing cancer cells of the blood, colon, pancreas and melanoma. The Project is now stepping up their trials with human subjects, and it won't be long until their initial results come in, with preliminary reports suggesting that the mechanism initiating apoptosis is switched on, where the cancer cells commit suicide, given that a major problem with cancer is that it switches off the ability to die and only reproduces.

Another bonus of dandelions is that they also induce autophagy, as shown from research at the University of Maryland, creating homeostasis (balance) by making sure all the damaged and dead cells are removed from all systems. So dandelions provide a double knock-out combo of inducing suicide in the cancerous cells, apoptosis, with no side effects to the human host, while at the same time inducing the very definition of health, this balance called homeostasis by cleaning out what has just been killed via autophagy and supplying all the required nutrients. Pretty cool, eh?

The anti-carcinogenic activity of the *Taraxacum* species was established at the end of the last century by Kyoto Pharmaceutical University in Japan (Takasaki et al. 1999), whose results suggested an extract of the roots could be a valuable chemopreventive agent against chemical carcinogenesis. Further studies by the same scientists suggested that the compound taraxasterol was at least partly responsible for this activity.

The Canadian research from the University of Windsor mentioned above, focusing on the cancer cell death aspect of DRE, was first initiated by looking at human melanoma cancer cells, successfully inducing apoptosis (Chetterjee et al. 2010). This work continued under another scientist involved in the first studies (Ovadje et al. 2012), looking at how DRE induces cell death in aggressive and resistant chronic myelomonocytic leukaemia, which it does, and that's all we need to know for now. The same year she also showed its effectiveness against aggressive pancreatic cancer cells, which has one of the highest mortality rates, with no damage to any other cells, or any side effects. Later research from *Oncotarget* (Ovadje et al. 2016) showed that DRE induces cell death in aggressive colorectal cancer cells, focusing on the cancer cells and taking out the mitochrondria, with no side effects from dandelions ever reported, while irrefutably improving the quality and lifespan of cancer patients.

Earlier work from Mexico reported in the *International Journal of Oncology* (Sigstedt et al. 2008) showed that DRE decreased the growth of breast and prostate cancer cells. Dandelion roots also showed their anti-carcinogenic activity on cancer cell lines in research at the Abdul Wali Khan University, Pakistan (Rehman et al. 2017), being seemingly potent against several human cancer cell lines. Additional anti-cancer potential from Dongguk University, Seoul, Korea (Saratale et al. 2017) was shown by taking the novel approach of using *Taraxacum officinale* for the synthesis of silver nanoparticles, which the team concluded was an original yet effective approach to cheap and eco-friendly antimicrobial and anti-cancer drugs. This was corroborated by research from Shanghai Jiao Tong University, China (Rasheed et al. 2018). Research at Kongju University (Lee et al. 2016), South Korea, demonstrated the anti-cancer effects of *Taraxacum coreanum* and a close relative *Ixeris denntata*, both of which had a powerful apoptic effect on human melanoma cells, being a contender for a chemotherapeutic against skin cancer. Rounding off the anti-cancer ability from DRE is work published in *Biomedicine and Pharmacotherapy* (Zhu et al. 2017), this time against gastric cancer, one of the commonest cancer forms worldwide.

The above findings represent only a small portion of what is an impressive amount of research that signifies dandelions as being an excellent anti-carcinogen, although surprisingly little comes from the UK or USA, the countries with by far the biggest cancer charities. Cancer Research UK had funds of £640 million in 2018 while the National Cancer Institute in the US is the biggest provider of funds for cancer research, receiving over £3.8 billion annually. Additionally the American Cancer Society has about a billion dollars. Total donations to the world cancer charities exceed £11 billion every year. A lack of cancer research on dandelions from the US or UK could just be an oversight, but with no research from these cancer societies focusing on the preventative or curative potential from any local plants, it's a little strange, but it is what it is.

Taxol is the main ingredient in chemotherapy drugs, and this comes from the bark of the yew tree, underlining the obvious fact that medicines come from plants—although, to be botanically accurate, it is a fungal colony living in the bark that produces taxol as a secondary metabolite. This yew medicine (coincidentally a highly revered tree of ancient European cultures) is extremely toxic, to the point of destroying the human immune system while it poisons the cancer—It's like a deadly game of balance, and which falls first: if it's the cancer the patient survives, if it's the immune system the patient dies. That is scary odds. Surely it makes sense in the early stages of cancer to try the dandelion root first, while bolstering the immune system with other nutrients to help the body heal itself? If this fails, then move onto the more invasive and dangerous strategies being currently practised. The beginnings to cancer treatment could start less aggressively, to see how the tumour, blood or skin reacts to a gentle nudge towards producing the right enzymes and compounds to assist the cancer cells in their own apoptosis, as suggested by the scientific research above.

* * *

Let's introduce some monetary values here, as ultimately in our economic world this is what it boils down to. Taxol commands around £3.8 billion annually at present, so naturally there will be many pharmaceutical feathers ruffled by suggestions of free or much cheaper alternatives. It should be said that this £3.8bn is chickenfeed compared to the amount spent on cancer medicines and machines, which topped £115bn in 2020. Back in 2006, two researchers at Chicago University estimated that a cure for cancer would be worth about £38 trillion, although with the caveat that it will only be as much if developed and sold by big pharma. It'll be worth far less if dandelions et al. get involved, and that is the biggest obstacle, this economic loss.

Here's the revenues clocked-up from the top 10 selling cancer drugs in 2021: Keytruda (owned by Merck) £14.3bn; Revlimid (Bristol-Myers Squibb) £10.6bn; Opdivo (Bristol-Myers Squibb) £6.2bn; Imbruvica (AbbVie) £4.5bn; Ibrance (Pfizer) £4.5bn; Tagrisso (AstraZeneca) £4.15bn; Perjeta (Genentech owned by Roche) £3.6bn; Tecentriq (Genentech) £3bn; Avastin (Genentech) £2.7bn; Herceptin (Genentech) £2.4bn.

If dandelions prove effective, these top firms are each set to lose over £15 billion in revenue from their cancer sector. Do you think they'll sit quietly and take that hit? Fifteen billion pounds in revenue is equivalent to the entire GDPs of Papua New Guinea, Cambodia or Afghanistan—all from the sales of drugs from just one corporation's subdivision of health!

* * *

It is the cancer *industry* after all. It's making a lot of money, turning the cogs of capitalism. The last thing they would do is to suggest making your own dandelion root tea. The cancer societies can't even bring themselves to state it's imperative to stop eating glucose-forming foods (the whites of sugar, flour and rice) and dairy when initially diagnosed with cancer, as suggested in countless reports demonstrating that cancer cells thrive on these glucose-forming foods, yet die when deprived—Otto Warburg was awarded the Nobel Prize for this in the 1930s, so it's rather old news yet still true today.

There's also no mention of the seemingly endless amount of side effects these highly toxic chemo-drugs cause. As if being debilitated by cancer wasn't bad enough, there is a high likelihood the drugs will cause more problems. As we discovered in the last chapter, the actual number of people suffering from ghastly side effects from their prescription for arthritis was sensational, not to mention the way the arthritic drug Diclofenac had put the health of the entire Indian subcontinent at risk from plague by wiping out the vultures. Sadly, the cancer drugs fare little better. Being a cancer patient is a scary position to be in, but so sewn up does the cancer industry have it that a patient will have to go on their drugs, with cancer laws making it illegal to even suggest an alternative method. This may account for why there are so few human trials with legislation in the way. Here's what the number one drug Keytruda causes as an addition to the patient's sufferance from cancer—up to 55% suffered anaemia, 47% fatigue, 30% nausea, 20% diarrhoea, 16% headaches, 14% insomnia, 12% back pain, 11% upper respiratory tract infection, 1% to 10% hypo or hyperthyroidism, 0.1% to 1% renal failure, partial seizures, pancreatitis, exfoliative dermatitis, hepatitis and Type 1 diabetes mellitus among a lot more.

The thing with dandelions and cancer is that by regularly eating them in the past, the body was being constantly cleansed, so the chances of the cancer becoming severe or even continuing long enough to be diagnosed were greatly reduced; this was assisted by the number of carcinogens in the past being fractional compared to today; and, coupled with this is the modern diet of fats, sugars and processed carbohydrates being a bonanza for cancer. So in the past, not many noticed any association, as they weren't getting ill—most of the ills of the past were from contagious diseases, often due to unsanitary conditions; most of the ills today are caused by the body reacting badly to modern chemicals, processed foods and civilization itself. Before WWII the wild plants of nettles and dandelions (and many more) were doing as they did best, acting as a preventative to the manifestation of serious diseases. By eliminating these preventative plants from our diet, the door has been opened to the diseases of civilization.

Duelling with the diuretics

Dandelions have been used for their diuretic effect for at least 2,000 years in Traditional Chinese and Indian Ayurvedic Medicines, a timespan probably also true for Europe as well, what with their traditional name of piss the bed! The fact that dandelions have been employed for the same ends across cultural, geographical and time divides adds veracity to the diuretic claims.

This veracity has been partially demonstrated by amazingly only one human experiment, which considering the money involved in the diuretic sector is stupendous; or, as mentioned before, perfectly understandable if a large slice of a pharmaceutical corporation's profits are threatened by free and readily available plants. No sane CEO wants such news broadcast, especially if it is from their paid-for research. This human clinical study carried out was a small one involving only 28 people (Clare et al. 2009), making it easy to downplay the results as there was no double-blind group, but nevertheless the results showed that dandelions acted as a significant diuretic over a five-hour period. This result echoed one carried out on mice, where the results from the dandelion-fed group were as favourable as those given furosemide, a pharmaceutical diuretic. So that's two ticks scientifically and a big one culturally suggesting dandelions are an effective diuretic. Dandelions are also seen as good enough for a German Standard Medical Licence and for the German Commission E to approve dandelions for diuresis. Anecdotally, last summer after eating about three leaves, after 30 minutes I was suddenly desperate for a pee, being out cycling at the time—quite a fuss was made trying to extricate myself from the tight spandex in time!

* * *

What's it all worth then, this diuretics market? For starters, that diuretic from above, furosemide (prescribed principally for oedemas), which dandelions did as well as, has global sales of £115m annually, with the total diuretic market today in the region of £1.2bn annually. Big money for a health assistant to help you piss. It is the first line of Western medicine's defence whenever a patient is diagnosed with hypertension, high blood pressure and oedema, often also for liver and kidney diseases and obesity. The theory is that excessive salt from the diet encourages water to be built up in blood vessels to dilute the salt, raising blood pressure, so by releasing the salt and water blood pressure is lowered … apparently.

However, the biggest problem with all these diuretics being consumed is that it isn't just the excessive sodium being flushed out. Potassium and magnesium are lost among many others, with none replaced, causing serious nutrient and electrolyte imbalances. Often potassium will be also prescribed, although this will be a denatured potassium supplement that is still missing all the other salts and minerals flushed out by the pharmaceutical diuretic. If these imbalances are left unchecked other, far more serious diseases can manifest such as chronic muscle weakness, seizures, irregular heartbeat, heart attack, kidney failure, pancreatitis, thyroid problems and depression. Pharmaceutical diuretics also leave multiple tissues vulnerable to free-radical reactions, an easy entrance for cancer cells.

Dandelions completely negate these problems as they are a net provider of potassium and magnesium (that is, more goes in than comes out after the diuretic effect), and the suspected diuretic agents such as abscorbic and caffeic acids, calcium, luteolin, magnesium and mannitol also conveniently play several other key health roles, such as being anti-inflammatories, antioxidants, vasodilators, hypotensives and nutritives, as well as all the other nutrients that have nothing to do with the diuretic effects, playing several essential roles in human health, cleaning and repairing from the cellular level right up to the tissue and organ level, as indicated by the research and clinical trials reported throughout this chapter.

Dandelions also avoid the side effects associated with pharmaceutical drugs, which thankfully are not as serious for this class of drug, with cramps, migraines and nausea being the most common, although usually the diuretic is only a part of a cocktail of drugs prescribed to a patient, often with several ailments and other associated prescriptions.

Dandelion may also help the liver regulate blood sugars to avoid hypoglycaemia (low blood sugar), often a cause of binge eating. The increase in bile flow stimulated by dandelion also helps to improve fat metabolism in the body. The diuretic effect removes more sugar from the body, which is good for diabetics as it lowers blood sugar as well as regulating insulin production, avoiding the dangerous troughs and peaks in blood sugar levels so debilitating for diabetics. The diuretic effect cleanses the whole of the urinary system, removing unwanted wastes, while the disinfectant properties of the phytonutrients in dandelions inhibit microbial growth. The increase in urination can help to lower blood pressure, the fibre is good to reduce cholesterol and the high potassium content replaces the sodium, creating a more balanced and healthy ratio of sodium and potassium. The diuretic effect also aids in the treatment of chronic disorders such as eczema, gout and rheumatism.

Before looking at more scientifically corroborated benefits associated with other diseases, to put in some balance or equipoise, here's the problems associated with taking a regular dandelion infusion, tincture or capsule.

Warning!

Dandelion has almost identical contraindications as for nettles, suggesting all of these are general suppositions and not factual information of real-life problems observed from drinking some dandelion tea and noticing and reporting a problem, unlike the incidence of side effects from taking prescribed drugs, as these are from actual reportage from consumption. Of what little information there is available, there is the proviso that for those taking antacids, dandelions may increase stomach acid, contraindicating any antacids consumed. For those taking blood-thinning drugs, such as warfarin, the high vitamin K content (a coagulant) would make them work less effectively, if at all. Similarly, those taking diuretics may find they run the risk of an electrolyte imbalance by losing too much fluid, as dandelions are themselves a diuretic. Also dandelions run the same risk posed by nettles of possibly making the side effects of lithium medication worse by not flushing the lithium out.

Taraxinic acid appears to be the most potentially allergenic constituent prevalent in dandelions, although there are no details available on what this might entail should you come into contact with some pure extract. Contact dermatitis and other allergic reactions are possible from the sap, with a really simple solution of not touching it again. As is the case for those allergic to marigolds and other flowers in the daisy and sunflower family, sufferance could be caused from merely touching the leaves—it's a very rare condition.

Tests on *Taraxacum mongolicum* (a near identical sister) suggests their consumption lowers the amount of the antibiotic ciproflaxin the body absorbs—although we should only be taking antibiotics in extreme circumstances. Probiotics would be better eaten most days as a preventative and, as we saw with nettles, their preventative effects would help to avoid the ailment manifesting to begin with and thus negating the need for the antibiotic. Dandelions do increase the flow of bile, so for those with gallbladder disease, gallstones or bile duct obstruction, it might be best to avoid, or at least consult with a professional medical advisor so they can take the blame if the stone dislodges and gets stuck somewhere more painful. The doctor might well 'tut' when you mention taking dandelions to try to clear your stones, irrespective of them being used as an aid for eliminating stones across the world for aeons.

Traditional Chinese Medicine and Native American Medicine both have a long history of prescribing dandelion root for breast cancer and leukaemia, as well as for kidney disease and liver problems, digestive ailments and skin problems, as was the case traditionally in Europe for similar ailments.

There is definitely no mention of death in the limited information available on consuming dandelions and their possible side effects, unlike all the pharmaceuticals. Appreciably it's in the early stages regarding dandelions, but this does not refute the proposition that when it comes down to Western medicine, it is 100% dominated by pharmaceutical corporations' output of drugs and vaccines. Unquestionably more pills are being popped today than ever before, while disease rates continue to rise, as do the side effects attached. Is that really what the Father of Western Medicine, Hippocrates, had in mind? Because it is definitely breaching the Hippocratic Oath all doctors once swore on—stopped being said, coincidentally, once the drugs took over late last century.

Nutrients

The anti-carcinogenic and diuretic capabilities are because of certain well-known nutrients and unknown phytonutrients being the entire make-up of these plants that presents us with a plethora of health potentials. I don't know what does what, and the experts seem somewhat vague as well, even after spending much time and money analysing individual constituents. What is certain is that the nutrient analysis of dandelions looks pretty good, and it has to be a combination of these and the phytonutrients that has something to do with feeding our bodily systems and acting in a positive way. Of the nutrients we are more familiar with here's what was found in dandelions—

Firstly, the vitamin A content, or rather the precursor that enables the vitamin A to be internally produced, carotene, ranges from 8,400IU to 14,000IU per 100g of raw dandelion. This is as good as its namesake the carrot, which weighs in at around 12,000IU (RDA is 5,000IU, IU being International Unit). As mentioned before, the nutrient content is determined by the soil, as well as the sunshine, wind and water, so all nutrient figures are only a guide as to the potential nutritious content of any particular part of an edible plant. Dandelion leaves from rich mineralised soil could even provide more than 14,000IU of potential vitamin A. They have even more vitamin K potentially available, in the region of 750mg (as much as 850% of the RDA from 100g), playing an essential role in clotting the blood as well as being seen as a new aid to bone health for the elderly. Not a bad start.

There's also going to be about 40% of the RDA of vitamin C and 10% of the required vitamin E and several from the B complex per 100g. Among the minerals there can be found roughly 10% of the calcium, manganese, iron, magnesium, 5% of the potassium, copper, phosphorus and zinc of our daily requirements. Be aware that eating 100g of dandelion leaf in one sitting is unlikely, unless as a soup, so these figures are a bit rosy, although dandelions do look like a valuable addition to bolster overall nutrition of the diet. The numbers aren't spectacular, matching nettles, and as with nettles it wasn't the numbers relating to the known 50 essential nutrients (vitamins, minerals, EFAs, water, fibre etc.), it was the little-understood phytonutrients that appeared to hold the key to unlocking the health padlock. With 25,000 phytonutrients discovered so far, dandelions have them in abundance.

Phreaky phytonutrients

As mentioned in the nettles chapter, these oddly named phytonutrients are only just being discovered and identified, while what they do individually and collectively is hardly understood at all. It's no easy matter to isolate these individual phytonutrients, and to do so requires some seriously high-tech equipment, such as a high-performance liquid chromatography and electrospray ionisation mass spectrometers. Which don't come cheap.

Exactly as with nettles, dandelions are throwing up new chemicals and phytonutrients. Among the ones causing interest in the pharmaceutical world are those ones we saw in nettles, such as the fabulous phenols being plentiful here, with several caffeoylquinic acid isomers, quercetin glycosides and new di- and triglycosylated flavonoids. As well as some others already seen there are beta-sitosterol (being the most abundant free sterol) and among the coumarins there is scopoletin (deemed so important this is standardised in some nettle capsules). Carotene, lutein and cryptoxanthin have been shown to protect against lung and mouth cancers, while the flavonoid zeaxanthin offers protection to the retina from UV rays.

Dandelion leaves and roots have a slightly different composition. The roots tend to have a greater proportion of bitter glycosides (its namesake taraxacin for one), tannins, phytosterols, triterpenes, choline, asparagine and inulin, while the leaves have more carotenoids (including lutein and violaxanthin), terpinoids and coumarins …

… and loads more, but they tend to get in the way of a good story as they hardly roll off the tongue. The appendix lists quite a few of those found in medical literature, with undoubtedly new ones being found daily around the world. These individual phytonutrients can be patented and made into drugs, so it is almost guaranteed that future pharmaceuticals will be coming from this avenue, even if at the expense of health, herbalism and holism.

The inconvenient truth

It would seem that it is most definitely an inconvenient truth for the drug corporations if the truth were revealed—with them being entrusted with our health and medicine for the past 70 years we appear to be generally worse off. Today our very death is because of our lifestyle, a disease of civilization—heart disease, cancer, iatrogeny, kidney and liver diseases, dementia, diabetes, depression, neurological disorders and obesity. Our health over the past 70 years has progressed down an increasingly slippery slope of sickness. Considering the third biggest killer of industrialised humans is an iatrogenic one—a word whose meaning has been so well hidden hardly anyone has heard of it let alone knows what it means, not even my paramedic friend. An iatrogenic death is one caused by medical intervention, notably prescription drugs, with only heart disease and cancer killing more of us—that's quite a surprise, isn't it? Not to mention all those side effects, debilitating millions of people every day. At least nettles give a sting as a side effect, dandelions can't even muster this.

* * *

The very inconvenient truth for big pharma is that according to well-funded independent clever bods, dandelions do all this—

They are a choleretic, in that bile production and flow to the gallbladder is increased, as well as being a cholagogue, contracting the gallbladder and releasing stored bile. These two associated benefits assist a plethora of ailments that are interlinked with bile, as sufficient bile prevents us becoming oversaturated with cholesterol and calcium and the formation of gallstones, kidney stones and sclerotic deposits within arteries, a leading cause of heart disease and high blood pressure. While the clearing and cleaning detoxifies the liver and kidneys of calcified stones, cholesterol and uric acid, the spleen is flushed also. Eating and drinking dandelion leaves and roots also assists constipation and haemorrhoids, lowers the risk of a colorectal disease and other gastrointestinal ailments, as well as promoting healthy gut biota. An infusion of dandelion root stimulated *in vitro* (*in vitro* means in glass, here a test-tube) the growth of 14 strains of *bifidobacteria*, the good bacteria.

The root is also a rich source of inulin, with as much as 35% of the root being inulin in the autumn, although as low as 2% in the spring when it has lots of levulose, a natural sugar. Inulin is a natural prebiotic, another aid to restoring a healthy balance of the micro-organisms in the gut. Inulin is a perfect source of energy for diabetics, with dandelions also being shown to stimulate the production of the hormone insulin from the pancreas, maintaining low blood sugar levels.

Overall bone health is improved thanks to a superb content and balance of calcium, vitamin C and the phytonutrient luteolin (the ethyl acetate fraction of dandelion flower extract contains a whopping 10% luteolin). Vitamin C and luteolin mop up free radicals, thus reducing the risk of cancer, as the vitamin C detoxifies while luteolin inhibits cancer cells from reproducing. These nutrients also protect from age-related damage, as well as assisting with optimum functioning of the liver, with another compound, carbon tetrachloride, shown in studies on mice to protect the liver against liver fibrosis and possibly being a good tool to assist in alcohol-related liver damage.

The latex sap of dandelion when directly applied has germicidal, insecticidal and fungicidal properties, all very useful in treating skin disorders caused by microbial and fungal infections from warts to nail fungus; be careful not to get it in your eyes as it'll really sting when it dries. The leaves can be used for eczema and acne, as well as aiding the healing of the damaged skin caused by acne. The rich boron content available from dandelions helps to raise oestrogen levels in the blood, which along with the calcium, phosphorus, potassium and magnesium, with some silicon as well, all contribute to overall bone health.

As well as all the above, dandelions also come with loads of alphabetical benefits attached. They can be seen as being an aperient, in that they are a gentle purgative and mild stimulant for the bowels. Further research will prove beyond doubt dandelions to be anti-arthritic, anti-carcinogenic, antifungal and anti-inflammatory. They ease the pain associated with rheumatism, making them an anti-rheumatic. They behave as an astringent, being a binding agent that contracts organic tissue, reducing secretions and discharges of mucus and other fluids from the body. They are a blood purifier and detergent, owing to their ability to cleanse wounds and sores. They are also a demulcent because they form a soothing film over a damaged or inflamed mucous membrane to relieve pain and aid recovery. We've already seen they are cholagogue, cholerectic, diuretic and a prebiotic. By promoting the bowels, dandelions can be called a laxative. They are a stimulant as they promote the functional activity of the tissues, requiring less energy and making more energy available for elsewhere. As dandelions have nutrients known to stimulate, tone and strengthen the stomach, they are a stomachic. By promoting or increasing perspiration dandelions are a sudorific and as the entire human organism is invigorated and vitalised by dandelions they also act as a tonic.

Dandelion leaves are a rich source of cichoric acid (as good as its namesake chicory). This acid is causing some excitement with results from *in vitro* tests showing it to be effective against free radicals, protecting collagen and possibly preventing HIV taking hold. Anaemia is assisted by dandelion's good iron and vitamin C content, working as co-factors for the best bio-assimilation of iron possible. There are also indications that dandelions inhibit the build-up of plaque on teeth, could limit neuron damage in the brain and are a probable galactagogue (milk production), as with nettles from the rich beta-sitosterol content and the intricate web of co-factors making it all work correctly.

Other studies focusing on food production and distribution have shown that dandelion would be an excellent addition to many local foods, being so easily and cheaply obtained, assisted by them being able to grow nearly everywhere, as well as being highly nutritious; they are especially useful in regions of devastation where transport is severely disrupted.

Clearly dandelions have the ability to do a lot of things, although from a herbalist's perspective it is the preventative qualities that are the most exciting. Once homeostasis has been broken and ill-health occurs, however it manifests, the remediation is very rarely a straightforward matter. There are many things that will help, such as the phytonutrients and other nutrients, although it usually takes individual one-to-one consultation to find the best potential curative path. It has become apparent that nettles and dandelions individually or in combination (also with the others we have yet to meet in this book) should be the first line of defence for the whole of the human race, as their preventative capabilities will become progressively more advanced the more they are studied. Most people on pharmaceutical drugs, from their first prescription, are addicted to the erroneous faith and assumption that modern medicine provides the pinnacle of healthcare in a pill. Yeah, right … just like the food industry provides us with the best possible food, the education system provides the best education, while the health system gives the best health possible and the political system delivers real democracy …

More than an opinion

There has been quite a lot of scientific evidence already presented throughout this chapter, mainly on cancer and diuretics. Let us add to these a brief selection from a few studies about other ailments, notably from China and Korea, both at the forefront of recent dandelion research—

A dandelion study from China had a 92% success rate for patients with metaplasia and hyperplasia, often seen as the beginning stage of cancer, characterised by an abnormal change in the nature of a tissue usually in size or type—this was from as long ago as 1992 (Liu et al.).

Taraxacum mongolicum from China was shown to have an important role in protecting from mitochondria-mediated apoptosis (Zhong et al. 2015), which has itself been shown to be a pathway in human diseases relating to immune disorders, neuro-degeneration and cancer. Heze Municiple Hospital, China (Zhao & Jian 2016) showed that for those with acute paraquat poisoning (a herbicide), when administered dandelion along with a Traditional Chinese Medicine, incidences of oral mucosa, pulmonary fibrosis, multiple organ failure and mortality were all reduced.

The triterpene taraxasterol was shown to have anti-inflammatory attributes, as well as protective ability for the lungs and bronchial epithelial cells from cigarette smoke by Chinese researchers from Jilin University (Xueshibojie et al. 2016). Taraxasterol was again focused on by more Chinese researchers, as reported in the *European Journal of Pharmacology* (June 2015: Piao et al.), where this time it was shown to be a useful agent for the prevention and treatment of osteoarthritis, as well as anti-inflammatory capabilities, with the added bonus of offering protection from lung damage (*International Immunopharmacology* journal Apr 2014: San et al.). China's Pharmaceutical University, Nanjing (Cai et al. 2017) showed potent liver protection from dandelion root's rich polysaccharide content.

The Molecular Medicine Reports (Apr 2014: Jia et al.) used *Taraxacum mongolicum* extract to validate claims of antiviral effects in attempts to potentially eradicate the hepatitis B virus while protecting the liver generally, concluding that the extract was indeed a potent antiviral. Impressive research from Inha University, Republic of Korea (Duan et al. 2017), showed that inflammation was a key contributor to vascular diseases, including stroke, myocardial infarction, hypertension, chronic kidney disease, autoimmune diseases and cancer. Using a methanol extract the researchers concluded that dandelion could be used in the treatment of endothelial inflammation (interior of blood vessels, very common in those unwell) and atherosclerosis.

In Korean traditional medicine, dandelions are given to improve general health and raise energy levels, so the Korean Food and Drug Administration (Lee et al. 2012) wanted to see if this was actually true. Their research showed there was indeed less fatigue, more glucose for energy, with less lactic dehydrogenase, a cause of muscle damage.

The Korean Food Research Institute, Songnam (Davaatseren et al. 2013) demonstrated the liver protection offered by dandelion leaf extract, as well as offering a promising approach for the prevention and treatment of obesity-related nonalcoholic fatty liver disease, with this hepa-protectivity further corroborated by other research across the world. Further Korean research (Yan et al. 2012), this time on pigs fed *Taraxacum officinale*, showed improvements in growth performance, feeding efficiency and reduced fecal *E. coli* concentration owing to dandelion's antimicrobial activity.

From the online publication Gut Liver (Sep 2017: Han et al.), Korean researchers carried out *in vivo* (in a living body) and *in vitro* experiments, demonstrating that *Taraxacum officinale* has anti-inflammatory properties that can deal with inflammatory bowel disease (IBD) better than the standard treatment of Sulfasalazine, which has a 33% chance of added headaches, anorexic symptoms, vomiting and loss of appetite—drug revenues for IBD are about £5bn annually. Dandelions come with the added bonus of also having significant regenerative capabilities, aiding in balancing all human systems, wound cleansing, dead cell clearing and aiding new tissue growth.

Research from Mexico (Diaz et al. 2017) identified chlorogenic acid from *Taraxacum officinale* as having significant anthelmintic activity (kills intestinal parasites), notably against the *Haeomonchus contortus* eggs and infective larvae that plague the sheep industry. *Taraxacum officinale* was also shown to have varying degrees of antibacterial capabilities depending upon which extraction method was used, such as aqueous, ethanol or methanol, and the type of bacteria (Diaz et al. 2018), suggesting an antibiotic usage for urinary tract infections in the very least from their findings. Antiviral activity was also reported from Mexico (Flores-Oceloti et al. 2018), with an extract of nettle and dandelion effective against dengue virus, spread by mosquitoes that infect 100 million people every year, with at least 25,000 deaths.

Taraxasterol was shown to be effective in improving liver function by Azad University, Tehran, Iran (Yousefi et al. 2017), as well as reducing crystal deposition, once again showing these plants to be a perfect preventative for a whole variety of ailments, quietly working away in the background, at least in the days when they were consumed as part of the diet.

Reported in *Biomedical Pharmacotherapy* (Oct 2016: Rehman et al.), Pakistani scientists showed a dandelion leaf extract was as effective against hepatitis C infection as the traditional drug Sofosbuvir, if not even better—this drug has significant global revenues, several billion £$ at least.

More broiler chicken research was published in *Veterinary World* (March 2016: Qureshi et al.), in this case from India, where dandelion leaves and fenugreek seeds were shown to have both liver-protective ability as well as having a beneficial effect on intestinal morphology, thus reducing the effects of harmful pharmaceutical drugs in the animal industry tainting the meat.

From the University of Sfax, Tunisia (Gargouri et al. 2017), *Taraxacum officinale* was shown to protect against lead poisoning in rats; they don't quite know how though, although it was clear the liver was strengthened.

Research from the University of Gafsa, Tunisia (Hfaiedh et al. 2016) concluded with the opinion that *Taraxacum officinale* leaf extract not only had definite liver-protecting capabilities, but it could also find potential therapeutic use as an alternative medicine for patients with acute liver diseases.

Antioxidant activity was noted to be better from the petals than leaves when Polish research (Jedrejek et al. 2017) looked at the phenolic content, the flavonoids to be precise, as a source of natural compounds with no side effects for oxidative stress diseases—oxidative stress diseases range from cancers to heart disease, dementia and more.

The Food Institute in Madrid, Spain (Gonzalez-Castejon et al. 2014) isolated the chlorogenic and caffeic acids in dandelion as the main phenolic constituents that could play a major role in obesity and the way lipids are metabolised. This is amazing, because if nothing else, it underlines the potential is multifarious and not to be confined to single uses, as chlorogenic acid was just reported to be a sheep worm killer in Mexico.

The Journal of the Society for Biomedical Diabetes Research (Summer 2016: Wirngo et al.) reported that Danish researchers had confirmed the vast range of ailments dandelions could benefit, as well as being a main go-to drug for diabetes; working across the board from lipid metabolism to insulin regulation, dandelions appear to work synergistically for so many conditions. Just as one would expect from a Human Plant.

* * *

So far, I trust you agree, the potential from the humble dandelion has been extremely impressive, all that food and medicine by the mouthful. Why are we neglecting these in favour of a chemically laced demineralised iceberg lettuce with a zero nutrient chemical cucumber? And calling that a salad! The daily Victorian salad contained up to 20 different foraged leaves, full to the brim with nutrients from rich soils, all seasonal from ramsons, jack-by-the-hedge, sorrel, dandelions, hazel, hawthorn and beech leaves, deadnettles, mints, clovers and flowers galore. This looks positively depleted compared to the salad preferred by James I (King, 1603–25) who was said to insist on at least 40 leaves in his salad—today most people can't even name 20 different plants, let alone edible leaves.

That's not all as dandelions have even more to give … Imagine a field full of dandelions, all gently bobbing in the breeze, when roaring out of the woods comes a trials bike, full throttle, tyres spinning, mud flying, as it hurtles past, cutting a path through the field of flowers. That's a clue as to the very unexpected use dandelions are about to be instrumental in.

Taraxagum

This particular innovation was fuelled by historical concerns about rubber supply, rubber being such an integral part in so many manufacturing machinations and transport. It was led by Russia initially, which from as early as 1931 devoted a reputed 67,000 hectares to growing a dandelion they found in Kazakhstan, *Taraxacum kok-saghyz*.

This produced an unusually large amount of latex (about 400% more than the identical-looking *Taraxacum officinale*). Germany, Britain, America, Sweden and Spain also began to cultivate dandelions for this very reason up until about 1950, when the Russians were managing to produce 200kg of rubber from every hectare. The fact of a more peaceful world and the reality that the yield from the rubber tree is 1000% more halted the dandelion-for-rubber cultivation and it was soon forgotten about.

Apart from those at the Fraunhofer Institute for Molecular Biology in Germany, whose work is now in collaboration with Continental Tyres, who have a prototype tyre ready to roll out as soon as they can secure the crop. Continental Tyres initially sank £26.5 million into the project because there is 40% natural rubber content in truck tyres and 20% in passenger cars; and, with increasing car sales, irrespective of whether these cars are electric or diesel, all in need of tyres, there is a vast amount of other rubber uses it can be rolled-out into servicing. Since originally writing this, the tyre now graces the Extreme E (electric) racing SUVs competing over jaw-dropping terrains around the world to raise climate change awareness, apparently.

* * *

Rubber currently comes from a very narrow equatorial band of growth from predominantly the Para rubber tree (*Hevea brasiliensis*). It was originally from South America, until in 1876 Henry Wickham smuggled 70,000 seeds out of Brazil, of which about 2,500 germinated. These were sent out to India, Sri Lanka, Malaysia, Indonesia, Congo, Liberia and Nigeria to begin new plantations, with Malaysia and Indonesia being the most productive today.

Continental clearly saw not just the PR coup but also identified a necessary alternative to a product that is contributing to the deforestation of the Indonesian rainforest. It is not just the Amazon under threat, a forest that tends to dominate the news, when more seriously and matter-of-factly is the devastation of the forests of Indonesia, where Sumatra, Borneo and Papua have seen their rainforest coverage fall from 145 million hectares in 1950 to around 70 million hectares today, with serious concerns that many parts only have as little as 20 years left. This rainforest has been primarily lost for the West's insatiable desire for cheap cooking palm oil, wood, minerals and rubber. To make matters worse, the rubber trees of Brazil have been devastated by a fungus, and there are no guarantees this will not happen in South East Asia, so it makes perfect sense to guarantee a dandelion rubber source that is not only local but also employs local labour and boosts local economies.

It won't be long when upscaling production takes place and there will suddenly be thousands of jobs needed on the land and processing plants. If just 10% of Continental's 135 million annual tyre output can be switched to Taraxagum … then there is the 2.5 billion made every year from 455 tyre factories across the world to think about. It just takes one, and now there is also Bridgestone looking at developing their own tyre … and so it begins.

Tyres use about 70% of the natural rubber harvested and produced, while other uses can be found in cements, adhesives, friction tapes, hoses, rollers, balloons, shoes, balls, cushions, durex, gloves, wipers, skateboard wheels, bicycle tyres, Lego and a million other things. Lego make about 320 million tyres each year, making them the largest individual producer of tyres in the world: they've got the *Guinness Book of Records* framed certificate hanging in their HQ to prove it, having already pledged to convert to hemp plastics for their bricks; hopefully it'll be Taraxagum tyres next. Bridgestone, Michelin and Goodyear produce about 180 million tyres per year each and Pirelli about 150 million, so if half of the companies begin to use dandelions there suddenly becomes a lot of raw material for many other industries from the waste, since rubber only needs the roots (also possibly the stems as these ooze latex). There will be sales of the rest of the plant, namely the leaf and flower, which can be used for animal and human feed, injecting essential neglected nutrients back into the human via pestos, sauces, dips, soups, seasonings and salads. Chlorophyll can be obtained and is used in food as a green colouring, E140, a probable super-antioxidant; other medicines, ethanol production and biomass electricity are also possible.

* * *

There is big money to be made from these ventures. Anyone at the beginnings of a new industrial sector stands to be handsomely rewarded if there is any longevity. The patents for these natural rubber products will have earned trillions if the industry is allowed to mature to fruition and continue into the twenty-second century. Morally these patents from Nature should be revoked after a short period, with all further proceeds going into a global fund to pay for initiatives to stabilise rainforest size and animal protection, employing millions to look after the environment, monitor and remediate local pollution, ensuring a clean, abundant and safe planet for generations to come.

With petroleum not only having an unknown future, it has a shady present with the spectre of the rubber trees being susceptible to a fungal wipe-out hanging over them, as well as the rubber plantations contributing to deforestation of in many instances primary and virgin rainforest, an alternative will be a good thing.

If rubber production can in the near future completely free itself from using petroleum—where currently a standard car tyre uses over 27 litres of petroleum in its production, while synthetic rubber is pretty much all derived from petroleum polymers—and limit natural rubber from South East Asia to only ethically sourced, while favouring locally grown and produced dandelions, which will have themselves created a whole new industry based around the dandelion, the future can certainly be golden.

As golden as goldenrod, which has a species known as *Solidago edisoniana*, named in honour of a species bred by Edison, with its 12% latex making a tyre in the late 1920s, and US troops in WWII benefiting from this novel rubber with water bottles. These were made from goldenrod, a great plant to grow with dandelions, both requiring little water or encouragement to thrive.

Whether these new economic sectors need to become a part of the current economic system is where some revolutionary innovations can step in. There is no reason why all new money generated by dandelions (and the other six plants) cannot be remunerated via a Green Currency, whose value is directly linked to the value of these new products, as money used to be backed by gold, even silver, only this is backed by crops—the Green Standard with a Green Currency. You'll have to wait until the final chapter to read about the jaw-dropping truth about our current economic system and the inherent failings of money.

Dandelions are proper cool

If all the medicinal, scientific and nutrient information wasn't impressive enough, the rubber potential should have tipped the scales towards the proposition that dandelions do have an important, sustainable and beneficial role to play in the future. We just need to recognise we have a disconnected viewpoint today. It was only three or four generations ago when dandelion juice was a favourite preparation both in official and domestic medicine, across Europe, Asia and all of North America. Local druggists would send out their collectors for the roots, then bruise and express the juice while still fresh, selling it as shots rather like the recent fad of necking a shot of wheatgrass, although dandelions are far better in taste and nutritional profile. This is reflected with the USDA ranking it in the top four green vegetables for overall nutritional value. Dandelions appeared in the United States National Formulary from 1888 until 1965 and are still listed in the United States Pharmacopoeia.

Internationally the modern Western medicine model has been very successful in gaining market share, at the expense of forcing out the traditional practices and healers. Russian and Indian medicinal cultures valued the dandelion highly as a liver tonic, for ulcers, tuberculosis, kidney disease and jaundice. In China, *The Chinese Herbal Materia Medica* suggests healers use leaves and roots of *Taraxacum mongolicum, T. sinicum and T. olepis* (all very similar to *T. officinale*) to reduce inner heat, counteract toxicity, bronchitis, colds, chronic gastritis, dysentery, enteritis, hepatitis, mumps, skin problems, stomach ache, swollen eyes, tonsillitis and urinary tract infections, often using dandelions with other healing herbs. I use the flowers as a skin cleanser, soaked in extra-virgin olive oil in a glass jar for about a month, placed in as much sun as the British summer provides; sieve the infused oil through a muslin cloth, bottle and apply it to eczema and other skin ailments.

The shape of the dandelion leaves channels rainwater and dew to the centre of the flower and down to the taproot. Dandelions are also a soil transformer, being an early coloniser species, loosening compacted soil by growing deep roots, adding nitrogen to the soil, creating a micro-climate and drawing earthworms in, who mix and turn the soil, bringing nutrients nearer to the surface, as well as adding their worm casing. Quietly and efficiently dandelions magically renovate the soil.

Akita Prefectural University, in Yurihonjo, Japan, with the aid of microscopes and super-slow-motion photography, studied the opening and closing of dandelions, concluding that the dandelion had superior functions of a plant from the engineering standpoint, maximising water efficiency and strength, right from the cellular level. The researchers were extremely impressed with the design, to say the least, proclaiming each single seed-pod as being at the absolute pinnacle of aerodynamics, maximising chances of effective and wide dispersal. The fluffy parachute-shaped head of each seedpod, called a pappus, can create a vortex just above itself as it floats through the air, which in the right conditions can hold the pappus in the air indefinitely by creating an anti-gravitational field. Each dandelion plant has in its 13-year lifespan the chance for 65,000 seeds to begin the cycle all over again.

Dandelions also radiate on the UV light spectrum, acting as a beacon drawing insects in to assist in their pollination. In addition dandelions emit ethylene gas as an aggressive act to preserve their bit of soil, preventing other plants encroaching. Ethylene gas stunts any neighbouring plants, as well as causing premature fruiting and flowering, hampering the chances of there being any next generation for plants getting a little too close to them. They can of course be utilised and processed for this ripening potential if required as well, isolating this ethylene for such purposes. Gaining an inroad to this ethylene industry may be rather difficult, as currently there is about 140 million tons produced every year, obtained from distillation of natural gas and crude oil. Most of this fossil fuel-obtained ethylene is used to make polythene, being worth in revenues every year somewhere in the region of £250 billion! Mainly to Dow DuPont, ExxonMobil, Sinopec and Total. There will definitely be significant hurdles in this sector, and even if dandelions could provide an environmentally safe and sustainable alternative, the chances for an outsider gaining a foothold are bleak.

Dandelions are already grown commercially in Germany, Poland, Australia, Canada and several States in America such as New Jersey, Texas, Florida, Kansas, Ohio and Pennsylvania. Which means there is already a font of knowledge about growing dandelions from across the globe.

* * *

All animals when needed will seek out plants to self-medicate, a process known as zoopharmacognosy, which is no different from the essential requirement of us eating nutritious food. There is no research on dandelions as fodder, although undoubtedly any trials will have positive results as a naturally nutritious plant should bestow health benefits. Given an opportunity, all animals, not just livestock, will prefer dandelions, yarrow, fumitory, cowslip, fat hen, plantain, nettles (dried and added to their hay), daisy, comfrey, chickweed, burdock, clover, marigold, sorrel and hundreds of other plants over boring grass any day. All these plants are also good for humans.

This shouldn't be surprising, as these plants have all been enjoyed as long as we've been on this beautiful planet. The animals/livestock know full well what they should be eating, but annoyingly for them they have to make do with what they're given, which is at best a field of grass, with perhaps some tasty wild plants and flowers around the edge, which are soon eaten. What would be a good idea for farmers would be to deliberately seed not just a strip at the edge, but entire grazing fields with wild flowers, as the livestock will then benefit from extra natural nutrition, their health will be improved, they will require less pharmaceutical drugs, their meat/milk will be tastier and more nutritious, with everyone a winner, even the soil.

Given half a chance, cows would rather eat foraged flowers and grasses as opposed to a mono-diet of industrial rye grass and processed grain and oil pellets. Some humans are amazed that animals have anything in the way of a food preference, or any intelligence for that matter—personally I'm amazed that any humans have any intelligence after studying the state of the toxic environment we have created and the food eaten!

With dandelions being such a feature of the British landscape it is to be expected a whole array of small birds, insects, bees, butterflies and moths all enjoy them and no doubt many small mammals also have a munch.

Foraging/Harvesting

The number one precaution when foraging for dandelions has to be what at first might appear a positive, because one of the reasons dandelions are so nutritious and healthy is that their taproots are so super-efficient at gobbling up everything in the soil—not just the nutritious beneficial minerals, such as absorbing 300% more iron from the soil than other plant roots, but they also hoover up environmental pollutants, heavy metals and pesticides.

This means it is vital to ascertain the purity of the local soil and water, particularly if you are looking at the health abilities of any plants, as a compromised plant that has pesticides entwined within its fabric is never going to resemble a health provider. As we saw with nettle's ability at phytoremediation, dandelions are an even better first line of biological defence against contamination and spillages at human sites of degradation and pollution, such as mines and industrial complexes, where waste can be filtered through them with their disposal via incineration for ethanol/electricity production (with suitable filtration).

Science has demonstrated that agricultural chemicals are carcinogenic, cause genetic mutations, respiratory failure and more—if you could see how much of a deadly chemical barrage your 'fresh' non-organic food had endured from before birth all the way to getting wrapped in plastic-leaching plasticisers into the food while stacked on a shelf, you'd think twice before buying. If we are going to bother foraging we might as well make the effort to get the best possible.

The British Pharmacopeia considers autumn-dug dandelion root to be more bitter, containing as much as 30 to 35% more of insoluble inulin, making November the best month to make extracts and tinctures from the root. Also remember that without the landowner's permission it is illegal to dig any roots on private land, so get your dandelion roots from your own unsprayed garden, a friend's or clean public land. When you are confident about the lack of contaminants, pick as much as you need. The leaves are less bitter in taste before the flower bud emerges, although, as we will see in the next section, bitters are good and if the leaf is bitter it can be tampered down with other salad leaves, tomato and cucumber. When going for roots and leaves, it's best to find land that hasn't been too trampled, as roots and leaves in such conditions tend to be considerably smaller, so go for those specimens with the largest leaves. Many from the garden will have small, low-lying leaves to keep under the blades of the lawnmower; these are best left alone and deserve your respect anyway for them being so clever, audacious and tenacious.

Culinary dandelions

Dandelions are considered to be bitters, which as the name suggests they have a bitter taste, which they do. The bitter taste is actually the taste of the abundant minerals, a flavour the modern food industry has deliberately veered away from, progressively producing cultivars that contain less bitters and more mild-flavoured carbohydrates. Not a good move as bitters help to restore our acid–alkaline balance, being alkaline, as well as aiding the absorption of the fat-soluble vitamins A, D, E and K and iron, plus being nutritious (full of essential phytonutrients)—all this is forsaken for carbohydrates. Food grown in permacultural plots has a fuller flavour because of the extra phytonutrients, a richer and more complex flavour, also healthier as the taste is the phytochemicals, which aren't in food after processing.

Bitters also stimulate the release of stomach acid, which triggers something known as intrinsic factor, a glycoprotein. Among other things this has been shown to help the body absorb vitamin B12, thus avoiding two very common modern complaints, namely B12 deficiency anaemia and pernicious anaemia. By stimulating the release of digestive juices, bitters improve the digestion of fats and proteins. With no bitters in the diet (that is, no bitter-tasting vegetables) the digestive system becomes sluggish, metabolism slows down, the person becomes also sluggish and weight gain is a natural result. This relatively recent diet of processed carbohydrates is a major reason why so many people suffer from acidosis, bloating, digestive weakness, heartburn, malnutrition, metabolic conditions and obesity. Bitters also reduce the appetite for processed white carbohydrates and sugars, helping to break the cycle of overeating sweet and processed carbohydrates to our detriment, which only distributes insulin and prepares the body for more of the sweet stuff.

Dandelions negate all this, as well as offering another dietary benefit, as the root is a rich source of what is known as insoluble inulin. This insoluble fibre, inulin, absorbs water as opposed to being dissolved (that'll be a soluble fibre, also vital) and acts as an important bulking agent to maintain a clean and healthy digestive tract and colon. It is definitely worth bearing in mind that the sensitive mucous membranes of the intestines and digestive tract contain as much as 80% of the entire body's immune system—this is going to be the new avenue for public health and associated products, as this little-understood digestive flora actually does a lot more than simply digest food.

* * *

Dandelions are still a part of many traditional dishes around the world, as diverse as Amish, Polish, Chinese, Japanese, Iberian, Korean, Cretan, Slovenian, French, Swiss, Sephardic Jew and Greek. In Greece a side-order of greens, usually dandelions, is called *horta*. This has very interesting etymological roots as Horta was the Etruscan Goddess of agriculture, food and health, so a dish named after Her should be super-healthy. Dandelions get used as an ingredient in these cultures for breads, pizzas, sausages and tomato sauces. In Japan and Switzerland the buds are battered and eaten with a sweet sauce or honey.

The dried and roasted roots have been traditionally made into a drink, which annoyingly is called a coffee substitute. It isn't that, it's just a similar colour as well as being very bitter. The effect of this drink is a little like that of caffeine, even though dandelions are caffeine-free—for my first trial I mistakenly put three teaspoons of ground roasted root in the espresso maker, realising too late that this was too much as for the next two hours my brain was a discombobulated whirr as if I had mainlined some amphetamine! The next day, one teaspoon had a much more satisfactory result, definitely comparable to coffee (only healthier) in effect, not taste, as it's more bitter.

The whole plant has been used since time immemorial to make a type of beer and wine, known as a health tonic, with dandelions also being an effective hangover cure.

Dandelion and Burdock, as mentioned earlier in the chapter, was a thirteenth-century healthy drink made from fermented roots, and continues today as a soft drink. A jam can be made from the flowers, which have also been used to make what is called May-Honey, a syrup as a substitute to honey (especially popular in Poland), although the large sugar content seriously impinges on its healthiness.

* * *

Without stretching any culinary abilities, here's some really quick ideas to get you started in incorporating some of these plants into your daily diet, as that is all that has to be done. Start eating some, and it will all snowball from there if you have any real interest; indeed the whole change of culture could start with the ingestion of a single foraged plant, as this reactivates neural and ancestral networks—

The easiest way if you have a handy clean supply of fresh dandelions is to pick a few leaves from a plant that has yet to send up the flower stem (new growth as late as October in south east England) and add these, chopped up, to a **salad**. Or the leaves can be lightly **sautéed** with a little garlic and lemon juice and eaten as a side dish. If you are feeling adventurous, a leaf or a flower bud can be battered and **deep fried**.

The best way of consuming quite a few leaves in one go would be to make the **pesto** from the last chapter, using whatever other leaves can be foraged seasonally. In the same way add a few dandelions along with the nettles for an extra boost of nutrition in a **hummus** or **soup**.

For the root beverage, called **dandelion coffee** (although Europeans have been drinking this for many millennia longer than coffee), dry the roots, in the sun if it's sunny for a couple of days, and then roast at about 170°C for 45 minutes to an hour, grind in a pestle and mortar or coffee grinder, and brew as coffee. It's worth experimenting with the oven temperature and duration as we all have different tastes.

Final word

It does appear that these first two plants, nettles and dandelions, hold within them the key to health—establishing homeostasis which is why they were incorporated into stories and folk tales about these plants, giving them lofty positions as symbols associated with the cultural pantheon of gods, goddesses and the sun, the ultimate provider of life. Today we might find it amusing that the Egyptians, Romans and other Ancients had so many deities; indeed it seems as if every plant and animal has its own one, as it should, because every living thing on this planet is a marvel and deserves veneration. Yet it is pretty much exactly mirrored in the Christian church with the Saints being the equivalent, fulfilling similar duties and functions such as Mark the Evangelist stylised as a winged lion is also the patron saint of quite a lot including painters, lawyers, pharmacists and people dealing with insect bites; St Christopher is the patron saint of travel; St Joseph for employment; St Anthony for recovering lost items, health and people; St Arnulf is the patron saint of beer and St Giles the patron saint of disabilities, as well as the fear of breastfeeding! The lost old stories and songs of folklore were once imperative in remembering how to look after ourselves.

There are now appearing hairline cracks from when our current culture plastered over our knowledge (of Mother Earth) and birthright (personal empowerment and sovereignty). These have opened up after millennia, and in some remote places of the world the plaster has either fallen off or it wasn't put on thick enough, exposing in glorious technicolour the world behind this plaster. The illusion of truth that is the fabric of the plaster of our culture is crumbling.

When I was formulating The Culinary Caveman's Green Gaia Goodness powder quite a few years ago, the first two ingredients that had to be included were nettles and dandelions. I had no idea at the time why these two presented themselves to me so forcibly, I just went along with it because I have the upmost respect for Mother Nature and was in no position to question Her interfering with my cerebral rationality. It can't be put into words quite how these seven plants in this book presented themselves to me, because of course they didn't speak to me in English, Latin or any words or speech, but there has been a communion, although sadly there is no way to scientifically quantify this phenomenon. From the subsequent research and knowledge learnt from this book, their inclusion became far more clear and obvious, being, as they are, two eminent Human Plants.

* * *

Let's move on to a less well-known plant, this one not only being sacred but also mythical and inter-dimensional. With a long history of folklore and traditional use, let's see what this one has to offer in this plan to remediate the unsustainable, disastrously desecrating cultural systems we are presently caught up in—

YARROW

(Achillea millefolium)

Armageddon

Not so long ago, I was spending some time with the lovely proprietor of one of the organic health shops that sell my herbal powders. We were in gorgeous woods where Winnie the Pooh once roamed and Eeyore moaned—it's way beyond cool. We had stepped out into the woods to find a parasol mushroom that my friend had noticed had just begun opening earlier, as we walked and talked about some local trees, I mentioned the plants in contention for this book. When I said yarrow was one, he asked did I already know about how yarrow will save us after a nuclear war? Well no, I hadn't come across that particular fact! So he went on, saying it was to do with the Doctrine of Signatures, as the inflorescence (flower-head, which is actually often between 15 and 40 flowers, arranged in what is called a flat-topped capitulum) resembled a nuclear bomb's cloud! I exclaimed that was fascinating, while at the same time being totally sceptical about how a 35 million-year-old plant decided to take its shape because in some distant future arrogant humans could potentially destroy pretty much everything with their science and deadly toys.

The Doctrine of Signatures is fascinating, yet scepticism is to be expected from the postulation put forward by Paracelsus (1493–1541). Born Philip von Hohenheim, but perhaps feeling this didn't sound lofty enough, he later became known as Philippus Aureolus Theophrastus Bombastus von Hohenheim. Apparently still not satisfied, he finally took on the name Paracelsus. This name translates as 'above Celsus', with Celsus being a second-century Greek philosopher, best known for his tome *On the True Doctrine*. Celsus very much objected to Christianity, or at least the illogical dogma he saw pretending to be Christianity. Paracelsus is therefore a somewhat cheeky, perhaps arrogant name to take on, but he appears to have been a fine fellow, probably with an elevated sense of wit, as he also improved the study of illness and medicine along with some fine quotes, such as "The art of healing comes from nature, not from the physician. Therefore the physician must start from nature, with an open mind." Pertinent to his own doctrine he also stated the visual appearance of a plant indicated how it may benefit the human patient: "Nature marks each growth … according to its curative benefit," be it red for blood or walnuts for the brain.

Giambattista della Porta continued the concept in his 1588 edition of the wonderfully entitled *Phytognomonica*. However, there is always that voice at the back of the mind chattering away, wanting to know why it is that a plant, existing long before us, would have any inclination to resemble those organs and parts of us that it is beneficial for.

The Doctrine of Signatures does sound like a human-centric view of the world. Although, flipping that concept on its head, it could have been the plants that planned for our evolution. Think about that for a moment. It is not as outlandish or flippant a proposition as initially assumed. All animal life has evolved from a state of matter on this planet that passed from the stages of mineral to plant to animal. The reptilian brain is accepted as being at the base of the human brain, normally explained away as a relic of evolution. In exactly the same way, the sympathetic nervous system is a relic from the plant phase—if it were removed from the body and allowed to hold its own weight upright, it is essentially a tree—with one thing connecting and communicating it all as a whole singular entity, namely acetylcholine. As we saw in the Nettle chapter, this first Human Plant has an enzyme allowing acetylcholine to function effectively. Our physical form has to have been created around the plants being eaten—there wasn't a conscious decision to start with a cell knowing it would become a human after millions of generations … or maybe there was ….

Anyways, back in 100-acre Wood, I thanked Dave for the mushroom—it was so large it made a mushroom risotto and veggie burgers—saying I'd investigate that connection further as it sounded captivating. That evening, after the smallest was put to bed, some research into this yarrow and nuclear bomb connection certainly had an ominous beginning, as yarrow means 'war' in the Language of Flowers. Although, it must be said, it is also sometimes referred to as meaning 'peace', which proves to be a common theme with yarrow, it having the ability to be both black and white and up and down, stop and start or open and closed. It also turned out after some probing that the claim was probably true: yarrow could well be the plant to save us after a nuclear holocaust, as we're about to learn (from the words in this chapter, not an actual bombing!), and quite a bit more besides. So, for this chapter it doesn't get any lighter. First the Apocalypse of words and a veil of truth, now an Armageddon—it's all beginning to sound more like a Monty Python sketch. I sincerely hope the Spanish Inquisition isn't about to turn up!

* * *

In all seriousness, of all the plants of this book, yarrow will prove to be the plant of the future, although its assistance is also needed today. In ten years' time, maybe less or maybe more, there are two very real situations that if allowed to escalate will push us further down the slippery slope of devolution and desecration. Above and beyond the host of serious complications we're about to look at, there are two bad predicaments that yarrow can ameliorate, having just seen the first menace, the nuclear threat, a peril not necessarily dependent on terror despots and itchy fingers. Consider the tragic travesty at

Fukushima, where the resulting despoliation from nuclear pollution and the ignorance of how to contain the damage is going to cause far worse, longer-term complications than dropping a bomb. It's as if they have dropped a bomb on themselves, scored the ultimate own goal, albeit a global one. As the constant leakage of unbelievably irradiated water continues to spread across the Pacific, it will soon to be added to, if not already, with the planned-for emptying of the contaminated stored water. This contamination is going to spread globally, and we can only wait to observe the consequences. There are no studies or contingency plans for what happens when millions of tons of irradiated nuclear pollution mixes across the world, becoming intricately entwined in the fabric of the food chain and the human cellular structure. Oh! It's a bit late then to realise that dilution was never the solution, which shockingly has been official government policy, here, there and everywhere since the 1920s.

It is just like plastic and what was forewarned in the 1980s—that it had to be recycled as it lays rotting for centuries, leaching out deadly chemicals (genetic mutating ones at that) into the soil and sea. These warnings fell on deaf ears for decades, as the seas have been awash with plastic for some time now, attested to by the islands of plastic in the huge gyres in the centre of the Pacific, Indian and Atlantic Oceans, as well as being found in ice cores from Antarctic and the Arctic after 1960. We are becoming saturated in plastic, from the tropical shores of Bali to the frozen wastelands of Alaska.

Sadly, it is just another man-made thing, where the only consideration being the amount of money it could earn. Corporations make billions, while hardly spending a penny to clean up the waste they created in earning such vast profits. There is no doubt that if we manage to get to the twenty-second century, we will look back at this time and gasp in amazement at how dumb, naive and arrogant we were from 1950 to 2025—spraying deadly chemicals directly onto the food we eat, polluting the air with transport and electricity production from fossil fuels, making plastic from petroleum and non-biodegradable genetic-mutating materials, wrapping food in it, cutting down vast swathes of the rainforest for cheap cooking oil, meat and rubber, laying barren the soil, and putting nuclear waste in the chemical-laced seas, while at the same time becoming addicted to fats, sugar, pharmaceuticals and phones. In the words of Handsome Boy Modelling School, "The World's gone mad, we've lost so much of what we had … The World's gone bad, we've lost the life we once had."

* * *

Yarrow is going to help with the present dangers of radiation poisoning from wi-fi, electro-magnetic radiation and possible nuclear contamination, as well as preventing a potential deadly invasion to remain unmentioned for now. Before then, it's worth spending some time looking at the details of this magnificent watercolour beginning to be painted by yarrow, with an eye for detail and an exuberant style on a par with the Pre-Raphaelites of Sargent, Waterhouse and Millais, or John Martin and his great triptych of *The Last Judgement*. Having gotten lost in those pictures in the past, in very much the same tone I've gotten lost and become full of appreciation of yarrow recently. The more one looks at the detail, the more one gets spellbound in wonderment at that sheer detail.

For now, let's go back in time a few millennia, because …

Them Greeks, they knew a thing or two

Yarrow, as indicated by its botanical name *Achillea millefolium*, can be traced back to the ancient Greeks, specifically to Achilles, a hero from the Trojan War who used yarrow extensively to treat the wounds of his soldiers. *Millefolium* means 'a thousand leaves' or the 'coming of a thousand leaves', a direct reference to yarrow's feather-like leaves, as indicated by its Greek root of *myriophyllon*, which means 'bearing numerous leaves'.

Whether Achilles was actually a real person, being the star in Homer's famous *Iliad*, is certainly open to conjecture. If he was, the events would have taken place some time around 1200 BC. There must be some truth to the stories other than yarrow being a superb wound healer, blood stauncher (styptic) and helping to prevent infection—considering that up until WWI, most fatalities from all combat for three millennia had occurred from infections caused by wounds not being properly treated with antiseptics etc.

Achilles' Heel is a real thing today—being a tendon above the heel area, the calcaneal tendon. I snapped my right one skating a few years ago. Pushing off a kerb, I heard a pop and a sensation as if a mouse had run up my lower leg, then I crumbled to the floor. The phrase is also a figure of speech, meaning a weak point amid overall strength. The 'real' mythical heel came about because Achilles' mother, Thetis, held him in the water of the River Styx by his left ankle to make him invulnerable to harm, although tragically she left the ankle where she held him unprotected (other accounts mention a vat of yarrow tea he was dunked in). It's also said Achilles covered his whole body in a yarrow tincture to make him invincible to arrows when in battle. Yet another legend variant has Achilles scraping the rust from his spear, which subsequently grew into yarrow, allowing him to heal the wounded.

Achilles was tutored in the arts and mastery of medicine, botany, pharmacy and archery by the centaur Chiron. Chiron is venerated by the supercar made by Bugatti, as one would expect a mythical hero to step out of such a sleek motor, although it's very low and probably extremely awkward for a half-man half-horse to get in and out of stylishly! With strong reservations about the reality of centaurs, although not of the healing power of plants, Chiron, if indeed a valued herbalist, would have taught Achilles about yarrow, because of the truth of the claims made of it as a plant that should not only be valued, also remembered—which is why it appears in so many stories.

History and names

The fact that most people aren't able to recognise yarrow today would have astounded people 200 years ago, if not less. It would have been unthinkable then not to know of yarrow and the stories attached to it, as is the case with all the plants in this book—a sure sign of our cultural amnesia.

The tradition of using yarrow as an insect repellant and aura cleanser may well span an astonishing 60,000 years, as well as spanning human species. The Shanidar Cave in Iraq is the site of a burial of several individuals who were laid out and revered in a similar way to other *Homo sapiens'* burials of the time and much later, except for one major difference—these weren't *sapiens*, they were *Homo neanderthalensis*. One of the burials was very different, with yarrow, horsetail, cornflower, groundsel and a few other flowers covering the body, perhaps singling out this individual as being the herbalist or shaman among the group, whom we ignorantly refer to as cavemen or troglodytes. We are stuck in a dogma that all of our ancestors, especially those from the dim past, were quite frankly just that, a bit dim—although recent archaeological findings and reattributions clearly show *Homo neanderthalensis* to be far smarter than previously imagined, in terms of clothing, music, artistic symbolism, and even possessing a larger brain. As the 'Culinary Caveman' is my alter ego, I see it as us today who are a little bit dim, with our toxic, unsustainable cultural lifestyle. We lived as hunter-gatherers—that is, all types of *Homo*, whether *Homo erectus, heidelbergensis, denisova, neanderthalensis* or *sapiens*—in apparent harmony with the planet, as demonstrated by it being by some distance our lengthiest and most sustainable cultural existence to date, with agriculture surviving for quite a few millennia thereafter. Tragically, modern monoculture and chemicals have managed in just 70 years to bring the world's food supply and much more to the brink of ruin and collapse.

* * *

It's hard to articulate clearly enough how different our relationship is with the flora of our country today than in any other period of our existence, with that term 'cultural amnesia' hinting at this disconnection from Nature.

Let's go back to the ancient inhabitants of Britain, who made amulets of yarrow and strewed it on thresholds to protect themselves and their homes from evil and to keep witches away. A yarrow plant hung in the house on St John's Eve (23 June) would protect the house from sickness for the coming year. A somewhat disturbing ancient Gaelic chant alludes to its preventative capabilities, with, as is so often the case, a woman saying "I will pick the green yarrow that my figure may be fuller … that my voice will be sweeter … that my lips will be like the juice of a strawberry … I shall wound every man, but no man shall harm me."

Dioscorides (Roman physician, AD 40–90) recommended rubbing the crushed plant directly onto wounds, which had been a standard practice with Achilles some time before and probably for long before him as well.

The Saxons were also known to wear amulets of yarrow to provide courage and protect against blindness, evil spirits, robbers and rabid dogs. Your average twelfth-century knight would have carried with him a pouch of yarrow leaves, fresh, dried or powdered as a part of his first-aid kit for wounds, which were far more commonplace when muscular, broad-shouldered men were frequently swinging heavy, sharp metallic implements around at one another.

The famous twelfth-century Abbess and Saint, Hildegard of Bingen had this to say about yarrow:

> After cleaning the wound, simmer yarrow in warm water. Once the yarrow has had time to simmer, drain the water, and apply the warm yarrow with a bandage to the wound. The warm moist yarrow relieves pain and infection to heal the wound. Repeat this process several times. As the wound starts to heal, remove the bandage and place the cooked, warm yarrow directly on the wound and it heals even faster,

> without complications. For those who have suffered internal injury, drink yarrow powder in warm water. As the person begins to heal, drink the powder in warm wine to accelerate the cure.

Not limiting yarrow to just these uses, Hildegard also advised its use to strengthen and heal the respiratory system and tract, reduce inflammation, improve the functioning of the glandular system and kidneys, promote circulation and purify the blood, prevent the development of varicose veins, as well as being effective for abdominal cramps and lower back pain. She was clearly, and quite rightly so, a fan of yarrow.

Albertus Magnus (1206–1280), in his *Book of Secrets*, suggested smearing one's hands with yarrow juice and plunging them into a river. Fish would be attracted to the scent and be able to be caught. It's a technique that the forager Ray Mears hasn't corroborated, but it might still be worth a go if you feel a bit peckish, near to a stream with some yarrow growing nearby.

John Gerard (1545–1612) used yarrow for "swellings of the private parts" in his controversial *Herball* of 1597. It is controversial because most of it was plagiarised from Rembert Dodoens's herbal, published in 1554. Gerard's edition was full of errors, its good reputation having earned a place in history through a later edition meticulously edited by Thomas Johnson in 1633. Johnson corrected the mistakes and even used the same woodblocks as in Dodoens's work. It is the Johnson edition that was a standard reference book for the next 200 years, not Gerard's, but somehow he still managed to be the only one remembered.

John Parkinson (1567–1650), who gave us *Theatrum Botanicum, or The Theater of Plants* in 1640, is said to have been the last great herbalist and the first great botanist. He used yarrow to stop nosebleeds, as well as suggesting that if drunk warm it was good for ague (any disease involving fever and shivering). Parkinson's contemporary (and plagiarist), Nicholas Culpeper (1616–1654), tells us "It [yarrow] is under the influence of Venus. As a medicine it is drying and binding. A decoction of it boiled with white wine is good to stop the running of the reins in men, and whites in women; restrains violent bleedings, and is excellent for the piles." Culpeper also recommends yarrow for sleeplessness, inflammations, ulcers, fevers, fistulas, menstrual problems, high blood pressure and poor circulation, as well as increasing cow's milk production. Carl Linnaeus (1707–78), he of the two-name nomenclature, recommended fresh bruised yarrow as a wound healer and a styptic.

* * *

One of the forefathers of naturopathy, the German, Sebastian Kneipp (1821–1897), said of yarrow: "women who resort to yarrow are spared much mischief". This evaluation is reflected in many of the names associated with yarrow from Germany at the time, such as tummy ache herb, wound herb, heaven's back herb, thankful wife's herb, mother's herb, haemostatic herb, the all-inclusive healing herb of the world and the equally positive salvation of all damage (obviously all in German, not English!). In modern German yarrow still has pertinent names to indicate its multi-purposefulness, such as *scharfgarbe*, loosely meaning 'sheep healthy-maker' after shepherds had noticed that their sheep healed much faster when they ate some yarrow; the shepherds themselves also ate yarrow, and in the process assisted themselves with digestive disorders, bleeding and colds. Frederick Weiss (Anglo-German botanist, 1865–1953) recommended yarrow for gynaecological problems such as leucorrhoea, as well as vegetative dystonia of the small pelvis.

Other names in English for yarrow include Allheal, Arrowroot, Bad Man's Plaything, Bloodwort, Carpenter's Plant, Carpenter's Weed, Death Flower, Devil's Nettle, Devil's Plaything, Devil's Rattle, Eerie, Evil's Nettle, Fernweed, Field Hops, Green Arrow, Herbe Militaris, Hundred-leaved Gradd, Knight's Milfoil, Knyghten, Milfoil, Nobel Yarrow, Nosebleed, Old Man's Mustard, Old Man's Pepper, Poor Man's Pepper, Sanguinary, Seven Year's Love, Soldier's Woundwort, Snake's Grass, Squirrel's Tail, Staunch Grass, Staunch Weed, Thousand Leaf, Thousand Seal, Thousand Weed, Warrior Plant, Woundwort, Yarrow Bloodwort, Yarroway and Yer.

The name yarrow itself may have come from the Greek *hiera*, holy herb, or a corruption of the Anglo-Saxon *gearwe*, meaning 'the healer' or 'the provider'—derived no doubt from the similar Dutch name, *gerwe* and Old High German, *garwe*. All of which casts no doubt as to yarrow's health capability.

Ethnobotanical

As with the dandelion, many commentators state that yarrow also originated in Europe and was spread to the Americas by colonialists. Again, this is extremely unlikely, with yarrow being already extensively utilised across the whole continent by the First Nation Peoples.

Tribes around the Pacific Northwest hung dried yarrow around the home, or rubbed the flowers on the skin and clothing to keep mosquitoes, flies and other insects away. The Abnaki infused the whole plant for children with colds, grippe and for fevers, while the Aleut used a similar infusion for stomach and throat pains, colds and the leaves as a coagulant to stem bleeding, as did the Nanivak. The Algonquin used an infusion for colds, respiratory disorders and a tea from the leaves and flowers to ease headaches. The Bella Coola used a poultice of the leaves for burns, boils and bronchitis, while the Blackfoot used an infusion of the plant both drunk and rubbed on the body to soothe gastroenteritis, liver ailments, swellings, sore throats, other sores, as a diuretic, a general tonic and to help expel the afterbirth as well as being drunk just before birth to assist the delivery, and as an eyewash for horses (and for other mammals as well, no doubt).

The Carrier used the above-ground parts for an infusion for colds and as a poultice for swellings, while the Chehalis used a decoction of the leaves to assist with bloody diarrhoea. The Cherokee also used it for haemorrhages, spitting blood, fever and bowel disorders, bloody urine, as a digestive aid and also drunk in a tea to reduce fever and provide a restful sleep, and also smoking the leaves for catarrh. The Cheyenne used an infusion of the fresh or dried plant for colds, nausea, chest pains, coughs, fevers, sore throats and to induce perspiration, crushed leaves in the nose for nosebleeds and rubbed on the body for respiratory diseases and tuberculosis. The Chippewa used a decoction of the root for eruptions on the skin, also chewed and spat onto the limbs as a stimulant, using a decoction of the leaves poured on hot stones for headaches and applied to horses as a stimulant. The Cialiam used the seeds for their fragrance around the home and the leaves for childbirth, while the Cowitz used an infusion of the leaves for a hair-wash and of the roots for stomach ailments.

Continuing, the Cree used an infusion as a compress for headaches and fevers, chewing the roots for toothache and fevers, while using the flowers as a bait to trap lynx. The Crow used a poultice for burns, boils and open sores and wounds, and the Delaware used an infusion of the plant for liver and kidney disorders. The Flathead used crushed yarrow leaves for aching backs and legs, wounds and an infusion for colds; the Gosiute also used an infusion for colds, headaches, a poultice for rheumatism and bruises and the leaves for digestive ailments. Pretty much these same uses were repeated among the Haisla, Hanaksiala, Hesquiat and the Iroquois, who also used yarrow as a wash for babies with convulsions and for children with worms, a digestive aid, for venereal diseases and almost everything else, even as a wash for someone unconscious. The commonest uses were also shared among the Karok, Kawalisu, Kialiam, Klamath, Kutenai, Kwakiutl, Lakota, Lummi, Mahuna, Makah, Malecite, Mendocino, Menominee, Meskwaki, Mewuk and the Micmac. The Miwok have used it as an analgesic, for head colds and during influenza epidemics, which have been devastating for indigenous peoples, wiping out entire communities from the sixteenth to the twentieth centuries.

Yarrow also had a central role in health for the Mohegan, Montagnais and Montana, while the Navajo held yarrow in such high esteem they considered it 'life medicine', as well as chewing it for toothache and pouring an infusion in the ears for ear ailments. Yarrow was also a general go-to among the Nitinaht, Ojibwa and the Okanagan-Colville, who also smudged (burnt) using the leaves and stems to deter mosquitoes. The Okanagon used a yarrow decoction for sore skin, snake and insect bites, with the Palute also having a wide range of uses, as did the Pawnee, Pomo and the Potawatomi, who used it as a smudge to repel evil spirits, witches and to revive comatose patients. The Potawatomi again shared many of the commoner uses with the Quinault, Saanich, Salish, Sanpoli, Shoshoni, Shuswap, Skagit, Snohomish, Squaxin and Swinomish. The Teton Dakota People called yarrow 'medicine for the wounded', with other tribes referring to it as 'warrior plant', in direct reference to its staunching and healing capabilities. Yarrow was also commonly used by the Thompson, Ute, Washo, Winnebago, Yuki, Yorok and Zuni, who chewed the root and applied the juice before fire-walking or fire-eating, pulverising the plant mixed with water and applied to any burnt areas as a poultice.

Summarising, yarrow was even more versatile than nettles, covering almost everything that could have ever ailed them—that is, before smallpox, influenza, measles, tuberculosis and Christianity were imported across the Atlantic.

With such a rich ethnobotanical spread of regions and uses, it is most improbable that this extensive pharmacological use only spread across the continent after the *Mayflower* docked. The new inhabitants would obviously have brought yarrow with them from Europe, as it was so useful, and this would have been documented. As time passes, such records become the oldest documented evidence, and it is easily misconstrued as it being the documenters who brought the plant to the American continent, as opposed to them merely bringing a European yarrow to add to the varieties already happily growing there.

Traditional

The list of other traditional uses for yarrow from the rest of the world is vast, and from all these it can be seen that yarrow fights bacteria and viruses and acts as a perfect preventative, especially when combined with other plants, such as ginger, plantain, dandelion, nettle, elderflower/berry, lemon balm, mint leaves and echinacea. It is also popularly used in conjunction with a small amount of tormentil for diarrhoea.

By itself yarrow has been used for detoxifying and benefiting digestion, with its bitter content increasing bile production and hence stimulating the appetite. With yarrow's drying action it is a useful decongestant, being effective against mucus formation caused by coughs and sinus infections. It alleviates allergies to dust, mould, dander (skin flakes and dandruff) and pollen. As a herbal first-aid, it is used for burns and ulcers and other skin conditions such as eczema. It relieves pain from arthritis and rheumatism, haemorrhoids (yarrow tea and poultice are effective) and acts as a menstrual regulator, stimulator and suppressor.

The flowers and leaves when eaten were used to treat gastrointestinal disorders, quell fevers and improve circulation. Also regarding the blood, yarrow can activate blood platelets and break up coagulated blood, helping it move more freely and easing strain on the heart. It brings blood to the surface, inducing sweating, with the possibility of lowering blood pressure as it does so. It is pertinent to note all the SARS family of diseases would be considerably assisted, from prevention to alleviation, with yarrow being the number one herb for blood (restoring and moving), the respiratory system and fevers. Blood clots in the lungs are the main cause of death from the SARS virus, applicable to a much smaller percentage of Covid-19 deaths as most people died *with* it (being so contagious), not *of* it—The Office for National Statistics gives a figure of approximately 2,000 UK deaths solely from Covid-19 (2020–22).

Chewing yarrow leaves provides good relief from tooth and gum discomfort—I can attest to the veracity of this claim. While at times the complications, extractions and infections have required a Paracetamol or two, more often than not yarrow, plantain, clove oil and ice-packs have worked admirably. During my most painful bouts I pondered on how we dealt with this problem before dentistry and painkilling drugs. The simple answer, from the experience within my own mouth, was that often an infection was actually derived iatrogenically (caused by medical intervention), because it was the filling that was the root of the problem as much as it was my unfettered indulgence in sweet-eating as a child, from the corrupt school dentists filling poor kids' teeth unnecessarily (all for the £12 fee) and an ability when older to hit trees and concrete while cycling. Teeth were and are a major cause of sufferance throughout our time in cities, although, as so marvellously proven by Dr Weston A. Price (1870–1948) before WWII, indigenous peoples around the world had almost perfect teeth, with a fraction of the incidence of cavities compared to urbanites—meaning that terrible teeth is actually yet another disease of civilization. This made me realise it wasn't how they coped with this pain in the distant past (and for sure they did, also with abcesses and sepsis never too far away), it was more pertinent to ask how we have allowed our teeth to get into such a deplorable state today and how the sugar manufacturers have not been held responsible for this calamitous cavity and diabetes pandemic. What I will do is prepare some yarrow root preserved in rum or brandy, to be chewed when required in the future.

* * *

In Ayurvedic Medicine, yarrow is an ingredient in the popular Liv52 formula, being the best liver support in the world apparently; the compound also contains capers, chicory, ferric oxide calx, black nightshade, arjuna, negro coffee and tamarisk, and if you fancy knocking one up for yourself the quantities are online.

Among the Highland clans of Scotland an ointment was made from yarrow for injuries and cracked skin from exposure to the bleak winter conditions. A tea made with yarrow from Orkney dispels melancholy, and it can easily be imagined such a brew being drunk by those living at Skara Brae over 5,000 years ago, or those who erected the other great Orcadian Neolithic sites such as the Ring of Brodgar and the Stones of Stenness. In the Hebrides, with their equally stunning and famous megalithic Callanish stones, second sight is believed possible by placing a yarrow leaf against an eye.

Matthew Wood, in his wonderful *Earthwise Herbal* books, speaks most highly of yarrow, calling it the master of both blood and fever, being able to move blood to either cool or warm areas. This ability suits bruises, lacerations, the digestive tract, fevers and chills, and any ailment involving inflammation and congested blood. It is also good, he says, for earache, toothache, bronchitis, pleurisy, diverticulitis, inflamed liver, varicose veins, measles, arthritis, bee stings and 'female' complaints from endometriosis to prolapse and sore nipples. Midwives, rurally and historically, carried yarrow with them to prevent patients suffering blood clots and haemorrhaging, and with menstruation yarrow either stops or starts, and healing if painful.

In Iranian folk medicine, yarrow is used for bleeding wounds, inflammation, pain and spasmodic diseases. In Germany, yarrow is approved by the official Commission E for menstrual cramps. Drunk as a tea or tincture, it has been shown to aid the body during invasive surgery and also to support the healing process. In Norway, yarrow has been used for rheumatism, with the leaves chewed to cure toothache.

Yarrow can be used internally and externally for hives. Externally too, it has been used for pink-eye and styes, skin diseases, chicken pox, damaged nails, black eyes, varicose veins (the oil or a poultice is good for these). It acts as a blood vessel tonic, improving peripheral circulation by dilating blood vessels, as well as assisting in lowering blood pressure and subsequently heart disease, as it also helps lower arterial cholesterol. Yarrow as a poultice also aids mastitis, a common infection in humans as well as livestock bred for milk. Simply lying in a bath with a few ounces of fresh or dried plant matter added, such as yarrow, lemon balm and comfrey, would be beneficial for ear infections, skin ailments, injuries, piles, hair loss and more.

The essential oil of yarrow can balance oily skin, stimulate new hair growth and is an effective skin and hair treatment. While calming inflamed skin, if mixed with basil oil it is effective against acne, psoriasis and seborrhea. It is also effective for irregular and painful periods and to reduce excessive bleeding. All the blue essential oils (from yarrow, blue tansy and chamomile) help to relieve burns associated with cancer radiation treatment—applied as a mist in an atomiser. The oil is also used for circulatory disorders, skin conditions, digestive ailments, asthma attacks and epileptic seizures.

Finally, with its balancing capabilities yarrow is an excellent remedy when experiencing times of transition, such as a change of employment or residence, or those highly emotional occasions after a row with a loved one, or other personal stresses. Yarrow brings with it a very strong inter-dimensional matrix of balance—as peculiar as that may sound now, by the end of the chapter hopefully it will sound more rational.

"It's all about the phytonutrients, darling!"

These phytonutrients, the soon-to-be must-have 'new' nutrient, so common among the previous two plants are also naturally abundant in yarrow. In fact yarrow has more than its fair share of what are known as terpenes, typically associated with conifer/pine trees—these aromatic, resinous and volatile essential oils contribute to the healing potential of the yarrow plant. Among these terpenes, eight are well known and researched for their healing capabilities—alpha pinene, borneol, camphene, camphor, chamazulene, cineol, limonene and sabinene.

Alpha pinene has been shown to have bronchiodilating properties, while the chemically similar borneol improves circulation. Traditional Chinese Medicine uses mugwort for this terpene, valued not only for its anti-carcinogenic abilities but also to help shift a foetal position to alter any breech presentation. Camphene and camphor are antimicrobial compounds found in yarrow, evergreen trees and rosemary, and are the main ingredient in Vicks Vaporub; they are also an effective insect repellant and a vasodilator, cough suppressant and decongestant. Chamazulene has anti-inflammatory capabilities, and is also found in wormwood and chamomile. Cineol is perhaps best associated with trees, and although also found richly in yarrow, this spicy aromatic substance is the principal compound found in eucalyptus oil, used for flavouring and as a cough suppressant. Limonene is an insecticide, and recent research has looked at its anti-carcinogenic properties, with ongoing studies already showing positive results. Finally, the bitter terpene compound sabinene, also found in oak and spruce trees, nutmeg, black pepper and tea-tree oil, is the subject of ongoing research trying to establish its varied beneficial workings.

* * *

As we will see in the next section, when science conducts studies on these plants to see if there is any veracity to the folk traditions, a resounding yes is usually the answer. It must be mentioned (again) that reductionist scientific methods of expected biological phenomena to be dependent on just a single compound are proving to be problematic, especially when several compounds all appear to do the same thing individually, but together the results are very different. As any holistic practitioner would expect, it is a combination of these phytonutrients that bestow health benefits upon us. Reductionism is only practised and dominating because it results in isolated patentable chemicals that can be sold for a profit as drugs.

Among the well-known minerals found in yarrow are calcium, chromium, cobalt, magnesium, phosphorus and potassium. These occur alongside an abundance of phytonutrients, including azulene, bornyl acetate and caryophyllene. It is rich in the flavonoids apigenin and luteolin, and sesquiperpenic lactones such as achilicin, achilleine and millefin. Alkaloids are represented by betaine, choline and stachydrine. Then there are asparagine, bitters, coumarins, eugenol, isoartemisia-ketone, linalool, salicylic acid (aspirin), sterols, tannins and yomogi-alcohol. Achilleine may be one of the chemicals playing a role in yarrow's ability to stop bleeding as well as being one of the antiseptic chemicals and compounds (see Appendix for a longer list).

Phytochemical studies have shown yarrow to be particularly rich in flavonoids and caffeic acid derivatives, and along with the others in this section, all have been shown to be antibacterial, anticarcinogenic, anti-inflammatory, antioxidant, antispasmodic, antiviral, gastro-protective, hepato-protective (hepa = liver), immunological, oestrogenic and more.

It does seem to be looking more probable that health is determined by not just a variety of known nutrients such as calcium, magnesium and zinc, and that phytonutrients will begin to play a more central role, as Western science begins to discover and name them. As mentioned in the Nettle chapter, dieticians, nutritionists and food scientists have predominantly been looking at the well-known 'essential' nutrients, with no mention of any essential *phyto*nutrients. This is not because these might or might not be essential and vital—they are not called 'essential' yet because science has little idea what they are all about, and focus has been on the macronutrients of water, potassium, protein etc. Health and nutrition do not end with these 50 or so 'essential' nutrients, but merely begins with them. They end with the interactions of thousands of phytonutrients among themselves and the already recognised nutrients. These phytonutrients act as tiny keys and jigsaw pieces, unlocking bits here and there, while also joining bits there and here. Unfortunately, they are also fairly delicate and are all but destroyed by the damaging processing of the modern food industry. Shame.

There have been many claims made already of the power of yarrow, but these claims are just words on a page—is there any meat to these claims? Surprisingly, there is quite a bit of scientific research for this undercover plant, which conveniently corroborates most of the traditional folkloric uses mentioned above.

Axis of evil leading the way (again)

The Iranians are certainly among the leading pack in terms of scientific herbal research, and here is some of what they found out about yarrow—

In 2013 (Jenabi & Fereidoony) a study carried out at Islamic Azad, Toyserkan Branch, looked at the efficacy of the traditional use of yarrow in treating post-menarche pain, dysmenorrhea, specifically the common affliction of abdominal pain radiating to the inner thigh. There should be no surprise to find out the conclusions to the study clearly demonstrated yarrow's ability to minimise the severity of the associated pain of dysmenorrhea. It's surely worth a mention that the revenue from drugs for dysmenorrhea is worth somewhere in the region of £1.9 billion annually.

Diabetic benefits from a 2014 study (Zolghadri et al.) at Shiraz University on rats, saw decreased blood glucose levels, with the all-important insulin-producing B-cells of the pancreas being protected by yarrow from the damage caused by alloxan. The rats not only showed reversal in their diabetic status, they also gained weight and overall health, with the suggestion it was the polyphenols doing much of the good work in protecting genes from damage. The yearly revenue from drugs for diabetes is a massive £41.5 billion, rising 6.5% annually—simply cutting out sugar from the diet of the general public and not replacing it with dangerous sweeteners would probably halve this figure within months.

A double-blind randomised controlled trial from 2015 (Miranzadeh et al.) looked at one of the common debilitating side effects suffered by cancer patients from chemotherapy, namely oral mucositis. This condition impairs drinking, eating and even talking, and is sometimes so severe the cancer treatment has to be terminated. A mouthwash was made to gargle and for those given the yarrow one, as opposed to the traditional one, the symptoms not only lessened but the oral mucositis was healed. The drugs that didn't work as well as yarrow earned in revenues globally about £750 million.

Another 2015 study, this time from Mazandaran University of Medical Sciences, Sari, looked at the protection offered by yarrow from ionising radiation created by various sources of radiation. All types of radiation have been shown to lead to genome instability and chromosome abnormalities—which is obviously not good for you, or any transmission of DNA to the creation of a new you. Samples of human blood tested with varying dilutions of yarrow extract all lessened the side effects from radiation, proving to be an excellent antioxidant with potent free-radical scavenging ability. Those flavonoids and azulene played a major role, making yarrow a perfect companion to traditional radiotherapy and chemotherapy.

Studies from Rafsanjan University of Medical Sciences in 2017 called for further investigation into yarrow's benefits for multiple sclerosis and the debilitating symptoms of inflammation and acute joint pain, following on from studies earlier in the year also done by F. Ayoobi on the use of yarrow for epilepsy, Alzheimer's, Parkinson's and stroke. Ayoobi was also involved in a 2014 study looking at yarrow's efficiency with encephalomyelitis, concluding that it lessened disease severity, inflammatory responses and demyelinating lesions—which could be massive for multiple sclerosis (MS), in both the prevention and treatment of its symptoms. The annual net worth of the drugs failing to curb the escalation of this disease is already in excess of £20 billion, rising significantly year-on-year. Ayoobi continued her research into yarrow and again corroborated the MS benefits in 2019, confirming that apigenin, luteolin and kaempferol all appeared to play a significant role.

MS is a mysterious illness, although from a dietary perspective an explanation for its rise could be the present diet having a deficiency of the plants from this book and their associated phytonutrients, leaving us exposed to deficiency diseases. The problem here is, of course, that we don't even know that we are dangerously excluding certain plants from our diet, completely missing the whole point of food—which is as a delivery system for as many nutrients as possible. Even the way we eat food has a lot to be desired from a nutrient absorption position—it would be best to graze throughout the day, consume more raw food and eat nothing several hours before bed and after waking.

Another study looked at yarrow's affect on a horrible infection caused by sandflies, leishmaniasis (noted in the Nettle chapter as nettle also proved to be an effective treatment). Tiny parasites called leishmania feast upon the skin, leaving large lesions and ulcers, as well as enlarging the spleen and liver while lowering the red blood cell count. About two million new cases a year are reported and up to 12 million people are currently suffering from the infection, with from 25,000 to 50,000 deaths every year in hot climates, notably south west Asia and southern Europe; it also affects many dogs and undoubtedly other mammals as well. There is no known cure, only prevention. This 2008 study, conducted in the Isfahan University of Medical Sciences, used mice and herbal extracts of yarrow, thyme and propolis, comparing efficiency against the normal treatment of cutaneous leishmaniasis with glucantime. The results showed conclusively that extracts from all three outperformed the glucantime, making them a superior anti-parasitic agent. The current treatment regime required for glucantime is pretty hardcore, involving hospitalisation and constant medical observation—it is not just a pill-and-go, with common side effects from glucantime including fever, irregular heartbeat, nausea, vomiting and pain in the abdominal area, drowsiness and headaches. Sanofi-Aventis make Glucantime, and unsurprisingly do not suggest yarrow as an alternative, with or without thyme and propolis. With annual revenues somewhere in the region of £75 million, a small amount compared to some of the astronomical numbers already seen, but it is still seventy-five million quid.

It should be pointed out that most research in Europe and US is funded by industry, with results focused on a profit-potential outcome enabled by a product, while in other countries, such as the former Soviet Union and clearly as shown above in Iran today, the government funds scientific research because that's the best way of finding out about really useful stuff. One of the first sanctions against countries such as Iran imposed by the West is banning the export of pharmaceuticals, forcing them to undertake their own research. So thanks, Iran, for some great scientific work—but no thanks for their terribly sexist and oppressive regime.

Here's more countries' opinions

Published in the *Journal of Ethnopharmacology* in 2012, a study on anxiety performed at the University Paranaense, Brazil (Baretta et al.) showed that the potential of yarrow as an anxiolytic (reducing anxiety) was comparable to diazepam and definitely worth future study. Importantly, there appeared to be no tolerance after repeated administration, not to mention negating the addiction and side effects from diazepam-type drugs, which as ever include mental confusion, heart attack, vomiting and quite a few more. The market for this type of drug has been out of control for some time—from 2005 onwards the US has been prescribing to at least 10% of its over-12-year-olds drugs for anxiety and insomnia. This is staggering, and it's not much better for the rest of the world, with prescriptions for this age group ranging from 3% to 5%. Diazepam is a drug from the benzodiazepine family with global revenues annually of about £1.8 billion—a drug shown to be not as effective as yarrow in a clinical study.

Yet another study, this time from Milan University, Italy (Giorgi et al. 2008), looked at the gastro-protection offered by yarrow leaves, namely the cytoprotection (protecting cells from dangerous agents), with a view to prevent or reduce mucosal necrosis caused by ulcers and other necrotising agents. These authors' conclusion was that the phenolic component had good anti-oxidant and cytoprotective properties.

Innocenti et al. (2006) offer more research from Italy, this time from the University of Padova, an institution with a fantastically rich history. Founded in 1222, it was one of the first universities to teach law and medicine, with many flocking here from across Europe to learn and be healed, and the university was instrumental in the history of medicine, the body, how it works and surgery. This study tried to establish which compounds caused the emmenagogue effect (increase to menstrual flow), which clearly presumes that the positive emmenagogue capability of yarrow was real and already accepted. The authors isolated nine compounds, with two of the flavones, apigenin and luteolin, proving to be the most important oestrogenic compounds that certainly warranted further research for their nutraceutical benefits—a great word we've seen a couple of times already, and a counterpoint to pharmaceutical with its toxicological methodology; this is a nutritional methodology, namely through food and herbs.

In a study from Hungary (Csupor et al. 2009), reported in *Phytotherapy Research*, five flavonoids and five sesquiterpenoids were extracted from yarrow leaf. These were then tested to see how effective they were at preventing human tumour cells from proliferating *in vitro* (*vitro* means glass, so test-tube laboratory conditions; *in vivo* is 'in the living', with a human study being the most relevant, a rodent one more common). Four extracts failed to have any effect, while three proved to be mildly effective and three, centaureidin, casticin and paulitin, were highly effective. This means when you eat a yarrow leaf you are also eating a minimum of three effective anti-tumour agents.

A 2005 study of four *Achillea* species, including *A. millefolium*, from Tokushima Bunri University, Japan (Stojanovic et al.), demonstrated that all four had similar antimicrobial activity against five bacteria including *E. coli*, *Staphylococcus*, *Salmonella* and two fungi, *Candida* being one of them. All of these are very common among humans, with the potential to be devastating, and definitely compounded by humans' overuse of antibiotics. This overuse removes what was an effective first line of defence, and we have gone full circle to the point where we have bacteria that have evolved resistance to antibiotics and pose a serious hazard (MRSA, for instance).

Yarrow is also specific against *Shigella*, as well as other bacteria responsible for infectious diarrhoea. *Shigella* is the leading bacterial cause of diarrhoea worldwide, and this is one of mankind's biggest killers, responsible for more than 200,000 deaths (some reports state it's more than a million) every year from around 150 million cases. A study from Mosul University, Iraq in 2011 (Hasson) demonstrated that an extract from yarrow flowers was an effective antibacterial agent against, among others, the Community Acquired Methicillin-Resistant *Staphylococcus* Aureus (the CA-MRSA drug market is worth at least £2.25 billion annually), *Salmonella*, *Shigella* and *Staphylococcus*. All these bacteria are mutating with drug resistance and deadly potential. Those last three studies—just in case you drifted off—suggest that yarrow is both anti-carcinogenic and antibacterial. There's no small fry here.

The Aga Khan University Medical College, Karachi, Pakistan (Khan & Gilani 2011), looked at yarrow's hypertensive, cardio-depressant, vasodilatory and bronchodilatory capabilities using a variety of rodents. The authors concluded it was as effective as verapamil, one of the World Health Organization's essential drugs, recognised as being in the safest category although it still has side effects. The study showed yarrow's efficacy for hyperactive cardiovascular activity and airway disorders, such as hypertension and asthma—where the value of asthma drugs and COPD drugs hit £32 billion in 2020.

Another study from the University of Karachi (Choudhary et al. 2007) looked at the anti-inflammatory properties of yarrow, concluding that seven of the compounds isolated within yarrow showed activity similar or higher than that of indomethacin, a nonsteroidal anti-inflammatory drug (NSAID), prescribed for similar problems as that bad boy Diclofenac we met in the Nettle chapter, for such ailments as arthritis, inflammation and pain etc. Indomethacin comes with the preliminary warning that it will substantially increase your risk of a heart attack if taken for a continued period; the other side effects were as we saw with Diclofenac, which are broad and varied to say the least—with 41% suffering some form of renal (kidney) dysfunction, 34% suffering nausea and 16% headaches; 1% to 10% experienced constipation, gastrointestinal bleeding, general bleeding from absolutely anywhere, depression and exhaustion, while 0.1% to 1% suffered acute respiratory distress, involuntary muscle movement, deafness, blurred vision, acute anaphylaxis, toxic epidermal necrolysis (that's your skin falling off, we saw that before), congestive heart failure, insomnia, mental confusion and renal failure. Remember, all these things have actually happened to patients who were prescribed indomethacin by a doctor.

* * *

And here I am, beside myself with worry if a client has so much as a fart out of place when taking the herbal powders I prepare for them! If yarrow has already been shown to be as effective, if not more so than indomethacin and verapamil, with the added benefits of being nutritive, anti-inflammatory, antimicrobial, and of course avoiding all the possible side effects, why in the name of reason, compassion and science has there been no advancement in replacing these clearly toxic drugs with a safe herbal alternative? Money?

Yearly financial figures and accounts have become much harder to get free of charge since researching my last book—now everyone wants more than £2,000 for their research on future markets and what's next to buy and what's hot. All this reportage and market analysis on how to create a market, diversify across territories and increase revenues from drugs all seems a bit arse-about-face, because surely it is the incidence and prevalence of an illness that should determine the territory and extent of the sales of the drug? It looks as if the pharmaceutical corporations are more interested in establishing market territories and finding customers through advertising and diseasemongery than actually looking for cures …

What all of the market reports indicate (from their free summaries in the introductions) are increases in revenues, which suggests there most definitely is no plan in sight to eradicate so much pain being felt all over the world. It's clear that any cure in any health sector would stop those huge revenues from the drugs. Economics is only really concerned with profits and cash flow, and it makes absolutely no difference if some drugs have in many instances less effect than yarrow, dandelions or nettles—if the revenue stream potentially threatened by a non-patentable, free plant does not generate a replacement revenue stream, the first stream will never be replaced. Commonsense, or health sense cannot compete with economic sense … yet.

OTC cash registers on fire

The over-the-counter (OTC) market for prescription-type pharmaceuticals is massive, with the USA spending the most every year at £16.5 billion; Europe is not far behind at £15 billion (UK £2 billion), while China and India both spend over £11.3 billion, giving a world total today in the region of £75 to £90 billion. Europe already spends £2bn on vitamin supplements, £5bn on cold and cough remedies, £2.25bn on skin treatments and £3.5bn on analgesics (pain relief) every year. That's big money, fought over hard to obtain and maintain by the duelling pharmaceutical companies, none of whom are going to relinquish their hold and admit that a home-made tonic made from a few freshly picked leaves and flowers is ever going to be as effective as their state-of-the-art, billions-spent-on-research-and-development pills that they say are more powerful and better than Mother Nature Herself. Because advertisers can and do say pretty much whatever they want, with the truth just getting in the way.

OTC is not just a significant market, it's a shop front, a drug dispensary to the public. In many instances the same drugs are obtainable OTC as those available from prescription, often not as strong. It's the first port-of-call for many suffering from a non-life-threatening injury, from plasters to Ibuprofen and everything in between. This makes it an ideal place to educate the public about the advantages of natural remedies, especially if recommended by the pharmacists and endorsed by celebrities and emblazoned across social media. In this case the public will have to take notice, as would the pharmaceutical corporations. This is of course what the pill makers did from the middle of last century—hire advertisers and create marketing campaigns, so today the biggest-selling drugs aren't the biggest-selling because they are the most effective. Far from it, those drugs have simply had the most money spent on promoting them. The economic world that revolves around the health industry has nothing to do with efficiency, merit or even necessarily health—it's all money-orientated. If, and hopefully when, these new green industries develop, the growers and the secondary and tertiary producers of nettle, dandelion, yarrow, plantain and hemp can all club some funds together for their own marketing campaigns.

While on the topic of marketing, a certain Edward Bernays (1891–1995) has to be mentioned. He is considered the father of marketing and advertising, hailed as a pioneer in public relations (aka bullshitting to the public—he is therefore aka the King of Lies), particularly when used as a psychological tool, as one would expect from Sigmund Freud's nephew. It would be fair to say that Bernays hugely influenced the trajectory of capitalism towards consumerism, responsible (or to blame) for the rise of lies to get sales, or, as he called it, 'engineering consent'. He is even responsible for the fluoridisation of water while in the employ of Alcoa (Aluminium Company of America), who were already in murky waters with their refusal to accept any possible danger from using aluminium for cooking pans. Bernays managed to sell the expensive-to-dispose-of waste from the aluminium industry at a vast profit to the water boards of the USA. Some commentators even say it was as a bet to himself to see if he could convince a whole nation that something bad for them is actually good. A quote from his *Propaganda* book of 1928, said to have influenced Joseph Goebbels, Hitler's Minister of Propaganda says it all:

> The conscious and intelligent manipulation of the organized habits and opinions of the masses is an important element in democratic society. Those who manipulate this unseen mechanism of society constitute an invisible government which is the true ruling power of our country. We are governed, our minds are moulded, our tastes formed, and our ideas suggested, largely by men we have never heard ... It is they who pull the wires that control the public mind.

Is it slanderous to call this man a bastard? The Guatemalans would definitely have been better off had he not got involved in their banana industry or government, or his marketing for the tobacco industry condemning millions to death by lung cancer. We definitely would have been better off without him, and his uncle also for that matter, a failed wannabe cocaine dealer and just a little too anally fixated to ever feel comfortable with. Another real danger is that once the public are convinced of cultural 'facts' it's harder to see the truth—as said by Mark Twain in the nineteenth century, "It's easier to fool the people, than it is to convince them they have been fooled."

* * *

The previous pages in this chapter were, I hope you agree, rather interesting, casting yarrow in a new light, which only gets brighter as the following section puts yarrow, the plant of war and peace, firmly in the spotlight. We met the nuclear and radiation threat (more on that later), while the other one that's been kept quiet is in reality not quiet, it's an irritating familiar high-pitched buzz. Hearing it once in your bed at night, or in the hotel on holiday, your mind starts to tell you that it's only here for one reason, and that is to get a suck on you. You can either fall asleep, no problem, let nature take its course, or you can turn the light on to seek and kill it before it gets a bit of you and maybe even infects you. I'm sure you guessed, it's a mosquito, and this might well be news to most people.

Aedes albopictus

Will be a name best remembered. Here our path has taken us to the edge of a forest clearing on the outskirts of a newly erected city, rising from the ashes of the forest in Asia like a shiny glass and metallic phoenix. Skyscrapers reflected in a few small puddles appear to be pulsating rhythmically in a mass of tiny wriggles. These are the larvae of the Asian Tiger mosquito, also known as the Forest Day mosquito, *Aedes albopictus*. As one of its names suggests, this critter is active during the day as well, and is also particularly annoying, in so much as it is a prolific feeder, biting as frequently as possible, with one survey noting biting rates of up to 48 bites an hour per mozzy, on their favoured victim, that soft-skinned, almost hairless mammalian human.

Worryingly, *Aedes albopictus* is a known vector (an intermediary in the transmission of pathogens between species) of dengue fever, yellow fever, Japanese encephalitis,

Chikungunya virus, West Nile virus and dirofilariasis. All are compounded by the fact that this mosquito will take blood from reptiles, birds or mammals, meaning the chance of these zoonotic diseases transferring to humans is increased with such a spectrum of animals being victims—only the females eat blood before laying their eggs, with the males gentle nectar-sucking vegans. Their most disturbing potential is for them to act as vectors for new future diseases, as they extend their geographical range. So don't go thinking these aren't going to bother you anyway, buzzing away over in the heat of South East Asia.

Oh no, don't go thinking that! That's why we are looking at this particular mosquito, because it's on the move. It has already arrived and is surviving in many parts of Europe, having hitch-hiked a passage around the world. Big groovy tyres take the brunt of the blame, with some South American countries, New Zealand and Australia—the latter two have always had very tight import restrictions—putting Asian tyre imports into quarantine, though with little effect on the migration of this mosquito. This can't be the only transport route, as there are all the containers full of consumables and on the boats themselves; the possibilities for little nooks and crannies are endless.

Aedes albopictus is already listed in the top 100 invasive species on the Global Invasive Species database held by the Wolrd Health Organization. What makes this particular species so potentially dangerous (along with its vector danger) is that unlike most of the other mosquito species, this one doesn't require water in the egg stage. It can lie dormant, hibernating and surviving in temperatures as low as zero degrees Celsius, and when the egg later gets covered with water it hatches. As the world gets regionally warmer and wetter with global warming, the population of this mosquito is going to go through the roof. Preferring temperatures not falling below zero degrees and rainfall not below 300mm per annum, with at least one month averaging 20 degrees, means much of the southern half of Europe is already ideal. They are also starting to be found indoors more often, with the possibility of overwintering and laying eggs in any small vessel and uncleared clutter around the house and garden.

With regards their global spread out of South East Asia, they were first sighted in Albania in 1979 and not again until Italy in 1990, with tyres sourced from China being blamed. Having established themselves throughout Italy, they spread to France by 1999, Montenegro by 2001 (I saw one there in 2018), Belgium by 2003, Greece 2004, Cyprus 2010; in 2016 eggs were found in England for the first time in Folkestone, Kent, then again in 2017 in Ashford, Kent.

Considering there aren't many in the public actively reporting to the relevant authorities when certain animals, plants and insects are spotted, it is probably quite safe to assume that this mosquito has already extensively spread across much of Europe, and with winters cooling their further spread is guaranteed. Continuing the story of their spread, they were reported in the US by 1983, Brazil in 1986, South Africa by 1990 and the Middle East by 2003. In other words, basically everywhere, with their ability, if not a propensity, to breed and lay eggs in tiny niches, and being left to get on with it pretty much undisturbed by the local fauna—they are a future horror movie.

These little blood-suckers (weighing about 5mg each) bring a whole new dimension to climate change. This species spreading deadly viruses is absolutely not what we want, although it is only a matter of time before we could be experiencing swarms of these mosquitoes during not only the day, also the whole year. This is not to mention the other 34 mosquito species already living in the UK, all of which will fully appreciate the increase in temperature and moisture offered by global warming in the UK. Although, as climate change kicks in, the increase in meltwater from the Arctic, particularly Greenland, pushing into the Atlantic will probably redirect the Gulf Stream further south to Iberia, making that a desert and giving the UK the same climate as Finland today.

Aedes albopictus does have a very distinctive black and white striped pattern, hence the Asian Tiger name—so keep your eyes open for this one (perhaps drop the Environment Agency a line to let them know). There could very soon be 35 species of mosquito living in the UK. With humans a popular snack, and a resistance to some of the deadliest chemicals known to man already, these insects pose a very real and extremely serious threat to public health. Cities are going to be dinner-time, all the time, for these little bastards.

It only takes one of these voracious biters to bite a sick chicken one day and a farmer the next, and for the farmer to go to the busy market at the weekend, sneezing and spluttering, to send a new strain of the so-called bird-flu into the air of the crowded market.

This species of mosquito can severely diminish the quality of life. It is so voracious that people could well develop more of a tendency to remain indoors, impacting the children, who may be less inclined to play outdoors, as it's not pleasant to be under attack repetitively, particularly with the itching caused by the bites. It will become yet another risk factor for obesity, a major risk factor in itself for diabetes, which is fast becoming the number one killer of humans.

* * *

I hope you are waiting for it, because no matter how diabolical the situation may become, there is a glint of light burning through the putrid ponds of pestilence. That chink in the armour of *Aedes albopictus*, its Achilles heel, is of course yarrow—the clue being the chapter it's in. Specifically, we are talking about the essential oil produced by steam distillation using yarrow flowers, which produces the most beautiful and deepest blue oil imaginable (occasionally greenish from yarrow species other than *Achillea millefolium*). This oil kills the mosquito larvae. Dead.

The larval stage is the best stage for eradication, although this species is more than happy with a wet corner somewhere, not just the vast crèches found in ponds and water butts of other species. Although due diligence and observation will assist in Britain, as currently this mosquito will not be able to survive through the coldest week of winter, where there are still a few days in a row each year where temperatures never rise above zero. They still require a water stage, no matter how small this wet nook may be, so changing and emptying any containers, localised spraying of a yarrow insecticide, vigilant observation and removal of distinctive long black eggs near to water containers will, with enough households doing their bit, lessen urban populations of mosquitoes in general.

Any sprayable yarrow insecticides need to guarantee there are no adverse propellant gases and ingredients used, with clear usage instructions. Caution is required, as the public are mostly unaware of the lung and respiratory damage being caused by many brands of plug-in air fresheners.

It's also nice to know that dragonflies are an excellent natural predator of both the mosquito larvae and flying adults. There shouldn't be too many complaints about there being loads more dragonflies, although there might be problems if they are allowed to take over ponds and breeding grounds uncontrolled, as when nymphs in the water they are a most deadly predator of all aquatic life.

Radiation

This is as of yet uncharted territory, with there being little concrete evidence that could be taken to a courtroom definitively proving that the radiation in our environment is a danger to our health and is solely responsible for any specific illness. It's just another pollutant, and considering how long it took to oust tobacco, with a similar scenario now being played out with the deadly agricultural chemicals such as glyphosate, the background of the human environment is so polluted today that it's becoming harder to isolate which carcinogen is causing which cancer.

Most people don't even know there is any radiation, or that it poses any adverse health affects. This radiation is the unseen pollution caused by our culture, our lifestyle. If radiation was a colour, the air would be such a mix of colours it would be an impenetrable dark smog. We have a current problem with electricity and electromagnetic radiation, and we will also have a future problem with nuclear radiation, from either more accidents such as Fukushima; or the contamination from the leaking of previously poorly dealt with nuclear waste, with steel barrels in the North Sea and Mediterranean full of nuclear waste, rusting away, with reports of the barrels that didn't initially sink in the Mediterranean being shot at to help them to sink; or a deliberate act of terrorism or war.

We have increasingly surrounded ourselves with electro-magnetic radiation (EMR) pollution during the past century. EMR extends from the stray electricity from home plugs, cables and the pylons delivering it, to mobile phones, wi-fi masts, transmitters, smart meters and a lot more. The danger from this source may not be as immediate and obvious as sticking your wet fingers into the plug socket, although it is quite possibly equally as damaging in the long run. There are plenty of professional scientists who consider EMR as a major risk factor when looking for a cause of many of the diseases of civilization, a danger understood in the 1930s but conveniently ignored for the prospect of filling our homes with metered electricity, electrical consumables and vast profits. It's madness to even have electricity pylons criss-crossing every country, and even more insane to have it in overseas hands. Intelligent cultural progress would have made all energy locally sourced, each house, farm or factory having its own generator from either solar, wind or even from the cool-sounding Tokamak generators using nuclear fusion, if these can be substantially downsized and proven to be fail-safe, preferably by independent experts.

The problem with EMR is all to do with harmonious frequencies and vibrations. Electricity should oscillate around us at 50–60Hz for it to be considered safe, because we are electrical entities ourselves, requiring constant electrical cellular pulses for all actions and thoughts. Any electricity outside of this range can biologically affect our delicate electrical workings. It can be ELF or EHF—Extremely Low Frequency or Extremely

High Frequency. EHF is the more dangerous of the two as this freely passes into, around and through us, being called 'biologically active', whereas ELF tends to bounce off. The EHFs today are ultra-high frequencies from 300MHz to 10GHz, and constantly interfere with our biological electrical system, negatively affecting us in a multitude of ways. None of which has been conclusively researched, although independent research strongly suggests it is in the very least a contributory factor in the incidence of many of the diseases of civilization.

Stray electricity, where it is not confined to the cabling and plugs, is surprisingly common, while wi-fi is literally everywhere. There are also estimations that anywhere from 5% to 35% of people suffer from electrical hypersensitivity, making them more prone to cancers, diabetes, fibromyalgia, depression, irritability, asthma and more from this adverse stray electricity reaction. Filters have been successfully fitted to 'sick' buildings where the incidence of cancers, ADD/ADHD and asthma inhaler use all decreased significantly in schools and institutional buildings after being fitted to plug sockets, preventing surges in the 4kHz to 100kHz range. There is already acceptance that living under electricity pylons increases incidences of leukaemia, depression, miscarriages, autism, Alzheimer's, addictions and MS, effects seen in sheep and cattle as well, apart from the addictions, of course. Dirty electricity also affects blood sugar levels, to such an extent that there could soon be a new class of diabetes attributed to it, Type 3, especially among those hypersensitive to electricity—that's if intolerances to agricultural chemicals causing sugar imbalance don't claim the title first.

* * *

There is no refuting that this electromagnetic radiation affects us biologically, as it does all animals, insects and plants. There is intercellular communication disruption caused by this proliferation of EMR exposure affecting us constantly today, when 100 years ago there was very little and 150 years ago, none. This disruption to intercellular communication is certainly worth pondering upon—that's all to do with the Second Law of Thermodynamics that is, albeit not heat dispersal. Coincidentally or not, there is a forever-growing menu of conditions and patients with neurological dysfunctions, most classed as auto-immune problems with no apparent cause (but seemingly defined by this neurological disruption). Common among these are MS, Parkinson's and myalgia, as well as other diseases of cognitive impairment. It seems logical to attribute at least a percentage of these incidences to those with electrical hypersensitivity, suffering a serious disease from the constant electrical disturbance causing intercellular failure—when the cells do not pass on the correct messages in their complicated neural pathways. It has got to be a consideration.

This could also be where acetylcholine plays yet another critical role and where its deficiency from the diet exacerbates present complications with man-made chemicals and man-made electrical frequency; the exclusion of these Human Plants from the human diet causes a domino effect on health. How do we know nettle neglect hasn't caused diseases? We have Human Plants with unique enzymes to activate acetylcholine (ACh) to enable motor and mental functionality, but their absence in our diet is made worse today as the processing of the food chain destroys the ACh enzyme as well, so it's commonly absent along with all the phytonutrients. Very few scientists have been looking at what these deficiencies caused by the drastic change in diet over the past few decades may be causing health-wise. With fewer minerals available from the soils and the majority of the diet consisting of processed carbohydrate-dense foods—has dietary change really not had any detrimental effects?

A quick look at an essential amino acid, tryptophan, essential as it needs to be ingested via food, will hopefully explain what is going on with EMR and why it is so dangerous biologically. One of the important roles of tryptophan is in the synthesis and production of melatonin, a hormone made in the brain. Melatonin is an antioxidant, which when we sleep, surges around the body and mops up free radicals, those cancer and cancer-promoting cells. It is also a surge of melatonin that causes drowsiness at night and allows us to switch off to sleep; yawning assists or is a consequence of this melatonin release. The pineal gland cannot produce this melatonin, or is at least severely hampered in performing its cleaning role, when disrupted by EHFs. These EHFs could easily be the result of any number of electrical appliances, stray voltage or wi-fi signals.

Linked with melatonin is the neurotransmitter serotonin, responsible for, among other things, our happiness and pleasure sensations. Disrupting this leads to many mental dysfunctions, from addictions to depression. From one basic biological process being disturbed, that of the amino acid tryptophan, the result is an increased risk of cancers, depressions, sleeplessness and a whole bunch of dysfunctions. There is a huge number of enzymes and cellular activity going on in the body, all of the time. If every cell has the potential of getting adversely affected by this sharp increase in electromagnetic radiation, particularly the higher frequencies, surely it is worth considering the connection and possible causal factor between this EMR and the correlated rise in cases of these diseases of civilization in the past few decades.

* * *

Can you smell that? There's that whiff of money again. Could the dangers of yet another industrial sector have been ignored or overlooked and mitigated by the massive revenues they can generate? It looks that way.

Companies supplying the electricity are also astronomical in size, with some of the largest being owned by governments such as State Grid of China with revenues of £383 billion (all figures here are from financial year 2020/21), or Électricité de France, EDF, with £75 billion or part-owned such as the Italian ENEL with £63 billion in revenues, or Japan's privately owned Tepco with £33.4 billion. Globally the electrical utility companies have a combined revenue in the region of £2.9 trillion, which makes them bigger than the world's seventh largest economy, France (£2.8 trillion). With that sort of money to back up any claims, it's easy enough to make EMR non-existent! They certainly won't be telling you that every 40 minutes enough solar energy hits the Earth to supply the world's power requirements for a whole year ... they'd also probably rather it was kept quiet that sulphur hexaflouride, essential for electrical transference, is over 23,000 times worse than carbon dioxide as a greenhouse gas ... Ssshhh!

These massive figures continue with the telecommunication companies, whose radiation is generally more accepted as being potentially hazardous. It recently cost the mobile phone companies for the use of the British airwaves, allowing them to saturate the air with their EHFs, a cool £21 billion when 4G was rolled out. I doubt there was much consideration to the possible health consequences or future medical costs caused by this desecration of the air we breathe, with yet another contaminant. It's invisible, though, how can that be a problem?

The top three telecommunication companies (according to Forbes 2022) all have annual revenues well over £75 billion, topped by the American AT&T with £135 billion. Total global revenues in this sector are about £1.25 trillion, just less than Australia's GDP of £1.4 trillon, the 14th largest economic nation.

From the day the first digital cellphone network went into service in the late 1990s across 122 US cities, all reported an unexplained spike in mortality of 10% to 25% lasting two or three months immediately after connection, as published by The Centers for Disease Control and Prevention. From this moment on, it was clear that despite any dangers, nothing could jeopardise future revenues. Remember those scare stories about tumours and other dangers of mobile phones? It's never changed. They haven't suddenly become safe, it's just been swept under the carpet. Remember how industry pays for a viral video showing how two phones can fry an egg? It gets debunked, of course, and somehow an association is made that all the stories about the dangers of phones can be categorised as 'fake news' alongside this discredited video, while still allowing the manufacturers to step up the frequency range, where it fills our air, freely flowing through us, affecting our sensitive electrical circuitry, which when combined with the endless array of other pollutants, carcinogens and genetic mutagens in everyday intimate proximity to us equals total biological chaos.

With such large revenues corporations have quite a loud voice when lobbying governments. Believe what you like, but a mobile phone or bluetooth is not good for your health. As the most recent research has shown, it heats up the brain behind the ear, acting as a neuro-excitor, with the future almost certainly going to see a rise in Alzheimer's-like symptoms and various tumours, as well as the constant disruption to the number of white blood cells produced (essential for immunity), making susceptibility to any illness higher. Quite how bad it gets we'll be seeing in the very near future, what with the proliferation of mobile phones among children, whose physiological and immune systems are still developing. The immune systems of both adults and kids are already working overtime to deal with this constant electromagnetic pollution, a situation compounded by the fact that we aren't feeding the body with the correct nutrients to deal with all the modern free radicals. It's not going to be a positive outcome, is it?

This is all without a mention of 5G, which could be the last desecration, the proverbial straw that breaks the camel's back. All I can say about 5G is, if it was such a good idea why has the capital of Europe, Brussels, banned it from being implemented across its own airwaves owing to radiation rules?

As is often the case these days, research on the dangers is conducted in-house by the polluters, which always conveniently ameliorates them, while no independent scientific evidence has demonstrated 5G as being biologically safe. Who to believe? This self-regulation of pharmaceutical and food industries and the money institutions has failed, as we have seen how unscientific and untruthful they can manage to make their research.

How liable are they going to find themselves when looking for the dangers of their own industry? 5G, anti-vaxxers and Covid conspirators all seem to have been thrown in the same 'fake news' bin alongside flat-earthers, moon-landing, Holocaust and 9–11 deniers. Check out Lloyd Burrell, who is a bit of an expert on EMR and 5G—if you want to know more about how EMR is a real pollutant, see his electricsense.com site.

* * *

Another form of radiation, soon to become more prevalent, is nuclear radiation, which finally brings it all back to yarrow, because of course yarrow is one of the solutions to EMR. The protection offered by yarrow is thankfully the same irrespective of the source of the radiation, be that electric or nuclear.

After the 1986 Chernobyl nuclear power station disaster in the Ukraine, the popularity of a volatile oil extract from three different yarrow species first took off, and after favourable clinical research the recipe was developed to include arnica and echinacea, sold today as Yarrow Environmental Solution (YES). Its intention is to strengthen and protect against electromagnetic radiation, X-rays, aeroplane flights, mobile phones, computer monitors, televisions, radiation treatment, contamination from nuclear power plants and fall-out from a bomb if there ever is one. For oncological radiation therapy relief, the oil is already being sprayed directly onto the skin. YES, as one would expect, is also very popular in Japan at the moment, where it was already being used to counter adversity from exposure to computers and a technology-led life. All it takes is just a few drops a day of YES or any yarrow extract. It truly could be as simple as this to either prevent or remediate biological radiation damage.

As we are about to see in the next couple of sections, one of the reasons yarrow's powers will remain hidden to modern science is that its superpowers actually work on levels beyond the visible—as humans only hear and see in a very limited range of available wavelengths, admittedly with technology pushing this wider, we are still very limited.

* * *

As we've been mentioning frequencies, here is a positive ELF. The harmonic of the resonant frequency of the Earth is as low as 8Hz. This is the sound of the Earth as it speeds at 108,000km/h through space. A low, distinct hum. A deep, constant Om. So when 'Om' is chanted, this is equal to Earth's voice, and when two equal frequencies come together, they cancel each other out, to form silence. The Buddhists chant 'Ommmmmmm' for a very good reason—it is a direct attempt to increase the peace.

Divination

Thou pretty herb of Venus' tree,
Thy true name it is Yarrow;
Now who my bosom friend must be,
Pray tell thou me tomorrow.

The above rhyme, repeated before going to bed after placing a sewn-up piece of cloth under the pillow with an ounce of yarrow in it, will enable a vision to enter revealing any future spouses. A close alternative was to pick some yarrow on May eve and this, placed under the pillow while repeating the following rhyme, will reveal the nature of a true love's affection:

> Good morrow, good morrow, sweet yarrow to thee;
> If I see my true love in white, his love for me is ever bright.
> If he appears to me in blue, his love to me is ever true.
> If he appears to me in black, his love to me will lack.

In Devon, a lady would rise at dawn, pick a spray of yarrow and put the flowers in her left glove or shoe, while reciting the following saying. By so doing one of the men met before the day is over will be a future husband:

> Good morning, good morning, good yarrow
> And thrice good morning to thee
> Pray tell me before tomorrow
> Who my true love is to be.

Cutting the stem across the middle and looking at the wound was said to reveal the initial of a future spouse. Small sachets filled with yarrow carried about the person were said to attract friends and family to you and to ensure at least seven years of love, as well as bestowing confidence, courage and protection from evil. The flowers were also commonly used in weddings and handfastings, ensuring love would last at least seven years.

Fact or fiction, it's hard to say, but the biblical story of Joseph relates that working one day he accidentally slipped and cut himself quite badly with one of his carpentry tools. The infant Jesus slipped out and found some yarrow for his dad to staunch the bleeding and aid the healing of the wound. This is why yarrow goes by the name *herbe de St Joseph* in France and carpenter's plant in many other countries.

The opposite is also true for yarrow, as it can cause bleeding, such are its extraordinary powers of movement and balance. Which is where another form of divination puts yarrow to use, by sticking a leaf up a nostril and reciting this rhyme:

> Yarroway, Yarroway bear a white blow
> If my love, love me, my nose will bleed now …

The Church would not have been impressed with the popular uses of yarrow for everyday divination or spiritual protection, such as the practice of tying yarrow to an infant's cot to prevent its soul from being stolen—the Church would rather a quiet prayer asking the Big Man to look over him/her. With the Church's intention to dispel these spells, this could well be the source of the corruption of some of the traditional names given to yarrow, such as Bad Man's Plaything, Death Flower, Devil's Plaything (a name also given to nettles), Devil's Nettle or Devil's Rattle—with names such as those, curiosity to dabble may well have been stifled. In the world of Magic, yarrow is sacred to the Horned God, and an old traditional usage of yarrow was in exorcism, to drive out evil from a place or person or even the Devil in cases of possession. Nor would the Church have been impressed with the practices of picking yarrow from graveyards from Sussex to Herefordshire, although it must be said the Church usually reappropriated sacred 'pagan' sites by plonking churches on them.

Drinking an infusion of yarrow can, for those of such a disposition, or when suitably relaxed, increase psychic powers, lulling the conscious brain and allowing psychic communication. As mentioned earlier, in the Hebrides, second sight is believed possible by placing a yarrow leaf against a closed eye. It is this otherworldly connection that is most interesting, and not just from European accounts—this position of yarrow as a mediating force between worlds is even more fundamental in China.

* * *

Notably, it is only the stalks of yarrow that were deemed worthy to be used in the Chinese ultimate tool of divination, the *I Ching*, originally known as The Yarrow Oracle. Attributed to the legendary emperor Fuxi, the 64 hexagrams were formalised in text by King Wen of Zhou (c. 1152 BC) and later became known as *The Book of Changes*. There is no documented evidence as to why yarrow stalks have to be used. I've asked *I Ching* experts, Chinese medicine practitioners and searched high and low, all to no avail. All the same, from the evidence in this chapter, there is a strong case to put yarrow forward as being a perfect candidate as an intra- and inter-dimensional harmonic translator. It is the ultimate adaptogen, altering itself for any situation. There's obviously a perfectly good reason for using the stalks, because if it were just a straight stalk required there are many other plants that would be better suited, such as the globally omnipresent willowherbs—although of course only yarrow has the harmonic resonance of yarrow.

The yarrow stick is said to represent the yin and yang forces of the Universe in perfect balance, and as we've seen it does have contradictory benefits, such as stopping or starting bleeding. The yarrow stick awakens the spiritual forces of the mind, as well as balancing the yin and yang forces to make possible the meeting of heaven and earth. And with that as an introduction, somewhere deep in the old quarter of Beijing, a wrinkled elderly Chinese man smiles, winks with a crystal-clear eye and leans over and sets out the 50 yarrow sticks on to a rattan mat, removes one and lets the others fall … I'd definitely want to hear what they had to say.

"Need ya aura cleansing or chakras aligned, Mister?"

There is much outside of the paradigms that confine modern science, such as the idea of the chakras, which to most in the West might seem amusing, but only because most of us are innocently ignorant. For most of the world except Europe, the existence of chakras and energy lines and qi/chi, today and far back into the past, are taken for granted. As with Indian and Chinese medicine, we in the West look upon it as a bit backward and old-fashioned, being so supremely arrogant as to look down at thousands of years of continual scientific evaluation by a significant percentage of the world's population. It's definitely extremely ill-advised to simply dismiss these things.

Just as it's hard to fathom and certainly to quantify the powers of yarrow, it does look as if the yarrow leaf can help to weave back the skin, while the higher notes of the flower do the same knitting, but on a higher vibrational network, one that influences the aura/human magnetosphere. This sounds a bit out there, probably because it involves frequencies not visible or audible to humans, and it is definitely not a field of current scientific study, being beyond scientific parameters and its relatively myopic scope of vision—but just because modern science cannot see it, this most certainly does not prove it's not there, as science has yet to tune into this new paradigm. Who are we to doubt this dimension beyond the field of our vision, blinkered as it is by our own culturally indoctrinated twenty-first-century lenses?

In the same light, what of intuition, gut feeling, a sixth sense, vibrating with someone, being on the same frequency, synchronicity, telepathy and a plethora of other psychic phenomena? Do these only really start to exist when modern Western science finally catches up and verifies their existence?

Even our perception of the colours we see is nothing but different vibrations, and to modern Western science this is all there is to it, just a vibration. So what of the three yarrow colours—white, pinky red and yellow—all with different vibrations and purposes, all on a level beyond our current perception, all used in the YES formulation, where within Eastern theology red is said to be for the root chakra, yellow for the solar plexus and white for the crown chakra? This talk of chakras might well have the sniff of patchouli oil to it, but as we are seeing with yarrow, dandelions and nettles, new scientific endeavours are corroborating traditional uses. This chakra imagery is ancient, so best not to reject this aspect just because we in the West have become desensitised—it does not mean such a connection is not true.

In biodynamic agriculture, yarrow acts as a frequency harmoniser, being an important compost preparation plant. Adding just a few leaves can bring about a sensitivity that enables the compost and subsequent soil to receive and regulate different forms of solar radiation. Yarrow harmonises the polarity of cosmic light and earthly form together, making it the queen of adaptogens. Snigger if you want, but this is how yarrow harmonises the human psyche, whatever the vibratory adversity, be that personal animosity directed against someone, radiation or geopathic stress—this is Earth's natural magnetism and energy lines, which can also cause adverse health effects.

The human psyche is very fragile within our modern culture, as this whole selfie and social media addiction feeds a fake ego, taking the human further away from the id and the soul, destroying all empathetic capacities along the way. Yarrow can return our sensitivities and receptivity while allowing for a healthy, non-dominant ego. It seems incredible that an occasional tea with a few grams of dried or a few fresh leaves, or a glug of tincture, can perform this magic. It's easier to dismiss it as absurd, instead of actually attempting to incorporate a little each day for a few days to gauge the effect. If you don't try it, you'll never know what you are missing.

Anything else?

As being a suspected anti-carcinogen, anti-inflammatory, possible agent in the curative path of many diseases, a nutritive, wound healer, blood stauncher, radiation deflecting and eliminating, mosquito killing, body soothing and sooth-saying … just isn't enough, is it? Unbelievably, there is more …

A 2011 study, paid for by BASF in the *International Journal of Cosmetic Science* (Pain et al.), found that yarrow improved the appearance of both wrinkles and pores. This anti-aging global market is enormous, being somewhere in the region of £50 billion annually. Clearly now is the time to put together completely safe skincare products and moisturisers using yarrow and other plants, not least because yarrow is also successful as a hair tonic, benefiting scalp problems and stimulating hair growth (excellent mixed with rosemary and nettle). It is an effective ingredient for a multitude of uses, and once the public can see the positive results from products containing no detrimental chemicals, a proportion will switch. The list and risks attached to so many of the scary chemicals commonly used in bathroom products means you are never being very beautiful to yourself by using the vast majority of these highly toxic products, even the rather paradoxically named 'beauty products'.

Applying yarrow tincture to exposed areas of skin acts as an effective insect repellant, with unconfirmed reports of the US Army using this, its downside being that it only lasts for an hour or two before requiring a further application, although this is still better than covering yourself in some of the dangerous chemicals found in insect repellants such as DEET (dimethyl toluamide), which amazingly is still available to liberally cover your skin with, allowing free entry to your inner workings by a seriously disrupting chemical. Yarrow mixed with eucalyptus, lemon or citronella would be both effective and aromatic. A valuable lesson from nature would be that many species of birds, swallows especially, include yarrow in the lining of their nests to inhibit parasite growth and as an insect repellant. Yarrow works. Full stop.

A gum resin called opopanax, also known as opobalsam, can be obtained from yarrow. It is used today in perfumes and in the past as a source for incense. The etymology of this word is fascinating, with the Greek root giving us *opos*, 'vegetable juice' plus *panax* or 'panacea', meaning all-healing. The most common source for this opopanax is the tree from which myrrh is derived—the resin from a small thorny tree, *Commiphora myrrha*—although yarrow still has it in harvestable amounts, especially if fields were to be cultivated in the near future. The healing fields.

A compound in the oil, called chamazulene is being regarded as one of the best components for wound healing. All the azulenes abundant in yarrow appear to impart a certain kind of magic, acting as an anti-inflammatory as well—azulene itself is a compound only formed as a result of distillation. The essential oils and sesquiterpenes are standardised for the drug industry by the European Pharmacopeia. By weight the volatile oils make up 1.4% of yarrow, and 51% of this is azulene, followed by borneol, terpineol, camphor, cineole, isoartemesia ketone, a trace of thujone and loads more.

Yarrow flowers and less so the leaves can be dried and added to a smoking mix, complementing other herbs such as mullein, sage, borage and mint, with a wide assortment of other herbs all interchangeable depending on the required outcome, be that attempting to give up tobacco, used in a joint with cannabis, or just to tamper down tobacco. It's obviously best not to smoke anything, although most natural plants will not be as damaging as the chemicals added to rolling tobacco and cigarettes.

Yarrow has also been shown to be effective against napalm, not that it was widely used after the barbaric, inhumane use of napalm against totally innocent populations in Vietnam, Cambodia and Laos by the Americans throughout the 1960s. Agent Orange and napalm were made by those caring folk at Monsanto and Dow. Unforgivable pure terrorism.

* * *

Moving swiftly on to agriculture, as I've spent years observing and researching farmers and their poison regimes, to keep it short and light for now as we'll get to agriculture and the simple solution to this devastation towards the end of the book. As it becomes apparent that the use of insecticides, pesticides, herbicides and fungicides is having and has had devastating consequences for not just the flora and fauna, but for us humans as well, more evidence emerges of safe, sustainable higher-yielding agricultural methods. If anyone, anywhere, takes the opinion that it is impossible to grow food without applying deadly chemicals, they should be the last person to be allowed to be growing our food, but instead this is the general indoctrinated consensus of farmers, the very people growing our food, some of whom I've had frothing at the mouth in exasperation at my permacultural ways. The problem is deep-rooted, to say the least.

As an insecticide, yarrow has a proven and continual use, stemming back those 60,000 years to our Neanderthal cousins. Effective sprays can be made from the volatile oils of yarrow, or a compost mix made, along with nettles and comfrey to make a stinky liquid fertiliser infusion. All these methods act as insect repellants, and when coupled with other sensible agricultural strategies, such as growing yarrow and companion planting to encourage natural predators of any crop's pests (poppies are great to attract blackfly to keep them away from beans, for instance), help maintain soil fertility and a healthy top soil with the all-important mycelium network kept intact.

Mycelium is the collective term for all the fungal filaments, from the near invisible to bootlace size. Those which form a symbiotic relationship with plant roots are known as mycorrhiza—*myco* means fungus and *rhiza* root. These are vital and manifest around the root to increase uptake of water and nutrients, establishing a symbiotic relationship. Gardeners and agriculturalists are only just waking up to the essential role played by mychorrhiza in the soil. They play an analogous role to that of the bacteria in the human gut, barely given any attention in Western medicine until very recently, but always vital, as long recognised in Eastern medicinal practices—with both the mycorrhiza and the gut flora making nutrients, making them utilisable and contributing to the immune systems of the soil and humans.

No poisons need or should be involved in growing food, certainly not the deadliest ones ever invented, which is what we have at present, being willingly sprayed everywhere. Healthy people need healthy food grown in healthy soil, but man-made chemicals have destroyed the mycelium and the health of the soil, just as human health is dependent on healthy gut biota, which has been destroyed by man-made chemicals (especially antibiotics, helped by the continual sugar/starch/highly processed carbohydrate diet).

It can't all be good, can it?

As with nettles and dandelions, unsurprisingly there are exactly the same possible cautions for yarrow, such as making the effectiveness of certain blood-thinning drugs like warfarin uncertain owing to their rich vitamin K content, a known coagulant.

With the anticoagulants global market worth about £22.5 billion annually, it's best to avoid stepping on the wrong toes.

There is also the possibility of yarrow making lithium drug side effects worse; those on stomach acid-reducing drugs like omeprazole should avoid yarrow as it increases stomach acids; yarrow might make you have too low a blood pressure if on drugs to lower this pressure; and if taking drugs for sleepiness, yarrow might get you too sleepy.

There is no evidence that yarrow causes any problems with pregnancy, however, because of its ability to move blood it's always best to be cautious when pregnant and avoid it. Some herbalists suggest its best use is during and after birth, helping to heal any internal wounds, acting as a restorative.

As with any plant rich in coumarins, there is the slight chance these coumarins in yarrow may produce an allergic reaction or sensitivity in some people's skin. The juice can also be phototoxic (reacting with skin when exposed to sunlight), which is more common at higher altitudes or where solar radiation is highest. Additionally there is the caution to those who may be allergic to the daisy (*Asteraceae*) family to avoid both internal and external uses of yarrow.

Foraging, growing and culinary curios

Yarrow's feathery leaves can be a year-round lawn companion, with new growth appearing quite early in the year, in meadows, pastures and waste ground, often blending in with the general green background. Yarrow prefers hot and dry conditions, with well-drained soils, favouring the top of a south-facing slope or hill, as opposed to the bottom of a shaded valley. It is not averse to a sandy soil and is quite happy to share its space with various mosses.

The best way to find and identify plants is to just stop, take in your environment, whether that be a green, unused bit of land at the back of a park, on the South Downs, a lush meadow, field or in a garden—and then stop looking, try to switch off. Take a deep breath, breathe out slowly and bend over and refocus back on to the greenery, try to see through the sea of green, observe the many shades of green, the different shapes. What was at first glance just grass often becomes a seething mass of various plants, with grass often becoming the minority upon closer inspection. Depending on location and time of year, the 'grass' becomes leaves and shootings of dandelion, plantain, mosses, speedwell, clovers, vetches, tansy, scabious, bird's foot trefoil, ragwort, dodder, burdock, docks, wild thyme, cornflower, knapweed, hawkweed, centaury, fat hen and various grasses. Once you see yarrow, it'll suddenly appear everywhere, with its feathery leaves cascading across smaller plants.

Not that we'll have the pleasure for a while to see a sea of yarrow, but an acre yields approximately 43,000 plants. This will hopefully become a common sight as entrepreneurs in the food, medicine, toiletry and health industries begin to utilise yarrow for its myriad uses. Imagine the pleasure this will bring to the bees, with fields of flowers, and no dangerous man-made chemicals, making glorious honey. The future could be mixed fields of yarrow, dandelion and the other five plants, feeding us, the soil and pollinators, butterflies, moths, birds and small mammals.

As with all these plants, they are easy to grow, readily inhabiting soils spoiled by humans. Once established, they prevent soil erosion with their leaf coverage and deep roots, redistributing minerals in the soil and providing valuable nourishment for grazing animals. It is not only the humans missing out on nutritious diets, the animals too are not being fed an ideal diet either, although it appears more thought goes into livestock feed than human feed, which is crazy although perfectly understandable from an economic perspective, as malnourished, sick livestock provide less meat to sell. Yarrow used to be one of the few staples of this food in fields along with clover, vetch and dandelion, before ryegrass took over and the biodiversity and natural immunity was lost. Yarrow was introduced to Australia and New Zealand for its drought resistance and rich nutrient content as a feed, although sadly it is no longer utilised and is now classified as an invasive weed.

* * *

If you want to grow some yarrow for yourself, it's best grown from splitting a root clump of some already well-established locally. Keep as much of the surrounding soil as possible (a pooh bag's worth, not a truck load!—with of course the landowner's permission as it is against the law to uproot and remove any plant) and replant somewhere sunny. Yarrow can also be grown quite successfully from seed, although it will be at least two years till it flowers as it's a perennial, which means it lives for more than two years. On the first year it lays down roots and leaves, and from the second year onwards it will flower some time around or after June. Splitting the roots every few years will see the plant thrive for decades.

Yarrow is a superb companion plant in allotments, with borders of yarrow, marigold, feverfew and wormwood all acting as insect repellants, while also attracting pollinators. Yarrow can also enhance the essential oil production—the secondary metabolites—of other herbs growing nearby.

Yarrow leaves are, as already mentioned, a great start for a compost pile, taking just a few leaves to act as a catalyst, helping the pile mature quicker. A compost pile with some yarrow leaves, pine needles, nettles, garden foliage and food waste (meat attracts rats so best to avoid) will provide excellent feed for any plants being grown when it breaks down. A good liquid feed uses just a little yarrow with some dandelions and loads of nettles and comfrey soaked in water. This liquid feed mix is not only extremely nutrient-rich it also acts as a natural insect repellant. This makes it both a fertiliser and a pesticide, with none of the horrendous stories attached to them, such as cancer, bee colony collapse, soil erosion, nitrate run-off, algal blooms in estuaries and even dead-spots in oceans where there is no oxygen and of course no life. Nature knows best.

Where is natural fertiliser being industrially applied? You'd have to look hard to find it—as we saw earlier, nettle fertiliser is now outlawed in France—but there is some being used, and it only takes positive results from research with increased yields, healthier plants and some marketing for the switch to happen. Every single online forum for gardeners, whether organic or not, mentions that nettles with other plants makes the best fertiliser for garden plants—so why not agricultural plants in smallholdings and larger? It could possibly be because the fertiliser industry is worth somewhere in the region of £113 billion annually (organic fertilisers are about £3.4 billion and growing fast), while the pesticide industry is worth at least £57 billion; the entire agricultural chemicals and fertiliser business is worth about £190 billion in annual revenues globally.

That's the answer right there, £190bn. This doesn't want to start talking about environmental and sustainable solutions—this might bring up awkward questions about whether the current fertiliser and pesticide industries are unsustainable or not environmental. That's why commonsense loses out. The agrochemical corporations are well aware that their biggest threat is from organic and they are spending a lot of money on slandering and discrediting safe, sustainable agricultural practices with their own fake news, for no other reason than to protect their profits, certainly spending more on protectionism than on R&D (true also for the pharmaceutical industry).

* * *

The spring leaf of the yarrow offers the best eating, becoming tougher and bitterer from June, when the flower heads start to develop and most of the energy goes to the flower. The leaves are still edible throughout the year, although the tougher larger leaves may have to be cooked as opposed to eating raw. There is usually plenty of fresh, soft new leaves growing year-round, with the flower also offering a raw treat. If you are to collect some leaves, make sure the land has not been sprayed with agricultural chemicals and wash the leaves before use because tiny bits of soil and grit easily get caught in the segmented leaves. The flowers and leaves can also be made into extracts, tinctures or dried for tea.

The food possibilities are exactly as we've seen before, although instead of it being the main ingredient in a soup, as with nettles, yarrow leaf is more a garnish, with just a few fresh and tender leaves being all that is needed, when added to a salad or sandwich, for instance. It is excellent as a dill substitute, or addition in gravlax and other vinaigrette-type sauces; it also acts as a substitute for tarragon in some recipes. Dried, it can be sprinkled as with other dried herbs on anything cooked, with the flowers added to salads. A few leaves can also be added to the hummus and pesto from the previous chapters along with any other foraged seasonal plants.

Tea is the easiest way of ingesting yarrow. Fresh leaves and flowers can be steeped in hot water, along with any other herbs you'd like to add, such as mint, rosemary and nettle. Just a few fresh or dried grammes can be added to a pint of hot water and topped up through the day—this will help to purify the blood and cleanse the kidneys, especially if the tea is incorporated into a daily regime.

In the seventeenth century and before, yarrow was a popular vegetable, used very much like spinach. Yarrow has been used around the world as a herbal addition to beers for much longer than hops, from as far afield as Africa; in Sweden, yarrow's name is Field Hop, indicating its former use in brewing, with Linnaeus stating that beer brewed from yarrow was more intoxicating than its hop counterpart. A once-popular European beverage was Gruit beer, made with a mixture of herbs, including yarrow and the potent marsh rosemary. These were all outlawed with the regulation of the brewing industry, beginning in 1516 when the Bavarian Reinheitsgebot was put into effect, allowing only three ingredients (water, barley and hops, with yeast added later when it was discovered as being an essential agent). This wasn't an international agreement, and its main purpose was to prevent grains getting diverted and squandered on beer instead of bread production. A 1710 Act of Parliament ended the brewing of beers using any herbs other than hops as a bettering agent. An ulterior motive for this was not because of any health concerns, but more so to prevent brewers from evading the new penny-per-pound hop tax, and from this point on hops took over.

Unrelated to yarrow, but still on the beer trip, here's a fascinating thing about early beer production. Up until the fifteenth century, beer was made by alewives, and was drunk in the main because it was a safe drink, not necessarily for inebriation, as most brews were low in alcohol. The water in cities gave more than just a quench of thirst, being a breeding ground for nasties such as hepatitis, gastroenteritis, cholera and smallpox. Think more kombucha than beer—who if selling beer in public wore pointed hats to be noticed and would have owned a cat to see off the rodent threat to their grains, which were used to make the beer in large cauldrons, using as their symbol of a beer-seller a broom hung over the door. Sound familiar? These alewives were becoming rich from the growing interest in beer, with many adding herbs for flavour and health. The men grumbled at this and began copying German and Belgium business practices, setting up shop themselves and forcing the women out of a profession they had dominated since Mesopotamian times (3000 BC). Then, as luck and timing would have it, the Spanish Inquisition arrived ("Nobody expects the Spanish Inquisition") depicting women with independence and power as Devil worshippers, borrowing from the alewife profession to stereotype what a witch looked like. Cheeky, eh? It was a little more complicated than this, with the fledgling male-dominated pharmacists adding to the malicious misogynist murder of thousands of innocent women and quite a few men for being, most definitely, not a witch.

Summation

There is much to be said for the saying "You don't know until you've tried it." Have you actually tried it? If not, the opinion is unfounded and hearsay, if not heresy. If you've an appropriate ailment that has already been highlighted in this book, why haven't you already bought or picked one of our seven plants and taken it for five days in a row to actually see for yourself the veracity of all these seemingly outlandish claims being made here about these plants?

Yarrow protects and strengthens us, within and without. It bolsters the immune, digestive and circulatory systems, as well as our unseen aura system. It protects us from radiation damage, be that environmental, telephonic, electronic, nuclear, geopathic or magic (malevolent and evil thoughts against us) and mosquitoes. It is also a probable anti-carcinogen and an anti-inflammatory. It is an incredible plant.

Here's the fourth plant, busting out to impress us—

PLANTAIN

(Plantago major)

Canine crisis

A few years ago we exchanged some pieces of printed paper with another human in order to mitigate in our minds the merciless separation of a golden retriever bitch from one of her extremely cute eight-week-old puppies. We justified it by saying that someone else would have had her and they might not have been as loving with her. Thank Gaia we did, because she has been by some considerable measure the best-behaved member of our household. She has also taught us much about animal emotion, intellect, health, companionship and love—as well as making it blatantly obvious these attributes are species-transferable.

Bella's 15 now, but she very nearly didn't make 11. One day she was having real trouble walking, dragging her rear legs as if she had done something to the muscles in her legs, possibly from chasing a deer through the woods (she has never harmed another living thing, she lives for the cuddle and hug). We gave it little thought for the rest of the day, but the next morning she was much worse. I had to lift and drag her outside to sit in the sun, while we fretted that perhaps she had had a stroke, having just heard her mother had suffered one. Leaving her to rest in the sun, I went for a cycle with an old friend not seen for 30 years. On my return I noticed Bella hadn't moved or drunk any water. Lying with her head to the side, she looked completely forlorn, and as I looked into her eyes, I sunk into them, pulled to her core, the message was clear: "I'm really not well." I pulled my gaze away, looking her over and gazing back into her eyes: she was falling into an abyss, all of her life-force was being drained, there was no doubt. I was watching our beautiful Bella die right before my eyes.

Stricken with fear, all I could think about was the problem with fastening the Little Van (small trailer that attaches to a bicycle) to the mountain bike (with only the one gear working), getting Bella into it and comfy and the five miles to the vet, guessing it'll be nearly an hour in total. Fortuitously, my wife appeared that instant and within 15 minutes we were sitting in the vet's reception, tears flowing and emotions never experienced before filling us, with Bella a bit too much of the literal dead weight on my lap. The vet was far more composed, assuring us he had a good idea what this could be. It had not been a stroke, and the scan confirmed it was a problem in her womb. Her final menses had not cleared itself fully, allowing the remaining blood to stagnate and then putrefy, a fairly common complaint for bitches who have not had puppies or had their ovaries removed.

In just three hours when we next saw Bella, she was walked out of the vet's room to meet us, looking worse for wear but she was at least walking. We were also a little concerned that she was allowed to walk so soon after a pretty serious operation, having assumed she would be on a dog-stretcher—I could not imagine a woman walking out of the hospital straight after a hysterectomy and catching the bus home! The vets who carried out the operation said during a subsequent check-up that Bella's was by some margin the thickest and largest, most pus-filled uterus they had ever seen, with so much pus it was backing up and systemic sepsis was imminent, with almost certain death that evening.

* * *

All very emotional, but animal communication and sad stories is not what this chapter's all about, it's about plantain, one of the most common plants found growing wild in Britain.

A common plant it may be, although for this little story it is specifically about a small patch growing in the garden that had presented itself to me earlier in the week while I was having a look at another plant it was sharing some shade with under an ornamental cherry tree. The other plant is called self-heal, although grandiose curative abilities for it failed to materialise on some preliminary research, so I looked into the plantain growing next to it, which did have the curative abilities self-heal proclaimed from its name. So, as soon as Bella was home I thanked the plants for introducing themselves a week before and assured them their leaves would grow back even fuller and more vital after some had been given to Bella. Blitzing a few leaves and feeding their juice via a pipette into her mouth was all Bella ate for the first two days. That first evening after the morphine had worn off and the pain was returning, all she could do was lie down, and although now there was life in her eyes, her body had just gone through immense trauma from a major operation. She'd need at least a week to recover some health and then, as with a human, six weeks without any strenuous exercise.

The other drugs the vet insisted she also take were clearly causing her serious discomfort, blatantly not what she should have to endure post-operation, but my family insisted she take them as well. I was giving Bella the plantain because it is one of the best antibiotics, antiseptics and anti-inflammatories of the plant world, while the vet, who admittedly had just saved Bella's life, then proceeded to give her drugs that had her spewing and excreting, causing considerable distress. The problem was the nonsteroidal anti-inflammatory drug (NSAID) meant to help with the pain, although its most common side effect was clearly emblazoned on the front of the box, that if it caused vomiting or diarrhoea it was to be stopped. So it was.

We continued with the plantain leaves and introduced a probiotic mix (to counterbalance the antibiotic Bella was taking) to try to stimulate her appetite as she wasn't eating, no doubt fearful it would just shoot straight out from either end, causing her to wince in pain as she kept moving and agitating her wound. The first check-up was after four days, and as the NSAID was rejected so violently, the vet gave her some Tramadol to calm her down and maybe help her eat. Tramadol! With my wife and daughter (a nurse) bemoaning my plant ways, Bella took some for a couple of days and completely zoned out and started to heal by not agitating herself from diarrhoea and retching. We continued to pipette her plantain and probiotic (realising the paradox of her taking both antibiotics and probiotics at the same time—she did take the antibiotic for five days), and she slowly started to eat a tiny amount at first. After two weeks she was enjoying her gourmet meat treats and her scar had all but healed.

Bella is great now, albeit elderly. She and I are also closer than ever, helped by the fact we got a cat soon afterwards to discourage the mice who were having it a little too easy (we live in the back of beyond, so it was their home before ours, but there becomes a point when enough is enough). My wife and youngest daughter are besotted with this carnivorous, cold-hearted murderer, so I spend more time with Bella, which is great. She also showed me first-hand the efficacy of plantain post-surgery, cross-species.

The people will not revolt. They will not look up from their screens long enough to notice ...

This chapter doesn't come portraying plantain as a multi-purpose, heal-all and build-all, a span-all-industries contender. It will follow a similar schematic to the other chapters, as hopefully the historical and traditional uses have been of interest, although the main crux here is going to be about scientific research, clinical trials and health. This far into the book, we have already seen a plethora of uses for nettle, dandelion and yarrow—there'll be a recap at the end of the book, as there was a lot to take on board (with loads more to come) as we flew through what will be a revolution in healthcare from these preventative marvels.

Benefiting the vast majority of the public—apart from the unfortunate less than 1% who have allergies to these plants—if made under safe and measured conditions, these plants can be transformed into tinctures, balms, powders and pills, servicing a plethora of ailments. It's a radically different approach compared to the current toxic drugs, causing, as we have seen, significant problems from their side effects, as well as laying the foundations for future diseases by damaging our genome in a multitude of malevolent ways. Thus far for these herbs there has been no side effects recorded, no toxic dangers of overdose or third-party pollution problems (as with Diclofenac wiping out the vultures) associated with any of these plants.

Plantain is another definite nutraceutical contender—not that you'd know it from the general lack of interest shown by the health industry with regards any of these plants. The pharmaceutical corporations are all perfectly content, having positioned us into the precarious situation of not only dependence on their pills and creams but also a blindness and misunderstanding has been installed—some would say indoctrination—that pharmaceuticals are the pinnacle of centuries of scientific endeavour in studying medicine and health. With no better alternative, we should be just grateful for it.

* * *

All the required information as to the potential of these plants in this book is already out there, it just needed to be compiled to demonstrate the overwhelming epidemiological evidence that these plants do help with a range of ailments, diseases, dysfunctions, deficiencies and detrimental lifestyles. These plants do work, not with everyone all the time, that isn't the claim; the claim is that these plants definitely act as efficient preventative nutraceuticals against a range of diseases. It's not just these seven plants needing to be eaten, the whole diet and food industry requires a reformation, as it can be clearly seen how unhealthy the population currently is, especially if compared to tribal peoples or our caveman ancestors.

Another claim is that the scientific evidence testifies to the veracity of traditional usages of these plants as being pertinent and correct. It has to be understood that the pharmaceuticals do not work all of the time, often only suppressing an irritating symptom, and, as we have seen, a significant percentage of users are stricken with added side effects, many of which are classed as diseases in their own right. This refers especially to neurological diseases such as fibromyalgia, which many sufferers coincidentally develop after several years of taking pharmaceutical drugs for other ailments. The word 'fibromyalgia' literally means 'muscle and fibrous connective tissue pain', which is prevented and quite possibly cured by these very plants. Just think back to nettles with their acetylcholine connections and anti-inflammatory capabilities. Is this a disease of nettle neglect and living in a polluted world? The medical profession simply washes their hands of these auto-immune diseases, claiming it to be effectively your fault you got ill as your body is the mutineer, when the cause for the vast majority is actually man-made chemicals, pharmaceuticals and industrial pollution.

These pharmaceutical corporations could well have a plan, as Aldous Huxley (an author very popular among the intelligentsia and upper classes, who might even have unwittingly planted the seed for this idea to manifest) pointed out in the 1950s when these pharmaceuticals first became popular:

> There will be, in the next generation or so, a pharmacological method of making people love their servitude, and producing dictatorship without tears, so to speak. Producing a kind of painless concentration camp for entire societies, so that people will in fact have their liberties taken away from them, but will rather enjoy it, because they will be distracted from any desire to rebel by propaganda or brainwashing enhanced by pharmacological methods. And this seems to be the final revolution.

It blows my mind every time I read that: any person who wants to end their turn on this mortal coil with LSD in their veins, as he did, more than deserves a quote in my book. Another visionary and luminary is George Orwell (whose French teacher at Eton was

Huxley ... we won't hold it against them!). From his portentous 1948 classic *1984* comes the even more relevant

> The People will not revolt. They will not look up from their screens long enough to notice what's happening.

A prediction of a future that we are living in right now and a book well worth a read or reread.

Species specific

This chapter is a bit of a cheat, as are five of the others, as the title of the book would be more botanically correct if it were 'Seven Genuses to Save the World', because it makes no differentiation between *Plantago major* (called Broad Leaf, Common or Greater Plantain, whose name and silhouette graces the title page of the chapter) or *Plantago lanceolata* (called Ribwort or Narrow Leaved Plantain), the two most common plantains to be found In Britain.

There are a few more, notably *Plantago media*, which is less common, also called Hoary Plantain, a smaller, downy version of *Plantago major*, which if repetitively mowed or disturbed will only grow small, adding to identification difficulty; *Plantago coronopus*, Buck's-horn Plantain (also called Buckthorn), was once a popular salad addition; and *Plantago maritima*, Sea Plantain—all share many of the beneficial compounds to be found in this chapter. There is little difference in the nutrient content across the *Plantago* species, beyond of course the variability of mineral content of soils around the world—some growing by rivers are super-fertile and mineralised, while older agricultural soil is usually barren, depleted of minerals, as would be the plantain growing on them.

The plant known as Water Plantain, or European Water Plantain (*Alisma plantago-aquatica*) is not related to these plantains, although it is another beneficial medicinal plant. Also what most definitely isn't in the plantain family, although is what most people erroneously assume is the plant I'm referring to, is the tropical banana-like Plantain (*Musa paradisiaca*). Why they share a name is a mystery—maybe it's the shape of the seed head, as the etymology uncovers nothing, at least no more than both are plants and indeed both are herbs, which surprises many, but it doesn't answer why these unrelated plants both have the same name—if anyone knows please let me know.

* * *

Others in the *Plantago* family found across the world share in most instances similar appearance and features as well as similar nutrient profiles—*Plantago coronopus*, mentioned above, is native to Eurasia and North Africa but can be found as far afield as America, New Zealand and Britain today. With a resilience to low temperatures, it is becoming more popular as a frost-hardy winter crop, sometimes sold under the name Erba stella. The seedpods are unmistakably those of a plantain, while the leaves are more similar in shape to a narrow dandelion leaf or shepherd's purse. I grew a few plants from seed, and they were very much enjoyed as a regular crispy addition to a sandwich and salad throughout the summer and early winter—they live for years as well.

Plantago asiatica, also called Chinese Plantain, is indistinguishable from the British Broad Leaf Plantain, and is used for the same ends medicinally and as food, as well as being the common go-to plantain in the Asian research discussed below. *Plantago amplexicaulis* has adapted to the harshest desert environments, with the seed being able to survive for as long as necessary. It has been shown from core samples from lakes that the Sahara bursts into life every 20,000 or so years, being last green as recently as 4,000 years ago, when the right combination of the Earth tilting and rainfall becomes more frequent and abundant leads to the sands springing to life magically within days, with *Plantago amplexicaulis* being one of the first to muscle in.

Plantago elongata can be found across North America and provides a good amount and size of seed; it is smaller in the leaves, and as with many of the other *Plantago* species, will cope in most terrains, being resilient and adaptable to even the most compacted of soils, prising them open to allow nutrients to flow through the soil. Adapted even to beaches, *Plantago maritima,* with a dense rosette of stemless leaves and the unmistakable plantain seed stem, is up there with samphire as the best raw seashore vegetable, found globally from the Arctic to South and North America, Europe, Asia and North Africa.

Plantago ovata, also known as Ispaghul, Desert Indianwheat, Blond Psyllium and Blond Plantain, is often the source of dietary psyllium. The mucilaginous seeds are used as an aid to digestive and co-rectal ailments, instead and alongside that of the aptly named *Plantago psyllium,* which along with *Plantago ovata* is cultivated across India and the former Soviet Union for the seed and the husk that protects the seed. Psyllium is the name given to the seeds used commercially for the production of mucilage.

I have pure envy for *Plantago leiopetala,* which looks similar to Ribwort Plantain, if a little smaller, as its preferred habitat, Madeira, in the Atlantic Ocean, has the perfect temperature of a fraction under 20°C average all year. This plantain might be clinging to the cliffs and coasts of these volcanic outcrops, but it is doing so while enjoying the best weather.

Without going on and on about multiple similar plants, North America has many Broad Leaf-looking species, with heaps of ethnobotanic usage outlined in a following section. The few species highlighted above merely demonstrate that *Plantago* has always been a companion to man, it being another Human Plant.

Which continues to fit in with the proposition that it is these plants and the phytonutrients they contain that we have evolved on for thousands of millennia. Thus far, we have *Urtica* species across much of the globe, all pretty much similar, just as there is *Taraxacum* and *Achillea* across the world sharing pretty much the same nutrient profiles wherever they are growing. These profiles are rich in exactly what humans need, as would be expected from plants integral to our evolution and continuance globally. They have all the phytonutrients that the scientific evidence presented throughout this book repeatedly shows perform essential biological operations on the quiet, whether anticarcinogenic, anti-inflammatory, digestive, diuretic, nutritive, wound-healers inside and out, cleaners, cleansers and repairers of not just humans and animals but the soil as well.

Names

Local names for both *Plantago lanceolata* and *Plantago major* are wide-ranging, with every district of Britain once having many colloquial names for them. Fighting Cocks, Hard-Heads, Ripple Grass, Buckhorn, Fireweed, Cuckoo's Bread, Englishman's Foot, White-man's Footprint, Broadleaf, Common Plantain, Great Plantain, Hoary Plantain, Rat-Tail Plantain, Waybread and Waybroad are just a few still remembered, with many more lost.

In England they were also called Soldiers and Sailors after a game of twisting the stem around the head and pulling sharply to hopefully send the head shooting off as if a catapult. In Scotland they are known as Carl Doddies, as is the game of using them rather like a conker to knock the heads off the stem. This name is a historical reference also, with Carl being a diminutive of Charles, and Doodies being George, a reference to the Jacobite Rebellion of 1745, where both Bonnie Prince Charlie and King George III would loved to have knocked each others' head off.

A more innocent name is Angel's Harps, because Ribwort and the Common/Broad Leaf (and probably all *Plantagos*) share an unusual attribute of ribs running down the length of the leaves parallel to the stem, making it possible to gently rip the leaf in half exposing the ribs as fine hairs, hence the name Angel's Harps. No doubt lost are other regional names involving violins and cellos, and there's even word around Liverpool where they were known as Beatle's Guitars for a short period.

Plantains also go by the name of Snakeweed, possibly from the resemblance of the seed spikes to a snake, or from the plant's ability to draw out snake venom from a bite. This was suggested by the nineteenth-century American herbalist Dr Taylor, who recommended an application of plantain after promptly cutting away the flesh around the bite of a rattlesnake or cottonmouth snake, as death can be as quick as six hours from a rattlesnake bite.

* * *

Another name, Whiteman's Footprint, is not because the shape of the leaf resembles a human foot, be that a white, brown, red, pink or black foot. The name comes from the notion that plantain was said to follow the white man around the globe, with the Latin for *Planta* being a foot. Not that it was literally following white men around, it's a reference to the plant being able to thrive in compacted soils, being tenacious and strong-rooted enough to break the soil apart and seek out nutrients. It is often one of the first plants to be established on compacted soils, being known as a coloniser species (along with dandelion, nettles and brambles). The name is more of a slur as it is the white man who 'comes over here', disturbs the soil with new tracks, compacting the earth over and over again with horse-drawn carts, as he snaked progressively across what was definitely not a United States, with the roads encouraging the *Plantago* to take a first and dominant foothold on this unnatural track made by white men.

Certain plant species would definitely have been brought over from Europe to North America, undergoing hybridisation with similar local species from the *Plantago* family, as well as more subtle changes caused by environmental adaptation. The early settlers would have taken plantain seeds with them to the colonies, as they did nettles, yarrow, hemp and dandelion, because they were both part of the foods eaten and were a part of their botanical and pharmacological knowledge at the time. If these plants were not deliberately brought over the Atlantic, there is the possibility of them being accidentally spread with seeds among the feed, soil and litter brought over with horses and livestock etc.

The whole confusion about the introduction of plantain probably falls at the feet of Henry Wadsworth Longfellow and his 1855 poem, *The Song of Hiawatha*, this being the first known mention of the term Whiteman's Foot for plantain. Longfellow was neither a botanist nor an ethnologist, and his poem has been taken perhaps a little too literally—he was after all a considerable wordsmith, giving us such sayings as 'Ships that pass in the night' and 'Footprints on the sands of time'. One also has to consider this statement and poem in the context that all the Native Indian peoples across both North and South America had experienced centuries of genocide, with many white Americans being fully aware of this annihilation. A false history eases the conscience, and within a generation there is no one left to counter the claims of the Americas being inhospitable and barely survivable, while the ravaged locals warmly welcomed the plants from the wise white man—this version was deemed better at the time than telling the truth of a shameful slaughter and extermination of entire cultures, which spanned both continents for millennia before. It was reprehensible, and this is why this erroneous history of the spread of plants is mentioned in all four chapters so far. I do not have a particular fixation on discrediting theories on global plant transmissions, but I do have a problem with people being unaware of the slaughter done in God's name only a few centuries ago wherever white Christians went.

Longfellow's time, from the 1830s to 1870s, could well be considered the height of the stereotypical Wild West of Cowboys and Indians, with the extermination of the buffalo and Native culture, and survivors resettled in reservations, disconnected and starved. Anyway, Longfellow appears to have been a good soul and a cultural icon in his time with huge sales of his work. Here is the relevant extract from the poem, which was little more than pure conjecture and artistic licence by the poet:

> Wheresoe'er they tread, beneath them
> Springs a flower unknown among us,
> Springs the White-man's Foot in blossom.

First Nation peoples loved them

Ethnobotanically, among the First Nation peoples, there is considerable evidence of utilisation of the whole *Plantago* family, from *Plantago aquatica* (American Water Plantain) to *Plantago aristata* (Large-bracted Plantain), *Plantago australis* (Mexican Plantain), *Plantago cordata* (Heartleaf Plantain), *Plantago maritime* (Goose Tongue), *Plantago ovata* (Desert Indianwheat) and *Plantago patagonica* (Woolly Plantain). They were all used, from the Aleut to the Zuni, across North and Central America for a whole range of ailments and uses, many of which are repeated with the two main *Plantago* of this chapter, *Plantago lanceolata* and *Plantago major*. Starting with the former, the Cherokee used a poultice of the leaf for headaches, burns, blisters, ulcers, insect stings and snake bites, an infusion for bloody urine and bowel complaints as well as to strengthen children as they learn to walk, while a decoction of the roots was used for dysentery. The Kawalisu used an infusion of the leaves to place in the ear for earache, while the Mendocino used it for its antibacterial and antiseptic benefits in fodder for livestock.

In Native American medicine, plantain is often regarded as snake or frog medicine—a snake as it can draw and pull venom and heal, with the powdered root often being carried around for this purpose, hence its name Snakeweed; and a frog because these have been seen to eat the leaves of plantain as a means of antidote after getting bitten by a snake.

Plantago major was also used across the continent by all Peoples, further detracting from the notion that given its name of Whiteman's Footprint it was in any shape or form introduced to the Native pharmacopoeia by white man. The Abnaki applied a poultice for rheumatism, swellings and general pains, while it was enjoyed as a green vegetable by the Acoma; the Algonquin used the leaves for a poultice for burns, wounds and contusions. The Carrier used a decoction of the plant for stomach problems, coughs and a laxative, as did the Cherokee, both eating the plant as a vegetable. The Chippewa used all parts of the plant for inflammation and rheumatism as well as carrying a pouch of powdered root as protection from snakebite. The Costanoan made a decoction of the roots for fever and constipation, while the Cree chewed the leaves to apply to burns and scalds. The Delaware used the leaves as a poultice for bruising and as a general tonic, and in combination with other plants for a range of female ailments.

The Hesquiat used a plantain poultice for drawing out pus from wounds, cuts and infections, as well as for headaches, with the Iroquois using an infusion of the seeds to regulate menses, facilitate digestion and reduce intestinal inflammation—boiled flower stems were applied to abscesses and a poultice of the leaves given for sores, arthritis and of the roots to purify the blood, for fevers, diarrhoea and for difficulty with breathing from lower chest pains and even for nervous breakdown.

Plantain's benefits for the blood, stomach, stings and bites, splinters, sores and swellings, a wash for eyes and skin, as an antiseptic and antibacterial and as food also figured prominently among the Isleta, Kawalisu, Keres and Kiowa—who used to make garlands and wreaths of plantain to symbolise health for dances and ceremonies—and the Kwakiutl, Laguna, Mahuna, Meskwaki, Mohegan and Navajo—who give the root the endearing term of 'life medicine', encapsulating its usefulness perfectly—Nitinaht, Ojibwa. Okanagan-Colville, Palute, Ponca, Potawatomi, Rappahannock, Shinnecock, Shoshoni, Shuswap, Thompson, Yurok, Zuni and no doubt many others also incorporated plantain into their lives long before the white man arrived.

History

Some of the oldest evidence for European use of plantain comes from several famous Iron Age burial remains in Denmark, notably Tollund Man, an amazingly well-preserved 40-year-old discovered in a peat bog in 1950. He had been hanged, either a sacrificial victim, a criminal, a victim, a suicide or something entirely different, but however he was taken he looks undeniably most serene and content in death. Dated to 375 BC to 210 BC, Tollund Man was so well preserved that from his internal organs his last meal could still be identified. This was a porridge or gruel that contained an astounding 27 seeds and grains, including chickweed, flax, oats, rye, barley, docks, knotgrass, shepherd's purse, mint, mustard, and plantain. His teeth and bones also provided a good idea of his general diet and environment through microscopic, isotopic analysis.

Another Danish peat-preserved body is Grauballe Man, only 30 years old and with even more varied seeds and grains making up his last porridge, before he had his throat slit. This goes to show that although the diet may appear limited, being seeds from plants from the end of the autumn (hence the suggestion of a sacrifice after the poor harvests were in), the nutrient spectrum from these two burials was tremendous.

What these preserved remains also show is the obvious fact that people at that time utilised a massive range of plants for food, always supplementing any cultivated crops (barley, rye and flax) with many more from the natural larder. Though berries, meat and fish were all absent, Grauballe Man also had traces of ergot, the hallucinogenic fungus that grows on grains, notably rye and barley. Whether he had ergot by design or an accident, as part of a ceremony, or the reason for killing him—owing to possible madness and visitations brought on by the ergot—will always be subjective conjecture.

The ancient Greeks and Romans both used plantain for herpes, skin infections and as an antidote to rabies. Plantain's extraordinary adaptability to draw sustenance from compacted soils is mirrored by its capability to draw foreign objects from the skin, as well as expelling out of the body from within that which should not be within. It is this 'drawing' ability that provides one of the oldest traditional medicinal uses of plantain, with the ancient Greeks using it as a drying agent, able to draw out pus as well as water. It was known to be soothing to the membranes, encouraging healing, by keeping the area free from infection and bacteria. A member of the audience at a recent talk I gave explained how she had made a tumour disappear from her breast in just a fortnight with a poultice of masticated plantain, although this is of course only deemed anecdotal, not scientific evidence, but she definitely spoke the truth.

Held in such high esteem by the previous inhabitants of *Olde* England, plantain was included as one of the nine sacred herbs of the Anglo-Saxons, as described in their tenth-century poem *Lacnunga* (Remedies). The other sacred plants mentioned are usually seen as mugwort, watercress, chamomile, nettle, crab apple, chervil, fennel and thyme—although some go with lamb's cress, cockspur grass or betony instead of thyme—with these nine certainly providing relief from a vast range of ailments. Herbalism grew in favour towards the end of the European Renaissance (roughly 1300–1600), when plantain gained a reputation as being the best plant coolant and was associated with illnesses such as scurvy, a disease caused by vitamin C deficiency.

Herbs continued to be in favour throughout the seventeenth and eighteenth centuries, as plant power gained public as well as academic support. Bear in mind that at this time genuine plagues were sweeping across regions, sometimes wiping out entire villages, so the public were very aware of the need for preventatives. Culpeper (1616–1654) certainly held plantain in high esteem, differentiating nine different types, and although his taxonomy is not botanically correct as of today's accepted genera, he was right in the main. One species was Dover Plantain (a smaller ribwort), which would have been growing wild in relative abundance across the South Downs of Sussex from the village of Isfield where Culpeper grew up (also my own local village, just outside of Lewes), across to his ancestral lands of Folkington close to The Long Man of Wilmington, 15 miles east, where an ancestor's plaque suggests spelling was not standardised as this was a Colepepyr.

His other species were Grass Plantain. Hoary Plantain, Ribwort Plantain. Sea Plantain, Water Plantain, Lesser Water Plantain, Welsh Plantain and simply Plantain. Culpeper also gave plantain a wide span of curative abilities, seeing it as a herb of Mars because it cures the diseases of the head, guts and bowels, eases menstruation, bleedings of the mouth, coughs, ulcers, jaundice, toothache, piles, inflammation and swellings of joints. He wrote that a powder of the dried leaves kills intestinal worms and benefits a "hot womb … and helps lunatic and frantic persons very much; as also the biting of serpents or a mad dog". Above all else Culpeper advised plantain as a wound herb, be that fresh or old wounds, sores both inward and outward.

William Salmon's *Botanologia or The English Herbal*, from 1710, maintained that plantain

> cures the Scurvy in the hot habit of body, heals the excoriations of the stomach and other bowels, stops catarrhs, and all sorts of fluxes in man or woman. It stops also all sorts of haemorrhages whether internal or external, the overflowing of the courses in women, spitting and pissing of the blood, and therefore is of excellent use in the curing of tuberculosis, or ulceration of the lungs.

Such diseases and ailments seem far removed from our culture of illness today, but in all probability physicians and authors of the time would have actually seen this curative potential of plantain first-hand, repeatedly (and of the other plants, particularly as the polluted background of the human organism and causative agents of illness were easier to establish before our current deluge of deadly man-made chemicals), during an age when such worrying unsavoury pestilences were rampant in the rapidly expanding conurbations. Why would authors report erroneously? Surely it would reflect badly on the author to exaggerate and distort such matters.

Alfred Vogel (1902–1996), nutritionist and herbalist, states that *Plantago major* was crushed in the Tropics and applied as a poultice to ulcers and boils. He mentions *Plantago lanceolata* as being particularly effective as a remedy for the ears, taken over a relatively long time (weeks to months): "it also sharpens the auditory senses … When the ear infection is accompanied by nasal catarrh and pharyngitis … put five drops of *Plantago* on cotton wool and place in ear, replacing every day." He praises Ribwort's ability to act as an antiseptic and antibacterial for many ailments including coughs, and favours an old custom from his native Switzerland of a spring-clean to cleanse the blood by drinking teas infused with leaves from berry bushes (blackberry, raspberry and strawberry), ribwort, coltsfoot flowers and cowslip in the morning and evening, as an effective tonic. He recommends finely chopping a couple of plantain leaves and adding them to a daily salad, especially for those suffering from respiratory problems, catarrh and sore throats.

The fantastic and much-missed Roger Phillips (passed 2021) tells us in his excellent series for Pan Books, *Herbs* edition: "The fresh juice, extracted from the whole plant and boiled with honey, then stored in the fridge, is considered a very good cough medicine." He agrees with the general consensus that the leaves mashed into a poultice make an excellent aid to soothing and healing wounds. American herbalist Matthew Wood brings us up to date with his fabulous *Earthwise*, finding plantain "useful with abscesses around the teeth, inflamed roots of teeth, or old lingering infections after root canal surgery … it will almost never fail to draw out the pus and stop the infection." Wood also gives it a wide range of uses from acne, Bell's palsy, bladder infection, bronchitis, cataract, cholesterol lowering, colitis, conjunctivitis, constipation, diarrhoea, dysentery, earache, eczema, emphysema, haemorrhoids, hepatitis, impetigo, insect bites, jaundice, kidney ailments, neuralgia, oedema, pneumonia, skin irritations and tuberculosis to vaginitis.

Traditional uses

There is a long history of plantain's use in Traditional Chinese Medicine (TCM) and among Indian Ayurvedic physicians, using the leaves and seeds of a variety of *Plantago* species as treatments for diarrhoea, haemorrhoids, constipation, urinary problems and more recently high blood pressure. In China the seeds of *Plantago asiatica* are used for ailments involving the kidneys, often as a kidney tonic, especially if inflamed, as well as being employed for water on the lungs, oedema and gonorrhoea, while the leaves are effective against bacterial enteritis. The German Commission E, the counterpart to the UK's Medicines Control Agency, endorses its use as safe and effective for bronchial conditions and throat inflammation.

Boiled plantain roots can be used as a mouthwash for sore and bleeding gums, as an eyebath, and made into an ointment or cream for piles, eczema and other skin ailments. With plantain's effectiveness in cooling and moistening of the mucosa it can be helpful with cases of bronchial infections and pneumonia. An effective preparation would use plantain extract with other plants, herbs and fruits depending on the severity of the bronchitis, such as thyme, marshmallow, blackcurrant, liquorice and echinacea.

Plantain is also a mild diuretic, a useful nutritive one, where more nutrition is gained from ingestion than is lost through extra fluid loss. Tannins kill harmful bacteria in mucous membranes, while the mucilage protects the mucous membrane from germs, with the rich silicic acid content helping the body build mucous membranes, cartilage and bone tissue.

Crushed plantain leaves directly applied to the area of insect bites or nettle stings work as one always hopes a dock leaf would do but always failed—plantain leaves actually curtail the pain in less than a couple of minutes. Plantain also has styptic abilities (stemming blood flow), used along with yarrow. The silica and tannin content also makes both plants useful in treating varicose veins, damaged lungs and haemorrhoids.

The range of ailments plantain is used as a treatment for is vast, with all the above, as well as benefiting asthma, coughs, cystic bronchitis, dysentery, gastritis, hay fever, inflammations, IBS, sinusitis, splinters/thorns, ulcers; it can be used as a vermifuge (parasite killer), while the Saxons used red wool to affix some plantain leaves to the head as a cure for migraine.

Two glycosides found in most plantains (not *Plantago coronopus*) are acubin and catalpol. Acubin has antibiotic properties, while catalpol, also found in dead-nettles and the figwort family, has been shown to stimulate the adrenal glands in animals that consume these plants, as well as inhibiting oxidative damage to the brain, meaning that catalpol could be a nutraceutical, for instance, with regards Parkinson's disease in the future. Used in conjunction with nettles, the preventative capabilities of plantain appear solid, and considering that the dementia and neurological revenues from this sector of medicine run in excess of £20 billion annually, this is not an inconsequential amount and definitely worth further investigation.

Gargling with a prepared plantain syrup has been used for stomatitis and pharyngitis, as an antiseptic, for insect bites, a demulcent, anti-inflammatory (the flavonoids have well-established anti-inflammatory capabilities), antimicrobial, as rich in the tissue-knitting allantoin (comfrey wins hands down for quantity of this miraculous chemical), expectorant, antiviral and reliever of sunburn. A tea made from plantain leaves is said to help with the quitting of smoking tobacco. It is great to eat during pregnancy, as a nutritive and preventative, and it helps to keep the digestive system and colon clear.

In the past I've also included some leaves and seeds of plantain in a balm to soothe and heal a friend's skin cancer sores, as well as it being a great first-aid balm to keep handy for any cuts, sores, stings and grazes.

* * *

Recently, I was fortunate to be asked to represent the plants at a Climate Change conference in the gorgeous Montenegro. After putting the case that there was definite potential available from these plants, listing many of the uses you've been reading of, one stuck in my mind for plantain when I returned home, namely insect bites. I looked down at my inflamed ankles after having given them a good itch at home the first night back, and I mean a really good itch! It was torture, so I got the torch, went out behind my house and found about 25 good-sized leaves of broad-leaved plantain growing by the edge of the field. I chucked them in the bath, squeezing them while soaking in it. After this night, I was still itching and was also bleeding from my rigorous itching. We counted 189 mozzie bites on just my legs, with a similar amount on my arms, head, face and body (many acquired when trying to photograph last chapter's Asian Tiger mosquito, as it drew me deeper into some woods, when I become instantly covered by a different species—and also failed to get a decent, non-blurry shot). After the second evening's bath and rubbing the leaves over me I had no more itching; all the bites had thoroughly cleared up in just 24 hours more. Here was living proof, and in a comparison with some other delegates who had also been bitten, their bites continued to menace them for at least a further four days. Next time it will be far more sensible to apply a natural mosquito deterrent to begin with, such as yarrow.

Firm scientific credentials

This has to be the most exciting bit, because after banging on for over thirty years about how our diet and lifestyle are devolving our genome and leaving us susceptible to disease, there is now plenty of scientific corroboration. I was initially made aware of this disparity in the health of the West when meeting with tribal peoples a few decades ago, as they were clearly far nearer the blueprint for a *Homo sapiens* than I could ever hope to be. At this time (1980s) there was little literature available, except Weston Price's exceptional *Nutrition and Physical Degeneration* and anthropological books such as *Man the Hunter* to go on; there was also very little, if any 'science', and this always lead to a brick wall in any discussions. Now with open source internet there's loads, and here's some that is plantain-pertinent—

Russian studies from 1989 (Karpilovskaia et al.) published in the *Farmakol Toksikol* journal, showed that the polyphenolic complex from *Plantago major* (a complex shared with nettles, yarrow and dandelions) was an inhibitor of carcinogen synthesis and reduced toxic damage to the liver, as well as decreasing tumour size by over 60%. Further studies (Markvicheva 2009) in the Russian journal *Biomeditsinskaia Khimiia* found tissue repair was accelerated when participants were given soluble extracts from *Calendula officinalis* (marigold) and *Plantago major.*

In Romania, the Institute of Medicine, Faculty of Pharmacy (Hriscu et al. 1990) showed that the polyholozidic fraction when separated from the leaves and seeds from *Plantago major, P. lanceolata* and *P. media* had significant gastro-protective action. Even older Bulgarian studies from 1982 (Matev et al.) reported that of 25 patients with chronic bronchitis, 20 of them showed a rapid response in symptomatic relief when administered a plantain preparation, demonstrating that the claims of *Plantago major* as an expectorant, anti-inflammatory and pain reliever were true for the majority.

In Serbia, the University of Novi Sad (Beara et al. 2009) looked at the flavonoids from a range of *Plantago* species, showing that luteolin, apigenin, rutin and quercetin were present across them in varying degrees, all with a potent antioxidant effect. Further studies reported in the *Journal of Pharmaceutical and Biomedical Analysis* (Beara et al. 2010) used extracts of *Plantago lanceolata* and *P. major* and showed *in vitro* (*vitro* = glass, *vivo* = living) plantain acted as a significant anti-inflammatory agent.

There were pharmacological properties reported in research on antimicrobial activity (Nostro et al. 2000), antioxidant and cytotoxic activity (Beara et al. 2012), anti-tumour activity (Herbert et al. 1991) and antispasmodic effects (Fleer & Verspohl 2007).

Researchers at the University of Debrecen, Hungary (Gonda et al. 2010) compared five *Plantago* species' (including *lanceolata, major and media*) spectrum of phytonutrients, finding only slight variations between the different species, and concluding that all of them were potential replacements of each other for any future medicinal applications.

* * *

The Department of Food Science at Korea University in Seoul (Park et al. 2007) reported in the journal *Phytotherapy Research* that there was no toxic effect from the plantamajoside concentrate in *Plantago major*, even when administered up to a huge 2g/kg of body weight—which would amount to a 70kg person eating 140g of pure plantamajoside with no ill effect. This basically means it's safe to consume plantain in what would be ridiculous portion sizes, and considering we're only talking about consuming a small handful in a soup or a few leaves in a salad and seasoning with ground-up seeds, there is never going to be any need for concern. So once again, the likelihood of side effects is incredibly remote and will again only affect those who are actually allergic to a certain plant species.

From the same university in Seoul, as reported in the *British Journal of Nutrition* (Chung et al. 2008), the essential oil of plantain was shown to be comprised mostly of monoterpenes, especially linalool, with the positive effect of reducing oxidised LDL cholesterol levels with additional cardiovascular protective benefits. As stated previously, cardiovascular disease is the number one cause of death, with the heavily overprescribed statins proving hazardous and ineffective in many instances—although sadly it is the biggest-selling drug sector of them all, and there will be no change contemplated to this non-stop cash machine of dangerous drugs and perpetual profits.

A 2012 Korean study from Chonbuk University (Choi et al.), published in the *Molecular Medicine Reports* journal, echoed the anti-carcinogenic properties found in the earlier Russian studies. In this study, Choi et al. showed that hollyhock (*Althaea rosea*) and *Plantago major* (both long used for therapeutic and inflammatory reasons in TCM) also significantly suppressed neoplastic cell transformation, concluding that plantain should be considered for chemopreventive and chemotherapeutic applications.

Kaohsiung Medical University, Taiwan (Chiang et al. 2002) conducted studies that looked at *Plantago major*'s antiviral capabilities, specifically against the herpes viruses (HSV-1 and 2) and adenoviruses (ADV-3, 8 and 11). The results of an aqueous extract of the whole plant had only a slight effect, whereas certain pure chemicals extracted from the plant, such as caffeic acid (one of those phenols) had the strongest activity against the herpes viruses, and chlorogenic acid was strongest against the adenoviruses. Results were very positive and showed that plantain contains significant antiviral compounds. The following year Chiang used the less effective aqueous extracts of both *Plantago major* and *Plantago asiatica*, with results showing anti-carcinoma (against stomach, bladder, bone, lung, kidney and cervix cancer cell lines), anti-leukaemia and antiviral properties, as well as modulation of cell-mediated immunity.

Plantamajoside was isolated from *Plantago asiatica* by the College of Veterinary Medicine, Huazhong, China (Wu et al. 2016), where it was shown to be a potential therapy for the treatment of pulmonary inflammation and lung injury. This plantamajoside was further studied at the Huaihe Hospital of Henan (Ma & Ma 2018), as reported in the journal *Inflammation*, with a view to using it for treatment of respiratory inflammatory diseases, which the authors concluded could be a definite possibility.

A joint Thai and Indonesian study (Kartini et al. 2016) published in *Pharmacognosy Magazine* used Indonesian *Plantago major* to test both the anti-inflammatory and the anti-carcinogenic capabilities of the plant. The authors concluded that the seeds displayed both anti-carcinogenic and anti-inflammatory activities, while the roots were especially anti-inflammatory. Human breast, cervical, lung and pharynx cancer cell lines were tested on, with the greatest anti-proliferation agent being ursolic acid. Ursolic acid, oleanolic acid and aucubin were also shown to be viable anti-inflammatories, demonstrating that benefits are available from plantain's seeds, leaves and roots.

An older American study in the *Journal of Phytotherapy Research* (Gomez-Floes et al. 2000) showed immune-enhancing properties from a leaf extract of *Plantago major*, concluding that plant extracts could be clinically relevant to diseases such as tuberculosis, chronic viral infections, AIDS and cancer.

The condition known as urolithiasis is uncomfortable at best and excruciatingly painful at times, and is caused 80% of the time by calcium oxalate, where the crystals form into a solid ball/stone that lodges in the kidney, ureter, urinary tract, bladder or the urethra. Clever researchers at Kebangsaan University, Kuala Lumpur, Malaysia (Aziz et al. 2005) discovered that *in vitro Plantago major* extract was better at inhibiting these crystals than potassium citrate or allopurinol, of which fewer than 1% of patients suffered liver failure when prescribed this. Leaving the side effect negativity aside, allopurinol garners somewhere in the region of £500 million in global sales annually, though actual figures on this particular drug would mean forking out £3,200 for a market report. AstraZeneca expects to hit a half-billion pound figure for a new gout drug, which allopurinol is prescribed for, as well as for kidney and gallstones as it lowers uric acid—but not as well as plantain lowers it. The same Kebangsaan University (Hussan et al. 2015) looked at the potential anti-inflammatory capabilities, and observed *in vivo* that *Plantago major* significantly reduced the pro-inflammatory cytokines, preventing extensive tissue damage.

An excellent study from the Department of Botany at Kohat University, Pakistan (Aziz et al. 2018), published in the *Journal of Ethnobiology and Ethnomedicine*, looked at the problem of the loss of traditional knowledge of plants from across the world and notably their own Pakistan in favour of Western medicine and pharmaceuticals. With the younger generation not taking on board any traditional knowledge passed down from generation to generation for millennia, this corpus of knowledge is today facing a massive threat of being lost forever. The authors also recognised that *Plantago lanceolata* was one of the most popular herbal plants used for medicine, along with *Caralluma tuberculata*, *Thymus serphyllum* (wild thyme), *Fagonia cretica* (aka Virgin's Mantle, used as an anti-carcinogen), *Periploca aphylla* and *Citrullus colocynthis* (bitter apple, desert gourd) and a few others.

A study at the Mazandaran University of Medical Sciences, Iran, published in the journal *Acta Pharmaceutica*, showed that plantain seeds were able to improve release rates from pharmaceutical tablets (Saeedi et al. 2013), and when mucilage from the seeds was used in the formulation a better tablet was made. From the same University of Medical Sciences (Sangian et al. 2013), the use of *Plantago major* and some other plants for antimalarial activity was examined, an issue that is becoming increasingly more imperative as the effectiveness of traditional antimalarial drugs is declining. Plantain proved to be effective and non-toxic, while the study also showed for the first time the usefulness of yarrow as an antiplasmodial—as we saw in the last chapter with the blue essential oil killing the larvae of *Aedes albopictus*, the Asian Tiger mosquito. A combination of both yarrow and plantain might well make the desired breakthrough in the search for the prevention and eradication of malaria.

Shockingly, diabetic foot ulcer (DFU) is a common complication among diabetes patients, and when left untreated (often not healing and becoming desensitised) gangrene sets in with amputation necessary as death would otherwise occur soon after. To see how effective plants may be with DFU, the *Current Drug Discovery Technologies* journal reported that the Mashhad University of Medical Sciences (Najafian et al. 2018) in Iran made a gel from *Aloe vera* and *Plantago major*, and this significantly reduced the ulcer, a promising avenue for what is a costly and very hard-to-eradicate problem associated with diabetes. Further Iranian studies, this time on *Plantago major*'s kidney protective capabilities, demonstrated improvement of functionality; this study was also from Mashhad University (Yazd et al. 2016 & 2018), as reported in the *Iranian Journal of Kidney Disease*. Other work from Mashhad University (Parhizgar et al. 2016) conclusively demonstrated that a *Plantago major* extract and vitamin E were able to improve kidney function and reduce oxidative stress.

The hepatoprotective (*hepa* refers to the liver) and anti-inflammatory activities of *Plantago major* were further studied by Yuzuncu Yil University, Turkey (Turel et al. 2008) and reported in the *Indian Journal of Pharmacology*; this study concluded that such activities were considerable when using a methanol extract of plantain seeds. This hepatoprotection was further demonstrated by the University of Mogi das Cruzes, Brazil

(Mello et al. 2012) in the *Journal of Pharmacy and Pharmacology*, as well as from *in vivo* studies at the University of Gaziantep, Turkey (Ozaslan et al. 2007), where a 1% extract inhibited the laboratory-created *Ehrlich ascites* carcinoma.

* * *

As seen with the Iranian yarrow research into oral mucositis in the last chapter, a study (Cabrera et al. 2017) in the *European Journal of Oncology Nursing* from the University of Barcelona used *Plantago major* instead, with similar results, working as well as the group using the traditional method of chlorhexidine, whose global worth in revenues exceeds £150 million.

It was those phenols that researchers at the University of Naples, Italy looked at (Ferrazzano et al. 2015), trying to control the cariogenic bacteria of *Streptococci* and *Lactobacilli* with a simple mouthwash made from a tea of *Plantago lanceolata*—which it did, as reported in the journal *BioMed Research International*. The authors also said that the effect was not solely owing to the phenolic compounds, with the flavonoids, coumarins, lipids, cinnamic acids and lignans all seemingly playing a role as well. Success was greatest against *Streptococci*—and may have been better had they not boiled the plants for quite so long to begin with! The scientists all agree that an alternative is required as the usual antibacterial agent, chlorhexidine, is allowing bacterial resistance and causing unwanted side effects—which include a staggering 50.7% of those who took it complaining of toothache, 28.4% upper respiratory tract infection etc. …

We've referred to the reports demonstrating the safety of these plants, even in unrealistically large doses. No dangers. Once they are rolled out as nutraceuticals and millions of people are prescribed them, there will be a number of contraindications, not least from the estimated 1% of the population who are unfortunately allergic to these plants. But it's never going to include side effects as broad and serious as liver and kidneys eating themselves, hearts stopping and skin and eyes failing as a consequence of a prescription.

The University of Seville, Spain, served up a smasher (Galvez Peralta et al. 2003), published in the *Journal of Ethnopharmacology*, answering a few questions and assumptions. Firstly, seven species of *Plantago* with *major, lanceolata, media, asiatica* and *ovata* all included, confirmed that similar health effects are transferable across similar species (exactly as done for this chapter and the assumption for dandelions, yarrow and nettles before). This is good to know, but their study was more significant than this, as the authors also showed that the *Plantago* species exhibited cytotoxic activity against all the commonest cancer cell lines. They also identified it to be among the flavonoid compounds, with luteolin in particular exerting this cytotoxic potential—luteolin is also found in abundance in the previous three plants. It is these phytonutrients that are among the first to be lost when food is processed, which amounts to most of the food that isn't in the fresh food sections of the (not so)super market.

The report certainly alerted the various cancer industries and researchers that something as simple and freely available as plantain might be worth looking into further. You would have thought Marina Galvez Peralta would have been inundated with research grants getting her to look at the potential of such easily found plants in the cancer arena—it was after all over 15 years ago now since she made this discovery. You would have thought so, wouldn't you? Yes but ... the cancer industry is worth somewhere in the region of £150 billion annually in revenues from the sales of pills, operations, scans, machinery, donations, research etc. If the plantain (and dandelion) prevention doesn't net £150bn there will be no change, as this would cause a significant dent in the GDPs of the medicated Western world. In our culture of monetarism, wealth always beats health. So, it's not going to happen fast, and even if it works medically, if it doesn't work economically, it will be prevented.

In the Swedish University of Agricultural Sciences (Zubair et al. 2016), the wound-healing capabilities of plantain were looked at and recorded as present, with an ethanol extract performing better *ex vivo* (outside of a living organism) than an aqueous extract, although both were effective, as reported in the journal *Natural Product Research.* These findings were corroborated by the University of São João Del Rei, Brazil (Thome et al. 2012), as reported in the journal *Experimental Biology and Medicine,* which showed wound healing was accelerated with a *Plantago major* ointment.

Brazilian researchers (Mello et al. 2015) used a water-ethanol extraction process when looking at how this protects mitochondria from oxidative damage, thereby preventing free radicals and the associated damage they may cause, which it did. This is potentially quite a big deal.

Iztapalapa University, Mexico had already established the antibacterial ability ten years before (Velasco et al. 2005) using methanol, hexane and aqueous extracts, all showing antibacterial activity against *Candida albicans* and *Escherichia coli (E. coli),* as well as importantly improving the functionality of the bone marrow and the spleen (for immunity), demonstrating hematopoietic activity (increased stem cells that become blood cells).

The Indiana University of Dentistry, Indianapolis (Shirley et al. 2017) seconded the Mexican research above, showing that *Plantago major* in extract form, as well as two of its active compounds, aucubin and baicalein, inhibited *Candida albicans,* proving the plant to be an effective fungicide with considerable potential in this field if it were to be used pharmacologically.

The Scandinavian Journal of Immunology reported on a Norwegian study (Hetland et al. 2000) where mice were offered protection from *Streptococcus pneumoniae* by a soluble pectin polysaccharide isolated from the leaves of *Plantago major,* proving it to be antibacterial.

Studies from the University of Pardubice, Czech Republic (Bajer et al. 2016), found 154 compounds in *Plantago lanceolata,* showing there was also a good quantity of these hard-to-read and easily forgotten compounds. Among the monoterpenes were isomenthon, limonene (the most prevalent), linalool, nerol, terpineol and thymol, and among the oxidated sesquiterpenes were caryophyllene, fokienol and valeranone. Then there were alcohols, aldehydes, aliphatic and aromatic hydrocarbons (rich in toluene), diterpenes, esters, fatty acids (rich in palmitic acid), ketones and phenols—just in case you ever needed to know (with plenty more in the Appendix).

* * *

Studies from the wonderfully named Zagazig University in Egypt by Ahmed Romeh established the potential of *Plantago major* for phytoremediation, initially in 2006 by removing the highly toxic and persistent insecticide imidacloprid from the soil, and again in 2014 with the equally toxic cyanophos, proving plantain's versatility in detoxifying contaminated soils. Plantain even manages to remove arsenic from polluted soils when most other plants just die. Romeh's work was corroborated by French studies (Jana et al. 2012) that looked at using plantain to clean up abandoned contaminated mines, as also demonstrated by Italian researchers in the same year (Malizia et al.). This environmental cleaning-up of industrial sites is a huge financial burden, taking years, often with dubious results and using vast quantities of only slightly less detrimental chemicals than the ones trying to be eradicated. Plant remediation prevents this, while offering many other positives, from being effective carbon sinks to being a source of energy from their combustion. This phytoremediation was further corroborated by Polish studies (Malinowska & Jankowski 2015) in the journal *Environmental Monitoring and Assessment*; the authors also point out the rather obvious hazard of pesticide uptake invalidating the medicinal potential plantain may offer, while demonstrating that both *Achillea millefolium* and *Plantago lanceolata* uptake heavy metals and clean the soil.

* * *

Wow! That's quite a bit of hard-hitting evidence, across the full spectrum of diseases and ailments, all shown to benefit from various extracts of plantain. There is a surprising dearth of research from either the UK or USA—it's very strange that the countries with the largest pharmaceutical corporations don't appear to be doing much research on plants the rest of the world has proven to have definite potential and efficacy. Then consider the billions of dollars and pounds the public in America and Britain donate to Cancer Societies, with the National Cancer Institute in the USA having spent £68 billion on research and treatment in the past 40 years alone. You'd have thought the scientific research coming out of these countries would be bursting at the seams, when in actual fact there is hardly any from thousands of studies available from NCBI and PubMed (peer-reviewed online scientific journal libraries), yet plenty from those countries that fall outside of the umbrella of Western pharmaceuticals, such as Iran, China and Korea. Just saying …

I appreciate it's all a bit boring and somewhat mind-numbing, showing the scientific efficiency of this plant and all that it can help with—but that is the whole point. It is the scientific evidence, not poetry or a fairy tale. Importantly, it has been conducted in the name of science, not commerce. Which is definitely not the case with much of the scientific research conducted in Europe and America, where it is industry directly funding the research.

Forbidden knowledge

With most, if not all plants that urbanites come into contact with, whether those tiny ones growing between the paving slabs in the street or a park full of trees and smaller plants, there is rarely any recognition, no one-to-one interaction going on, and no connection. This is even though, as just shown from the four plants thus far, the potential of these plants as providers of human health is phenomenal, as is to be expected from Human Plants. It might well be worth reconsidering the respect and familiarity we bestow on all plants, because apart from herbalists and a few others, most of the population see plants as a quaint, ineffective, old-fashioned, weak and unscientific way to consider health. This perspective, it must be added, has been wrought from years of marketing propaganda, and it's no fault of ours that we are merely products of the cultural system we were born into and programmed by—that's just the way it is.

Part of this reawakening to the natural world will be this new appreciation of the plants. They are all astonishing, although as an introduction and not to overwhelm the reader, seven of them easily set the scene and can begin to heal the separation, particularly as these seven are so easy to become acquainted with. We need to relearn that plants are incredibly effective, their powers are omniscient, from killing you with a couple of leaves, to fusing a broken bone, flushing your digestive system or preventing dementia, cancer and heart disease or even opening doors to new mystic and spiritual worlds. It was of course from plants that most of the pharmaceutical drugs we use today originated. We've merely been indoctrinated by our cultural memes against anything free or old, and most definitely have been swerved away from any participation in any embodiment of personal empowerment—particularly in feeding or healing ourselves. The hunter-gatherer, living as far removed from our culture as it is possible to get, when approaching old age (and many do), doesn't suffer from obesity, dementia or high blood pressure, isn't riddled with cancer or atherosclerosis and remains a productive and cherished member of the community till death. These diseases of civilization the Westerner is suffering from at an ever-decreasing age should be the beginning of our questioning of the definition of 'civilization' and whether it really is best for us to continue with this particular version of civilization.

* * *

Plants are becoming a forbidden knowledge, again. There have been herbalists for millennia, and they have been persecuted cyclically. There is currently a new persecution, as this time the herbalists aren't threatening the notion of the Lord being the only saviour, when plants with curative and spiritual abilities made them a serious threat to the Church's monopoly on health, wellbeing and all matters theological or spiritual. Now herbs are a threat again, this time to the dominance of Big Pharma, whose monopoly on health and wellbeing is being questioned, and rightly so, because the drugs are not always working but are also causing added sufferance for millions via their side effects. It is as clear as day to observe that more and more people are getting sick, more are taking drugs yet the cycles of illness are getting larger not smaller, as should be expected if the drugs were working.

The pharmaceutical corporations are desperately trying to suppress this knowledge about the power of plants, as it is the ultimate threat to the noose-like hold on health they have, an economic threat, and it is this threat to profits that is the biggest obstacle to plants gaining a foothold. The same reason is why it took so long to begin to utilise wind and solar energy, not because it wasn't a good idea. It was always a better idea, but as it threatened the economic stability of the petroleum industry, which effectively wrote the laws revolving around capitalism and consumerism, there was never going to be an alternative to petrol for decades. We can't wait for decades to repair the food and health industries, it has to start now, but as the pharmaceutical corporations are making so much money nothing much will change.

When the pharmaceutical industry reached its initial peak in the late 1950s, it wasn't simply a case of the public being thankful and embracing the new pharmaceuticals, replacing herbs and roots with modern science, white coats and microscopes, devising drugs for all our ills—it also took quite some marketing as well. The rise in size and revenues for the pharmaceutical corporations has been astonishing, meriting respects for their business prowess, creating a monopoly on health with their pills, vaccines and creams. Most people thoroughly believe there is no other or better way to health than the modern pharmaceutical way, in exactly the same way many people know that milk is a good source of calcium and is good for their bones. They *know* this, yet these are no more than 'cultural truths', a lie disguised as a meme—being the result of years of marketing, sponsoring, advertising and celebrity endorsing by the Milk Marketing Board making people assume it was good for bones. Counter-intuitively the countries consuming the most dairy also have the highest incidences of osteoarthritis—read my *The Seven Deadly Whites* for the dirty lowdown on dairy, and how profiteering has connected all of us together in a web of not actually real cultural truths.

The pharmaceutical corporations have gotten lost in power and wealth and have forgotten about providing health. Whatever their skewed arguments may be to rationalise their affairs, they are causing immense suffering and, coupled with the agricultural chemical corporations (who coincidentally are often the same corporations), their man-made chemical pollutants are by an enormous distance the biggest and most disgraceful desecration ever perpetrated against the planetary systems we are dependent on. Climate change will become a serious problem for sure, flood plains will be inundated and available agricultural lands will dwindle, so terracing on higher ground and growing quinoa and rice will work, as would the proposal later in this book, but those chemicals aren't even seen as a threat or danger by the populous currently.

It is no coincidence that the rise in fortunes of the pharmaceutical corporations has gone hand-in-hand with the advertising agencies and food industries' increase in wealth as well. The revenues generated by the pharmaceutical industry globally have more than trebled in this century alone, rising from £294 billion in 2000 to a huge £753 billion by 2014 and somewhere around £1 trillion by 2024.

War was a big initial paymaster last century, as the apothecaries of early Pfizer and Glaxo (formerly Beechams) became producers of pills, inspired and financed by injured soldiers, and painkillers (morphine-derived usually) and antiseptics found widespread use. The 1950s saw a significant increase in global pharmaceutical revenues, with antibiotics, polio vaccines, anti-psychotics and general anaesthetics all bringing the science and fashion of drugs to a new level. The 1950s also saw the beginnings of fast food with the first take-away restaurants, supermarkets, science-fiction films, spaceships and visions of universal conquests. The 1950s is also when devolution sets in unabated.

Prescription drug revenues rose from 32% of all revenues from drugs/remedies in 1929 to 83% by 1969; there are no figures from today, but it must be a minimum of 90% by now, what with the meteoric rise in revenues in the past twenty years, mostly publicly funded. It's clear that in the 1950s these relatively small pharmaceutical companies' only interest was in providing a healthy service to customers, justifying any riches forthcoming. This changed to a much more aggressive way to make money through devious tactics, notably advertising and pseudo-science throughout the 1970s and 1980s, where today the only real concern is to post greater profits year-on-year. They are certainly not providing a service any more where their success in curing is reflected in their profits. Greater profits actually are easier and bigger to come by when more money is spent on marketing. Today, vast sums of money are made irrespective of efficiency, and it's clear to see that more and more people are taking pills for longer and longer for more and more diseases but no-one is actually getting any better. When the public see this deception, herbalism and the plant world are indeed going to be the death of Big Pharma, which is why the Pharma Corps want to eradicate all public knowledge of plants before this happens, by using the law.

The law or legislation is the new threat to herbalism. We have already seen the French law prohibiting the use of nettle fertiliser, which also restricts even talking about it! Yet more troubling is Directive 2004/24/EC, which holds potentially severe implications for the future of herbalism. Having caused confusion already as it sanctions against all herbalism and herbal preparations other than those given by a doctor in member states (and for those countries wishing to join the EU they have to comply with it), there is some unspecific wording indicating there is also a €75,000 fine for anyone picking or passing on knowledge of wild plants. Mother Nature has been made forbidden knowledge by the law itself. Directive 2004/24/EC will prevent any herbal preparation being made by anyone who is either not qualified or has used a plant that does not have at least 30 years of continued medicinal use. This immediately discounts any use of medicinal cannabis (the money involved in this might well cause an amendment allowing its exemption) and any new preparations from wild plants, irrespective that recent medical research has proven them to be effective against a range of diseases. The law now says no.

This directive will almost certainly cause repercussions, putting the stoppers on any dandelion root extract being further investigated for any anti-carcinogenic abilities as it clearly hasn't had 30 years of continual medical use, even if it has been used for millennia by herbalists and with recent worldwide science corroborating these claims. It looks very much like the product of lobbyists from the pharmaceutical corporations asserting pressure to get legislation in their favour, as is the norm. With herbalism already having the laws in place to eradicate it in Europe, as a herbalist myself this could be devastating, although as with Brexit and Covid-19, not much is very clear, but many laws have certainly been passed in their wake.

Psyllium (silly om)

Plantago psyllium is a really close relative of *Plantago lanceolata* and *Plantago major*, in appearance a bit like an amalgamation of the two. Often recommended for those with IBS and other lower digestive issues, it is better known as psyllium husk, which is actually just the seeds and outer coating of the seed. The seeds work so efficiently because up to 30% of the shell of a seed is formed of a water-absorbing substance called mucilage. When powdered and soaked mucilage swells significantly, giving it a gelatinous texture. When eaten this cleanses and scrapes through the digestive system, further swelling, picking up waste matter and increasing bulk, hence its description as a bulk-laxative.

In the case of diarrhoea, psyllium absorbs the excess fluid in the intestinal tract, giving some bulk to the stool. In the event of constipation, it lubricates the colon as well as pressing on the colon wall, triggering the contraction to open. The cleansing action also provides some relief from haemorrhoids.

It should be understood that it is not always the seeds from specifically *Plantago psyllium* that are found in products claiming to be psyllium husk. Psyllium is a sort of trade name as it also includes seeds from *Plantago ovata*, grown throughout western and southern Asia and south western America as a seed crop.

Taking a teaspoon of the husk, soaked in warm water for 10 minutes, three times a day for a 12-week study, as reported in the *Journal of the American Medical Association*, lowered cholesterol by 5%, equating to a reduction of 10% in the heart attack risk. Market analysts predicted a £17.5 billion increase in global revenues from cholesterol-lowering drugs from just 2018 to 2021, a near doubling of the revenues in just three years. Irrespective that this demonisation of cholesterol may well be wrong and that the statins etc. might not be as safe or effective as marketed, there is massive money to be made here. Sales of Lipitor (made from a chemical, atorvastatin), the biggest-selling drug, pulled in a staggering £125 billion in sales from 1996 to 2012.

The wonderful thing about eating these seeds regularly is that by repeatedly cleaning the digestive tract in its entirety, the build-up of all toxins is never allowed to happen, ulcerative risk factors are purged and hypoxic regions are avoided, preventing cancer from gaining a foothold. This is such a mind-blowingly simple preventative measure for a whole bunch of digestive ailments. The only caution is that it is important to always have sufficient water, otherwise what is known as a 'psyllium plug' may occur, being the polar opposite of the benefits achievable by sensibly administering psyllium.

This rather nicely shows that it is indeed the whole *Plantago* family that has so many health attributes, with each species offering a slightly different capability here and there, but all positive.

Farmers need to become the guardians of the soil

I have an asparagus farmer for a neighbour, who is somewhat stuck in his ways. We've exchanged harsh words, with him screaming at me about how on earth I thought all the food could be grown to feed people without pesticides! Unsurprisingly, he was a sales representative for Monsanto and the like before he switched to farming. Rest his soul, he passed away last year.

This use of pesticides and the like is the consensus of most farmers, however, the supposed guardians of our soil. They are so brainwashed that they can see no danger in growing our food while spraying the most deadly toxins invented by man on their fields! I hope you can see the insanity in this statement. This farmer's spraying hospitalised one of my children, who had to be rushed to hospital and placed on a ventilator because the pesticide being sprayed had done exactly what it was designed to do in wartime—knock out the respiratory system of a soldier. So of course those beetles don't stand a chance, as neither do the other insects, birds and bees. Nor did my two-year-old son.

So, this spraying is very personal to me but what's it all got to do with plantain? Well, in mid-April the crop sprayer covers the fields with a ghastly mist and within a couple of days the ground turns a deathly yellow where the poison saturates the soil. It's such a sad sight, this hue of death just as the asparagus starts to poke up. A fortnight after the death-sprayer cometh, a plant is already emerging through the yellow crust of slaughter, forming a protective barrier between the poisoned soils of the farm and the surrounding land—it's plantain. In complete chemical defiance and abundance, it grows in straight lines, a foot soldier of Gaia, a first line of defence, taking over areas of distress and bordering them, containing them, putting down a barrier that prevents the poisonous human desecration from spreading too far into adjacent soil. Phytoremediating. I love it.

We'll return to the soil in the final chapter, as it, beyond anything else, is the determining factor to a functional future.

* * *

These four plants—nettle, dandelion, yarrow and plantain—are probably the commonest 'weeds' in Britain. The same word is also the commonest nickname for the following—

CANNABIS

(Cannabis sativa)

Human Plant numero uno?

Cannabis could be considered the odd one out of the seven plants, in that it is already well known, with many advocates, being utilised and recognised as holding many of the answers to enact a reformation of our cultural systems. It can be used as food, medicine, clothing, paper, building material, oil, plastic and as biomass for power stations, with future uses including batteries and carbon fibre. It pretty much does it all. If all the rhetoric and hyperbole being thrown about hemp and cannabis is to be believed, it could be argued that it alone was the plant to save the world. So, let's take a closer look to see what all the fuss is about.

It's been nearly forty years since Jack Herer's seminal *The Emperor Wears No Clothes* was published (1985), in which the historical evidence of hemp and marijuana's uses and legislation issues was collated. Hemp is the male plant, marijuana the female, and both are *Cannabis sativa*. With legalisation now spreading across the world, it must be seen as a victory for the trail-blazers in the legalise hemp and marijuana movements, although in many respects it feels a hollow victory, as the real winners will be the same feckless corporations we've already seen throughout this book, from pharmaceutical to petroleum, plastic and chemicals.

It has been very much a two-pronged effort to re-establish legalisation for both hemp and its assorted industrial uses, and another from the medical marijuana angle, while quietly following in their wake will be the almost guaranteed massive income and taxes available from recreational marijuana. The market potential for the recreational use is as big if not bigger than alcohol, and irrespective of the fact that up to 30% of adults may already indulge occasionally in their leisure time, historically and dogmatically, recreational legislation has never been viewed as a 'vote winner'. So, by over-egging and enlarging the industrial potential of hemp and the lives saved from medical marijuana, the topic is given an ethical and moral sheen that helps the other 70% of the public to forget all the anti-marijuana propaganda erroneously and repeatedly espoused to them over the past century.

Mary Jane, what's in the name?

Mary Jane is one of the plant's well-known nicknames, heralding from North America. It refers to the dried, smokable parts of the female plant—the bud and leaf—and is a play on words with marijuana. Some of the other popular names for cannabis, some varieties and its paraphernalia (among thousands) include Bhang, Blonde, Blunt, Bong, Broccoli, Bubble-hash, Bud, Buddha, Cabbage, Camberwell carrot, Charas, Cheeba, Christmas tree, Chocolate, Chronic, Citrol, Combustible, Cryptonite, Dab, Dagga, Ditch weed, Doobie, Dope, Durban, Ganja, Grass, Green, Green goddess, Hash, Hashish, Herb, Hot knife, Hydro, Jay, Jazz cigarette, Joint, Joy stick, Leb, Loco weed, Kief, KGB (Killer Green Bud), Kush, Mr Green, Loco weed, Nugget, Phatty, Pollen, Pot, Primo, Ragweed, Railroad weed, Rasta weed, Reefer, Reverend Green, Rocky, Shatter, Sinsemilla, Skunk, Slate, Smoke, Solid, Spliff, Squidgy, Swazi, Thai stick, Torpedo, Trees, Wacky backy, Weed and 420. After a Blast, Blaze, Hit, Hot box, Puff the dragon, Session or Toke you may be Baked, Buzzed, Fried, Roasted, Stoned, Travis or Wasted!

The language and words vary around the world, and the lingo rapidly changes as cannabis culture is very different in countries and states where it is legal and freely bought and enjoyed in public, as opposed to where it has to remain clandestine with penalties being a loss of freedom or even execution.

Some of the most traditional, popular producing countries of smokable cannabis would have to include Afghanistan, Hawaii, India, Jamaica, Lebanon, Mexico, Morocco, Nepal, South Africa, Thailand and Turkey. Many countries argue they produce the best, with some small regions growing some particularly potent strains covertly, and a tremendous amount of varying quality and strengths being grown in backyards, roofs and cupboards across the world as well.

Whether or not *Cannabis sativa* should be split into several closely related species has been hotly contested. The commonest one that many argue is a separate strain is *Cannabis indica,* a nomenclature that was used for some time to indicate a hashish plant from India, which tends to be shorter and bushier; it is botanically classed as *Cannabis sativa* ssp. *indica*. There is also *Cannabis ruderalis,* containing only a trace amount, about 1% of the part that gets the user high, of the psychedelic tetrahydrocannabinol (THC); this is officially classed as *Cannabis sativa* ssp. *ruderalis,* and grows as a wild species throughout Asia and southern Europe. Variations in distinguishing features of the plant, wherever grown, such as altitude, shape and size of leaf and aroma (linked to the phytochemical terpene family, of which THC is one), are in the main adaptations to climatic conditions and geographical location, and from a botanical perspective both are different varieties of *Cannabis sativa*.

The original homeland of *Cannabis* is also hotly contested, although agreement allows for a swathe of western Asia from the Black Sea around present-day Kazakhstan and Turkmenistan, south east across northern Afghanistan and India, southern China and South East Asia. There is, however, no botanical proof of this, with only scant archaeological evidence about the global spread of the plant being utilised, often as cordage.

Etymologically there are a few interesting words that come from cannabis. Notable among them is the word cane, especially when as a reference to a plant growing with a strong, straight stalk, called a cane because hemp was the number one crop of old. Its botanical *Cannabis* name derives from the Scythian for the plant *kanab*, which is probably where the Greeks got their name of *kannabis*. Another etymological link is from a light fabric made from hemp by the Dutch, which went by the name of *canefas*, again derived from *Cannabis* and from which the English word canvas is derived.

The actual name of the plant, *Cannabis sativa*, was first used in 1542, 211 years before Carl Linnaeus, in an illustrated herbal written by Leonhart Fuchs. Linnaeus collated a lot of extant names when deciding on his nomenclature—as clever as he was, he wasn't responsible for making up all the actual names, usually just shortening the multiple names of a single plant down to two words. It certainly wasn't as simple as this, though, in the cataloguing or in getting his system accepted by the various global horticultural and scientific fraternities, but it stuck eventually.

In the Old Testament, in its Aramaic translation, hemp is referred to as *q'neh bosm* or *kaneh bosm*, which in traditional Hebrew is *kannabos*. With the root for *kan* being reed (long, straight, stem) and *bosm* being aromatic, it means an aromatic reed. Seeped in controversy, as so much is to do with Jesus, the word 'Christ' means the anointed one, while the Christians are the anointed ones and an anointing was called a *chrism*—with cannabis being argued for some time now as being an ingredient in the anointing oils. Were Jesus and the Apostles actually spreading the word of cannabis oil? *Chrismation* was a way of administering healing balms and oils, so has there been a corruption or melding of these words, christened and chrism? Was the son of Joseph actually a famous anointer of the time, the best at *chrism* in *Chrismdom*?

According to many sources the name assassin comes from hashish, with Persian (Iranian) assassins committing their deeds while intoxicated after eating hashish. It's a nice story, but probably no more than that. The Assassins referred to may have come from a sect of Shia Islam, whose leader Hasan-i Sabbah, in AD 1090 infiltrated a castle and ousted the king of Daylam in a bloodless coup. The followers of this Hasan were called the Hassassin, and they did commit assassinations, reputedly with a golden knife, but they were very unlikely to have done so after eating hash as Hasan-i Sabbah was strictly against all intoxicants. The hashish-eating assassin stories were probably made up by the hashish-eating early Victorian European writers, as incredibly have a lot of 'facts' about the past, all being just elaborations and fabrications, often by drugged-out writers trying to make stories appear a little more sensational and accessible by mismatching real events.

"Pssst! ... you wanna buy sum smoke?"

Smokable cannabis comes as either hash or weed. Hash is a solid agglomeration of the resin obtained from either rubbing the female plant or drying, sieving and pressing it. Hand-rolled charas from north India is a fine example of a hash, with every female plant when mature having some resin to rub and collect, although being such a laborious and time-consuming method, hash is more commonly made with the entire plant shaken and sieved and compressed into blocks, with a higher quality obtained from a very fine-meshed sieve using the best buds (the female flower trichomes, the same word as for the individual stinging part on a nettle). Morocco supplies by far the most hashish today, although sixty years ago north India, Afghanistan, Pakistan and Kashmir were the leading producers. Not many countries make hashish from their cannabis plants, tending to use the female flower bud and leaf, as in Thailand, South Africa and Jamaica—this is weed or marijuana. Some outdoor wild plants growing in these three countries can grow to over eight feet tall, with over 500g of high-grade marijuana harvestable, while some farmed cannabis from California can achieve up to 3kg of top-grade bud per plant when grown in an optimum sunny climate. Here, the terpenes and secondary metabolites require the strong sunshine to develop better than if grown under artificial lights, although in London and other urban areas, legality and poor weather makes indoor hydroponic growing favourable, with a yield as low as 25g per plant.

The THC content ranges as a percentage of the total dried weight of the buds, from 1 to 4% in the wild to up to 35% from some growers in the legalised States and countries. The hydroponic varieties that have now begun to spread out, taking over local regional varieties—the landrace strains—are often called skunk (because of the potent smell), with thousands of different strains, going by some interesting names such as Buttercream Gelato, Cobalt Haze, Double Banana, Girl Scout Cookies, Gorilla, Hindu Kush, Northern Lights, Purple Haze, Shiva Skunk, Wedding Cheesecake, White Widow and Zkittlez.

The traditional smoking method is in either a hand-rolled joint mixed with tobacco or smoked by itself in a pipe. Vaping has become widespread more recently with THC oils and dried herb being smoked in a variety of electronic pen-looking contraptions. Water-bongs have been a popular way of smoking for some time, with the smoke being slightly cooled as it passes through the water; both hash and weed can be smoked this way, similar to a Middle Eastern hookah. There are many other types and methods, although it would be true to say hashish is smoked by a minority today, while these buds and leaves known as weed, once from a plant in Thailand, Jamaica or Eswatini (Swaziland), are now more likely to come from a roof in Walthamstow or a polytunnel in Devon.

Hamp or hemp

I live in the county of Sussex, in the south east of England, an area once famed for its vast hemp production. It's an image you definitely wouldn't get from a quick drive through the present landscape, or a thorough ramble across it for that matter either.

Not helping in understanding the position of hemp in the past in England is the fact that the etymologists are strongly divided in opinion about what means what, determined still, it seems, to eradicate this plant from history. Take, for instance, the county of Hampshire, as 'hamp' is often documented as meaning home, said to derive from ham used to signify a small village or settlement or just a home, and 'shire' being a rural area with an elected council. It is a useless name for a county, disclosing no information or individuality whatsoever, meaning 'home in a rural area'. However, Hampshire was particularly well known for its hemp, and the words for hemp in Anglo-Saxon and Celtic are *hamp* and *hanf*, and in Swedish it is *hamp*.

Taking into account the location of Hampshire and the fact that this county and the whole of the south coast of England was growing a considerable amount of hemp primarily for the naval industry, it makes perfect sense to assume Hampshire means 'the shire of hemp'. We've forgotten, or rather the history books forgot to mention, how central hemp would have been to many societies. A ship would have had several tons of hemp on board by way of the ropes, sails, sacks and uniforms, and easy access to this hemp was an absolute priority. It wasn't until researching this book that I had given any thought to the quantity of ropes used in the past and what these ropes were made from. Considering that in the eighteenth century each of those large ships had over 100 tons of hemp rope and sails on board, it's a surprise Roper is not a more common surname. As of 2020 there were 11,919 Ropers in the UK, being the 860th most common surname; by way of comparison there are 124,146 Millers and 139,740 Bakers, making them the 53rd and 40th most common surnames respectively. Smith (from being a blacksmith) is the most common name, although surely never the most common profession, with 747,960; second is Jones with 558,125 and third is Williams with 419,845.

With so few clues remaining, it is surprisingly difficult to prove that hemp was a crop of such prime global importance in the seventeenth to early nineteenth centuries. The few relevant place names around Britain still surviving tentatively suggest the way it once was so important, such as Hempsted in Gloucestershire, with sted or stead meaning place—hence Hemel and the several Hempsteads around the south and east of England. I have a Hempstead Lane and Hempstead Wood not far from me, clearly stating where a 'place of hemp' can be found. Hempton is a name in Norfolk and Oxfordshire, where ton means a farmstead, enclosed village or a manor; then there's Hemp Green in Suffolk. Scotland offers more survivors, with Hemphill, Hempisfield and Hempland being self-explanatory, while Hemp-buttis, Hempriggs and Hempy Shot leave a bit more to the imagination, but all attesting to its previous commonality.

In France, the local names for hemp have been preserved in regional dialects across the entire country, underlining the plant's importance a little over a century ago and for millennia before. There is far more preserved evidence linguistically from France, which is unsurprising as the climate throughout much of France is very accommodating to growing cannabis. The smallish south west central region of France, forming the Charente-Maritime provinces, has over 80 villages with hemp in their names. There is also evidence from France that cannabis was more popular than tobacco until as recently as the beginning of the twentieth century, no doubt because it is intoxicating and could be grown so easily and freely with no import duties.

Cannabis history

History is the period of documented evidence, be that carved on stone, or written in books, so roughly the past 5,000 years. It would be true to say that history needs to be viewed with an open mind, as the legendary Jean-Jacques Rousseau said: "The falsification of history has done more to impede human development than any one thing known to mankind."

Here's a true history of Cannabis presented as a wordy timeline—

BC

Cannabis, as with nettles, are nitrophilic, insomuch as they adore nitrogen-rich soil, particularly spoil heaps of human and animal excrement. Cannabis and nettles knew long ago where they would rather be growing, so it wasn't much of a step for the humans to discover them, as it was more a case of these plants discovering us.

The oldest evidence of humans utilising hemp is from Taiwan from the 12,000-year-old excavated village site of Yuan-shan, also one of the oldest evidences of pottery in Asia. The excavated pottery had been impressed with cordage to give the vessel some texture and decoration, with microscopic analysis revealing these impressions to be made from hemp cord. Also at this archaeological site a beating stick was found, used to pound the retted hemp in the preparation of hemp fibre.

From at least 8,000 years ago, the amazing site of the oldest large town/city in the world at the time, Catal Hoyuk in south east Turkey, where the remains of a burned house helped to preserve the remains of a baby wrapped in a woven hemp cloth. These two sites conveniently define the region of hemp's homeland, although it is only a matter of time and luck before more archaeological evidence is unearthed showing even earlier human utilisation of hemp from southern and eastern Europe.

The fashion of impressing pottery vessels with hemp cordage was still being practised throughout the fifth millennium BC, as attested to by the Chinese imprinted Yangshao pottery over 6,000 years after the Taiwanese pots. Hemp being utilised for cordage, twine and textiles is undoubtedly much older than this, but these dates stand until new evidence predates them.

In 2737 BC the Emperor Shen Neng, the Father of Chinese Medicine, a most genial fellow and expert farmer by all accounts, commissioned the first medical compendium, the *Pen Ts'ao*. This is also the first documented evidence of the medicinal use of cannabis, in which it was identified as being beneficial for treating constipation, beriberi, rheumatism, gynaecological disorders, absentmindedness and malaria. So impressed was Emperor Shen Neng with marijuana that he classed it as one of the *Super Elixirs of Immortality*. It was an affection that continued for millennia, whether that be for superior strings on bows to kill from a greater distance, producing the paper to hold all of their knowledge on, building the roofs to shelter and clothes to protect them or the seeds to fill their bellies, hemp was everywhere in a country the Chinese themselves called 'the land of mulberry and hemp'.

A cup of charred seeds was found in an archaeological dig in Romania, dated to the early third millennium BC, where a tradition of smoking cups emerges from the East (often tripod bowls, referred to as 'incense bowls') and had evidently spread to Romania soon after. The hard casing of the cannabis seeds often protects them in a fire, becoming carbonised but still recognisable, whereas the rest of the vegetative matter burns—as it was intended to do.

Some interesting thoughts can be had from the conclusions of extensive analysis of teeth from over 200 people buried in a cemetery north of Naples, Italy, Gricignano d'Aversa, dated to the Early Bronze Age (2500 BC to 1800 BC). It is vital information that could have been so easily missed if not for a few fortuitous microscopic-sized fibres of hemp recovered from calcified tooth plaque. These fibres matched perfectly the striated wear of 28 people's teeth, mostly women, who used their teeth as a tool in hemp preparation, as can still be observed among several hunter-gatherer and tribal groups today, as well as among nettle weavers in Asia.

* * *

Some time in the third millennium BC, the ancient sages of India came together to share and amass all knowledge held about health and plants. This knowledge remained an oral tradition until documented in two notable books, the *Charaka Samhita*, with its 600 drugs from plant, animal and mineral sources, and the other from ground-breaking surgeon Sushruta, who put parts of the oral tradition and his theory of surgery into writing sometime in the sixth century BC, with his *Sushrata Samhita*, which explained 100 different kinds of surgical instruments and 650 drugs. *Bhang* makes quite a presence (it being cannabis, whether smoked or eaten), being prescribed by Ayurvedic physicians for anorexia, asthma, bronchitis, constipation, cold, diabetes, gout, leprosy, malaria, menstrual pain, rheumatism, spleen disorders and tuberculosis (in Sri Lanka cannabis has never been illegal if purchased from an Ayurvedic practitioner). Northern India also has a rich tradition of sadhus, ascetic monks who have rejected the world to follow a religious path, usually of Hindu or Jain background. The image of a smiling sadhu, with his long beard and smoking a charas-filled chillum is a stereotype dating back to before the *Rig-Veda*.

In the *Rig-Veda*, a book of vedic Sanskrit hymns, dating from about 1500 BC, *bhang* is referred to as 'the healing herb'. A little later, from about 1200 BC, the Hindu sacred text *Atharvaveda* (*Science of Charms*) mentions the 'sacred grass' *bhang* as being one of the five sacred plants of India, used medicinally and ritually as an offering to the god Shiva. There is a further mention from about 700 BC of *bhang* from the Zoroastrian ancient Iranian religious text, *The Law Against Demons* by Zend Avesta. *Bhang* was used extensively throughout the entire region—hence Bangladesh, which translates as 'the land of *bhang*'. When hemp was criminalised and cut down the country began to experience catastrophic flooding and landslides owing to hemp's ability to prevent soil erosion and stabilise water tables, coupled with extreme rainfall and the onset of global warming.

Bhang was used in Tantric religious yogic sex acts, consumed an hour before to heighten the pleasure. The Hindu spring festival Holi sees *bhang* being heavily consumed, with a *bhang lassi* the most popular method—a *lassi* is a drink prepared a thousand ways, often with yoghurt or milk, spices, fruits and mashed-up buds and leaves of marijuana; *bhang* is also mixed with ghee (clarified butter) and sugar and made into a purple halva, chewy balls or even a *bhanga* chutney. The first Western account of *bhang* consumption comes from Thomas Bowrey, a merchant sailor and writer who travelled around the region in the 1670s, later writing *A Geographical Account of Countries Round the Bay of Bengal*. It would be true to say that the use of cannabis in India had been institutionalised for thousands of years, and still today more than anywhere else it is central to the orientation of their culture.

* * *

The ancient Egyptians also made use of cannabis, with mentions in the *Ramesseum III Papyrus* (1700 BC), the *Ebers Papyrus* (1600 BC) and the *Berlin Papyrus* (1300 BC) for both its fibre and medicinal qualities, such as a remedy to cool the uterus, use as an enema, for sore eyes and a medicated bandage. Pollen was found on the mummy of Rameses II, who reigned 1279–1213 BC, and even though no inebriating effects are documented, and the medicinal references are limited, there is running contrary to this the results of hair analysis from mummies dating to 1070 BC, all showing high levels of cannabis, which indicates significant consumption, and suggests it may have been the preserve of the rich. Cleopatra was given cannabis when she gave birth to Julius Caesar's son, Caesarion in 47 BC; his name means Little Caesar and is not a reference to either him or his father being born via Caesarean section, because they weren't. Also putting the plant to other uses, the Egyptians made a concoction of plants prepared into a solid wax to be rubbed on the body, unlike their Mesopotamian contemporaries who preferred oil.

Mesopotamia is the name of the region that runs roughly along the courses of the Euphrates and Tigris Rivers from south east Turkey to the Persian Gulf—in Greek it literally means 'between two rivers'—and has housed various civilizations from what is touted in the West as the first civilization, the Sumerians (c. 3100 BC) to the Assyrians, Hittites and Babylonians (ending c. 600 BC), all of whom appeared to make a lot of use of hemp and marijuana, with numerous depictions of a 'Tree of Life', which might or might not be a large cannabis plant with many ripe fruity buds growing from it. By 680 BC Mesopotamians were burning the flowers and making an oil from them for spiritual and medicinal purposes, including to help in the treatment of diseases of the chest and lungs, stomach issues, skin lesions, against lice, and for inflammation and other ailments. An interesting ancient Assyrian use of this oil was for a disease called the 'Hand of the Ghost', considered to be epilepsy today, bringing it in line with what has been rediscovered about the uses of this plant, as epilepsy and tremor conditions such as Parkinson's are diseases medical marijuana excels with. Several cuneiform inscriptions dating from the reign of the famous Assyrian King Ashurbanipal (668–627 BC), who ruled from Nineveh in the north to Babylon in the south, describe sacred rites, medicinal applications and recipes for cannabis incense, or *qunubu*, as it was called.

Further to the north east of Mesopotamia in Iran, from the land of the ancient Magi, the prophet Zoroaster (628–551 BC) had a religion named after him. The Zoroastrians are associated with the enigmatic 'soma', believed to be *bhang*, possibly mixed with opium or other psychoactive substances. Zoroaster would quite likely be called a shaman today, particularly as, according to the theologian Mercia Eliade (who himself sounds like an ancient philosopher but was actually born in Romania in 1907), Zoroaster gave himself to ecstasy with cannabis, and this is when he hears the voice of the one god, Ahuru Mazda. It is also around this time that hashish was either invented or at least became popular and culturally consumed across this part of the world.

* * *

An archaeological excavation of burial sites in north west China uncovered the oldest complete female plants of cannabis with flowering tops, thirteen in total, laid across a man with Caucasian features who died at about 35 years of age, 2,500 years ago. The site,

Turpan, was an important stop on the Silk Road, with over 240 burials excavated from a cemetery and some also having cannabis seeds and powdered leaves. For an unknown length of time cannabis was associated with Chinese shamanism, where serpents were carved into the stalks and used as wands for healing ceremonies. From the classic ancient Chinese *Greater Lord of the Long Life*, from about 300 BC a quatrain reads "First a yin then a yang, No one knows what I do, Jade buds of holy hemp, for the one who lives apart." Hemp paper was invented sometime in the first century BC—a significant cultural tool China held secret for a few centuries.

The Scythians, who ranged from Kazakhstan in the west to Siberia their homeland, and Mongolia in the east, as well as causing a ruckus in Macedonia, Mesopotamia and more, were very fond of marijuana. These horse-backed marauders were appreciators of the finer things in life, as exuberantly displayed by a 2,400-year-old elaborately decorated gold bong, still containing a resinous mix of cannabis and opium (Nepal used to produce in small quantities a hash called Temple Ball that was peppered with opium and available until the early 1990s).

The Scythians were little known apart from a few stories, with Peter the Great (reigned in Russia from 1682 to 1725) becoming besotted after having seen some of their unearthed gold, with the first major archaeological finds centuries later in the 1940s in the Tien Shan Mountains of Kazakhstan. Much of what was known about the Scythians until then had come from the Greek author Herodutus (c. 485–425 BC), as well as one piece of misinformation about the Scythians, in reference to them adding hemp seeds to fires to create a feeling of euphoria. Many archaeologists believe this burning of the seeds bit, corroborated it seems by all the charred hemp seeds found fairly extensively.

I once had a disagreement with a professor when studying archaeology/anthropology, who was trying to uphold this erroneous idea of burning seeds. I asked him if he had actually ever tried to burn the seeds as it was not the best smell and definitely instigated no euphoric effect. Coincidentally, or scientifically, just a couple of days prior while having a sauna after a squash match, I put a couple of grams of marijuana on the hot coals, resulting in not only a heavily euphoric effect, but also sending an incredibly strong smell into the air conditioning system, which quickly spread throughout the entire Student Union building, causing its evacuation! This hot coal method was employed by the Scythians, as correctly pointed out by Aeschylus (525–456 BC), who tells us the Thracians threw the entire plant on the hot coals—the Thracians were a Scythian clan from north of the Danube, inhabiting Thrace, now Romania.

Cannabis was used by the Scythians extensively in their shamanic rites, with them spreading the plant throughout their travels, getting as far as Berlin by 500 BC. A continuation of a culinary dish, said to be from the Scythians, is a soup called *semieniatka*, served in Poland and Lithuania, made from hemp seeds and eaten on the night before Christmas to assist the souls of the dead to visit their families. The Scythians are so called because their Goddess used a scythe as her weapon, an agricultural tool used to harvest the plant best associated with the Scythians, cannabis, adopted by the Grim Reaper who was only ever Mother-Time in disguise.

The earliest record of hemp being used in Britain comes from accounts of a Celtic Princess, Cambri Formosa, who was known to have taught hemp weaving and sowing in around 373 BC. Although not to be elaborated upon here, there is a convincing argument in favour of there being an intimate connection between ancient Goddess worship and the intoxicating female of cannabis, having already seen how Jesus was an anointer of the oil, with the associations extending as far and wide as Shiva, Kali, Freya, Cybelle, Dionysus, Aphrodite, Hera, Seshat, Asherah, Ishtar, Ishara, Kundi, Ninurta, Bushyasta and more ...

AD

Regarding those anointing oils that brought Jesus and his Apostles such fame. Cannabis was certainly being grown throughout Palestine, Lebanon, Jordan and Syria at this time, if not for millennia before. It is implicated when God instructs Moses in Exodus, chapter 30, on how to make a holy anointing oil with myrrh, cinnamon, *kaneh bosm*, cassia and olive oil. There's more to this anointing malarkey than merely splashing some water on the forehead.

Both the navies of the Greek and Roman Empires used ropes made of hemp because of its strength and it not stretching when wet, which would spell disaster for a sailing ship caught in a storm. Pliny the Elder (AD 23–79) makes mention of uses of the seed and root for health, also noting its versatility as a rope, which the Greeks got from the famed quality of Rhodanus (Rhône River valley, near to Lyon) and Colchis on the Black Sea (present-day Georgia); the Romans got a good supply of theirs from Babylonia.

Dioscorides, the physician from Nero's army (first century AD), listed cannabis in his most excellent *De Materia Medica,* differentiating the sexes and quite rightly stating: "The root being sodden, and so laid on hath ye force to assuage inflammations and to dissolve Oedemata, and to disperse ye obdurate matter about ye joints. Ye bark also of this, is fitting for ye twining of ropes." The Greek physician, Galen (AD 130–200), a century later was prescribing cannabis, noting in his *De Alimentorum Facultatibus* the Roman habit of eating pastries made of marijuana at banquets to promote hilarity, proving its use was not unknown, even if its documented evidence is scarce.

Hua-T'o, a Chinese physician around AD 220, understood the anaesthetic and analgesic effects of cannabis popularly administered via a wine, and he then performed quite complex surgeries, not surpassed in many respects until the twentieth century.

In France the Frankish Queen Arnemunde, in AD 570, was buried draped in hemp, gold and silk cloth. The Vikings extensively used hemp for their vessels' sails, ropes, fish-line, nets and undoubtedly clothing as well. It could have been the Vikings who introduced hemp to the North American east coast around AD 850, a period currently shrouded in mystery, although as further evidence continues to surface it is being shown that the American continent appeared to have been repeatedly visited and even inhabited from nations as far apart as Scandinavia, Babylon, Egypt, Polynesia and China.

From about AD 850 hash becomes increasingly popular in the Middle East, causing some consternation among the clean-living Muslims. Not long after, by about AD 875, paper is being made from hemp in Baghdad, helping to spread the word of Islam with the *Qur'an,* introducing hemp paper to Europe in around 1150 through Moorish Spain. The *Canon of Medicine* was written by one of the greatest minds from the Islamic Golden Age, Avicenna (aka Ibn Sina), in about AD 1000. *The Canon* makes many mentions of cannabis, hashish and *bhang* as an analgesic and an ingredient in tonics for many ailments; his was a standard reference book for over 650 years (even reprinted in 1973), although by the sixteenth century the popularity and abuse of hashish had tarnished its image.

The Middle East led the way with cannabis during this period. The Arabian collection of stories compiled during the ninth to thirteenth centuries, *1,001 Nights*, mentions the intoxicating and aphrodisiac effects of hashish, appearing in several of the Nights. On one particularly amusing night, a beggar falls asleep in the bath-house after eating hashish, sleeping till noon, dreaming of being in the arms of a young lady, only to be woken by laughter. The beggar is somewhat relieved he hadn't gotten to the deeply sexual part in his dream but is still rebuked: "Art thou not ashamed, O Hashish-eater, to be sleeping stark naked with stiff standing tool?"

Cairo was celebrated in its day for cannabis cultural congregations, as was Amsterdam recently. When the Moorish botanist Ibn al-Baytar was passing through Cairo in the thirteenth century on his way to Mecca he noted the Sufis' habitual use of hashish had enfeebled their minds. It was in a popular hashish centre in Cairo, the Garden of Cafour, that in 1253 the army invaded and cut down all the cannabis plants, burning them in an enormous pyre. The plant's popularity was unabated, then in 1378 the Ottoman Emir Soudan Sheikhoumi issued not just an edict against hashish eating, but had all the plants searched out and destroyed in Cairo and its environs, imprisoned or executed the farmers and pulled the teeth out of known users as a public spectacle. This still did not stop the use of this repeat-felon known as kief or kif, graded by bud quality and sieve size, with a proverb stating "Kif is like fire, a little warms, a lot burns."

* * *

The awesome twelfth-century Abbess, Hildegard of Bingen (1098–1179), wrote quite extensively of hemp in her medical work *Physica*: "whoever has an empty brain and head pains may eat it and the head pains will be reduced ... gentle and profitable in the stomach." Also a tradition with a long continuation was a cloth soaked in hemp and dried and used as bandages and sheets, as hemp is both hypoallergenic and antibacterial, with Hildegard suggesting "a towel prepared out of hemp laid upon sores and wounds works well since warmth is contained within it". The tradition continued until at least 1480, when a midwife's recipe for the bandage to be used after a Caesarian was a plaster made from three eggs, hemp cloth and Armenian earth (a type of clay).

In early 1303 Richard Puddlicott discreetly sowed hemp seeds outside of Westminster Abbey, doing so as he knew the plant was fast growing. As it grew it hid his antics of making a tunnel into the crypt, where he stole a part of the crown jewels, some of which turned up among local prostitutes, some in the Thames, causing an investigation after which Puddlicott was caught and hanged in 1304.

It appears that the Vatican chef to Pope Martin V (Pope 1417–31) cooked with marijuana, coating chopped leaves and flowers with breadcrumbs and frying them. By 1484, so incensed was Pope Innocent VIII with cannabis that he banned herbalists from using this unholy sacrament of satanic rituals, making no difference whatsoever to the public popularity of cannabis.

When Christopher Columbus brought future ruin to tens of millions of inhabitants of the Americas in 1492, the ships *Santa María*, *Pinta* and *Niña* each would have carried several tons of hemp by way of ropes, sails, cloths and clothing. Which is nothing compared to the vast quantities required by the sailing battleships a few centuries later, such as HMS *Victory*, commissioned in 1765. This ship took a staggering 60-acre forest of oak, about 2,000 mature trees (some sources suggest as many as 6,000) to build, with an astounding 53 miles of rope weighing in at well over 100 tons, and 4 acres of sails divided into 37 sails weighing easily 20 to 25 tons. In all there was about 150 tons of rope and sails, all having to be replaced every few years, at a time when the Royal Navy had about 300 large ships (not on the scale of *Victory*), adding to the countless private small fishing vessels around the country.

There is mention of Columbus introducing cannabis seeds to America, which he almost certainly didn't do, although he would have taken some, given the importance of always maintaining a source of ropes and sail—the inference from a couple of accounts is that hemp was probably already established throughout the continent. The earliest account of hemp in the Americas was from a Florentine explorer in 1524, Giovanni de Verrazano, who commentated on the hemp growing wild in Virginia as well as later observations of it growing in Canada.

The dearth of ethnobotanical knowledge and uses of cannabis among the First Nation Peoples suggests it was a novel plant to them, unlike the ample pharmacological and cultural evidence from the previous four plants and the two to follow. This either means the plant has never been utilised as it didn't grow there or the records have been deliberately censored. All those pipes of peace have (from residue studies) shown varieties of tobacco, with marijuana and other substances only found post-contact, adding weight to the late introduction to the Americas, although it doesn't discount a Viking or Chinese introduction a few centuries before Christopher Columbus got so terribly lost. The thesis of this book about humans evolving in localities of these seven plants is not diminished by the absence of cannabis from the Americas, as there has never been a claim (by anyone as far as I can find) of human evolution taking place in the Americas.

* * *

Roman and Anglo-Saxon finds attest to the continued use of hemp in Britain, although around 1000 AD pollen analysis from sediments of lakes suggests the agricultural practice of growing hemp hit a trough, with the grains of oats, wheat, barley and buckwheat all taking centre stage as staple food crops. It was not until hemp re-emerged as a vital naval ingredient that it reached its pinnacle again in the sixteenth and seventeenth centuries, being instrumental in forging the British Empire. All those sails and ropes had to be locally sourced initially, although as the British Empire and the number of ships grew so rapidly it was soon more cost-effective to source hemp from Russia, until the question of national security came up, as reliance on another nation for such an essential item was clearly putting the entire nation at risk, particularly in the days of potential naval invasion. Henry VIII (King 1509–47) had made it compulsory in 1533 to grow a quarter of an acre of hemp or flax for every 60 acres under cultivation of other crops to maintain supplies for shipping. However, the opening up of the Empire combined with the inhumane trade in slaves enabled Britain to source large quantities of cheap cotton from the late seventeenth century onwards; within a century cotton had replaced hemp.

The French priest, scholar, doctor and lawyer François Rabelais (1494–1553) deviously devoted three chapters to cannabis, calling it Pantagruelion in one of France's first great satirical novels, *Gargantua and Pantagruel*, telling the tale of two giants; he heavily influenced Jonathan Swift's 1726 tale *Gulliver's Travels*. Rabelais declared life unimaginable without this plant, with no work possible for the writer or lawyer with no paper, no water drawn from the wells or church bells rung without hemp rope, no bed clothes, curtains, sheets or boats sailing people, goods and armies—his father was a hemp farmer and wine maker so perhaps he was a little biased, although hemp was clearly important at this time in history.

With little to no mention of the inebriating qualities of cannabis in Britain up until the sixteenth century, the same could not be said for hashish's continuing popularity in Western Asian culture, as indicated in the epic poem written about 1550, *Benk u Bode* by Mohammed Ebn Soleiman Foruli of Baghdad, where the two most popular intoxicants of wine and hashish engage in a form of dialectical battle. With this and several contemporary physicians of the Islamic world all verifying the medicinal benefits of cannabis, it truly was flying the flag. Because during the 1600s hemp was indeed utilised to make flags, among a range of other things such as napkins, sheets, tablecloths, towels, curtains, sacks, horse blankets and bedding. The English King's horses have hemp bedding, for the same reason that hemp makes superior fabric, as the hollow fibres allow for clothing and bedding that has better insulation.

Robert Burton's *Anatomy of Melancholy* of 1621 suggests marijuana may be able to cure depression, clearly alluding to utilising the female buds. This period is characterised by an abundance of hemp, with fields upon fields a-swaying in the breeze of a green and plentiful land, and it's unfathomable to imagine that Mary Jane wouldn't have been as energetically consumed as Harry Hempton was in being grown and made into rope and cloth. There has been mention that Shakespeare (who would have read *Gargantua* and *Pantagruel*) was a marijuana smoker from pipes found in his garden and around Stratford-upon-Avon, which is circumspect at best, with Shakespeare dying in 1616.

This is the same year Nicholas Culpeper (1616–1654) was born, and Culpeper was a man who very much embraced smoking a pipe and can be relied upon to have been aware of what was fashionable at the time. He mentions hemp seed as an aid to clearing up dry coughs, and also tells us the Dutch made an emulsion of the seed oil "for continual fluxes, cholic, humours in the bowels, [it] stops bleeding in the gums, mouth and nose and will destroy worms in man and beast". This defence was taken by a French man in 2010, who claimed he fed the cannabis to his ducks to rid them of parasites and for their general health, "Honest, guv!" The ducks were indeed healthy and he got off lightly with only a 500 Euro fine and a month's suspended jail sentence for possessing a hefty 5kg of marijuana.

During the sixteenth to nineteenth centuries, explorers, adventurers, writers and anthropologists travelling south of the Sahara commented on the extensive hemp eating and smoking of the tribes, known as dagga, from the Zulu and Bantus in the south to the Pygmies (no offence intended by using this word) in central Africa. The Pygmies claim to have smoked cannabis since the beginning of time and for some groups it is their only crop, with the rest of the diet foraged. Sadly, their numbers have been decimated in the past 40 years with repeated massacres committed against them. Interestingly, the name for cannabis in the Republic of Congo is mbanga, strongly suggesting an Indian influence, where it's known as bhang, possibly introduced by Indian merchants. The antiquity of sub-Saharan African hemp is a bit of a mystery, although it is probably far older than the currently accepted 700 years or so, dated from fourteenth-century smoking pipes found in Ethiopia.

* * *

With cannabis appearing in many dispensaries and pharmacopoeias throughout the eighteenth century, it began to be seen more as a medicinal plant, with the State of Georgia offering free hempseed to farmers in 1767 to encourage more plantations. Not long afterwards, confirming the quality of hemp paper, the draft of Thomas Jefferson's Declaration of Independence in 1776 was written on it (the finished article is written on hide as was the tradition). Jefferson (1743–1826) wrote that hemp "is of first necessity to the commerce and marine, in other words to the wealth and protection of the country".

A Jefferson contemporary, Thomas Paine (1737–1809), a political troublemaker from Lewes—he was actually from Thetford, Norfolk, but it was in Lewes he coalesced much of his philosophy—helped with the Declaration of Independence, and was confident America could win a war against the British because 'hemp flourished here'. The words were said from memory, as Paine would have been very familiar with the sight of fields of hemp across the Sussex countryside before fleeing to America.

The first US President, George Washington (1732–1799), was an enthusiastic supporter of the native Indian hemp, growing several varieties around his estate and being fully aware of the industrial uses of the fibre and oil. There is some debate whether the female plants, which he complains vociferously of not always separating from the males, were ever used recreationally. He wasn't such a keen supporter of the Indians themselves, however: "The immediate objectives are the total destruction and devastation of their settlements. It will be essential to ruin their crops in the ground and prevent their planting more."

It's a shame some of the third President Thomas Jefferson's words were not adhered to: "The best hemp and the best tobacco grow on the same kind of soil. The former article is of the first necessity to the wealth and protection of the country. The latter, never useful"—although an awful lot of money was made from it, not to mention the hundreds of millions who have died from lung cancer as a consequence of these tobacco profiteers. Some of Jefferson's other words displayed a darker side, notably his views of the indigenous inhabitants: "This unfortunate race, whom we had been taking so much pains to save and to civilize, have by their ferocious barbarities justified extermination."

By 1840 Persian pharmacies are selling hashish in America, with it added to the US Pharmacopoeia in 1850; until 1915 it could be purchased in pharmacies and general stores in various medicinal concoctions from syrups to tinctures.

Meanwhile, back in Europe, Napoleon Bonaparte was so incensed upon finding out the French troops returning from Egypt had brought back the tradition of hash smoking with them, that he imposed an immediate ineffective prohibition, as hash smoking took off, becoming a very popular social drug. It was also indirectly hemp that led to Napoleon's ultimate downfall. As it was logistically not possible for him to invade

Britain given the superiority of the Royal Navy, the next best endeavour was to stunt Britain's navy by cutting off the hemp supply from Russia. It wasn't just hemp, and Bony attempted a European-wide trade blockade against Britain, with some limited economic success. However, in 1812, against his better judgement, Bonaparte led an unsuccessful mission to Russia, being defeated catastrophically, as is so often the case, by the Russian Grandfather Frost.

British medicine saw cannabis reintroduced in 1842 by an army surgeon, Sir William Brooke O'Shaughnessy, who had become acquainted with it while serving in India. In Victorian times cannabis was extensively used in Britain to treat convulsions, epilepsy, menstrual cramps, tetanus, rabies and as a sedative.

Physicians continued to write about its benefits throughout the nineteenth century, one example being Dr Jacques-Joseph Moreau's *Hashish and Mental Illness* in 1846. Moreau (1804–1884) also supplied the Parisian Club des Hashischins with a hash paste, leading no doubt to some extraordinary discussions among some of France's leading luminaries, such as Baudelaire, Hugo, Dumas, Balzac, Delacroix and Gautier. These top-notch French cultural ambassadors, dressed-up in Arabic clothing, added a paste made of hashish, cinnamon, cloves, nutmeg, pistachio, sugar, orange juice and butter (common ingredients used in *bhang* across India) to their coffees, helping to well and truly lift the lid off their minds.

Edmond de Courtive published his doctoral thesis, "Haschish", a couple of years later in 1848, using himself as the subject for a range of psychological assessments, at a time when it was not just the French but many throughout the European intelligentsia had taken up cannabis appreciation. In 1857 Fitz Hugh Ludlow published *The Hasheesh Eater*, a sincere psychological autobiography of the effects and addiction, a work often overlooked or seen as the poor cousin of Thomas De Quincey's earlier (1821) *Confessions of an English Opium-Eater*, which is more bravado and hedonistic, although both clearly demonstrate the popularity of eating materials that alter one's state of consciousness throughout the nineteenth century.

In 1890, while the Greeks are making hashish illegal, Queen Victoria's private doctor, Sir Russell Reynolds, wrote of cannabis "that when pure and administered carefully, cannabis is one of the most valuable medicines we possess". True to his word, he prescribed it to Queen Vic to relieve her menstrual cramps, most likely in the form of a tincture, with just a few drops being an effective dose. This is becoming popular again in the States that have legalised recreational cannabis, especially the edibles, but it has all been done before, with the wonderfully named Gunjah Wallah's candied hash sold in America in the 1860s being not too dissimilar to the THC jelly babies on sale today.

With continued scientific work showing positive results and repeated recommendations from doctors across Europe throughout the mid to late nineteenth century, by 1880 and for the next twenty years cannabis surfed a wave of accolade and popularity. This was helped in no small part by the fledging pharmaceutical companies, all making medical-strength oils and tinctures readily available to the public, such as Merck in Germany, Wellcome in Britain, Bristol-Myers Squibb and Eli Lilly in America. Fluid extracts, powdered extracts and gelatine-coated one-gram extract pills were given for anything from babies' looseness of bowels to colds and coughs, cystitis, depression, menstrual and nervous disorders, mental exhaustion, migraine and neuralgia. They were sold under a plethora of names, from Cannabis Compound, Infant Relief, Indian Cannabis and One Night Cough Syrup to Dr Poppy's Wonder Elixir. Cocaine (invented 1859), opium and heroin (invented 1874) tinctures and tonics, it must be said, shared a similar breadth of uses, popularity and producers at this time also.

In 1876 Philadelphia hosted a huge Centennial Exposition that included a Turkish Hashish stand; with wafts of hashish hookah smoke filling the air, the effects would have been interesting to observe from what was a very popular stand. By 1883, every major American city had legal hashish-smoking parlours, from New Orleans to Chicago and New York (with an incredible 500 parlours). These parlours continued throughout the alcohol prohibition, actually outnumbering the famous yet illicit alcohol-serving 'speakeasies'. Who would have known?

Into the twentieth and twenty-first centuries

In the first quarter of the twentieth century there was little indication to suggest that things would turn on the cannabis plant in quite the way they were about to. So at ease did hemp fit into life that in 1914 a $10 bill was issued that not only depicted hemp being harvested, it was also printed on hemp paper. On the other hand, that very same year, the 1914 Harrison Act outlawed marijuana, a measure made as a response to the Mexican Revolution of 1910, which saw an influx of Mexicans cross into the United States, bringing with them their custom of smoking marijuana.

Britain weighed in with the 1925 Dangerous Drugs Act, with its main concern being control of opium, heroin and cocaine, and only a tiny but pertinent mention of *Cannabis sativa* by way of "Indian Hemp and all resins and preparations based thereon". Hashish was only on the agenda at the Convention because it jarred with Egypt and Turkey's interpretations of Islamic law and consumption of intoxicants, so recreational use of marijuana became prohibited in Britain. In 1937 the US Congress passes the Marijuana Tax Act, making both hemp and marijuana illegal. This Act more than any other was instrumental in global legislations against Cannabis.

Q. Why was *Cannabis sativa* made illegal?
A. Industrialists protecting their money.

Stories had begun to appear in the US press about the dangers of this new drug marijuana a few years before the 1937 Marijuana Tax Act, with a whole hotbed of racist propaganda about the dangers of the black man's smoke, who will get your women into jazz, share a hit of the evil weed and then lure them to have sex. Which is exactly how the newspapers equivalent to *The Sun* and *Daily Mail* of their time in America reported this weed menace for over a decade, led in no small part by the newspaper owner William Randolph Hearst, who spread continual smear campaigns against marijuana, lifting the word marijuana from the Mexican slang for the plant. Hearst had a vested interest in demonising cannabis, owning timber and paper-making businesses. Hemp was not only more economically viable it was also definitely a better environmental alternative than chopping down loads of trees, but Hearst didn't want this to become public knowledge, and as he ran several papers it was easy to spread malicious gossip. All this media propaganda against marijuana came to a head in 1936 with the release of two films, seen today as comedies. The first was *Marihuana: Weed with Roots in Hell*, with advertising posters depicting needles, wild parties and a woman getting injected! The other was the classic *Reefer Madness*, advertised as weed from the Devil's garden. Both were released with absolutely perfect timing, as it was undoubtedly these films that brought the topic to the public's eyes and smoothed the route to it being sanctioned.

The 1937 Marijuana Tax Act made it illegal to sell or grow any hemp or marijuana without a licence, which was never publicly available, and with this, in one fell swoop, the plant became criminalised. A part of the testimony for Congress to pass the 1937 Act was a compilation of articles that had appeared in newspapers owned by Hearst—the same as *The Sun* being primary evidence in a court of law today! The Hearst livery today includes 49 newspapers, 335 magazines and stakes in television companies around the world, with a famous granddaughter socialite and actress Patty Hearst who has an interesting story of her own of kidnappers, urban guerrillas, bank robbery, jail and a pardon from President Carter.

The Act had been orchestrated by the head of the Federal Bureau of Narcotics, Harry Anslinger, who had been fired up by the nonsense stories made up by Hearst and his newspapers, so in 1935 he set about legislating against it, culminating in the 1937 Act. Anslinger had been writing anti-marijuana articles for *American Magazine*, with one, *Marijuana Assassins of Youth*, released as a film just before the Act was passed. This was all done while ignoring the American Medical Association, who saw no evidence that marijuana could ever be termed a dangerous drug—rather they said future studies could show it to have a substantial range of medical uses. Pointing out this threat to the pharmaceutical industry was probably just another nail in the coffin for *Cannabis sativa*.

Cannabis was beginning to pose a serious threat to not only the paper and pharmaceutical industries but also the building, clothing, food and oil industries, as well as the imminent public release of petroleum plastics. A lot of old-school industries were to get their feathers seriously ruffled if hemp was to be rolled out as planned. Conveniently, the media played a critical role with false propaganda—easy when the people believe the news they read as being factual, and never more so than today. Which brings to mind a quote by the great man Malcolm X:

> This is the media, an irresponsible media. It will make the criminal look like he's the victim and make the victim look like he's the criminal. If you aren't careful, the media will have you hating the people who are the oppressed and loving the people who are doing the oppressing.

Not long after being criminalised, various U-turns saw hemp being utilised for the American WWII efforts, with each new naval vessel requiring a minimum of 20 tons of rope, over 6 miles in length, with hemp still offering the best solution. Of course, the Tax Act resulted in a supply shortage, so the USDA (US Department of Agriculture) made a *Hemp for Victory* publicity film encouraging farmers to grow hemp for the war effort to make all the necessary rope and even parachute harnesses, one of which reputedly saved the life of George Bush Snr, who somewhat ironically fought so hard to eliminate it as President. Although this *Hemp for Victory* film is available on the internet, years later the

USDA vehemently denied ever making it, no doubt a little embarrassed when accused of duplicity.

After WWII the US imposed a strict ban on cannabis on all defeated nations. In this way it could not only undermine any future attempts to remilitarise (selling it was an easy way to raise funds), it also opened the door to America's new weapon, economic subservience, by way of the World Bank, International Monetary Fund and consumerism.

* * *

A nation hit hard by this American drive against marijuana was Japan, with strict legislation in place today only thinly veneering a recent past far more dependent on cannabis. In 1948 there were 25,000 cannabis farms, today there are fewer than 10. Japan required vast quantities of hemp for their navy and shipping, as well as all the clothes made from hemp, as were parts of the Samurai outfit and the twine for stitching the leather and metal together. With the death of the last Samurai in 1877, the arrival of cotton and the US insisting on a ban in 1948, hemp fell by the way. However, a curious point was made when Emperor Hirohito died in 1989 (he was Emperor from 1926), as his son Akihito had a special hemp gown made for his crowning by Shinto farmers who had clandestinely grown the hemp for such an occasion. Hemp to the Shinto Buddhists is a symbol of purity, with all brides once wearing a veil of hemp to signify just that. Continually rocking the boat, Emperor Akihito married a commoner and was the first Emperor to abdicate, as well as possessing, like the Dalai Lama, a truly comforting smile. Emperor Akihito's gown was a message of continuity, sadly missed by most.

Japan was once a big cannabis-smoking nation, with a rich history of some very elaborate *kiseru* pipes—a name derived from the Khmer word for cannabis, *ksher*, used predominantly for smoking marijuana in. The wild cannabis growing abundantly in some regions of Japan has a high 4% THC content, equal to the cultivated varieties from Thailand, South Africa and Jamaica, where the wild strains only have 1% to 3% THC. It seems likely that Japan was enjoying the inebriating qualities from the beginnings of the Jomon period 10,000 years ago, when hemp was utilised for ropes and impressing designs onto pottery vessels. In an eighth-century collection of poems, *Manyoshu*, there are several mentions of cannabis. The Japanese even had similar social customs, dances, rhymes and school songs as other parts of the world, alluding to growing up straight and strong like a hemp plant and encouraging growth. This continued use came to a crashing halt a few years after WWII with US restrictions and a UN convention (1961), converting a cultural continuance of several millennia into a strict social taboo.

Underlining hemp's importance historically in Japan and the artistry of the Japanese language, there is a 320-year-old house, which is not only made of hemp, it was also used for the preparation of hemp, and is now rather fittingly a hemp museum; it is found in the village of Miasa Mura, which translates as 'beautiful hemp village'. Surreptitiously and forward thinkingly, in a secret field in Japan a thousand plants are grown every year with the seeds saved to guarantee a quantity of seed for when it is once again legal.

* * *

A seismic modern cultural impact was made by marijuana in a sixth-floor hotel suite in Manhattan on 28 August 1964, when Bob Dylan met with The Beatles, and behind a door with a 20 policeman barricade, he pulls out a bag of weed and changes their musical trajectory with a few shared joints, continuing a subcultural movement and helping to shape the rest of the 1960s by laying the final foundations of the hippy era begun with the Beat poets in the 1950s.

The powers-that-be in Britain were definitely becoming worried that drugs and pop stars held far more public power than the elite of the politicians and rich. This peace and love "Tune in, turn on and drop out" message was amplified by the leading musicians and writers of the time, who even had the audacity to place a full-page advertisement in *The Times* newspaper calling for marijuana to be legalised for medical research and personal possession; the letter was signed by The Beatles, 2 MPs, 59 physicians, clergymen, scientists and writers. The government responded by setting up an Advisory Committee, which concluded that cannabis was less damaging than all other drugs.

This was the last straw for the stuck-in-the-mud elite, who completely ignored Baroness Wootton's report and stepped up the gears to contain this drug menace from freeing too many of the public's minds—the word 'government' after all does mean to *govern*, 'control' and *ment* as in *mentasm* 'mind', control your mind. Arrests rose from just 4 marijuana busts in 1945 very slowly to 235 in 1960, but after that publicised hotel session in 1964 cultivation was made illegal in Britain busts increased to 626 in 1965. From then the figure went crazy, with a steep rise to 14,119 busts in 1973 continuing to a stratospheric 72,392 in 1994! At least 85% of these convictions were for personal use, with tens of thousands actually going to prison for possession, in many cases for as little as 10g. How much has this cost in public money and lives shattered from loss of employability? Families were split apart because an adult chose to smoke a plant that has been with us for millennia, as shown by the extensive worldwide history above, instead of the socially acceptable, violence-provoking, tax-creating drug of choice, alcohol.

* * *

With the surge in popularity of marijuana from 1965, many young Westerners started to travel to Morocco, India, Nepal and Afganistan to smuggle some hash home. With stories of Westerners even taking metal sieves all the way to Afghanistan to make their own hash with, such antics soon attracted the authorities' attention and by 1973 hash was made illegal in Afghanistan, while Nepal banned exports with a $4 million 'gift' from President Nixon.

This would have made a small dent in the cannabis varieties available on the menu from Amsterdam's first coffeeshop, Mellow Yellow, which opened its doors in 1972. Today there are over 200 coffeeshops in Amsterdam, although sadly Mellow Yellow isn't one of them; quite where they managed to source their cannabis was always a mystery, as importation has always been very much illegal and carries hefty prison sentences in the Netherlands.

Throughout the 1980s Morocco became the largest hashish exporter, although with demand increasing so quickly from American interest the quality quickly decreased (with additives from oil, henna and even old 78rpm records) as demand far outstripped supply.

This scenario was repeated in northern India as charas supplies ran out in the late 1990s, with too many Israelis (that's a fact, not antisemitism, because straight after their army service many young Israelis went off to Goa and tried to reprogramme themselves with trance music and charas) and European hippy entrepreneurs starting to take half a kilogram home with them to sell to raise the money to return for more trance dance and chillums.

Today, Morocco supplies around half of the world's supply of hash, employing nearly a million people in this illicit trade. However, because of the paid-for purge during the Reagan and Bush administrations' 'War on Drugs' in the 1980s world supply of Moroccan hash was severely affected. As a response to this, Dutch horticulturalists came up with 'skunk' weed with its very high THC content and the bonus of being able to be grown in Holland hydroponically, minimising the business threat to their booming coffeeshops. It is a rather ironic twist to the War on Drugs, although what the psychological effects might be to a generation smoking what is essentially a very different plant from the traditional sun-grown *Cannabis sativa* eaten and smoked for millennia have yet to be scientifically researched. The THC content has risen from 2–4% in 1990 and before to 30% in 2018, although you don't get 10 times as high on the 30% THC skunk, while the 3% Thai weed or Moroccan hashish can get you just as stoned and susceptible to a psychotic episode.

In 1985, nabilone, a synthetic THC made by Eli Lilly, received Food and Drug Administration (FDA) approval and appeared on the market, the first cannabinoid-based medication. A year later dronabinol got FDA approval for cancer patients as a relief from the severe sickness caused by chemotherapy; its use was extended in 1992 for HIV/AIDS patients suffering from severe anorexia. Two drugs with FDA approval and still marijuana is classed as a Schedule 1 drug as it has "a high potential for abuse with no accepted medical value". Erm! This duplicity is doubled, if not quadrupled, when one finds out that the US government actually holds the patent (number 6,630,507) for some drugs made from any non-psychoactive part of the plant (the CBDs)!

* * *

In 1914, California was the first State to ban marijuana and in 1996 was first again, this time to relegalise it, initially just medicinally for patients of cancer, AIDS and other serious illnesses, upgraded to fully legal today. In 2003, Canada became the first country to have countrywide approval of medical marijuana. At the time of editing this in 2024, States such as Idaho, Indiana, Kansas and Nebraska were the only ones to still have a total marijuana prohibition on recreational, medical and cultivation, although they had relented on CBD and will soon decriminalise further. Other States allow possession and/or medical use and 18 States allow small cultivation (from 3 to 12 plants) and possession (up to 4oz). This is soon to change, no doubt, as it is always the money making the most noise, loudest of which is the taxes from marijuana sales, totalling across all legal States over $1.4bn, with varying rates from State to State. Highest is Washington at 37% and the lowest Maine at 10%, and overall averaging out at about 25%. This sector already employs over 300,000 people in the US, a figure that will rise significantly as more States decriminalise.

The rest of the world has also been active, with the following countries passing laws that allow cannabis use for either medicinal and/or recreational use and/or cultivation: Argentina, Australia, Austria, Barbados, Belgium, Belize, Bermuda, Bolivia, Brazil, Chile, Colombia, Costa Rica, Croatia, Cyprus, Czech Republic, Denmark, Ecuador, Estonia, Finland, Georgia, Germany, Greece, Ireland, Israel, Italy, Jamaica (virtually fully legal), Lithuania, Luxembourg, Macedonia, Malta, Mexico, Moldova, Netherlands, New Zealand, Norway, Paraguay, Peru, Philippines, Poland, Portugal, Romania, Russia, San Marino, Slovenia, South Africa (nearly fully legal), Spain, Sri Lanka (has always been legal from an Ayurvedic practitioner), Switzerland, Thailand, Turkey, Ukraine, United Kingdom, Uruguay (nearly fully legal), Vanuatu and Zimbabwe.

That certainly is quite some history, included here to show what an ass the law can be and how it can be manipulated by money, with more instances shown later. However, with much still to cover in this chapter let's move straight on to looking at some of the touted industrial uses from hemp, with in excess of 25,000 reported—don't worry, I'm not going to list them all!

Building a better world—Hempcrete

In many respects, it is the industrial applications of hemp that generate such enthusiasm among its advocates. It is this premise that makes up the bulk of Jack Herer's book *The Emperor Wears No Clothes*, with its main contention being that the industrial uses are worth potentially billions for a plant that has definitely had unfair prejudice thrown at it, from both government and industrialists, consequently stalling development and uptake as well as missing out on the environmental benefits. It's the same scenario with the unbelievably slow uptake of the transition from fossil fuels to solar, wind or Nikola Tesla's mysterious magic boxes for sources of power. With the biggest players in the economic world being the petroleum corporations, their pollution is deemed necessary and unavoidable. If ethics, health and longevity mattered, none of the industries polluting our beautiful planet would still be in business, but instead they have left our soil, air and water a cancerous, chemical cauldron—because ethics and morality have no place in a business world ruled by the myopic, disconnected pursuit of short-term profits and the erroneous and corrupted use of the word 'economics'. We need a return to *oikonomia*, 'good household management', and we literally could build loads of houses and a whole better world with hemp.

When it became 'fashionable' to be carbon-aware as a way of tackling climate change earlier this century, the construction industry must have taken a long, deep breath waiting for the reforms to be published. They just happen to be one of the largest emitters of CO_2, with nearly 10% of global emissions coming from the concrete industry alone. The reforms never materialised, but quite a few showcase buildings and estates were built with remits of environmental sustainability and low carbon as their focus.

For example, in 2006, 1,000 cubic metres of cast in-situ hempcrete and 90,000 hempcrete blocks were used in the construction of the walls at the Adnams brewery's innovative distribution centre in Suffolk. Hempcrete was deemed the best material for not just the general construction, it also provided the best, cleanest and safest environmental conditions to store vast quantities of beer in. It is a superb building, currently the biggest in the UK made from hemp, as well as having the biggest living sedum roof in the UK at 0.6 of a hectare. The energy savings alone for this centre were in the region of £49,000 for 2006 (which would have amounted today to over £500,000 in savings) as well as the construction saving of at least 450 tons of CO_2.

A few years after in 2012 (planned and built from 2005) Marks & Spencer built a flagship store in Cheshire, UK, using a type of hempcrete (called Hemclad) for cladding the walls, as well as incorporating hemp insulation. The same year also saw The Science Museum build a Hemclad storage facility to house archives from the Science Museum, National Railway Museum and National Media Museum, including paperwork, models, microfiche, films, hard drives and steel. Hemclad was chosen as it created the most stable conditions for storage, as well as being the best thermally efficient building material, providing clean air and being environmentally sustainable in its construction and future.

A development in Swindon, completed before the M&S store, saw a collaboration between Kevin McCloud's (presenter of *Grand Designs* on Channel 4) Hab Housing and housing group GreenSquare build 42 town-houses with all external walls being made from Hempcrete in 2011. The development had many environmental and social ideas implemented, with all residents given a share of a small public space and a copy of an *Edible Landscape Manual*, achieving a good amount of press attention at the time. Sadly, the company has since gone into receivership.

There has been a very slow rate in developments, with thousands of homes built in Europe and America from hemp already this century. Hempcrete only made a (re) appearance in the mid-1980s, being initially patented in France as Isochanvre, having caused the innovator, Madame Pierre, many hurdles and barriers. Even today she can build no more than 300 houses a year because of legislation and the French government having a monopoly on hemp production. That said, France's first public building made with hempcrete blocks has recently been completed, a sports hall in the town of Croissy-Beaubourg near to Paris.

A Dutch company have made what appears to be the first economically competitive solution (i.e. it's cheaper), with a house offering better insulation, durability, cleaner internal air, carbon sequestering and being earthquake proof—not such a problem in the Netherlands unless fracking starts there! Creating a perfect flat-pack house, ready to send around the world for not only relatively cheap housing, it offers a long-term solution of relief for victims of natural catastrophes. A tent is cheaper, but not a real solution, even if canvas was once always, and still can be made from hemp.

An Australian company looks like it might be on the path to innovatively incorporating every stage from harvesting to a finished 3D-printed hempcrete-panelled house. They haven't as of yet built one, but with the dropping cost of hemp as more countries start to grow it, a super-cool building can now be printed and erected within a fortnight. Hopefully, by 2025 there should be some seriously interesting eco-buildings being built, especially now Elon Musk has invested in Boxabl and their foldable $10,000 homes.

Notwithstanding the celebrity endorsements and the environmental benefits, there has still been an extremely slow uptake, and even though the UK showcase examples all achieved better than expected positive sustainable environmental results, these remain pretty much one-offs in the UK. Clearly, the main problem is that the mainstream construction industry has no intention of changing the systems that are in place—an industry worth in the region of a staggering £8 trillion every year globally, with concrete alone being worth £600 billion of this.

* * *

In an attempt to take the *con* out of concrete and replace it with hemp, hempcrete is made in exactly the same way as the nettlecrete from the Nettle chapter, with the obvious difference being the substitution of hemp hurds. The hurd, also known as the shiv, is the airy, woody, pithy core. It takes just three ingredients to make a hempbrick, hempblock or hempslab—hemp shivs, cement or lime and water. Almost any form (mould/shape) can be used, and with a proper framework and supporting column (a rebar traditionally made from steel, with an even stronger hemp rebar recently patented) hempcrete walls can even be poured.

The possibilities are virtually endless, and all it takes is roughly four parts hemp hurds, broken into 1cm to 2.5cm lengths, with one part lime binder and one part water—with variations in the mixes depending on the use and finish required. You mix it up, press it into some largish brick moulds, leave to harden and remove from moulds and leave to set solid for six to eight weeks. Build walls. Simple as that.

The concrete industry is itself stuck in the past, just as with the energy production and transport industries, being still caught in the aftermath of the Industrial Revolution. There has been very little change in concrete production since 1824 with Joseph Aspdin's Portland cement (mainly a mix of clay, gypsum and lime) or Joseph Monier's 1849 steel-reinforced concrete invention.

Concrete is far older than this, of course. The floors of the royal palaces at Tiryns, Greece in 1400 BC were concreted, as were both the Colosseum (AD 80) and the Pantheon in Rome (AD 126 still the largest unreinforced concrete dome in the world), with the Mayans at Uxmal, Guatemala also using concrete in AD 850. A lime mortar with straw added was used when affixing the limestone casing stones of the Great Pyramid at Giza, and sticky rice was added to the lime mortar when putting the Great Wall of China together, all of which have unarguably stood the test of time.

Without getting bogged down in details about cement, concrete and lime, all that needs to be known is that a new, more environmental way is possible. This will still involve lime but being produced at a lower temperature of between 600°C to 900°C as opposed to the 1,400°C used to make Portland cement, thus saving a lot of energy. But the real advantages of using hempcrete with a more traditional lime ingredient, as apposed to cement based, occur after their final use.

Among these advantages is the continual uptake of CO_2 by the hempcrete, as it slowly returns itself to its original calcium carbonate rock form, sequestering in a century as much CO_2 as that released by its production. An average hempcrete three-bedroom house can sequester as much as 8,000kg of carbon—it takes about one hectare and a 14-week harvest cycle to produce enough hemp to build a 135-square metre house and a small shed. Therefore, for every 1,250 houses built this way 10,000 tons of carbon can be sequestered; 125,000 houses would sequester a million tons of CO_2.

Hempcrete with lime plasters is weatherproof but not waterproof, which is actually desirable as modern houses are designed incorrectly in regards to them being sealed, when in fact a building needs to breathe to regulate humidity. Hemp and lime plaster create a healthier building, with maximised airflow and all surfaces acting as effective sluice gates for moisture, allowing this circulation to prevent mould and damp. This is why new houses fail, by being so well insulated, quickly creating a mouldy, damp building, as the building sucks up ground moisture and has no way of dealing with it, compounded by the condensation forming from human and animal breath, the drying of clothes and towels etc. The vapour in the air inside has nowhere to go with modern paints, plasters and walls, including conventional building materials continuing to off-gas for many years, which is when those glues, resins, binders, dioxins and VOCs (volatile organic compounds) are released into an unventilated, impermeable new home.

New houses also have low humidity, as there is so much talk of damp being the danger, but buildings with an internal low humidity also have higher incidences of breathing complaints, asthma, allergies and viral infections, as opposed to a traditional house. Tragically, along with the cleaning chemicals and air fresheners, the insides of most homes are today many times more dangerous to be in than outside in the polluted air of the city, trapping exhaust and other noxious industrial fumes inside. This was exposed by a 2019 study by CleanAirDay, in which houses on average were 3.5 times worse, with 560 times being the worst tested!

Hempcrete's other advantages are that it is extremely unlikely to be affected by insects, especially certain ants and termites, known to destroy all types of materials from cob and adobe, straw bale and wood. Hempcrete also burns much more slowly in case of a catastrophic fire event, requiring higher and sustained temperatures. Hempcrete also has a better insulation R rating than traditional concrete, which equates to a 50%-plus saving on heating efficiency, as demonstrated by all those hempcrete buildings thus far. Along with this improved insulation ability are the acoustic insulation benefits, be that loud music not going out or the outside noises not coming in. Finally, but definitely not lastly, during the production process of hempcrete far fewer harmful chemicals are required from field to finished use.

* * *

Here's some quick concrete facts: 4.1 billion tons (bnt) of cement is produced each year globally, while total concrete production is somewhere in the region of 10bnt to 17bnt per year, depending on the mixes used, as some have 1 part aggregate, some 4. The enormity of this figure of one year's concrete is that it is far in excess of the total weight of all plastic ever produced (9bnt), and also twice as much as all the other stuff, such as glass, bricks, carpets, toys, cars, paper, furniture, clothes, everything, twice. Or, in terms of the biggest concrete structure ever made, the Three Gorges Dam weighing in at about 40 million tons, there is the equivalent of 425 of these dams' worth of concrete poured every year onto the Earth's skin!

Concrete is globally worth in excess of £600 billion annually, set to rise to nearer £800bn by 2030, even higher than the astronomical figures already uncovered in this book. This equates to a significant amount of protection in place to prevent—heaven forbid!—a better, cheaper more sustainable replacement getting a voice, let alone some foundations. That's 600 billion reasons why it's never going to be an easy ride for any opposition. With that sort of collateral backing any lobbying, educational institutes, pseudoscience, publications, research funding or whatever you can name, the building industry giants can pay for it.

* * *

The best thing about hempcrete is the fact it is a perfect companion revenue stream to the next section's textile and fabric production. Hempcrete uses the inner material, the 'meat', while fabric uses the outer fibrous 'skin'. That's two revenue streams from one plant, which can also have other revenue streams, none compromising the other, depending on when the plant is harvested. Leaves or seeds can also be harvested, with thousands of uses for these from food and feed with no shortage of other industries serviceable. Then there are the roots, with huge medicinal potential, or as filler in plastic production using other plant wastage (plastic gets a fuller mention soon). Which is why the massive potential of hemp has been evangelised for many centuries. But it just hasn't gained any traction, not because the idea is flawed, but because we are repeatedly seeing in the economic system, and within the industrial system's inbuilt protectionism—often called a lawyer.

Fabric

Looking at clothing from a forager's perspective, a large mammal's skin is an awkward proposition, particularly compared to weaving some fibres together from all manner of omnipresent upright grasses and plants—especially when the stalk of one lying around after its leaves have been torn off to eat, was trodden on and retted, to be one day got pulled apart by an inquisitive individual and hey presto! Fibre to twine and weave was discovered ...

Can you imagine having no clothing or sheets at night? No blankets or curtains? No string, twine or rope? Clothing goes hand-in-hand with all of these. We might think we were tougher in the past, but we would rather have been comfortable just as today. We could have had fine woven clothing for 100,000 years, yet as the archaeological record itself is biased towards articles that are non-biodegradable, further skewered by the term 'hunter-gatherer', which manages to squash hundreds of millennia of existence into an era where men hunted, humans ate meat and wore skins. Today, wild-food-gathering societies forage over 90% of their diet from leaves, fruits, insects and roots, with rodents being the main meat source. The danger with using limited archaeological evidence to deduce the truth is the same as deciphering the pattern of a textile from just a single dyed thread.

We will probably never know the extent of fibres used for clothing from the past. Thus far we have evidence of flax being woven from Georgia 30,000 years ago, while even more remarkable is the use of tree bark to make cloth over 40,000 years ago in southern France by *Homo neanderthalensis,* our supposed dumb caveman ancestors, who are repeatedly being shown to have used their larger brains for burial rites, music, art and much more. This is not forgetting those needles found in Siberia used by *Homo denisova*, 50,000 years ago, though we will never know what they wove.

Ethnographic evidence from first-hand accounts and actual garments made from hemp in the past century can still be found. Any clothing over 500 years old is extremely rare, with nearly all the garments worn and the people who wore them all being completely composted by now. Fabric that has survived this long, such as the bandages from mummification in ancient Egypt, have had surprisingly little research on them. Ramie has been identified, a member of the nettle family *Urticaceae*, used, it is said, for its antibacterial properties and excellent strength, attributes shared with hemp and nettles.

Of the very few garments that have miraculously survived the past couple of millennia, nettles, hemp and flax predominate, followed by wool, leather and silk. The choice of hemp for clothing enjoyed a long continuance up until the twentieth century, with the last surviving remnants being found in traditional folk costumes and fabrics from around the world.

In Japan, the kimono is customarily made from cotton or hemp, called a *Futomono* for everyday wear. As we saw, the Emperor made his point about the longevity and wealth attached to the history of hemp—there has always been a small plot with hemp growing on it reserved exclusively for the Imperial family on the island of Shikoku. In the Nagano prefecture on the main island, hemp is still grown to make the bell ropes and curtains for Shinto houses of worship. Cannabis has held a continuance in Shinto ceremonial use with Shinto priests still wafting bundles of hemp to scatter malevolent spirits. Today most of the hemp fabrics available in Japan originate from Nepal where the lower wages and cost of living allow for the required intensive labour. China utilised three main plants for clothing—hemp, ramie and an invasive vine known as kudzu—while silk was invented around 5,000 years ago it has always been the preserve of the rich. The Hmong tribe of Vietnam and parts of Thailand still use a back strap loom to weave hemp tribal garments, a continuation of practices millennia old.

Over in Europe, Hungary has a long and rich tradition of hemp clothing and rope making, growing perhaps the most of all European countries, in an environment conducive for hemp growing, where still throughout the country *Cannabis sativa* spp. *ruderalis* grows wild. The traditional folk costumes were a diverse array, certainly not everyday wear anymore, from Poland to Ukraine to Serbia, Bulgaria, Georgia, Latvia, Slovakia, Czechoslavakia, Lithuania, Russia and Romania, where people were once clothed traditionally in hemp (and nettle). As were the Chinese army and many peoples across Asia, parts of Africa and the Americas, all utilising hemp for blankets, carpets, curtains, sacks and shoes.

Fabrics made from hemp require far fewer detrimental chemicals in their growing and processing, also needing half the water of cotton to grow in such density that weeds cannot compete. It is possible to accrue three times as much fibre from an acre of hemp than an acre of cotton or flax, although it does take longer to process. Before economics dominated, hemp was always the choice for a stronger, durable and more absorbent fabric, irrespective of yields and cost. Hemp fibre is two to three times as strong as either jute or sisal—which is why it was the choice for ropes, cordage and natural carpeting, with the plastic backing now also being able to be made from hemp and other plants.

Another bonus is that any fabric with 50% hemp in it will also prevent all UV radiation from penetrating through.

It was not news in the past to proclaim hemp fabric superior, it was just known to be. That is why clothing and hard-wearing sails and ropes were made from it, as were some nets and fishing lines—industries that can be revived today. This is undoubtedly why the 1895 USDA's Yearbook had this to say about hemp:

> the value of all other fibres as to strength and durability is estimated by it ... The plant is an admirable weed killer ... puts soil in condition. The value of hemp fibre ... and oil would seem to make its cultivation a very profitable one.

Hear, hear. Before then, in 1840, John Baxter wrote that "hemp will bleach whiter than flax or cotton, and make the finest fabric, from lace to cambric, to good shirting, and far cheaper than either". (Cambric is a lightweight closely woven fabric, said today to be of linen or cotton, when in Baxter's time it was made from hemp.) No one had a gun to their heads to say what they did, but from all of the plants available as a source of fabrics, hemp kept being chosen because it was the best.

Before Baxter's time it is apparent from eighteenth-century local newspapers on both sides of the Atlantic that hemp shirts were highly prized, given their repeatedly reported theft in those newspapers. While going though my three-foot high t-shirt pile the other day I noticed the three from THCT, as old as 30 years, were still in good condition after hundreds of wears and washes.

* * *

As we saw with the building industry, it's the same scenario with the first foray into utilising hemp again in the fabric industry. Having laid down some very impressive foundations, demonstrating what can be achieved durably, environmentally and economically, proponents of hemp are finding the already existing competition complaining and lobbying to prevent hemp from gaining the necessary inroads.

Some fashion and high street clothing brands have made the initial foray into utilising hemp, notable among them being Adidas, which make a cool 420 skate shoe (manufactured since 1995). This has been done for PR reasons more than environmental concerns, although everything adds to the kudos and inevitability that hemp becomes progressively the go-to for shoes to clothing and everything else. The process continued in 2022 with Converse launching a Stussy collaboration made from hemp canvas. Proving its strength and versatility, a kickstarter (8,000 kicks) for a waterproof hemp rucksack underlines the current interest in this area, as the desired pledge of $10,500 saw a return of over $450,000, thus guaranteeing the production of a significant quantity of a quality product, serving to raise the profile for hemp products a little higher.

Fashion label Ralph Lauren had secretly been using hemp in its clothing lines since 1984, giving the garments a unique sheen and feel. This was only found out when in 1995 the *New York Times* interviewed Calvin Klein, who let the cat out of the bag by suggesting he would be using it in his future lines just as Ralph Lauren had been doing. This had been a trade secret until then ... ooopsie!!

It makes common sense to make the switch, particularly as hemp has up to eight times the tensile strength and four times the durability of cotton. But, of course, a t-shirt or shoe that's going to last for longer means less will be sold, which makes little current business or economic sense in a global clothing industry worth over a trillion pounds a year.

Paper production

I remember being told at some demo in the school-hall, in what must have been 1980/81, that computers were going to make paper redundant, the number of trees felled would halve and the environmental benefits would be felt for generations to come. It proves you shouldn't believe all you're told at school—I was also told not to worry about English too much, as I was obviously never going to be a writer!

Paper is taken for granted today. It's everywhere, and nearly half of all trees cut down every year are sacrificed to make it. This has gone on since the mid-nineteenth century; before then paper was made from hemp and cloth rags—worn-out clothes, rope etc.—all mashed to a pulp and used to make paper, hence the term 'daily rag' for a newspaper.

When Gutenberg's presses started rolling in the 1430s, it was hemp paper that held the ink that spread the word of God. It was the same centuries earlier when the Qur'an was written on hemp paper. Without paper, the pace of change and dissemination of information would have been very different. Paper certainly ranks as one of the most significant inventions ever, from China and/or the world, facilitating colossal changes to cultures worldwide.

The oldest dates for paper made from hemp all come from China, with fragments of a probable map from Fangmatan dating from 179–141 BC, with other hemp paper fragments surviving from Dunhuang dated to 65 BC. Somewhat confusingly, Ts'ai Lun is attributed in many books with inventing paper in AD 105. The story goes that Ts'ai Lun, so distraught at having his paper dismissed by the leading bureaucrats, devised a plan to draw attention to his invention. His idea was to fake his own death: he would be buried with a tube to breathe, and when some of his paper was burned, he would arise from the dead. It worked and secured the popularity of his invention. Rather than inventing paper, perhaps Ts'ai Lun came up with a recipe of hemp and mulberry bark, resulting in a thinner and stronger paper, or he just started the tradition of burning paper money at funerals.

Papyrus was used for a writing surface in the ancient Mediterranean world (oldest record from Egypt in 2500 BC), although this is not recognised as real paper even if somewhat paradoxically the word 'paper' derives from the Greek for this plant, *Cyperus papyrus*. Nor was parchment—animal hide, also known as vellum—considered to be paper, although it has proved a durable and long-lasting medium, which could very well predate papyrus and paper. In China bamboo strips or bone were stitched together before paper, but as with papyrus this process was cumbersome and prone to split; they also used silk but only rarely. True paper begins as a pulp, when plant material (particularly cellulose) is broken down and laid out in thin sheets, which harden when dried.

China successfully guarded its paper invention for a while, although by the third century AD Vietnam is known to be producing paper. Korea followed soon after in the fourth century, and Japan in the fifth century. The technology spread to India by the seventh century, with the Islamic world finding out about it in the eighth century. By the eleventh century it is known in Moorish cities in Spain from where it spread across the rest of Europe and America. Up until 1890 all history was written on it, as 75% of all paper products were made from hemp.

Paper made from hemp has less of a tendency to turn yellow and disintegrate as wood pulp paper is very prone to do. Much of the hemp paper still in existence is still in good condition, even if hundreds of years old, provided it has suitable safe storage. Any important paperwork was always printed or written on hemp. Moreover hemp can be made into every grade of paper—whether cartridge paper for painting on, fine stock for stationery, index cards or corrugated cardboard or newspapers. Hemp is still used in many brands of cigarette papers today and the super-thin paper found in many bibles and dictionaries. It is the paper currency of Slovenia, and financial security papers such as stock certificates and wills can still be found made of hemp paper.

The paper industry is the fifth biggest consumer of energy, using about 4% of the entire world's produced energy. Manufacturing of paper has risen to about 400 million tons of paper every year, a doubling since the initial popularity of the home computer in the mid-1980s. Breaking that 400mt huge figure down a little to a more manageable and comprehensible ton, it takes on average 17 trees (45-foot pine trees being the most common), 25 tons of water and over 2,500 litres of petrol to make one ton of finished paper. This ton comprises a pallet of 40 cartons, with one carton being 10 reams (500 sheets/ream, 5,000 sheets/carton), totalling 200,000 sheets at 5g an A4 sheet, for what is confusingly called 75g paper. Therefore one tree on average makes 23.52 reams or 11,765 sheets.

Of course, paper production is a lot more than A4 sheets—there's all the printed books, magazines, newspapers, wrapping and toilet paper, packaging, cardboard and take-away coffee cups, not to mention furniture, clothing, even shoes and building panels, which can all be part paper. All of this vast output of paper after a short life goes to make up over a quarter of all landfill sites, where it breaks down, releasing methane and CO_2.

This is why recycling is so important, especially as up to 80% less energy is needed to make a ton of recycled paper than virgin paper. With recycling being such an important process, it's good to know that hemp paper can be recycled eight times, whereas wood pulp paper manages only three or four times. With two crops a year from hemp being possible—unlike a tree, which takes nearer to twenty years to mature—four times as much hemp paper can be made per acre compared with an acre of trees. Using hemp for paper makes as much sense today as it did a century and more ago.

* * *

European and American regulations involving polluting discharge from paper mills have improved significantly in the past 20 years, although much of the rest of the world's environmental attitude has yet to catch up. In mitigation, the buyers in Europe and America leave such small profit margins for the producers that they can't afford to implement any environmental policies, which should be paid for by the West's larger profit margins. Paper mills' discharge runs heavy in dissolved organic lignins, alcohol, inorganic chlorine, nitrates, benzenes, methanol and heavy metals, including lead and mercury, with a significant amount of sulphur dioxide, nitrogen oxide, carbon monoxide and hydrogen sulphide pumped into the air (we can produce in our tum the last one of these, released as an egg-smelling fart!).

As hemp has a much lower lignin content, it is far easier to pulp than wood, not requiring any harmful chlorine to break it down. This reduces the dioxins created as a by-product—dioxins are recognised as one of the most dangerous and persistent of man-made chemicals, causing havoc with marine life as it is regularly accidentally and/or deliberately discharged into local streams and rivers. Hemp paper can be made from a far gentler and cleaner process using soda and bleached with hydrogen peroxide or sodium hydroxide, not the chlorines needed for wood-pulp paper. China is currently in the lead in using non-tree paper, with over 75% of their paper made from various crops such as reeds, bagasse (the pulpy waste from sugar cane), bamboo and the stem straw waste from wheat, rice, flax, cotton and hemp.

The bonus with growing hemp is that it can be used for phytoremediation, meaning it cleans and remineralises industrially damaged soils, such as those ruined by the paper industry. Hemp can absorb copper, cadmium, lead and mercury from the soil, and these are all common pollutants from paper production. As the pulp and paper industry has a market value annually somewhere in the region of £64 billion, there are close to 64 billion reasons why hemp will only be considered an ingredient for paper when the current industry is positioned ready to utilise it.

We've already seen that a guaranteed supply of hemp as a raw ingredient has been a major hurdle. Another hurdle to expanding the hemp paper industry, or any other industries hemp can service, has also been the initial labour intensiveness of the processing of the hemp plant, predominantly in separating the hurd from the shiv. The machine touted as solving this problem has been—

The decorticator

This machine has always been perceived as a vital component if hemp were to achieve a competitive market presence. The 'modern' version of a decorticator was invented in Bologna in 1861 and was last produced in 1950, with little indication about how busy the 'scavezzatrice' had been for the half-century before. Making hemp illegal in 1937 was an obvious hurdle to rolling out a potentially profitable machine, although there had yet to emerge a decorticator that could cope with the tough, superior fibres of hemp on a grand scale.

The situation today is that after a few years of legalisation there has been a small swing by a few farmers towards hemp growing, along with a few innovative machines being produced, from manual mangles to mobile factories—whatever your price budget there is a product out there waiting to empty your bank balance or put you in a lot of debt.

The reason the decorticator gets its own section is not to concentrate on the technology, but to begin to address a bigger issue, as here lies a fork in the road to the future—we can either continue down Unsustainable Avenue using the same agricultural system with a few new plants, or we can turn down Longevity Lane for a safe, sustainable solution.

* * *

The modern system, monoculture, is only cost-effective if the vast expense is spread out over thousands of acres and is indeed only viewed through the lens of profitability. The huge, expensive machinery compacting the soil, the destruction of the interconnected mycelium network (part of the soil's immune system) and dependency on fertilisers and chemicals was not factored in the calculations, yet they have had significant repercussions on the life and health of all planetary organisms. A 2019 report from the International Food Policy Research Institute stated: "the cost of the damage caused by agriculture is greater than the value of the food produced". That has to be worth a reread. They also worryingly suggested that the annual £560 billion (nearly £1m/minute) of subsidies given out worldwide to farmers was actually driving the climate crisis and destruction of wildlife.

With the current model being unsustainable, it is only the facilitators of monoculture—the seed, pesticide and machinery producers—who continue to argue for it being the best and only viable solution to feed the world. Meanwhile they continue to pollute, poison and make the soils progressively more barren, often with government encouragement and those subsidies (which is public money, our money). The countryside of every country has been desecrated by monoculture, while the city folk have been kept completely blind as to what has been going on during the journey taken from seed to plate.

Yet another product of our broken economic system was that somehow, somewhere, it was determined that the cheapness of the food on this plate was more important than the quality of the food on it. Cheap denatured, processed food was never going to be the long-term solution, which must be real food from real soil.

There is absolutely no proposition here to continue with this proven unsustainable agricultural system, as by placing cannabis within it will only sour the potential benefits it could reap. I'm sure by this point in the book it's becoming clear that it is the very systemic machinations of our capitalist culture that are so desecrating and unsustainable, and therefore it is these that are in need of reforming. By using a plant with billions of dollars of potential, the underlying problem of the unsustainable system is simply plastered over, until after a brief period of short term-profit the cracks reappear. Words are easy to throw around in any string of sentences, and can make everything appear possible, although actions will always speak louder in the long run—those actions and how to enact a systemic revolution are surprisingly simple, as outlined in the Solutional Summations chapter later.

All this means the direction we need to be headed for is unarguably 'small and local', the absolute opposite of the 'large and global' mantra of big business today. Which is just as well, because the truth of the matter with hemp processing is that it hasn't actually changed all that much in the past 10,000 years. The majority of the world's hemp production in China, Nepal and northern India is harvested and processed in exactly the same way as it has been for millennia. There is little reason for this to change, as long as there is the potential to grow an increased yield if demand increases. There will be a dramatic rise in hemp supply soon as the law is changing and more and more smallholdings will begin to contribute to sizeable total harvests that can with processing enable three possible revenue streams from the seeds (oils and plastic), hurd (hempcrete and paper) and bast/shiv (fabric), with any waste going to either plastic production or biomass energy, and any further processing raising local revenues and employing more people locally.

It's as if the hemp industry is waiting for a repeat scenario of cotton after the jig was invented by Eli Whitney in 1793. This machine revolutionised the cotton industry, giving it the near monopoly on fabrics it still holds today, as hemp, nettles and other plants could never keep up with the forever dropping price of cotton. The hemp industry needn't wait for a bigger, better decorticator, but rather needs to change its mind and consider positioning itself at the cottage-industry level where there are already sufficient decorticators, ranging in price from less than £1,000 up to £15,000. From the decorticator producers I spoke to from South Africa, the US and Australia, irrespective of the colourful brochures and shiny corporate videos and profit forecasts, the feedback is that sales have been very low for any above this price. This lack of sales suggests that keeping it small to begin with might well be a safer, more practical suggestion, as the way forward is hopefully going to be lots of smallholdings collectively making many tons, exactly as has been successfully demonstrated in China, with their extensive network of co-operative smallholdings. There is no future and certainly no longevity in the current monocultural system, especially as it has been shown to cause more harm than good.

Plastic, bioplastic and oil

Making plastic from plants isn't the first thing that springs to mind, especially when looking at a cannabis plant, although it's no stranger than the fact that most plastic made so far started as petroleum, a finite resource, as opposed to the plethora of plants. Petroleum can be replaced by hemp and other plants for not just plastic production but also its primary purpose as fuel for locomotion, from cars to ships and planes.

Let us begin with plastic, a story steeped in controversy, confusion and deception. Plastic's invention was not a simple single step, with Parkesine being perhaps the first man-made plastic albeit from natural ingredients, not petroleum. Parkesine was developed in 1856 by Londoner Alexander Parkes, with a few others soon to follow, although the very high production costs and limited practical uses only ever saw bankruptcies. Celluloid, invented in 1869, was the first successful plastic product, finding its fame eventually with early cinematography a few decades later. The first petroleum plastic was made in 1907 by New Yorker Leo Baekeland; this was Bakelite, now a fashionable antique. The following year Cellophane was invented. This is still in use, as is Polyvinyl chloride (PVC), invented in 1920, with the word 'plastic' being first coined in 1925. Clingfilm was invented in 1933, although it was not until the massive surge in consumables after WWII that many of the earlier inventions found uses. Many consider the typical plastic as perhaps a Lego-brick—this is actually acrylonitrile butadiene styrene (ABS), invented in 1945—and by the end of the 1970s all these types of plastic (except Bakelite) were omnipresent, from food wrapping to toys, televisions, credit cards and a million other things.

Plastic production uses about 3% of global petroleum consumption, while its manufacture uses about 2% of total electricity. Every year since 1950 plastic production has risen worldwide, from 2 million tons (mt) in 1950, rising to 50mt/yr in 1972; thereafter the production curve gets steeper, with 100mt/yr passed in 1987, 200mt/yr by 2000 and about 400mt by 2020. It continues to rise, being worth somewhere in the region of £500 billion/yr at time of writing, with figures set to double by 2050 at current growth. There is of course a lot of talk at the moment about cutting production, with manufacturers slowly waking up to their responsibilities, although it appears it's only when it starts to affect their profits that an alternative is looked for, with little said by governments.

Only 18% of the plastic produced is recycled these days, which clearly is not good enough, but is at least double the average of 9% (810mt) of all the plastic ever made (9bnt tons). With around 8mt ending in the sea every year, it is estimated that by 2050 the weight of plastic in the oceans will be greater than the total fish biomass! Each resident in Germany uses 38kg of plastic/year, UK weighs in with over 30kg and the EU average is 24kg, while India and Indonesia use about 10kg/person/year.

What may be considered worse and more shocking is the fact that humans are consuming about 5g/week of microplastics in our food and drink, as shown from studies from Newcastle University, Australia (Palanisami 2019). This is a worrying ¼kg every year, adding to other recent revelations of breathing in airborne plastic particles constantly. The majority of this 5g/week came from drinking water, both tap and bottled, and 35% of this was from washing polyester fabric, predominantly clothing.

* * *

What is particularly aggravating is the fact that the plastic producers, such as INEOS, BASF, DOW and ExxonMobil, pay little more than lip service to remediating the pollution created from their profits. They should be made to pay to help clear up the present mess and pay for far more efficient recycling plants, employing more and recycling far more. Not least because up to 60% of all recycled plastic has actually been wrongly sorted or contaminated and gets incinerated. Although of course the plastic polluters would argue they shouldn't pay for recycling as this is direct competition to their plastic production and will impact negatively on sales, a double loss to a corporation such as INEOS who make it known they have a worldwide workforce of 26,000, with annual revenues in excess of £50bn. The deafening din caused by economics and profits means all that's heard is the fact that recycled plastic costs about £200/ton more than virgin petroleum plastic.

It's also worth considering that one-sixth of the total weight of petroleum plastic is chemical additives (such as BPA, fire retardants, phthalates, pigments, dyes and plasticisers), comprising very harmful carcinogenic, hormone-disrupting DNA-mutating chemicals. Which means of the 8 million tons estimated to end up in the seas every year, there has also been 1.33mt of these chemicals—which equates to a very worrying 1.5 billion tons of man-made chemicals from all the plastic ever made, going nowhere fast as they're all persistent poisons; this means they don't break down, remaining biologically active in the sea.

A notable persistent poison is BPA (Bisphenol A), which began its life as a hormone-replacement drug at the infamous 1930s German firm I.G. Farben, which made Zyklon B, the Holocaust gas. This specifically was invented by a Jew, Fritz Haber, whose wife committed suicide when the ammonia gas he invented earlier was used in WWI. It is the Haber-Bosch method, involving turning nitrogen from the air into ammonia, which became the fertiliser industry, so there are some who say Haber saved billions from starvation, even though he was evil; ultimately fertilisers have caused far more long-term trouble and damage than any short-term benefits.

When BPA is free from its bonds in the plastic composite, it is allowed to run its course of genome- and endocrine-disrupting, gender-bending complications, with heritable dangers affecting yet-to-be-conceived generations, as well as being linked to childhood asthma, neurological problems, cardiovascular disease, diabetes, metabolic disease and Parkinson's ... oops! BPA is still used to line some cans to stop them corroding, yet it has been clearly shown to leach into the contents, especially if acidic, such as juices. Of more concern is the fact BPA was mainly used as a plasticiser from the 1980s to the 2010s to make products more flexible such as babies' dummies and teats for bottles, where many people claim it has indeed caused genomic mutations and gender confusion. This has all been kept quiet (consider the omnipresence of clingfilm), apart from the science, such as the conclusion from *Environmental Research* in 2004 (Levy et al.) where results showed clear evidence that BPA induces feminisation.

With around 2mt produced every year and a £10 billion global market value, it is being slowly replaced, although the safety of these replacements, such as Bisphenol S, is already disputed. The American Chemistry Council spend over $8 million a year on lobbying expenses—which somehow makes it safe—while those trying to expose the risks posed by BPA and other food additives such as aspartame usually only have the truth on their side but little to no funds to spread it. Sadly, the truth doesn't always sound quite as convincing when said by a selfless, passionate individual with a megaphone, billboard, fold-up table and leaflets in the local shopping mall, as opposed to the magazine articles, educational institutes, videos and periodicals produced by the polluting industries with the deliberate intention to confuse, ameliorate and ill-educate.

* * *

Bioplastic sounds like it's what we've been waiting for, and it even comes with 'compostable' emblazoned on it, whether a spoon, fork or cup—what's not to love? Unsurprisingly, it's not that simple, nor is bioplastic really compostable or for that matter even plant-based. We'll get on to that soon. To begin with it's important to make it clear that plastic can be made from an incredible diversity of plants, utilising either the cellulose, starch or protein, or it can be fermented to make organic polyethylene, then mixed with water, vinegar, glycerine, an alginate here and an acid there—plastics are easily producible. Petroleum plastics have had over 70 years of constant research with various synthetic polymers providing substantial versatility in the petroleum plastic arena. Bioplastics, on the other hand, are still very much novel, and irrespective of them being older there has been little industrial investment or continual innovation.

But this is all about to change and future developments will see most plastics being plant-, algae- or fungal-based. Most have already been invented and can utilise all agricultural crop wastage from corn to sugar and rice, with hemp already being used for this end, as can of course our other six plants. Even marine algal blooms causing hypoxic (oxygen-free) dead areas in many estuaries can now be harvested for plastic production, and another lucrative avenue will be chitin, found in fungal cell walls and the main component of prawn, crab and lobster shells (up to 30% chitin) and the exo-skeletons of insects such as cockroaches. Chitin is the second most abundant plant plastic polymer on Earth, the first being cellulose. Chitin plastics will soon prove to be not only more efficient than other plastic wraps and packaging, they can also be fully biodegradable. Which is where some of the problems begin to arise regarding bioplastics.

Firstly, dealing with the issue mentioned above, currently a product can call itself a bioplastic even if it is made from as little as 20% plant matter and be 80% petroleum-based—bioplastic does not guarantee petroleum-free. Secondly, the compostable claim is applicable if the bioplastic fork/spoon/straw/cup gets recycled to a special industrial composting unit—if it is mistakenly put in the home compost it will not break down with the grass and food scraps, or if the compostable bioplastic ends up in the plastic recycling it is considered a contaminant and the recycling batch gets incinerated, not recycled. If the compostable plastic ends up in the sea, it stays there, adding to the problem. The mantra should be: reduce, reuse, refill and recycle. What is required is biodegradable bioplastic. This is still at an experimental stage in many respects, but what is coming in this field is very exciting, and if the bioplastics can stop being so easily corrupted by big corporation green-washing, the future for plastic will be plant-derived.

* * *

When this door is finally opened to a substantial new revenue stream from plant plastic, another revenue stream from the abundant oil in seeds and nuts is also opened to replace petrol in cars etc. Hemp seed (just the kernel) contains about 32% by weight in oil, flaxseed and coconut about 40%, sesame, sunflower and castor seeds about 45%, peanut 50% and walnut 55%. A century ago these and other plant oils had several uses from food to paints, varnishes, lubricants, soaps, hair and skin tonics for humans and animals as well as of course for combustion engines, now all petroleum.

Bearing in mind there will be a huge move away from not just petroleum engines but all engines of this nature, consequentially in no more than a decade the demand for petroleum for locomotion will be substantially reduced, avoiding the oft-reported scenario of food shortages, rising prices and starvation for locals caused by human food getting diverted to energy production. There is no intention to directly swap from petroleum to plant/seed oil, certainly not at anything like the ridiculous rates of consumption of today. A better solution is to replace the machines, making them all battery-, solar-, wind- or something new-driven. It's somewhat of a surprise to discover the first internal combustion engine, invented in 1802 in Switzerland by François de Rivaz, actually used hydrogen and oxygen; thereafter the intention of other engines was never to run them on petroleum as with Rudolf Diesel's engines and vegetable oils.

It's already well into the third decade of the twenty-first century and it has to be about time we started making it appear as if we really could make a better future. All the technology has already been invented for a fully sustainable and community-sufficient future. We have barely touched the surface at harnessing solar and magnetic energy, Nikolas Tesla invented some energy systems that have been most likely deliberately hidden, while sonic energies haven't even begun to be looked at—instead we have been trapped in the Industrial Revolution, still using steam to create energy, which is all nuclear energy does: boil water to steam.

Surely the advantage of having the source of plastic, oil and energy all locally available and producible from local plants, providing local wealth with the bonus of being sustainable and environmental makes far more sense than reliance on countries prone to irrationality. Doesn't it?

Hempire

One of the first prominent modern advocates of hemp products in Britain must have been the ethical and environmentally forward-thinking Anita Roddick with her Body Shop, founded in Brighton in 1976. One of their best-selling ranges is the hemp range, growing since 1998, with some amusing anecdotes about staff getting arrested for supplying marijuana in Paris and the Bahamas when first stocked. Apparently every 9 seconds The Body Shop sell one of their hemp hand protectors, and definitely not because it's a gimmick.

There is already quite a surprisingly large market for hemp animal bedding using the shiv (the inner core), being sold in massive 180 litre bags for horses to 500g sachets for mice bedding, as it has long been known to offer the animals warmer and more absorbent bedding.

The Hemp Shop based in Somerset UK has been providing an ever-increasing range of hemp products from hats to sacks, rope and clothing, beauty products, foods and oils since 1998 (thehempshop.co.uk). Ethical clothing and accessory company Thought (wearethought.com) are incorporating 13% hemp in all clothing and have been utilising hemp since 1995. There are many more such companies, and as the price drops there will be more and more, because in the end quality will win out. Currently on Etsy it is possible to buy 10–20 metre balls of yarn for about £3, whether that be nettle, wool or hemp, so in the last couple of years there has been an increase in the supply of raw materials and a drop in price. The ball has started to roll.

Being incorporated into trainers, Puma as well as arch-rival Adidas use hemp fabric in some of their shoes, while new shoe company Dope Kicks claim what they tout as being the first waterproof hemp shoe, including recycled old soles and sustainable cork for the insoles (wearedopekicks.com).

Cars use hemp as well, with the very fast and expensive Porsche Cayman 718 GT4 sports car having body panels made from hemp. The current C-class Mercedes have up to 20kg of hemp in them, with the company already using more than 500,000 tons of bioplastic a year; the BMW 5 Series all have hemp in their door panels; and Audi, GM and the Lotus Elise all feature hemp and bioplastics. Actually, as we'll see in the final chapter, it's all been done before, when in 1941 Henry Ford unveiled a car made of and run on hemp.

Bast fibres of the hemp plant have been used in the biocomposite industry as a substitute for glass fibres, producing a lighter and stronger material than polypropylene. There are some ultra-modern composite companies out there, such as the Dutch Nabasco (Nature Based Composites), using hemp, soya bean oil, balsa wood, cork and linseed to produce some really cool and curvy car and building panels. Lego have also promised to switch their entire production to hemp plastic by 2030, which will require a considerable harvest, opening doors for any industries reliant on the waste from this industry.

* * *

The above uses are far from exhaustive, yet demonstrate the inroads already accomplished. The following are still speculative, but as soon as the raw materials become readily available they will slip into various industries—

Hemp fibre is not just limited to clothing textiles, and depending on any after-treatment it can be moulded into any shape or used for a vast range of uses and composites for anything from pallets to panels of cars, box trailers, trucks, boats, trains and buildings, even 3D printing. The bast and shiv can also be made into edible food containers, as well as coarse paper, hemp tiles for roofs and MDF (medium-density fibreboard), while soundproof walls and insulation are superior in all respects when made from hemp fibre.

Then there are also all the food, colourings and medicinal uses: the plastic, wrapping, packaging, bottles and bags and clothing can all be made from sustainable and non-polluting biodegradable hemp plastic (and other plants), negating the forever choking of the planet with petroleum plastic from bottles to micro-beads, nurdles and their despicable chemicals. There are also exciting advancements in developing batteries using the hemp shiv, transforming it into a graphene-like product where it can replace the horrendously polluting mining of Earth's finite natural resources to make batteries today.

At worst is the potential of just burning hemp for biomass energy—alongside all the nettles used for phytoremediation. Last and not least, all hemp products can be fully biodegradable, and most recyclable, with those touted 25,000 uses probably being quite an underestimate. It might be a bit of a slow uptake at the moment, being the formative period, and if hemp is to help reform/supplant traditional fossil fuel plastics, petroleum, pharmaceuticals, building, paper and forestry, because these corporations do not control hemp production currently, there will be only slow and stuttering inroads into hemp becoming mainstream. As soon as the agricultural and chemical corporations secure patents for hemp seeds, create specific fertilisers and pesticides, with other horizontally owned industries benefiting from new machinery, then it will become economically viable and hemp will be rolled out monoculturally. Which is exactly what is happening and is exactly what shouldn't be happening, as it's the wrong way to keep going if we want to secure a future.

The seed—health

There is a deficit in the evidence for historical hemp seed use as a common food item, although we did see how the health practitioners of the day used hemp and marijuana for many ailments; we also saw how much had to be grown to maintain the British navy's requirement in rope and sails alone. Surely with hemp being such a common crop for centuries, with swathes across most European countries, there had to be a ready supply of seeds (billions of) for future harvests. It's hard to believe the edible aspect of the seeds was ignored or that the psychoactive properties of the dense female buds were unknown. Alarm bells start to ring loudly when any information on it being grown extensively is so hard to come by from any literature, which strongly suggests there may have been censorship and deliberate interference in the transference of cultural information and knowledge.

There is a single piece of folklore widely spread across the British Isles, with local variants associating the hempseed with divination, from Sussex to Devon, Oxfordshire, Norfolk and Yorkshire. This involves scattering hempseed on either Valentine's Day, St Mark's Eve (24 April) or Midsummer Eve and reciting "Hempseed I set, hempseed I sow, The man that is my true love, Come after me and mow," while walking and upon turning around there will or won't be a spectral image of their future love, or a coffin indicating a young death. This at least proves hemp was being grown across the entire country.

For 4,000 years the Persians/Iranians have valued hemp seed as a roasted snack, calling it *Shah daneh*, or King of Seeds. This was a crown this two to five millimetre powerhouse of a seed could soon regain, with its potential to literally seed the revolution. In Russia it was common to make artisan hemp seed pastes, superior to peanut butter both nutritionally and in taste, but its major flaw in our economic world dominated by wealth not health is that hempseed butter has a short shelf life as the delicate and super-healthy oils can go rancid quicker than those in peanuts.

Undoubtedly we have always understood the health benefits of hemp seeds and the psychoactive properties of the female's trichome-rich buds, as indicated across time from one of the oldest cookbooks in existence, the 1475 *De Honesta Voluptate et Valetudine* (*Of Honest Voluptuousness and Health*): "A health drink of cannabis nectar: carefully treat food and divide for the stomach and head, finally remember everything in excess may be harmful or criminal." Recipes continued through the centuries up until 1884's *Frammento di un libro di Cucina* (*Fragment of a Cookbook*), with a tortellini recipe of bacon and hemp flowers, which emphasised the importance of cooking the flowers in the bacon—as this makes the THC more bioavailable, a process known as decarboxylation.

As we saw in the history section above, the entire plant was being used by not only the common folk but also Queen Victoria, American Presidents and Popes. Even the Buddha is involved in the story of hemp, as allegedly for six years before illumination, Siddhartha Guatama ate just one cannabis seed per day, after which time his enlightenment was complete. The veracity of six years is questionable, and we all know how stories get exaggerated—it may only have been six months or even six weeks, and with time the inflated boasts abound—some say the Buddha also had one grain of rice per day as well, making it a more balanced diet at least!

The Buddha was of course already clever enough before his enlightenment to realise that of all the possible seeds only hemp seeds would suffice, as the hemp seed provides the perfect balance of essential fatty acids (EFAs) and is also the only culinary seed to contain GLA, which we'll get to in a moment. These EFAs, namely the omega-oils, have had considerable health news coverage for many years now, no doubt kick-started by Udo Erasmus's excellent 1986 book *Fats that Heal, Fats that Kill*.

There are two EFAs of concern here. First is Omega3, also known as alpha-linolenic acid (ALA), the best source of which is linseed and hempseed. Fish oils are another excellent source, touted as unique by the fisheries industry as EPA and DHA, for instance, are absent from linseed, although understandably so because the human body makes the eicosapentaenoic acid (EPA), which breaks down into series 3 prostaglandins, an anti-inflammatory from ALA and hence why Omega3 gets labelled the good Omega. Docosahexaenoic acid (DHA) and stearidonic acid (SDA) are also made from ALA in the gut. As humans get older the ability to make enzymes and to break these acids down further can diminish, which is why supplementing with these oils may be important later in life. So it's a bit of a fishy tale by the fisheries, particularly as getting your Omega3 from hemp or linseed negates ingesting any of the increasingly prevalent toxic marine chemicals.

The other essential fatty acid is linoleic acid (LA), Omega6, whose richest source is sunflower seeds. The body breaks this down into gamma-linoleic acid (GLA)—found richest naturally in borage, blackcurrant seeds and evening primrose oil—and arachidonic acid (AA), which helps in the production of series 2 prostaglandins, said to be inflammatory and why Omega6 receives such bad flak. Further flagging the Omega6s is the fact the EFAs need to be ingested in a balanced ratio to be healthy and efficient. This ratio should be somewhere near to 3:1 (Omega6:Omega3), whereas the modern diet with its abundance of processed vegetable oils has a serious imbalance—the ratio is commonly nearer to 20:1 and higher.

Mention has attuned to these EFAs because it is among them where we encounter the very foundations of health, the very beginnings of any organic functionality, the cell. Each cell, which is a marvel beyond words, is its own world encased in a phospholipid membrane; across this membrane wastes filter out and nutrients enter, but not if the cell has been made with the wrong oils and the body can only use what is ingested. Considering that we are an amalgamation of anywhere between five to sixty trillion cells, if these aren't up to scratch to begin with, optimum health can never be achieved. That's why Udo Erasmus is such an advocate of these oils, and quite right he is. The magic with the hempseed is that it has this perfect ratio of omega-oils—as one would expect from a Human Plant—with that GLA getting converted to the protective hormone prostaglandin PGE1, which regulates hormonal balance. GLA-rich foods have been seen to benefit those suffering from ADHD, diabetes, heart disease, MS, obesity, skin allergies, high blood pressure, breast pain, PMS and rheumatoid arthritis.

The seed also contains Omega7 and Omega9. The ratio of these omega-oils found in hempseed is approximately 56% Omega6, 22% Omega3, 7% Omega7 and 15% Omega9. The seeds also have a fantastic amount of manganese, magnesium, phosphorus, zinc, copper, potassium, iron, protein (31% by weight), and good quantities of vitamins B1, B3 and B6. The good chromium content and other appetite suppressants in hemp, along with the high fibre content of the seeds, promotes satiety (feeling of fullness), which in turn promotes weight loss. Hemp is high in both soluble and insoluble fibre, providing essential bulk, feeding the probiotics in the gut and keeping the gastrointestinal system regular and in shape, even in as low as doses of 10g (a small tablespoon) of hempseed oil. The oil also assists with lubricating and repairing nerve endings (the myelin sheath), which is being shown to be a serious problem and probable causative agent in many neurological disorders, especially as most processed supermarket vegetable cooking oils actively damage the ends of these essential myelin sheaths.

Another exciting action of Omega3 oil was published in the *PNAS* journal, from work at the University of Illinois (McDougie et al. 2017). This showed that the ALA (Omega3) is used by the body to produce endocannabinoids, natural cannabinoids, the same as the ones found in the other parts of the plant, although these can be bio-assimilated without any psychedelic effect. This is a significant boost to the proposition that a dietary deficiency of the seeds is a major risk factor in many diseases. These natural cannabinoids produced from the Omega3 help to support an optimum inflammatory system, and maintain blood vessel dilation and blood platelet aggregation.

It isn't that Omega3 is simply the good oil because Omega6 also plays a role in this anti-inflammatory drama with GLA, even if vilified as the 'bad' inflammatory oil. That is not true as the answer lies in the balance of ingestion. Too much Omega3 for an extended period can cause severely dry skin and hair loss, basically the opposite of any lubrication benefits it can bestow when consumed in balance. Hempseed gets the balance right in regards human health, and no doubt for many other mammals, birds and insects eating them.

* * *

Deficiency of essential fatty acids definitely plays a correlated role in the manifestation of most, if not all of the diseases of civilization. It might not be the causative agent but its deficiency doesn't help with any remediation, which its balanced intake could well accomplish, as well as acting as a perfect prevention. Of all the seed oils it appears that hempseed oil is alone as the one that supplies the EFAs in the perfect ratio, although it shouldn't be heated as this converts the beneficial omega-oil ratio to potentially harmful ones.

Eating and/or applying the oil acts as an excellent skin moisturiser, which rehydrates and increases skin elasticity and water retention (useful for conditions such as psoriasis and eczema) and treats dry hair, which is why it is already being used in some bathroom products. When eaten as a ground-up seed it can help to regulate blood sugar levels and is fantastic for intestinal flora, providing ideal fibre for them to thrive. Buy a 100ml bottle of hemp seed oil from a local health store and give it as a present to either yourself or a loved one, and have 10ml a day. You can feel the difference.

The above applies only to the seed and the oil derived from the seeds, with only cold-pressed being safe and healthy for human consumption—surprising as it may sound, the refined oils on the supermarket shelves of sunflower, corn oil etc. have no scientific evidence of safety let alone healthiness. In fact it has been known since the 1930s that these refined vegetable oils are detrimental to human health and are considered carcinogenic. Only consume cold-pressed organic vegetable oils from coconut, olive, even palm and rape if they are organic and cold-pressed (remembering that palm oil has desecrated the Indonesian rainforests).

The Russian Vavilov Scientific Research Institute of Plant Industry in St Petersburg has one of the largest collections of hemp seeds in the world with about 500 varieties. Worryingly, it's a dwindling collection, as it numbered 1,400 varieties just 70 years ago. Within these seeds there would have been varieties excelling in various desirable qualities, with various THCs and CBDs, unique phytonutrients, different amounts of oil in the seed, thickness and length of the stems, size of leaf, or more shiv or fibre. These seed banks are imperative if we are to return to a Natural future.

This is all about the oil from the hemp seed, and it is important to know that it shares nothing with the other oils obtainable from the plant, hence why hempseed oil has always been legal. What hasn't always been legal are the cannabinoids—

CBD

Clearly Mother Nature has no time for petty human legal affairs, although it does help to substantiate the claim of cannabis being a Human Plant we evolved on. This disrespect of the law has manifested biologically as an extensive endocannabinoid system, evolved within us to utilise ingested cannabis throughout our bodies, as well as converting some from Omega3. The endocannabinoid system, comprised of receptors spread throughout the immune and nervous systems, has the ability to affect key body processes including appetite, memory, mood, inflammation, insulin, fat, energy metabolism and pain, as well as helping the human organism to achieve homeostasis by these receptors always measuring and balancing hormone requirements and production. The full extent of their usefulness is only just beginning to be uncovered, although this endocannabinoid system is undeniably integral to the human organism, or, as Jack Herer said, "This points to an ancient symbiosis between the plant and people ... these neuronal pathways are the product of a pre-cultural relationship between man and cannabis."

* * *

CBD is an abbreviation for cannabidiol—it is not an abbreviation for cannabinoid, as is so often wrongly reported. There are well over a hundred cannabinoids of which cannabidiol (CBD) is one. It's worth having a quick look at a few of these cannabinoids to appreciate their role within the human.

There is CBDA (cannabidiolic acid), which when exposed to light and heat undergoes a process called decarboxylation, which removes the acid to become CBD. CBDA is available in the raw, unprocessed plant and may well have different health benefits from CBD, possibly anti-inflammatory and anti-carcinogenic.

CBD (cannabidiol, with eight different distinct compounds isolated so far) when bought as an oil is derived from the hemp plant as this has a very low THC content, although CBD is found in higher concentrations in the female buds. The separation of the CBD and THC is not straightforward, especially to achieve the near-to-zero THC legal limits (0.3% THC in USA and 0.2% in UK).

CBC (cannabichromene, with eight different compounds) is also broken down from CBDA and has some exciting results from studies involving pain, tumour growth, inflammation and acne treatment.

First isolated in 1975, CBCV (cannabichromevarin) still has very little known about it. CBDV (cannabidivarin) could well be key in cannabis' anticonvulsant and anti-epileptic capabilities; CBDV also appears to be an appetite suppressant, an important distinction as the stereotype is of cannabis giving 'the munchies'.

CBG (cannabigerol, with 17 different compounds) is actually another precursor to many other cannabinoids such as CBD and THC. Ingestion before it breaks down suggests it may benefit anxiety and depression, and it may also be this CBG that helps reduce intraocular pressure associated with glaucoma; research suggests it could also prevent tumour growth and have antibiotic properties against MRSA.

CBN (cannabinol, with 10 compounds) results after heating and oxidation (as well as simply aging), this time from THC; it is very slightly psychoactive and appears to be a sedative, similar to diazepam.

There's also CBDL (cannabinodiol), CBE (cannabielsoin, five compounds), CBL (cannabicyclol, three compounds), CBND (cannabinodiol, two compounds) and CBT (cannabitriol).

CBGV (cannabigerovarin) is being touted as a cancer treatment, while CBGM (cannabigerol monomethyl ether) is more prevalent in plants grown in north east Asia. All in all, considerable canna confusion!

With a few THCs identified (all are cannabinoids), the best known is the psychoactive delta 9 THC (tetrahydrocannabinol, with 18 compounds); there's also delta 10 THC (with two compounds) and a delta 8 as well. THCA (tetrahydrocanna-binolic acid) breaks down to THC upon heating, and when consumed raw possibly acts as an anti-spasmodic and with anti-proliferative effects on tumour growth.

Finally there is THCV (tetrahydrocannabivarin), which may play a role in obesity-associated glucose intolerance.

In total 113 cannabinoids have so far been identified, undoubtedly soon to be increased as the discovery of the endocannabinoid system itself didn't occur until 1992. The cannabis plant has in total about 545 different known compounds (phytochemicals/nutrients) within it.

* * *

Cannabis is still a Schedule 1 drug in the USA, defined as a substance with a high potential for abuse, with no current accepted medical treatment with a lack of safety for use under medical supervision. This is as laughable as it is wrong, as demonstrated from the previous history, traditional uses and the scientific research in the next section; it's also somewhat ironic when one considers the devastation caused by the legal drugs of tobacco, alcohol, fentanyl and OxyContin!

Irrespective of the law, the US government holds the patent on the non-psychoactive cannabinoids (no. 6,630,507), with the others divided up; here US firms dominate, including Abbvie, Sanofi-Aventis, Merck, Pfizer and Bristol-Myers Squibb, while GW Pharma are the top non-US company along with the Hebrew University of Jerusalem—so many medical patents for a plant demonised and criminalised for nearly a century and still today in half the world.

Still illegal in Britain (with medical exemptions), there is unsurprisingly similar duplicitous behaviour being displayed on both sides of the Atlantic, although Britain's duplicity comes with a large slice of nepotism. The story goes that British Sugar (which grow most of the country's sugar beet) was having a torrid time with their tomato sideline they had expensively set up. Its managing director Paul Kenward managed to turn Europe's largest greenhouse, the size of 23 football pitches (45 acres) into the biggest cannabis grow room in the world, instantly becoming also the world's biggest supplier of legal medical marijuana. It was an exceptional achievement, especially when the plant was still fully illegal and licences are nearly impossible to come by.

Oh, yeah, I forgot to mention—Paul Kenward is married to the Tory MP Julia Atkins, who was the Drugs Minister and managed to preserve her husband's monopoly by refusing to allow cannabis legislation, rejecting the claims of any efficiency from medical marijuana as lacking any clear evidence. But it's her husband, and she might not have realised what he did, in just the same way that Kenward also managed to secure a single buyer for all his bud, to the then Prime Minister Theresa May's husband Philip, or rather to GW Pharmaceuticals of which he has a 22% stake and whose CEO just happens to be a significant Tory party donor. Roughly 100 tons a year of bud is being grown for immense profit by Kenward, securing 46% of the world's medical cannabis, and they've made many tens of millions of pounds from this self-imposed monopoly.

* * *

On the other hand, someone who should only be congratulated for his current wealth through determination, perseverance and a solid belief that the world would be a better place with loads of cannabis being grown, is the vision of Paul Benhaim. A hempreneur who has been shouting about hemp since 1993, and who was responsible for the cheekily named 9Bar mixed seed snack bar in 1996, he has helped with hemp build-ups globally, been involved with utilising it for building and as a plastic and now runs The Hemp Plastic Company, Hemp Foods Australia as well as being the CEO of Elixinol, a CBD company with stakes in many other companies as well.

This all amounts to little more than a drop in the ocean compared to the outstanding wealth creation centred on cannabis in the past decade. As of the end of 2019 the largest firms in the cannabis sector had experienced incredible year-on-year growth. Ranking number one and heavily invested in by musician Snoop Dogg, with the largest market capital of £17bn in 2019, Canopy Growth from Ontario experienced a 313% yearly growth rate from 2018–19, although it slipped to a 2022 worth of 'only' £1.75bn. Another with massive initial growth and subsequent drop was Aurora Cannabis from Vancouver; its peak was £9.5bn in 2019 and a growth rate of 305%, more recently reporting worth of only £440m. Putting these to shame was Arena Pharmaceuticals, which experienced an eye-watering 45,544% growth rate from 2018–19, with a market capital of £2.2bn from California, rising to £5.3bn in 2022. iAnthus Capital Holdings of New York must have been a little disappointed with their comparatively meagre growth rate of 4,176%!

Britain's GW Pharmaceuticals, out of Cambridge and founded in 1998, grew from £3.8bn market capital in 2019 to £6.1bn in 2022, with their Sativex and Epidiolex being worldwide big sellers, all from highly questionable nepotism and monopoly. There are plenty of other large companies to choose from, all with enormous growth rates not seen since and maybe even exceeding the silicon and dotcom frenzies of the late 1980s and 1990s. As expected there was a crash after the initial hype, with the massive growth rates proving to be unsustainable, and with too much product on the market many companies fell off the cliff from 2020 to 2022.

Former heavyweight boxing champion Mike Tyson has been reformed thanks to cannabis, helping him to get off the highly addictive opiates to help with his constant pain. He now grows and runs Tyson's Holistics, producing CBD oils, and wants to become a bigger advocate of cannabis by creating a festival, a college for wannabe growers and even a cannabis theme park—if he manages to find the time among smoking several ounces of premium bud every week, more even than Snoop!

CBD from cannabis and Taraxagum tyre manufacture from dandelions could well be the bigger industrial uses that allow for many other revenue streams and industries to be established from the quantity of raw material available after these initial uses. The extraction of CBD allows the waste material to be used in plastic, paper, fabric, ethanol, biomass, food and the supplements industries. It does look like some serious money is already being sucked up by the pharmaceuticals, so let us see what the world's scientific community has to say about *Cannabis sativa*—

Some science

Left to just the sensationalist media headlines, it could easily be assumed that cannabis is a new wonder drug with unrivalled omnipotence. Although, as we've seen from the extensive history above, cannabis was used for a range of ailments, albeit definitely limited, as corroborated here—

The epilepsy potential was scientifically established in 1970 in a study conducted by Raphael Mechoulam in which two of four epileptic patients stopped having seizures after cannabis was administered. Raphael discovered the active component of marijuana, the cannabinoid THC, in 1964 with Yechiel Gaoni—he has also uncovered some ancient connections with religious practices and cannabis, but that's another story.

Another human trial with a few more people, but still too small with only eight patients, a 1980 report in *Pharmacology* magazine (Vol. 21), showed that in a CBD blind experiment, four of the eight patients remained almost free from convulsions, with only one showing no improvement at all.

This 50% success rate was again demonstrated in the *Journal of Neurology, Neurosurgery and Psychiatry* (Stockings et al. 2018) for epilepsy and reducing seizure frequency, nearly fifty years after the Mechoulam research. Another study from the Paediatric Neurology Unit, Tel Aviv, Israel (Hausman-Kedman et al. 2018), with 57 patients, again gave a 50% success rate.

In the case of refractory epilepsy (refractory means conventional drugs are not working) a Mexican study by Dr Aguirre-Velazquez showed a CBD-based treatment decreased the frequency and severity of seizures, improved sleep and diet, with up to 80% seeing a reduction in convulsions. CBD has to be considered as a useful adjunct, particularly when other drugs have failed as in the case of refractory epilepsy, whose rates appear to increase over time—that is, the longer the conventional drugs are taken, the more they tend to not work. With up to 30% of those suffering from epilepsy being refractory, it makes sense to at least give a go to an alternative with 50% effectiveness.

On a personal level, a good friend's son, who is both autistic and epileptic, was suffering from absence seizures, previously called *petit mal*, when he'd just 'switch off' for a few seconds, as frequently as 30 times a day. After studying his diet it was seen his low-sugar cordial was sweetened with aspartame, a known neurological-excitor. With this eliminated from his diet he now only suffers a tonic-clonic seizure, a *grand mal*, every few weeks as he had always done, as is more 'normal' for an epileptic. It's a good example of a man-made chemical, aspartame, causing additional suffering for years, because his mum thought it would stop him getting fat and be better for the teeth of her other children, never realising that it could ever be responsible for a life-threatening seizure 30 times a day.

* * *

A rare piece of British research came out of Porton Down Science Park—definitely not an amusement park—where the UK's secret chemical and military facilities are as well as GW Pharmaceuticals' labs. GW first used nabiximols from cannabis in Sativex, the first market-approved cannabis derivative, in 2010; they also have Epidiolex (2018) for epilepsy. It is a company with both colourful and shady developers and partners, such as Sam the Skunkman who helps with the growing and Bayer who distribute them. From their own research GW found that Sativex aided sleep for those suffering from multiple sclerosis, rheumatoid arthritis or intractable cancer pains, with no tolerance allowing dosages to be kept minimal. Their conclusions were then reported in scientific literature, muddying the knowledge as the actual results from GW's research on Sativex's use as an analgesic (pain reliever) were actually both negative and positive, with severe side effects and drop-outs by participants; results beat the placebo for pain control, but were far from conclusive.

The recent press coverage of young children and epilepsy, with calls for legalisation in Britain, is on the one hand a good bit of press and exposure to a possible curative route, but on the other hand it also has the stench of an orchestrated marketing ploy, known as diseasemongery, by the pharmaceutical corporations. Putting pressure on the UK government to allow sales of their new 'wonder weed drugs' at the same time as extensive press coverage of patient pain, focusing on children being denied their CBD/THC medicine to garner public support—coincidence?

* * *

The anti-inflammatory ability was looked at for any benefits it may have with rheumatoid arthritis and liver inflammation by the Hebrew University Medical Faculty in Jerusalem (Haj et al. 2015). This used a novel cannabinoid they synthesised, HU-444 and reported good results, being recommended as a potent oral drug in the therapy of rheumatoid arthritis.

From the University of Medical Science, Tabriz, Iran (Rezapour-Firouzi et al. 2014) research looked at how multiple sclerosis patients fared on hemp oil, evening primrose oil (EPO) and olive oil. There was favourable activity among the enzymes involved with inflammation for those taking the hemp seed and EPO, while the olive oil group saw an undesirable increase in EDSS (expanded disability status scale) levels.

As with nettle and the other previous plants, it appears cannabis enhances the effects of pharmaceuticals, having the bonus that this synergistic ability will allow fewer toxic pharmaceuticals to be consumed, as well as opening the door to the possibility that these plants will act as an effective replacement. As reported in the journal *Progress in Neuro-Psychopharmacology and Biological Psychiatry* (Sales et al. 2018), this synergistic capability was acknowledged, as well as the antidepressant effects, which were noted as being dependant on appropriate serotonin levels (bananas and cocoa being superb for this).

From the *Frontiers in Oncology* journal (Yasmin-Karim et al. 2018), American researchers found CBD, alongside radiotherapy, effective at battling pancreatic and lung cancer cell lines, with expected further improvements with better localisation and delivery of the CBD to specific affected bodily areas, when used as an adjunct with enhanced efficacy potential from a bespoke THC/CBD ratio. This was also concluded in the *International Journal of Pharmaceuticals* (Fraguas-Sanchez et al. 2020), demonstrating that CBD was effective as an adjunct to the current breast cancer treatments of paclitaxel and doxorubicin.

CBD was shown to be a potent antioxidant (Hampson et al. 2000), with neuroprotective benefits, and reported as being best for prevention but feasibly also for current neurodegenerative disorders (Ferdandez-Ruiz et al. 2013), such as Parkinson's (Lastres-Becker et al. 2005).

Epidemiological work from the University of São Paulo (Zuardi 2008) looked at all the research over the past 45 years, noting that initial research was on the epileptic and sedative effects in the 1970s and that it was not until the turn of the century that the range got extended to include Parkinson's, Alzheimer's, diabetes, arthritis, cancer and inflammatory diseases—all the latter are mostly awaiting confirmation from clinical trials, it must be noted.

Research by Manuel Guzmán (another leading researcher in this field) from the Department of Biochemistry and Molecular Biology, Complutense University, Madrid, Spain, in the *Journal of Cannabis and Cannabinoid Research*, looked at the problem of there being little homogeneity in the extracts, preparation and consumption of cannabis and its components, even among the already approved drugs of Marinol (THC/dronabinol), Cesamet (synthetic analogue of dronabinol, nabilone) and Sativex (nabiximols)—all being approved as antiemetic (anti-sickness), anticachexic (anti-wasting), analgesic (pain relief) and anti-spasmodic medicines, each with a range of side effects. For all the millions self-administering cannabis there is no understanding or compilation of effects, uses and effectiveness. What is required is better understanding of bespoke THC/CBD ratios for individuals and ailments. More work by Guzmán with Abrams, in 2015, found the benefits were limited by its psychoactive effects, even if effective for anorexia and chemotherapy relief etc.

Research from Poland from the Poznan University of Life Sciences (Sledzinski et al. 2018) showed that THC had the propensity to both prevent and encourage proliferation of certain cancer cells—a clear warning and call to obtain better understanding of which cannabinoid is best for which ailment and which contraindicates. The only human study on record thus far was of a very limited group of nine patients with recurrent glioblastoma (aggressive brain tumour); in two patients the tumour growth decreased with THC, which is good but hardly concrete evidence of effectiveness.

The general consensus among conclusions on the overall efficacy of hemp and marijuana agrees with the findings of Rasche et al. (May 2018) from the University of

Bonn, Germany: … "In most cases, today's assessment of cannabinoids relies on studies that are classified as low evidence. Therefore, further studies which involve more participants and evaluate long-term effects are needed." Although being in the health/supplements world I come across a lot of customers and sick people, many of whom have sworn that either CBD or a THC/CBD mix (full spectrum) oil has been their saviour and the only thing to have worked; these people predominantly had muscular pains, sleep and anxiety, with some success and also failures with cancer.

The same conclusion was reported from the Division of Rheumatology, McGill University Health Centre, Quebec, Canada (Fitzcharles et al. 2015). The research team looked at the efficiency of cannabinoids (nabilone and THC/CBD) for patients with fibromyalgia, osteoarthritis, rheumatoid arthritis and chronic spinal pain, finding there to be insufficient evidence to recommend cannabinoids for these conditions—conditions that we have already seen can be ameliorated by the previous plants in this book. There is far more agreement that cannabinoids do have definite efficiency with conditions such as epilepsy, depression, muscular discomfort, appetite enhancement and all the other benefits of being 'high' (if on a THC medicine) as opposed to struggling with aches and pains.

* * *

Unsurprisingly, as profits are being adversely affected, there does appear to be quite a bit of research looking at the uptake of medical marijuana in favour of other prescription drugs, whether for anxiety, migraine, pain and sleep, done by the pharmaceutical companies themselves—no doubt to gauge how much they can prepare to lose before changing direction and joining in with the marijuana jamboree.

One of these studies involving nine researchers (Piper et al. 2017) was an epidemiological study, which concluded that as many as 77% of people taking opioid medications would or have taken medical marijuana as an alternative—72% for those on anti-anxiety medication said they had or would consider it; 42% as a substitute to alcohol and 38% as opposed to antidepressants. Over half of all authorisations for medical cannabis use in Canada are for those suffering from painful arthritis, which will have an impact on NSAIDs sales.

This switch from pharmaceuticals to cannabis was studied at the Center for Medical Cannabis Education in California (Corroon et al. 2017), noting that 36% switched from narcotics/opioids, 14% benzodiazepines and 13% antidepressants. This study involved 1,250 people, and if these figures were extrapolated nationwide and globally, the earnings from medical marijuana are potentially astronomical, as are the loss of profits for the traditional pharmaceuticals, which since the 1970s has been far more focused on profits rather than health (as said by the CEOs themselves). This is why these corporations with their vast finances use a strategy to ascertain all the associated patents, eradicate the competition by either buying them, undercutting them or slandering them (if not all three), construct a marketing plan, form media reportage of pseudo-scientific reports and 'news' stories, raising fear and anxiety levels, putting the drug and diseases in the limelight and then offer a solution, irrespective of the veracity.

Even with incidental evidence of veracity from CBD oils etc., disruption is still being caused to the revenues of the pharmaceutical sellers, notably a significant drop in prescriptions for ailments linked to chronic pain, depression, appetite enhancement and sleeplessness. Legal cannabis sales in the US in 2015 were £5.4bn, increasing to £22bn in 2022 and expected to exceed £37.7bn by 2025. As long ago as 2013 medical marijuana saved Medicare in the US $122 million, and if available nationwide the saving would be closer to half a billion dollars, with more up-to-date figures unavailable. Saving money should not be the goal, although it's a useful first step as it creates an incentive, especially where still illegal; the goal should always be to improve all aspects of health and education, providing a longer, healthier lifestyle as ailment-free as possible.

Even though half of the States of America have legislated in favour of medical marijuana, to conduct any trials on it applications have to go through the FDA. It would be reasonable to assume the FDA has been lobbied by pharmaceutical corporations to maintain as many barriers to research as possible, until they are able to secure their own patents and monopolise this new drug, applicable to several sectors of the health industry in the form of depressants, sedatives, appetite enhancers, anti-epileptic, anti-convulsive and possibly arthritis and cancer sectors as well. With cannabis still being an illegal substance federally, this acts as a severe impediment to applications, and irrespective of an individual State's legal status, the omnipotent Drug Enforcement Administration (DEA) still have it classed Schedule 1.

Side effects

Unlike the previous four plants, it would be wrong to argue there are no side effects from cannabis, or that all it does is potentially interfere with or even improve prescribed drugs. From hemp as such there are no side effects, apart from the positive ones from ingesting a nutritious plant, whether the leaf, root, oil or CBD. For marijuana with the psychoactive THC there are many side effects, and when it's smoked there are of course all the negative associations of tobacco, or the tars and lung damage if smoked neat in a pipe, or the as yet unknown damage to the respiratory system caused by the new vapourisers or the potential future psychotic side effects. No science yet.

Side effects reported from cannabis include possible addiction, asthma, rapid beating of the heart, low blood pressure, muscle relaxation, bloodshot eyes, dry mouth (xerostomia), gum disease, slowed digestion and metabolism, increased appetite, nausea, diarrhoea, depression, short-term memory recollection problems, disorientation, lack of motivation, dizziness, drowsiness, hallucinations, lethargy, OCD symptoms, slower reaction times, and decreased testosterone in men and increased prolactin levels in women. Much of this is well-known and commonly known as being 'stoned', which has a subjective rising scale from being slightly inebriated all the way to where mental facilities temporarily malfunction or are heightened to all manner of stimulus and awareness, be it auditory, smell, touch or visual, being nothing short of hallucinatory. Which might well be the state desired for a night out at the club with your friends but certainly not appropriate for the school-run or to get through the day because the pain and lack of appetite from cancer is debilitating enough, and being stoned all day may well be highly undesirable. This psychedelic effect is only experienced from the THC, although CBD in high doses can have a soporific effect.

Smacked out or stoned?

An incredible number of adults are taking painkillers at the moment, both OTC and via illegal distribution, with numbers constantly rising. It was reported in the *British Medical Journal* (2016 Fayaz et al.) that there are at least 28 million adults in England suffering persistent chronic pain. In 2015 119 million Americans (of 12 years and above) had taken painkillers, tranquillisers, stimulants and sedatives that year; it's definitely more now, so pretty much half the population of over-15s.

For most of these pain sufferers it literally is a toss-up between being smacked out on the opioid drugs or stoned on marijuana. It is perhaps worth a look at a couple of these opioid manufacturers to help assess where the faith of the patient should lie—

Insys Therapeutics have to be mentioned first, because they made the controversial fentanyl (there are other pushers as well). This causes the most deaths per year of all legal or illegal drugs (except tobacco and alcohol), and is continually rising and, alarmingly, it is spreading out from America. There, as many as 35,000 will die directly from fentanyl alone by the mid-2020s (it could be as diabolically high as 100,000, which is twice the number of US troops who died in Vietnam between 1965 and 1974). It's out of control in terms of production and addiction. Insys no longer make fentanyl, or Subsys, their sublingual spray, where for far too many people a-spray-a-day led to a crippling addiction; meanwhile fentanyl pills are sold for $5 and ground up and smoked just like heroin on the street, although this dragon is reputedly 50x stronger.

Fentanyl was approved for severe pain caused by cancer only, although Insys bribed thousands of doctors to prescribe it for general pain. The boss, John Kapoor, was not only arrested but also convicted and given a five and a half year prison sentence for racketeering—the charge they get gangsters on. He served two years and was released from jail in June 2023. The case is staggering, and the drug made Kapoor a billionaire while ruining millions of lives and killing hundreds of thousands directly—overdoses directly attributable to fentanyl were 20,100 in 2016 in the US, up 540% in three years and staying at this level still, where more people aged 18–45 died of fentanyl overdoses in 2020 than Covid, car accidents, cancer and suicide combined. The stories are so tragic, it's like an episode of *Narcos* or a gangster movie, and this drug has affected such a range of people, many of whom wouldn't even have dared to smoke a joint at college, fearing drugs to be bad for them—it's a travesty that has been kept fairly quiet in the UK for the most part.

Insys sales in 2012 were $8.6 million, and after their illegal tactics of bribes, lies, talks, conferences and lots of dinners and strip-clubs for doctors, this had risen to $329m in 2015. It kept rising until Kapoor's arrest in 2017. In 2019 he was sentenced with agreement to pay a $225m fine, only to declare bankruptcy straight after. Deplorable hardly describes him. He and the doctors should actually be on trial for mass murder, because that is what it has been. Insys started the craze (it was actually first made in 1960 by Paul Janssen in Belgium) and now it's being made by gangs in China and Mexico and continues to flood American streets, with it even set to replace heroin, creating droves of addicts daily, even finding its way into MDMA and ecstasy pills in Europe. Hopefully, just as with methamphetamine and the like it manages to mainly stay over the other side of the putrid pond.

Insys recognised the danger posed by cannabis, so not only did they develop their own cannabis-based drug, at the same time they hedged their bets and 'donated' £370,000 to an advertising campaign damning marijuana before Colorado had a public vote on the matter of public consumption in 2016. There has definitely been a swing to marijuana from opioid drugs, and as a consequence opioid deaths were down 25% in States where medical marijuana was available. It is too late to bring back musicians Michael Jackson, Prince, Tom Petty, rappers Lil Peep and Mac Miller and Justin Townes Earle, among many thousands more, including a State's execution recipe, and the gas used by Russian forces in a Moscow theatre siege in 2002. In this siege it was used to knock everyone out as quickly as possible, killing 130 of the 850 hostages. It would be fair to say, albeit with an addictive potential there have been no deaths from cannabis, so it has to be—Opioids 0: Cannabis 1.

* * *

Fentanyl deserves the top spot, but the podium is also shared with OxyContin, a weaker type of smack that manages to cause just as much addiction and harm, without the ever-present roulette of death that comes with fentanyl. OxyContin, made by Purdue Pharma, has been peddled so much that the owners, the Sackler family, are now worth about £10 billion. Arthur Sackler (who died in 1987) was no stranger to mass public drugging, having been responsible for marketing valium decades ago as the modern housewife's medicine. The law has now caught up with the company's aggressive marketing (racketeering again), flooding communities and basically making hundreds of thousands of Americans addicted to opioids and causing numerous suicides. Purdue was caught bang to rights but still guaranteed no jail term, only a massive fine, somewhere around £7bn (could rise to $10bn), causing the company to file for bankruptcy! Opioids 0: Cannabis 2.

Purdue should receive court approval for a reorganisation to transform the company into a 'public benefit company', run on behalf of the communities devastated by the opioid crisis. This is exactly what should happen to DuPont for Teflon, Monsanto for glyphosates, Ineos for plastic and other mass desecrators who should be up for crimes against humanity and the environment—ecocide—and instead of getting tax breaks, they should incur environmental taxes. There are books, a film and even a television series about Purdue now, so the information is seeping out to the public.

Pharmaceutical companies collectively pay out every year well over £3 billion in fines globally, usually for false claims and a failure to disclose safety data—that's called an off-label promotion, which makes it sound very minor, but it definitely isn't. From 2000 to 2021 pharmaceutical companies were fined a total of £70.4bn, from 1,006 violations, a staggering figure for a legal business. Individually, in the 2020–21 period Johnson & Johnson were slapped with £12.2bn from 65 violations, Merck £8.6bn (81 violations), Pfizer £8.2bn (75 violations), Purdue £7.3bn (10 violations), GSK £7.2bn (39 violations), and on and on. By way of comparison during the same period the world's oil and gas companies look like angels with just £44bn in fines from 5,905 violations; utilities and power generation came in third with £25bn in fines; chemicals unsurprisingly is a big violator with £17.5bn in fines from 4,738 violations, with bad boy Bayer paying nearly £11.4bn of that from 141 violations; healthcare services don't look too healthy either with £15.5bn in fines; aerospace and military contracting have been fined £5.7bn, with Boeing well ahead of the field at £3.2bn (which is more than covered by their subsidies, which have totalled over £12.2bn). None come close to the pharmaceutical corporation's transgressions. That is wrong, isn't it? Constant repeat offending to still be given tax breaks and subsidies ... Moving on—

* * *

In the vast majority of the larger publications from the bigger publishing houses there is no mention in any of the herbal/pharmacological/drug books that cannabis has any benefits. It simply is not in the books at all (apart from a 1970s Rodale publication). This exclusion can only lead the reader to assume it is because cannabis has no pharmacological or other use. All those book burnings of the past very effectively censored any information being passed down generationally, and more recently being excluded from books and censored for a couple of generations means that omitted plants soon become forgotten about, on the basis that it is only the plants documented that are of any importance. If cannabis had no role in the past, why did up to 60% of remedies from headache to sickness and pain have marijuana as their main ingredient at the turn of the twentieth century? This is a fact, that cannabis was the main drug, but since legislation the plant has become redundant for all purposes, and all history of its use becomes harder to unveil as less is written of it. Was this intentional?

Before our health was dominated by pills and profit-driven corporations, the foods, brews and views that actually worked had a continued use owing to their healthiness and efficacy. It would make no sense to continue using plants just because folklore dictated it (as is the dogmatic world we are caught in today); continual use was because of the proven effectiveness, all being a part of our timeless inherited wisdom—our humanness—now nearly lost.

Cannabis sativa would have been used as an important source of food, clothing and cordage, being central to the life of many people. Its ritualistic uses would have been contentious for the monopoly of the Christian churches on being the sole provider of rituals and spiritual endeavour, with everything else labelled as Devil worship—although in reality it had been the Inquisition, the Conquistadors and missionaries who were often the real devils, spreading death through violence and disease. As is the case today with terrorism, we call 'them' terrorists, when the truth is, 'we', our cultural system is the real terrorist. Terrorism is a word we stole from the French in 1795, as was the word guillotine (a contraption actually invented in Halifax, England in the thirteenth century) after the French Revolution, where *terrorism* meant being governed by fear and terror—which we are of course ruled by today, driven by the media, a world's media controlled by just a few global corporations such as Walt Disney, Comcast, Time Warner, Google, Facebook, Viacom, Bertelsmann and News Corp.

Wrapping it up

After nearly a lifetime of research into the role of psychoactive plants in human history, the late Harvard University Professor of Ethnobotany Richard Evans Schultes surmised that:

> Early man experimented with all plant materials that he could chew and could not have avoided discovering the properties of cannabis (marijuana), for in his quest for seeds and oil, he certainly ate the sticky tops of the plant. Upon eating cannabis, the euphoric, ecstatic and hallucinatory aspects may have introduced man to the otherworldly plane from which emerged religious beliefs, perhaps even the concept of deity. The plant became accepted as a special gift of the gods, a sacred medium for communion with the spiritual world and as such it has remained in some cultures to the present.

It is evident from the scientific research thus far conducted on cannabis that the results have not been overwhelmingly positive, unlike the previous four plants (or the two to follow)—not that this has stopped the money wolves from lapping it up. Little has been learnt thus far from the experiments that have taken place, although the traditional uses have been corroborated, ranging from convulsions, tremors, sleep, pain, appetite and digestion; it also appears, as with the other plants, that cannabis is an excellent adjunct to other treatments.

What hemp and marijuana do have are some very helpful tools. The seeds provide, for instance, the perfect ratio of omega-oils acting as precursors and controllers of inflammation, which will play a critical role considering that both the early stages and the latter stages of many diseases are characterised by inflammatory symptoms. These omega-oils are a perfect preventative, and with a lifetime's consumption many diseases would therefore have been prevented from manifesting. To pin any hope on a preventative acting as a curative is not being suggested here, especially when the human organism after years of maltreatment/life in a city is a very polluted entity. One also has to be wary in today's climate of marketing, as the new panacea of CBD and THC has to come with the warning of a very large fashion flag.

There is no conclusive evidence that an oil made from the female buds is any more remarkable than the seed oil from the same plant, although both are undisputedly excellent. The conclusion from this chapter is that if each of the seven plants were to have a unique selling point, that for *Cannabis sativa* would be the seed oil's dietary inclusion as a preventative, benefit and possible cure for inflammatory diseases. The potential of the cannabinoids is certainly excellent in the palliative sector, for pain relief, as well as helping with sleep patterns for the chronically ill, and an appetite enhancer for those too sick to eat, while its historically important role with epilepsy will hopefully play out with Parkinson's also. As more research is conducted and human trials are done, the truth about the veracity of these plants' health capabilities will have to come out in the open, or not if they fail to live up to the expectations.

If the money is there it is almost guaranteed that the industries will follow, and as legalisation spreads across the world every industrial sector will begin to utilise hemp as a raw material, especially if it truly is of a higher quality. It will become more cost-effective to do so as it becomes increasingly legalised, whereas when Jack Herer wrote *The Emperor Wears No Clothes*, it was but a pipedream in the middle of a war on drugs with no chance of even ascertaining any raw hemp to initialise any industrial uses.

It's seen as a failure if hemp cannot conform to a monocultural, chemical-led agricultural regime. Yet hemp does fantastically well on the cottage level with a multitude of smallholdings acting as co-operatives. This is where the future of hemp and cannabis growing lies, and this is where the raw materials will be grown for what will be well over 25,000 industrial uses.

* * *

A final point noticed from managing to uncover far more information than expected about cannabis, is that the hippies weren't the first to go all arty-farty and peace and lovey from smoking some pot. The hippy era is the only one in living memory. The smokers in the parlours in New York in 1900, the Hashisch Parissean club of the 1840s, the Saddhus and Buddhists, Sufis and Zorastarians of old all follow a similar thematic in many respects. It could be argued that the ingestion of the *Cannabis sativa* plant creates a particular frequency that is naturally drawn towards flowery symbolism, earthy matters and artistic appreciation, allowing the positive harmonious vibration of the plant to manifest as a subcultural wave of oneness, peace, love and harmony … man! As much as I'd love to pass it all off with a shrug and put it down to hippy shit, it's not, it's much bigger than that. All plants, every aspect of Nature is divine, and they are all about to shine.

Cannabis sativa is the plant that can earn billions from its growing and utilisation across all economic sectors from health, food and industry, and it can also retune the human organism to a more emotional, harmonious and caring frequency. People are not tuned in right now. Tuned in to their phones maybe, but the answer is never going to be in there; there is no app or algorithm to change cultural systems, just as the revolution will not be coming from Facebook, X (Twitter) or TikTok. The answer is out there—all we need to do is tune in and let the plants awaken us. Civilization has detuned us, and when this is realised, the reharmonising can begin.

Which is all well and good, but it's time to move on to our sixth plant, a tree, which is still a plant. It has a role to play with our biggest killer and to teach us a little about historical subjugation and segregation to begin to really stir up the reformation along the way.

HAWTHORN

(Crataegus monogyna)

Barbs of wisdom

A few years ago, while helping to thin a friend's hedge between two of their fields, everything went quite smoothly until we got to the last section. We were confronted here with a tangled and contorted, convoluted mess, with a few low horizontal branches sending up twisting arms, creating an impenetrable barrier of randomly protruding branches, each concealing a formidable armoury. It was possibly two hawthorns, maybe even three, being extremely hard to locate the trunks. Even with a chainsaw it was a tremendously difficult task, with more hacking and jabbing going on as opposed to cutting, but after a while enough was removed to establish a more reasonable height for the hedge, although this was far from the end. It was the end for the chainsaw, though, as it spluttered its way to the recycling centre.

In trying to clear the felled branches, I was taught countless lessons by the thorns, being left with plenty of reminders from torn clothing to the hidden extras in the tips of the thorns. Each one piercing through the skin created a puncture wound that left a searing pain with heat. So twisted and gnarly was each branch that it had to be carried away individually or else it got caught and snagged on everything it could, even the grass. The branches pulled this way and that, across me and inevitably into me for one final time, as I irritatingly exclaimed: "What is the point of this fucking monstrous plant?" It highlighted my ignorance at the time, although instant karma had been heaped upon me, repeatedly. There is of course a plethora of other points to this plant—or is that barbs of wisdom?

Hawthorn (Crataegus *sp.)*

Of all of the trees that surround us, the mighty oak, elm and ash, beech, birch and sycamore, why choose the hawthorn? The truth is, as already mentioned—all the plants chose me—or perhaps the hawthorn infected me from the above attack, of which I was the clear perpetrator. From that moment on, after uttering that ignorant, vulgar statement, I see the world more from the perspective of the hawthorn. This shift in view has unravelled what it can offer us, all of which was certainly somewhat unexpected.

Other local trees may have had a more illustrious role in the past, as well as offering some form of hope for the future. From a culinary and medicinal perspective elder laid a good claim (especially its anti-flu potential), while the oak certainly held a lofty and revered position from cultures past, as did hazel and my favourite edible nut, the sweet chestnut, but only the hawthorn resonated. The hawthorn represents so much more than merely a tree, a global companion to man, while we have certainly been kept ignorant of some of the significant and shocking historical facts it exposes.

A Hawthorn Berry

How sweet a thought,
How strange a deed,
To house such glory in a seed
A berry, shining rufously,
Like scarlet coral in the sea!
A berry, rounder than a ring,
So round, it harbours everything;
So red, that all the blood of men
Could never paint it so again.
And, as I hold it in my hand,
A fragrance steals across the land:
Rich, on the wintry heaven, I see
A white, immortal hawthorn-tree.

(Mary Webb, 1881–1927)

Botanical

The commonest hawthorn tree found throughout Western Europe, especially Britain, is *Crataegus monogyna*. This botanical name derives from *mono*, meaning one and *gyna* meaning female, which may allude to its single carpel (the female reproductive part of the flower), although *monogyna* also only has one seed, while many of the other *Crataegus* genus have five. The word *Crataegus* derives from the Greek *Kratos*, meaning strength and *akis*, sharp.

It's not a rule of thumb by a long shot, but the plants named with a Greek root are more likely to be a medicinal plant. This also applies to medicinal words in general, often Greek yet wrongly assumed to be Latin by a public that neither speaks nor learns Greek or Latin as a part of their syllabus, excepting of course those attending Eton and the like.

As with the other plants from the book, the hawthorn prefers disturbed soils rich in phosphorus and nitrogen, so by the way we spoil the soil, they would have found us initially, allowing us in turn to discover them.

The hawthorn tree is a member of the extended rose family, *Rosaceae*. This family contains a lot more than roses, including a range of trees, shrubs and small plants (covering a known 4,828 species and 92 genera), from apples and pears, plums, cherries, almonds, strawberries and blackberries, to meadowsweet and dropwort. The fruit of the hawthorn is botanically classed as a pome, and such a plant as this could have been the original progenitor of the apple family long ago, with quite free and easy cross-pollination and hybridisation potential, not to mention the almost endless grafting possibilities between trees of the same genus. This offers the interesting possibility of plum, cherry, peach and apple (all *Prunus*) being able to be grown from one tree.

The variation in hawthorn fruits, called haws, is quite large, ranging in size from 0.5cm to 4cm in diameter, maturing into hues of red (by far the most common), yellow, black and purple through the autumn. Every *Crataegus* produces an edible fruit, some more palatable than others; notable among the 200 or so separate species is *Crataegus arnoldiana*, which produces a sweet, 2.5cm-diameter, soft and mealy-fleshed haw, found in many nurseries in Britain, possibly heralding from America. *Crataegus azarolus* has been infrequently cultivated in the Mediterranean for its 2.5cm fragrant fruit, sugary with a slight acidic-kick; when grown in more northerly climes, the haws tend to be less sweet. *Crataegus baroussana*, most commonly found in Mexico, is one of the more shrubby of the genus, reaching little more than 5ft tall, producing a sweet, 2.5cm haw, while the rest of the genus average from 15 to 20ft tall, occasionally 30.

Crataegus missouriensis has a haw about 3.5cm in diameter (similar in size to a cherry), providing a sweet, comparatively juicy fleshed haw; also found throughout Britain, its species name suggests it has been introduced. Britain's most common hawthorn, *Crataegus monogyna*, has a slightly appley, dry fruit, about 1.5cm in diameter. *Crataegus schraderiana*, a South European species, has been claimed by the appropriately botanically named Ken Fern as the most delicious fruit he'd ever eaten. One species found around the world is *Crataegus macrosperma*, aka Bigfruit Hawthorn, obviously a favourite for that very reason. As is *Crataegus succulenta*, weighing in with the largest fruit from the genus at up to 4 to 5cm in diameter, and living up to its name with a sweet, juicy and pulpy flesh, although as seen below, the size and flavour of the fruit is not all it excels in.

The *Crataegus* family comes with a range of thorns to match its range in fruit: some provide double-trouble with barbs on the thorns, entangling and puncturing, while some also carry toxins as an extra treat. This tree has certainly evolved quite an arsenal to protect itself, which only says one thing—"there is something worth protecting here"—shrinking the leaf size while putting considerable energy into those thorns. For instance, *Crataegus succulenta*, the most wide-ranging of hawthorns in North America, has a profusion of 6cm-long purple spikes, with added barbaric spikelets for good measure—no doubt because this species' fruit is one of the best to eat, and extra protection would be expected from a highly evolved and intelligent plant.

Thorns from most of the *Rosaceae* family can cause synovitis, a type of arthritis causing inflammation of the lining of joints, while far more common is the painful wound left behind by many, blackthorn and roses especially, where the flesh turns purple around the puncture and can hurt for several days. The tip can also snap off as well, leaving a painful splinter, which takes the body's defences some time to slowly break down, if at all. Not all the hawthorns have thorns, with varieties of *Cratageus orientalis*, *C. pontica* and *C. aronia* and a few others being thornless; I have witnessed one on a recent walk and happily managed to fill a small bag with haws.

And if the thorns weren't bad enough, some species have a hard, sugary biofilm covering the thorns containing various bacterial colonies, with anthrax, dysentery, tetanus and gangrene all found on *Crataegus aronia* (native to the USA). Not only that, there's more—some thorns also harbour fungi, which when entering the body through the puncture wound can lead to very serious diseases such as Fonseca's disease, resulting in elephantiasis! Fortunately for most of us, these are only found on *Crataegus* growing in the tropics.

Folk and ethnographic

There are many historical folk stories and traditions about the hawthorn, clinging by a thorn just below the surface of our current cultural knowledge, suggesting this tree once had a deeper connection with us. We might have forgotten, but that doesn't change the fact that hawthorn finds its tangled form intertwined with the human drama on many levels, from history, agriculture, politics, religion, paranormal, literature and subjugation—all from a tree most people know next to nothing about or could even identify today.

In Staffordshire, if someone from outside of the family placed a bunch of hawthorn flowers 39 days after Easter Sunday (Ascension Day) in the rafters of a house the family would be protected from witches, spirits and storms. In Herefordshire, a hawthorn globe was hung in the kitchen all year from New Year's Day with a replacement burnt at 5am by the men in the field sown first the year before, while the women made a replacement to ensure good crops.

In Devon, similar traditions prevailed, notably on May Day when blossoms cut that day were fixed to houses and cattle sheds as protection against lightning, witches and evil ghosts. Blossoms were also woven into the crowns and wreaths attached to the Maypole, and a small spray carried about the person offered insurance against ill-luck.

In Suffolk, any servant who brought hawthorn flowers into the house on May Day would be rewarded with a dish of cream. A sprig fastened to a cowshed on May Day was said to safeguard the milk supply. In Cornwall and Devon, May Day is also known as Dipping or Ducking Day and for those found not to be wearing a spray of hawthorn flowers they were either dunked or had water thrown at them. Bathing in hawthorn dew on May Day was done to increase one's beauty, as attested to by Elisabeth Pepys, Samuel's wife, in 1667 and as this old rhyme suggests—

> The fair maid who the first of May,
> Goes to the field at the break of day,
> And washes in dew from the hawthorn tree
> Will ever after handsome be.

The sheer amount of different local names for the tree, leaves and flowers—remembering that an unknown number of other names have already been lost—suggests a greater familiarisation in times gone by, with names for the tree including Azzy-tree, Hagthorn, Holy Innocents, May, Mayblossom, May bush, Mayflower, Maythorn, Maytree, Moon Flower, Mother-die, Quick, Quickset, Quickthorn, Thorn-bush and Whitethorn.

Or for the fruit, the haw, local vernacular also calls them, among others, Agags, Agald, Agarve, Agasse, Aggle, Aglet (which is also the name given to that little bit of plastic or metal at the tip of a shoelace), Agog, Arzy-garzies, Asogs, Awe's, Azzies, Bird's eggle, Bird's meat, Boojuns, Chaw, Cuckoo's beads, Hag, Halve, Harve, Hazle, Heethen-berry, Hegpeg, Hippertyhaw, Hogail, Hogarve, Hoggan, Hoggosse, Hopperty-haws, May-fruit, Pig-All, Pig-berry, Pixy-pear and Thornapple. The familar name Bread-and-Cheese strongly suggests it may once have been commonly eaten.

Then there are all the towns and villages, starting from the obvious Hawthorn in Hampshire, Durham and Rhondda, Hawthorns in Staffordshire, Hawthorn Hill in Berkshire, Hawthorn Corner in Kent, Hawthorpe in Lincolnshire, Hawton in Nottinghamshire, Hawstead in Suffolk, or Haws Bank in Cumbria. Add to this the profusion of names with Thorn in, often in reference to the hawthorn, from Thorney Island in West Sussex to the one in London we'll see in a bit and a plethora of others across Britain from Thornend to Thorney, Thorneygreen, Thorneyham, Thorneyhill, Thorney Toll or Thorney Wood and well over another hundred others, not to mention thousands of street names, all attesting at the very least to a greater recognition of the hawthorn in the past.

* * *

With a reversal of migratory routes and claims of the other chapters, as for this chapter it was of course the actual boat that took some of the earliest settlers to North America that was named after the hawthorn, *Mayflower*. Also, in the case of the hawthorn, the flow of species has been very much in the reverse direction, with several brought back from America by exploring botanists in the past few centuries. Examples include *Crataegus persimilis* and *Crataegus crus-galli*, introduced to Europe for their ornamental splendour and vivid autumnal leaf colorations, with largish berries, making a popular addition to driveways, roadsides and parks.

The hawthorn tree was already very much established as a cultural and dietary fixture among the First Nation tribes across the whole of North America long before the acorn that grew into the tree, that got sawn into planks, that made the oak hull of the *Mayflower* in 1620. There is still a remembrance across the continent of when they were at times a crucial part of the winter larder. Even the thorns have been employed in a variety of uses, from removing splinters to an awl for sewing tough materials such as leather. Hawthorn is one of the hardest of woods and rarely rots, making it ideal for tools and handles, fence posts and wood engraving, all of which were employed by the original inhabitants.

The *Crataegus monogyna* of this chapter is known in America as the English Hawthorn today, being rare there, but the very similar other members of the family do figure prominently in First Nation ethnobotanical studies. All the Indian nations already mentioned throughout the book, and those apologetically not mentioned, ate the berries as a food source, with some commenting on its importance during times of famine.

The Chippewa used the roots to offer some protection and curative abilities from tuberculosis (brought to the New World from Europe and Africa), female weaknesses and back pain. The Comanche chewed the inner bark as a gum, while the Cherokee used the bark to improve circulation. All species of hawthorn on the continent were utilised, whether *Crataegus calpodendron* used by the Fox Indians to treat bladder ailments while the Kwakiutl in British Colombia made a poultice for swellings from it, or the Cherokee using *Crataegus spathulata* for cardiac problems. The Chippewa, Kutenai, Kwakiutl, Meskwaki, Omaha-Ponca, Ojibwa, Potawatomi, Thompson and Winnebago all utilised either the fruit, bark and root to treat gastrointestinal disorders, bladder ailments and diarrhoea, and the leaves as a poultice for swellings, as a food source and as a general tonic.

The Okanagan-Colville used the thorns rather like brutal acupuncture, scabbing an arthritic area that clears as the wound heals, as well as the bark and roots to make a stomach medicine, while the Thompson used the thorns for fish hooks, piercing ears, axe handles and digging sticks, as would many other tribes. The Iroquois also carved the roots of the *Crataegus submollis* into what we might call 'voodoo dolls', as well as the shoots and bark to stop menstrual flow. The toughness of the wood also lends itself to the making of a bow, and with its high tensile strength a very small bow can be made that still maintains plenty of power, making it an ideal travelling companion. With the arrival of horses by the Europeans from the sixteenth century (there is much debate about the date of introduction) these bows along with their quickly finessed acrobatic horsemanship were perfect partners, with the Mohawk and Seneca being particularly adept exponents of this deadly equine skill.

The Aztecs—an OG First Nation Indian tribe, who unfortunately found themselves exterminated by the Spanish before their cousins further north in North America were decimated by the British and other European colonisers a little later—used the roots of the *Crataegus mexicana* to make a strong tea known as *texocotl*; its modern name in Mexico is *tejocote*, with up to fourteen species of *Crataegus* used to brew it. There is also a traditional punch-type drink, *Ponche Navideño*, mixed with sugar cane, cinnamon, guava and served at Christmas, for the Day of the Dead celebrations (which itself harks back to the Aztecs), as well as making candied sweets and rosaries from the dried haws, used as offerings to the dead and as altar decorations.

Bubbling up from the depths of time

Quite possibly the oldest evidence for human utilisation of the hawthorn is from a site that also happens to be the oldest evidence of using grapes for wine. This is the Jiahu site in China's Henan province, dating back to 7000 BC, with analysis of the dried remains from an unearthed vessel showing it once to have contained a fermented liquid of grapes, haws, rice beer and honey mead ... which sounds lovely.

The ancient Greeks carried blossoming boughs of hawthorn in wedding processions, especially the bride, who would also wear hawthorn flower garlands. The altar of Hymenaios—God of marriage ceremonies, where the word hymen originates, as Hymen was expected to attend every wedding—was bedecked in hawthorn blossoms. The song sung at the ceremony is given the same name, Hymenaios.

Both the ancient Greeks and the Romans who followed them (plagiarising much of the Greek culture and those around them, such as the Etruscans) linked the hawthorn tree to hope, marriage and fertility, with the Romans also putting the leaves in babies' cradles to ward off evil spirits, in direct opposition to the superstitions looked at in a few pages.

* * *

The reason it's called Mayflower, May Bush and May Tree among others is because the hawthorn is synonymous with what the month of May once stood for. As the ultimate symbol of fertility and rebirth, signifying the beginning of summer (when temperatures tend to stay above 60°F), it was a collective cultural sigh of relief that the yearly cycle had safely turned again, with the return of the sun and longer days, fertility and regrowth, and also the time to sow buckwheat, onions, sunflowers, potatoes, and corn (the last two after the late sixteenth century in Europe) and many other essential vegetables.

Celebrations of the old May Day, also known as Beltane, were one of the most important festivities. Hawthorn was known as the May Bush by the Celts, who tied cloth and mementos to them, with special reverence for ones near to ancient springs. That's why it's been named after a month (no other plant is named after a month or vice versa, with the word May coming from the Greek fertility goddess Maia), because it is May, as it signifies with its blossoming Mayday and the Maypole. Maypoles have an uncertain history, probably a remnant of tree worship from our previous culture, when we decorated boughs of trees and celebrated and partied in honour and praise of Nature's larders being replenished once again. This was particularly in the latitudes when seasonally night becomes day and the sun allows life to flourish again. An interesting coincidence, in the name at least, is the short story by the American Nathaniel Hawthorne, *The May-Pole of Merry Mount*, published in 1836. It is a fascinating story about the Puritans of the eighteenth century and the end of a free-thinking and merry way of life; the Maypole is cut down and the people are forced to join the congregation, becoming the sheep, no longer the wolves.

Of all the plants, trees and nouns that did not find themselves as letters in alphabets of old, the hawthorn finds itself represented as a letter in not one, but two ancient languages. It is represented as a letter of the old Runic scripture, which was once the written language for most (if not all) of North and Western Europe, where it first makes an appearance sometime around AD 150 (although probably much older), surviving for a millennium until replaced with Latinised scripts by 1100. The Runic letter/symbol for thorn looks like a *p* and a *b* joined together, representing the letter 'th'—that's the hawthorn. The hawthorn also finds itself being represented by the sixth consonant of an old Irish language, Ogham, from some time around the third or fourth centuries AD (again, probably much older). The name of the letter is Uath, meaning terror and fear.

Making words and language even more entwined around the hawthorn is the fact that when the insightful Hildegard of Bingen was writing her herbals and music on vellum in the mid-twelfth century (paper doesn't make an appearance until the fourteenth century, when it would have been made of hemp) she used hawthorn ink. This was made by cutting the branches in spring and leaving them to dry throughout the year, then the bark was removed and soaked in water for a further 10 days. Next it was boiled and reduced down with wine added until thick and black, then poured in bags and left in the sun to dry. When it was dried, wine was again added along with iron salt and heated, and it was ready to be used as ink.

Sticking with literature, more recently Robert Burns (1759–1796) often made mention of the hawthorn and birch, while the famous French writer Marcel Proust (1871–1922) prosaically mentions the hawthorn at some length in his masterpiece *In Search of Lost Times*, envisaging a hedge in full bloom being comparable to chapels with altars of bouquets, with the spectacle of the flowers being equivalent to seeing a new painting by your favourite painter for the first time. Siegfried Sassoon (1886–1967) wrote a touching short poem called *The Hawthorn Tree*, about dying in France and a chance of a last look at the hawthorn in bloom to wipe away the horrors seen in the trenches during WWI. Coincidentally, Sassoon also has a commemorative plaque on his house, awarded by the Thorney Island Society, in Tufton Street, a most significant tiny part of London, as shown in the next section.

In Robert Graves's part-poetry, part-history, part-mythology *The White Goddess*, from 1948, the May Tree plays a central theme of many legends. This book has caused some confusion as many have mistakenly considered it to be factual as it is represented in this manner, when in fact it is only Graves's interpretation of old myths and the Ogham alphabet.

And finally, because I like football, the club West Bromwich Albion's emblem is a hawthorn leaf, with a haw and a thrush sitting on top. They also play their home matches at The Hawthorns, in acknowledgement (I wouldn't go as far as to say veneration) of the fact there were hawthorns on the site before it was levelled to be made into a football ground.

Which is all very interesting, but it gives little indication for the following highbrow cultural positioning of the hawthorn, where it is lofted high, as a symbol for the uppermost echelons of our belief systems, adding to the notion that we are currently suffering from cultural and botanical amnesia. This loss of memory has been seeping away for possibly many centuries, as pointed out by the co-editor of *The New Testament in the Original Greek* in 1881, Bishop Brooke Foss Westcott, who stated "the restoration of man and nature are placed side by side".

Jesus Christ and the soul of London

A curious fact about hawthorns is that they can procreate via a virgin birth, known as agamospermy—asexual reproduction, in which seeds are produced from unfertilised ovules—possibly another reason why the tree was seen as so special. This form of virgin birth is interesting for its obvious association with the Christian virgin birth of Jesus Christ, where it should be mentioned there have been numerous visions of Mary around hawthorns. Another word for this virgin birth is parthenogenesis, where lizards and reptiles can create an embryo from an unfertilised egg; some fish also manage this, even sharks, as well as scorpions, small water fleas and minuscule nematodes. This process has not been ruled out for humans, as scientists have already created human stem cells parthenogenetically—which brings it back to the ancient Greeks, because the Parthenon in Athens is a temple dedicated to Athena, who herself was made from the forehead of her father Zeus via parthenogenesis (which means in Greek, virgin birth/origin).

This whole virgin-birth story certainly has some extremely deep roots. It could even allude to a time when the gods did indeed make the human race via parthenogenesis, using themselves to be effectively cloned from, hence the saying 'made in His image'. That's what all the old stories from past civilizations say happened anyway, be that the Sumerians, Egyptians, Mayans, the Dogon of Africa and the Hopi of North America, among others.

* * *

The cultural significance of the hawthorn extends even deeper into our theological belief systems, as it is argued that sprigs from this thorny tree were fashioned into no less than Jesus's crown of thorns, and the cross itself may even have been made from the tree. On the other hand, this slander could have been an attempt by the Christian Church to discredit the hawthorn, to reverse the popular reverence for the tree by suggesting it was a bad tree causing suffering to Christ, the saviour.

The Jesus associations don't end with the crown of thorns either, as the following account knots the hawthorn and Jesus a little tighter together. In The Bible, the person who cut Jesus down from the cross, took him to the cave and tended his wounds was a kindly soul going by the name of Joseph of Arimathea, assisted by Nicodemus. He comes with quite a story attached, but to keep it short, Joseph soon scarpered the Holy Land, having clearly drawn attention to himself, although it seems not enough, as he also took quite a souvenir with him, the Holy Grail, which allegedly now resides in Valencia. Finding himself in southern Britain, Glastonbury, Wearyall Hill to be precise, Joseph decides to push his staff made of hawthorn into the ground, from which a hawthorn tree blossomed. Quite a special one at that. Unfortunately, what remains today of the Glastonbury Thorn is not this one, although probably a direct ancestor, considering a hawthorn lives for no more than 400 years and the one standing on Wearyall Hill in the 1640s was burnt down in the English Civil War by Cromwell's Roundheads, who were so set against any idolatry and superstitions. It is thought that the original tree had pieces taken from it and grafted onto other hawthorns to continue the barkline, hence why there are many trees across Britain claiming direct ancestry from the Glastonbury Thorn. A good job there is, because in 2010 the one on Wearyall Hill was cut down. The vandal who chopped it down could have been a Christian unhappy with the pagan associations, or a pagan unhappy with their tree being kidnapped by the Christians for their own devices and propaganda or even an anti-royalist.

Anti-Royalists? Obviously Cromwell wasn't too keen on the undeserving monarchy (being instrumental in the beheading of Charles I in 1649), but what's it got to do with the royal family today? Well, the Glastonbury or Holy Thorn, *Crataegus monogyna* spp. *biflora*, as its name suggests, flowers twice a year (around winter solstice, 21 December and May Day, 1 May) and since the time of James I (reigned from 1567 to 1625), a flowering sprig of this tree has been sent to the English monarch every Christmas, with the last incumbent apparently placing it on her table at lunch. The sprig today comes from a cutting taken from what's said to be the original, although it is more likely the one burnt by the Roundheads as it was grafted onto a tree from the nearby St John's churchyard in Glastonbury.

The ancestors of the cuttings taken throughout history from the Glastonbury Thorn have themselves been revered across the entire country, such as the hawthorn at Appleton, Cheshire, where on midsummer it is adorned with ribbons and flags, known as 'bawming the thorn' (interestingly, this word bawming dialectically means anointing in Cheshire). This was no doubt once a common ceremony with all cuttings from the Glastonbury Thorn nationwide, becoming blurred with Maypoles centuries later. In Kilmeford, Argyll, a hawthorn can be found that has had coins bashed into its trunk, rather like a wishing well, albeit a somewhat aggressive act of worship.

Crataegus monogyna spp. *biflora* almost certainly has its origins in the Mediterranean. As it would have been from there if it was indeed one from the Holy Land that originally found itself on Wearyall Hill around 2,000 years ago. The legend of the hill also incorporates King Arthur, who is reportedly buried there (as well as a hundred other places!), where he died on his quest for the Holy Grail—could the hawthorn be the Grail?

There is also the confusing issue of the switch from the Julian to Gregorian calendar in 1752 where 11 days were 'lost', adversely affecting the conjunction of flowering and May Day traditions and beliefs, many of which did indeed disappear thereafter, unbalancing local and regional belief systems for the whole year, while trying to replace them with mainstream/elite views of religion, governance, laws and social division. In 1753 people congregated at the Glastonbury Thorn and other *bifloras* around the country only to note that they neither bloomed nor blew on the new Christmas Day, but instead did so on 5 January, using this as proof that the new calendar was incorrect. If you have any interest in this most baffling switch of dates it's worth further investigation, although be prepared to get your mind caught in a most confusing calendrical conundrum.

* * *

Incredibly, there is one more piece in the jigsaw, which clearly imbues the hawthorn with the highest level of veneration—this time it's not a person, it's a piece of land that just happens to be the capital city's roots.

The landscape of London has changed dramatically in the past two millennia, if not just in the past couple of centuries, with land being drained, river embankments built, new areas created and old rivers and islands built over. One such island was Thorney Island, named after the hawthorn grove once found there. Thorney Island was once flanked by two arms of the Tyburn River as it joined the Thames, and is the part of Westminster housing two of the most important buildings in England. First is the Houses of Parliament, and while the current building dates to only 1888, this site has been the seat of parliament since 1265. Second is Westminster Abbey, with most of what remains dating from 1556, after fire destroyed Henry III's 1245 building, which itself replaced the church built by Edward the Confessor in 1065—Edward must have been quite a chap, being also the only king to be canonised as a legit saint.

Thorney Island was also home to one of the first churches in England, established sometime around AD 604, just seven years after St Augustine landed at Canterbury and became the first Archbishop there. One would assume that the location of the initial churches would most definitely have been highly symbolic and would have occupied an already significant place of worship for the old congregation, perhaps a congregation with a goddess tree-worshiping-orientated reverence, practised by priestesses in sacred groves of hawthorn. Was this the land upon which it was deemed auspicious and appropriate enough to build a capital city and honour the greatest ancestors of the nation? Or is it just a coincidence that the centre of power of the nation, the Houses of Parliament and the Abbey, where the royal and famous are buried, just happens to be a former sacred hawthorn grove?

From a time whence no buildings survive, King Canute (1016–1035) ruled over England, Denmark and Norway, and as kings go was a rather good one, argued by some historians as being the best of the Anglo-Saxons; he has been somewhat slandered for far too long as being so arrogant as to assume even the tide would halt for his supreme power. No doubt the soon-to-follow Normans were more than happy to spread salacious stories about competing royal houses. What Canute was actually pointing out was that he may well be king, but he was nothing special, as the real power lay with God. The location for this pious act has been placed variously at Winchelsea, Bosham and Southampton, although it probably occurred outside of his royal palace on Thorney Island, as the River Thames is of course tidal.

Westminster Abbey has christened, married and buried many of the royal family (only the royal family, members of the Order of Bath and those living in the precinct of the Abbey can get married and christened here), and a good proportion of the most famous of British people are buried here, including most of the monarchs since Edward the Confessor (1042–1066). In addition, far too many undeserving nobility to ever want to mention, loads of Prime Ministers and a few of the more popular figures, such as Isaac Newton, Charles Darwin, Stephen Hawking, Robert Stephenson, Charles Dickens, Geoffrey Chaucer, Laurence Olivier and Rudyard Kipling, with many others buried elsewhere but still commemorated in the Abbey with a plaque or statue.

All those famous people buried in a known area where folklore stretching back aeons suggests this might well be of assistance to their soul's passage back to the Great Soul Soup in the Sky.

Adoration of a tree: how quaint it may look, but it is immensely practical and cultural, just from a previous culture and state of mind once shared by all, which has been incessantly vilified by our present culture's programming, so we erroneously assume trees, plants and fungi have nothing whatsoever to do with us … ooops!

Vibrations, Green Men, fairies and pixilated superstitions

As we saw with yarrow, one of its abilities among many others was that it delved into the supernatural, by harmonising and allowing transference of energies from the Earth, universe and the netherworld—which I'm fully aware sounds somewhat odd, but it is what it is. The abode of the hawthorn, although rooted to Mother Earth, does also appear to act as a portal to another dimension not usually frequented by humans. A rationale for this could be that the vibration of the hawthorn tree (everything, right down to the infinitesimal-sized DNA has a vibration, which can be amplified by quartz and recorded as a frequency) harmonises the immediate vicinity and allows, perhaps with an alignment of Earth's magnetism or other planetary external forces—and here's the really 'odd' bit (meaning it doesn't fit comfortably within our current accepted mental or scientific framework)—for some people and at the right time to visit a place that is out of the ordinary. A step into the extraordinary is possible, into another realm, co-existing on a different frequency alongside our own, or as scientists would say, a different universe within the multiverse. Or not … but there is an awful lot of anecdotal evidence to suggest there is something going on here. One thing is for sure, and that is attempting to put it into words that are readily understandable tends to make it sound even more suspect.

That said, we've already seen how this hawthorn tree is intimately entwined with our past, from celebrating the regrowth and continual survival for the rest of the year, to being associated with the highest rung of our current cultural life. It's not coincidental, and there is a reason why the hawthorn is such a potent symbol, in the same way that the Chinese *I Ching* could have used any straight stalks, but only the yarrow would suffice, as only yarrow has the frequency of yarrow, just as only hawthorn resonates at the frequency of hawthorn, which is measurably different from the frequency of an oak or walnut tree, bumble bee, wildebeest or you.

* * *

Around a millennium and a half ago, the hawthorn also found itself being closely associated with the Green Man. Of the leaves often depicted in the foliage of Green Man sculptures and paintings, those flowing from all of his orifices and hair, hawthorn is more frequently used, particularly in the older depictions. Oak leaves and acorns appear also, being similar in shape, although an oak leaf is much larger with a wider end to the leaf, as opposed to the hawthorn leaf being wider towards the stalk end.

It appears that the Church adopted the Green Man as a badge of familiarity, especially during the seismic cultural transitional period when Christianity became the dominant belief system and story to believe in. It could be argued the Green Man represents the patriarchal Church's version of Mother Earth. The Green Man was more in touch with the Anglo-Saxon polytheism, including tree worship, allowing a friendly face, albeit a somewhat green and gnarly one, to be associated with the new Christian religion. From around the sixth to the eleventh centuries the Church adopted this pagan symbol to lure the flock, while paradoxically at the same time forbidding tree worship, often actively burning sacred groves to the ground. The Green Man iconography reached its pinnacle in artistic flamboyancy when incorporated into the elaborate Gothic and Romanesque architecture of the twelfth century. Having all but disappeared by the nineteenth century, there has been a revival during the past century, with a number of pubs and the wonderful Welsh festival The Green Man, helping to keep Him in the mind's eye.

Sticking with the Welsh theme, a wonderful vision is conjured up by the Welsh Goddess Olwen, the White Goddess of the Hawthorn. It is said She once walked the empty universe, and where She trod, hawthorn petals shone, illuminating Her path through heaven, which we see today as the Milky Way.

* * *

In Ireland, the hawthorn is known as *sceach gheal*, with *sceach* meaning thornbush and *gheal* being bright/radiant. The tales of sacred hawthorns and fairies resonate much more clearly from the past in Ireland, where many of the legends and stories remain nearer the surface of their current cultural consciousness—although you'd be well advised not to be using the fairy word, which is seen as derogatory and could well antagonise the little folk, who'd prefer to be addressed as the Sidhe (pronounced Sh-ee).

But what are the Sidhe? Not the colourful, Disney-cute, smiley and kind ones of the silver screen, nor the flower fairies of the gorgeous Cicely Mary Barker books, that's for sure. These are more menacing, some say goblin-like, although others call them the good people—having never seen them myself, I have to admit to having a belief in them, although more as some form of an inter-dimensional nebbish. I also can accept that the vibrational resonance of the hawthorn could well create a vortex to another dimension during specific energetic periods, or that hawthorns growing on Earth's energy networks (also known as Dragon lines) may act as a conduit or a portal. There has to be some truth to these doorways to other dimensions, otherwise why did NASA set up its Magnetospheric Multiscale Mission (MMS) in 2015, tasked with studying these mysterious portals to gain a deeper understanding of them?

Everyone remembers the gull-winged DeLorean car from the film *Back to the Future*, a car that also managed to go through a portal to another time dimension. Most also know that the company itself failed, mired in some foggy something or others. Well, guess what, there's mention it was the meddling of the fairies in revenge for the desecration of cutting down one of their trees while building the car plant in the late 1970s in Dunmurry near to Belfast that brought about its demise. The company heeded the warnings at the time, acknowledging they had done wrong, as they replaced the tree with a 'similar-looking one'—which would be the same as murdering an elderly tribal shaman and replacing him/her with a similar-looking old person! The failings of the company are inexplicable, and perhaps DeLorean was stitched up, caught in someone else's game of petrol economics. It could well have been due to the fairies, although the fact it was a terribly built car beyond its attractive exterior didn't help, and nor did the unpaid bills to Lotus or the cocaine smuggling charges!

Throughout Ireland it is not uncommon to see a lone hawthorn tree in a cropped field, suggesting the farmer didn't want to test the superstition, perhaps being backed up with first-hand accounts of the bad luck heaped upon those who were foolish enough to cut them down. Even roads in Ireland have to be planned around any significant hawthorn trees. In Downpatrick, Ulster in 1964, for instance, a local revered Fairy Thorn was cut down to make way for a road. All the workers refused to perform the sacrilege, leaving the foreman to do it, with legend telling of hundreds of white mice running from the hole as he cut into it; and while removing some of the soil away his leg was crushed by the horse. More recently the construction of a new motorway at Ennis, County Clare was delayed by nearly a decade throughout the 1990s, until being finally slightly rerouted, the motorway avoiding the cutting down of a hawthorn said to be the meeting place of the fairies of Munster. I can still remember news reports from the time and thinking how incredible it was that no one was doubting the presence of the fairies, from the policeman, mayor or construction worker, with the story attracting international news crews' visits. One would hope the fairies of Munster don't mind having to conduct their meetings today just five metres from the side of the motorway.

* * *

Hawthorn is considered a safe tree to sit under during a storm, as indicated from this old rhyme:

> Beware of an oak, It draws the stroke,
> Avoid an ash, It courts the flash,
> Creep under the thorn,
> It will save you from harm.

Having said that, caution needs to be given about sitting under a hawthorn on or around May Day, as it appears at this time of year it's possible to be whisked off to the netherworld with no return, or not till years in the future. This was the claim of the thirteenth-century Scottish poet, Thomas the Rhymer, who said he had met the Fairy Queen by a hawthorn tree. She led him to the Otherworld for what he assumed had only been a short visit but which turned out on his return to be seven years later. Such an event in the past centuries may have been called being 'pixilated'—to be led astray by pixies, which might just have been a lame excuse of very drunk country folk falling asleep in a hedge and blaming the pixies the next morning!

In Serbia or Croatia, if you wanted to kill a vampire, only the wood from the hawthorn would do. The most powerful of witches' wands are said to be those made from twisted hawthorn limbs, with the blooms of the hawthorn being used in spells for fertility, happiness and good luck in fishing.

In Germany, hawthorn wood was once used for funeral pyres, as only the smoke from the hawthorn was considered suitable for lifting the soul to begin its journey. Considering the famed Germanic sensibilities and technical proficiency, as well as fitting in nicely with the practices from England and Thorney Island, there probably is an element of truth to it.

The hawthorn tree is also said to guard wells and springs, an absolutely vital staple for life, with many sacred sites being no more than a water source. Water is the determining factor to life or death, where water from a river, well or spring was literally a lifeline in the past. Hawthorn's presence at many stone circles, seemingly protecting them, may be of note.

* * *

The blossoms in spring, after the leaves have begun to grow, colour from white, pink to red, with an aroma that has been placed somewhere between sex and death. The smelly chemical is triethylamine, which is not only present in semen, it also attracts pollinating flies by mimicking rotting flesh as this is one of the first chemicals released when the human body ceases to exist (the smell is fairly subtle from *Crataegus monogyna*). It could be this deathly smell that has given rise to superstitions about bad luck being attached to bringing boughs and blossom inside the house, perhaps harking back to the smell of the dead and dying from days of plagues etc., when the dead were apparently left in the house after death for a few days.

As we've just seen, Queen Elizabeth II had her royal sprig accompany her on the dining table, and many churches bring them inside for decoration, as did the Romans, even in babies' nurseries, with presumably no concern for retribution or ill-luck. It may have been a form of reverse psychology, a way to stop public veneration of the tree, while actually continuing such reverence themselves.

There is also an association with the Catholic Church, where the hawthorn blossom is known as Our Lady's Dowry, and to celebrate spring it adorned statues of Mary, a May Altar. So, in the 1540s when the country turned anti-Catholic, to continue this custom was a badge of Catholicism, which became, with or without any hawthorn present, punishable with death.

Just a quick pause to clarify the rather faint sex/death aroma from your average *Crataegus monogyna* blossom before moving on, in all likelihood it's not even the *Crataegus monogyna* species that is being referred to, but the similar-looking *Crataegus laevigata*, also known as the Midland or Woodland Hawthorn. This is because in the past, in the time when traditions, folklore and superstitions were laid down (when living a life more in tune with nature and seasonality), Britain was a different place, in which without the vast field networks of modern agriculture the Woodland Hawthorn would have predominated. This species not only blossoms a little earlier than the other species of hawthorn it also definitely has a far more pronounced deathly smell, especially from freshly picked flowers. So it's perfectly logical it was this species that was being referred to, although it's still a hawthorn.

Traditional uses

Dioscorides cited hawthorns as being good to stop stomach outflows and the flows of women in his first-century *De Materia Medica*, a tome that formed the core of the European pre-modern pharmacopoeia, enjoying well over 1,600 years as a medical textbook—it must have held some substance.

In Europe throughout the Middle Ages hawthorn was better associated as being beneficial for nervous conditions, although it would also have been employed as a remedy for gout, fever, pleurisy and insomnia.

Our old friend Culpeper used hawthorn as a singular remedy for kidney stones and dropsy (oedema). He also prescribed bruising the seeds, then boiling and drinking the water to obtain some relief from tormenting pains of the body, as well as soaking a cloth in this seed brew so when placed over a splinter it draws it out—it'll be great to verify this in super zoomed-in, hi-definition stop-frame slo-mo. To add a few words on Culpeper, who often bears the brunt of derision, being portrayed as somewhat of a fantasist, combining astrological perspectives with herbs and questionable ingredients such as the powder from an unburied skull! In real life Culpeper was actually a very knowledgeable herbalist, who with his *English Physician* made the vital connection between the plants and illnesses they assisted in, his book being made available for the first time for just a few pence, instead of the several pounds for other books on medical literature—more than the yearly wage for most. He was subsequently severely slandered by the doctors and apothecaries, being a threat to their medical monopoly, with most of the stories about him being pure fabrication and vilification, far from the distinction he deserved.

There are many interesting parallels between what may appear disparate times, the mid-1600s and today, not least the disconnection of many of the people and the scientists with Nature, or the constant battles between the herbalists and doctors, who put their faith in new quackery science over old-fashioned natural remedies, exactly as the discord between herbalists and pill peddlers is panning out today.

* * *

The nineteenth-century herbalists known as the Eclectics used hawthorn for angina and dropsy, as well as a sedative for chronic insomnia. The famous German medical herbalist and Chairman of the German Commission E, Rudolf Weiss (1895–1991), confirmed that hawthorn is one of the most widely used heart remedies. In Germany hawthorn extract is used to avoid complications for the elderly if already suffering from secondary conditions such as influenza, pneumonia and a multitude of heart conditions.

Hawthorn was mentioned in the first Chinese state-approved pharmacopoeia, the *Tang Ben Cao* from AD 659. In Traditional Chinese Medicine (TCM), *Crataegus pinnatifida* (*Shan Zha*) is used to relieve digestive problems from over-indulgence in meats and greasy foods, as it increases gastrointestinal secretions and enzymatic function. As hawthorns enter the stomach and spleen the phytonutrients stimulate the blood to improve digestion and remove accumulated foods. It has also been given to children who do not thrive and have frequent diarrhoea and dysentery, as well as being used as a clearing agent for postnatal abdominal pain, menstrual pain, hernias, testicular pain and swellings. TCM practitioners prescribe up to 500g of the fruit for tapeworm and other parasites.

Only more recently has hawthorn been used for high blood pressure and high cholesterol in China, in part because these are relatively new afflictions, as the East becomes progressively ill-fed and malnourished following the lead of the West, with its predilection for gorging on refined, highly processed carbohydrates. Hawthorn is still popular, though, a continuation suggestive as evidence of it being effective, with 36 digestive problems hailed as being curtailed with the Major Crataegus Pill (*Da Shan Zha Wan*) alone. Among more than fifty other OTC patented medicines with hawthorn as their main ingredient are some great-sounding remedies, such as Scorched Three Immortals (*Jiao San Xian*), Preserve Harmony Pill (*Bao He Wan*), The Great Tranquillity Pill (*Da An Wan*) or the Brain Vitalising Pill (*Nao De Sheng*).

As we saw above, the First Nation Peoples enjoyed an extensive pharmacological ethnobotanical use of all the species of hawthorn for their entire residence. Following their demise, tinctures of *Crataegus* were still sold throughout Europe and America in the nineteenth century. In contrast, by 2012 Americans spent only $1.5m on hawthorn capsules and tinctures from the leaves, flowers and fruits, an average of about 4 cents per person.

It has to be recognised that today the vast majority of the population has been eating a deleterious diet for most of their lives, with the opposite of preventative capabilities attached to repeated and excessive consumption of the dominant processed food ingredients. There is already firm evidence of the efficacy of hawthorn, which is why standardisation has occurred for some of the compounds, such as oligomeric proanthocyanidins and the flavonoid vitexin as found in OTC medicines from Lichtwer Pharma, Weiss, Willmar Schwabe, Nature's Way and many more.

There needs to be a paradigm shift in how we approach health before progress can be achieved, especially the bias in the West that their pills and science are the best, while the rest of the world's medicine is painted as antiquated, even though Western health is only just beginning to catch up with what real health is all about. Health, with its three aspects of mental, gut flora and physical, has seen Western medicine focus on the physical aspect of ailments by suppressing these and ignoring mental health as the pills are often contraindicative to this while destroying the gut biota with antibiotics etc., and that's before we even look at poor diets and man-made chemicals.

There can be no denying that before WWI, medicine, health, potions, creams, salves, pills and thrills *all* came from the plant kingdom. It has to be appreciated that our whole ethos regarding health and medicine has been completely indoctrinated and controlled by the pharmaceutical corporations. This may appear perfectly acceptable, that the makers of suppressers and cures dominate the realm of health, but as we've seen throughout this book, the side effects are enormous and can't keep being swept under the carpet. Then there is the fact that even with all the drugs available, most diseases are increasing unabated in incidence—so there has to be an element of 'the drugs aren't working' as well as the fact that our polluted background, from the water, air, soil and body load, has gone through the roof. This is further compounded by the stark reality that it is profits that are driving pharmaceutical corporations (as for all industry), while more people are getting sick and for longer, many on drugs for the rest of their life. More drugs, more illness, with no end in sight. That's not right.

* * *

A brief digestion of some of the ailments hawthorn benefits and has been used for traditionally includes the treatment of cardiovascular problems, such as heart failure, hyperlipidemia, hypertension with myocardial injuries, angina pectoris, arrhythmia and atherosclerosis. In addition, it has been used for improving blood circulation, as well as blood stasis elimination (when blood flow is impeded and stagnation occurs). Hawthorn has also been used for the treatment of gastrointestinal diseases, stimulation of digestion and promotion of stomach functionality. Moreover, hawthorn had uses in the treatment of indigestion, epigastric distension (pain below the ribs), abdominal pain and diarrhoea. In the European tradition, hawthorn is also used as an anti-spasmodic, cardiotonic, astringent and diuretic agent, while a poultice of the leaves is effective for boils, ulcers, splinters and sores.

Heart disease

In 1603 the physician to French king Henri IV, Joseph Du Chesne, announced that a syrup made from hawthorns was good for the heart. According to the Doctrine of Signatures, the red of the haws clearly indicated they were good for the heart and blood.

Today, heart disease is the number one killer of inhabitants of the industrial, allegedly 'civilized' first world, although it's soon to lose this ominous crown to cancer as cancer rates continue to rise. This is from not just a never-ending list of new cancers, cancer is also now affecting year-on-year a younger demographic, although obesity and diabetes are fast on their heels for this crown.

To treat such a prolific killer there has been an explosion in the drugs available for heart disease, which itself has been split into a huge list of conditions. It is similar to the scenario revolving around diabetes and the range of symptomatic diseases it can present itself as—in the 1930s it was well known that diabetes (imbalance in blood sugar, which has to have as a significant risk factor the ludicrously excessive quantity of processed carbohydrates, forming 80% of the average diet of today) if left to persist would either result in heart disease, cancer, liver and kidney diseases or depression (the sugar blues). When the symptomatic diseases were split into separate diseases, it was conveniently forgotten that the cause was the imbalanced blood sugar levels (diabetes). This still holds true today, even among the turgid background of various pollutants encountered every day, and uncontrolled blood sugar levels are a major risk factor in all aspects of heart disease, today even more so than in the 1930s.

* * *

Of the many subdivisions in the heart disease industry, angina is a fairly common complaint, increasing in prevalence with approximately two million people in the UK diagnosed with the condition, affecting 14% of men and 8% of women, with an estimated 96,000 new cases each year. Kerching! go the pharmaceutical cash registers. It is such a common disease as civilization itself is a major risk factor for this one, predominantly that'll be lifestyle, whether diet, social activities, exercise levels, stress at work and at home, all contributing to what is a human organism that finds itself very much at disease with civilization. Hawthorn has been shown to have excellent results with angina in medical trials when taken by itself, usually as a tincture, a condition characterised by reduced blood flow to the heart causing sharp pain. Hawthorn has also been shown to increase the efficacy of conventional drugs, as well as lessening the side effects—which, as mentioned before with the other plants also sharing these benefits, if the pharmaceuticals were lessened or even done away with results could be better, an option that is never trialled for, not least the malpractice laws in place to prevent this avenue of research.

Some of the drugs prescribed for angina include aspirin, beta-blockers, calcium-channel blockers, isosorbide mononitrate, verapamil and glyceryl trinitrate ointment, spray and tablets. There are plenty more, whose net worth in revenues collectively is somewhere in the region of £7.8bn annually. Some of these drugs are prescribed solely for angina, although with such a range of ailments adversely affecting the heart and blood pressure along with the side effects, often the angina prescription will be just one in a medley of pills popped by patients with poor hearts daily, probably for the rest of their lives.

Natural plant, herb and food-based remedies are numerous for the heart. In this instance, to regulate blood flow as well as benefiting angina symptoms, a nondescript plant (unremarkable in look, not effect) whose root has been successfully used for heart and respiratory conditions in Ayurvedic medicine for several millennia is coleus (*Coleus forskohlii*). More familiar, yet still very effective are garlic, ginger, ginkgo, khella (also known as bishop's weed, a member of the carrot family), the alliums such as onions and wild garlic, and of course hawthorn.

Hawthorn extract improves the energy processes in the heart, such as oxygenation and energy metabolism, and reduces the waste product of muscular usage and the cause of much heart muscle pain, lactic acid. It also strengthens the walls of arteries, acting as an antioxidant, anti-inflammatory and cholesterol reducer—all of which have been observed from a variety of ingestion methods, including from a tea made from the leaves and berries, tinctures of various extraction methods and standardised capsules.

Angina is merely a small part of the arena of symptomatic heart diseases. Another would be cardiac arrhythmia, with more drugs being considerately made available for this, and undoubtedly one of the ones just mentioned above would be deemed suitable by differing prescribers. There are also plenty of natural products that could benefit arrhythmia, with the following shown to have the best effectiveness: chamomile, hawthorn, kava-kava, reishi mushroom and valerian, but who is taking these for weeks on end to enact a curative outcome? Not enough, that is for sure. Because we've lost the faith, a faith that has cost the pharmaceutical industry a lot of money in propaganda and marketing to wrest us away from any natural answers.

It is the same for one of the biggest-selling drugs prescribed for yet another symptom of the heart at dis-ease with civilization—high blood pressure—with ACE inhibitors, alpha-blockers, beta-blockers, calcium-channel blockers, diuretics and others all being prescribed. With sales of the hypertensive drug Diovan peaking at £1.5bn in 2011–12, the global sales of pharmaceuticals for this sub-section of heart disease must be worth well over £20bn annually. Or there are the timeless remedies from Nature's medicine cabinet to control blood pressure—coleus, dandelion, garlic, hawthorn and reishi. High blood pressure is sold to us as an inevitable consequence of getting old, so why do the elderly in tribal and hunter-gatherer societies not get high blood pressure? They also very rarely suffer from cancer or dementia. Both of these are traditionally considered diseases of the elderly in our culture but are now beginning to affect an increasingly younger demographic.

Related in the CVD (cardiovascular disease) family is the condition intermittent claudication, which is a type of pain in the legs caused from blockages in the arteries from atherosclerotic plaques, and is often most severe after exercise. Hawthorn extract helps with this, by stabilising and strengthening collagen, the substance holding all the cells together, thereby lessening pain. It also helps with Raynaud's disease, as hawthorn assists with general circulation, which is also a benefit for those at risk from stroke as well as helping with the entire cardiovascular system, even varicose veins. Hawthorn is also a general tonic, revitalising and maintaining the heart as it ages, slowing the degeneration of the myocardial muscle from its constant use. An excellent tea for increased blood circulation would be dried hawthorn leaves, peppermint and yarrow.

Also included in the heart disease campus is the whole shambles of cholesterol control (you'd have to refer to my last book, *The Seven Deadly Whites*, for the lowdown on the cholesterol myth). The vast majority of any cholesterol dysfunctionality is definitely lifestyle-derived, diet mainly, but of course those caring drug corporations can help, with sales of the number one class of all drugs, designed and marketed to lower cholesterol, statins, bringing in revenues worth around £23bn a year for this class of drug. All the heart disease drugs combined are worth annually about £115bn to GDPs around the world (mainly the US and Europe). So, when looking at heart disease, there's basically 115 billion reasons why hawthorn is going to be kept as quiet as possible by those with a tight monopoly on the supposed 'health' industry.

* * *

With quite a bit of science available regarding hawthorn and heart health, let's start with Iran, a country at the forefront of plant research. Studies conducted at the Institute for Stem Cell Biology in Tehran (Pahlavan et al. 2018) showed that leaf extract from the *Crataegus pentagyna* worked as a nutraceutical with antiarrhythmic (steadies the heart beat) potential.

From the *Anatolian Journal of Cardiology*, a Turkish study (Alp et al. 2015) on poor rats with induced arrhythmias from digitalis toxin (from foxgloves) saw their blood pressure steadied after administering an alcoholic extract from *Crataegus oxyacantha*, subduing the arrhythmia as well as increasing life expectancy. The study concluded that if human patients reacted in a similar way when administered digitalis for their heart condition while suffering arrhythmia—as quite a few must do if looking at a way of preventing it in the first place—then hawthorn extract could well be used as a treatment.

A Mexican study on rats, reported in the *International Journal of Molecular Sciences* (Cuevas-Duran et al. 2017) used a variety of treatments looking at the response after an induced heart attack, using combinations of *Crataegus oxyacantha* and *Rosmarinus officinalis* (rosemary, now *Salvia rosmarinus*) extracts. They found that when the hawthorn and rosemary were combined, morphological and functional muscle necrosis caused by interruptions to the constant flow of blood supply was lessened, as the extract both reduced oxidative stress and improved the balance between vasoconstrictors and vasodilators—not to mention all the benefits that were not being studied or measured, as rosemary is superb for the brain and hair, as a powerful herb that loves to grow around us and can bestow superb health.

In TCM, haws known as *Shan-Zha* have been employed to improve digestion for millennia. With the recent global increases in certain cardiovascular diseases, attention has shifted to the haw's beneficial potential with these ailments, particularly with the significant change in diet of urban China, where highly processed, carbohydrate-rich snacks have taken over from the traditional snacks of fruits, nuts and seeds, with haws once a favourite. The haws from *Crataegus pinnatifida* were shown by the Tianjin University of Science and Technology (Hu et al. 2016) to have a suppressive effect on the mechanisms laying down damaging arterial lipids and cholesterol around the circulatory system. These authors chose a particular cultivar not traditionally used by TCM, which provided larger, earlier berries with better drought resistance.

Hawthorn's heart and circulatory benefits were further demonstrated by the Tianjin University of Traditional Chinese Medicine, China (Dong et al. 2017). Using flavonoids extracted from the leaf, it was shown these slowed down the development of atherosclerosis by reducing cholesterol, a major risk factor in heart disease. This study corroborated an earlier study reported in the *Journal of Atherosclerosis and Thrombosis* (Zhang et al. 2014) in which feeding haws to mice saw increased levels of antioxidant activity recorded and improved hypo-lipidemic activity with reduced cholesterol.

Similar results were garnered from the College of Science, Seoul, Korea (Yoo et al. 2016), using a hawthorn extract from the *Crataegus pinnatifida*; the authors also improved the lipid profile of rats and decreased oxidative stress, improving the body's defence system against possible damage, particularly cardiovascular stress caused by oestrogen deficiency.

The science conducted by the Academy of Chinese Medical Sciences (Fu et al. 2013) also corroborates the improvements possible from acute myocardial ischaemia (similar to angina but worse, involving complications manifesting from a reduction in blood flow) by administering extracted flavonoids from hawthorn leaves. An extensive Chinese epidemiological study from the Chinese Medical Sciences University, Beijing (Wang et al. 2013) seems quite explicit in the positivity of hawthorns and strongly suggests further research to identify the compounds with the best benefits as soon as possible, particularly as heart disease is a rising star in China's recent acceleration of ill-health.

The oligomeric proanthocyanidins and antioxidant bioflavonoids found within hawthorn have been shown to nourish the tissues of the heart and blood vessels. Hawthorn also naturally possesses compounds remarkably similar to the ACh (acetylcholine) inhibitors, and the beta- and calcium-channel blockers. These compounds carry none of the side effects and are excellent for long-term use, as these are preventatives firstly, not a quick cure. It's worth considering that the current pharmaceutical drugs usually come with little curative potential, usually just symptomatic relief, which hits in hard and fast, thus confusing the patient about what is actually relief and what is curative, as it is the body that 99.9% of the time enacts its own cure.

A German study from 1994 with 132 patients showed hawthorn extract to have similar results with increased exercise tolerance, and less fatigue and dyspnoea (shortness of breath and trouble with breathing). This corroborated an earlier placebo-controlled randomised double-blind study by H. Leuchtgens, reported in *Fortschritte de Medizin* (July 1993), where 30 patients showed improvements in exercise tolerance, with a lowered blood pressure. Another German study (Schmidt et al. 1994) showed the effectiveness of the Li132 preparation, widely available as an OTC. With 600mg administered daily the results were statistically significant, reducing systolic blood pressure and improving all clinical symptoms of the 78 patients in the study who had suffered chronic heart failure.

A 2003 study involving 102 patients, given 900mg/day of hawthorn extract for two months, saw improvements equal to a low dose of captopril. Captopril is often prescribed to improve survival after a heart attack, used to lower blood pressure and to relieve symptoms of congestive heart failure. Captopril is an ACE (angiotensin-converting-enzyme) inhibitor. Bristol-Myers Squibb, the makers of captopril, had profits in 2019 of about £13.1billion, with captopril earning somewhere near to £750 million at least. Ssshhh!! Don't mention the hawthorns.

* * *

The fact that so many of the traditional folklore uses have been proven true by modern science when independent research is done, does weigh towards hawthorn having a significant future role to play, particularly with the lack of side effects and the added bonus of being nutritive. The biggest danger from hawthorn is once again the economic threat—they are free, or can at least be cheaply and locally produced commercially.

We've already seen more than enough about the side effects from pharmaceutical drugs, and there is unsurprisingly a whole slew of side effects associated with heart disease drugs considering that so many people are prescribed them. All we need to know is, hawthorn is potentially as effective as the heart drugs and with none of the side effects.

Furthermore, hawthorn fruit possesses tonic effects on the heart, with several studies showing that it could reduce some cardiovascular risk factors, such as hypertension and hypercholesterolemia (high cholesterol in the blood). The heart is cared for by the good utilisable content of flavonoids, which can dilate the coronary and peripheral arteries, refining the heart's blood supply, improving the pumping force and correcting arrhythmias, while the procyanidins slow the heartbeat to strengthen it. Hawthorn can be deemed an adaptogen, as the fruit has the ability to both lower high blood pressure, as well as raise low blood pressure, depending on the individual's needs. It is shown to be a remedy for angina and heart cramps (stabbing pain in chest with breathing trouble), while the leaves make a fantastic herbal tea addition (blackthorn leaves can also be used), and they've been shown to help with Raynaud's disease.

Another use is for atherosclerosis, assisting with hardened and constricted coronary arteries, as well as helping with cholesterol deposits on artery walls; it can be mixed with arnica for better potential results. The biggest problem/side effect so far reported is that too much hawthorn may cause drowsiness, with a drop in blood pressure, so may contraindicate some blood thinners and blood pressure drugs resulting in possible fainting. Caution is required if already taking medication for a bad heart. Quite what too much is, isn't specified anywhere, probably too much extract, as it'll be a near-on impossible task to overdose on the leaves, flowers or fruit.

There is of course a possible deception while researching and reporting of only the positive trials to reinforce an opinion or theory, making hawthorn look amazing, giving a false impression by avoiding an assortment of negative studies. Common as that is, it's definitely not the case here, as corroborated by The Cochrane Collaboration, a reviewing organisation of epidemiological medical research, based in Oxford, with a collective globally of over 10,000 health experts. They said that hawthorn extract showed definite signs of efficacy and should be used alongside conventional treatments, and they don't say things like that lightly. It must be said that some see this Cochrane group as plugging their vested interests, which could well be true for some areas of treatment … but hawthorn? The non-existent hawthorn-growing society or union hasn't set aside money to pay for some lobbying and marketing firms in order to promote hawthorn products … yet.

* * *

Could hawthorn leaves and fruit really make a tincture that replaces the current statins market, worth globally around £23bn annually? The above shows that most aerial parts of the hawthorn, leaves, fruit, flowers and bark, all exhibit this definite potential with regards heart disease and certainly adds to the argument in favour of their preventative capabilities from regular hawthorn consumption.

So let's have a deeper look at what the scientific world has discovered to corroborate the extensive traditional uses of hawthorn for ailments other than the not insignificant heart disease—

Global science

Let us start at the University of Monastir, Tunisia (Mustapha et al. 2016), where the potential of hawthorn as a natural treatment for an aggressive form of invasive melanoma was looked at, particularly as there isn't one currently available. Using both *in vivo* and *in vitro* methods, a hawthorn extract showed anti-tumour activity, prevented oxidation and inhibited tyrosinase-mediated melanogenesis, which can also lead to the not so scary sounding, but still debilitating and far easier to pronounce, vitamin D deficiency. It was also found that it could be used safely as a natural skin de-pigmenting agent in skin care products, where astonishingly the global market for skin lightening products (predominantly detrimental toxic bleaches) is in the region of £6.75bn annually.

An article in the *Medical Science Monitor Basic Research Journal* from August 2017 reported research undertaken by the Lebanese University (Kallassy et al. 2017) that looked at the antioxidant, anti-inflammatory and anti-proliferative activity of the leaves of Lebanese *Crataegus azarolus*. The authors found that ethanol and methanol extracts exhibited excellent therapeutic efficiency and can be used in treatments relating to oxidative stress, inflammatory reactions and uncontrolled cell growth, as well as exhibiting excellent scavenging potential against free radicals. This is significant as oxidative stress, inflammatory conditions and cell proliferation nearly encapsulates all the diseases of civilization (cancers, heart disease, dementia and degenerative conditions).

A study undertaken by the Poznan University of Medical Sciences in Poland (Krajka-Kuzniak et al. 2014) used an extract of the bark from *Crataegus oxyacantha* to observe that the procyanidins acted as an antioxidant, with detoxifying effects observed in the tissue cells of the liver—clearly suggesting that the bark of the hawthorn has liver-protecting and chemo-preventative qualities. More bark studies from Wroclaw University of Environmental and Life Sciences, Poland (Wyspianska et al. 2017), this time using the same procyanidins from the bark of *Crataegus monogyna*, again yielded evidence for antioxidant abilities, as well as anti-inflammatory responses to add to the liver protection above, and all from just the bark. Quite conclusive evidence, therefore, that it is not only the whole plant that has beneficial properties for us it is the whole species from around the world.

From the *International Journal of Polish Pharmaceutical Research* (Bernatoniene et al. 2009) a fluid extract of hawthorn, maidenhair tree (*Ginkgo biloba*) and motherwort (*Leonurus cardiaca*), showed all three to be effective preventatives and additional remedies in the treatment of diseases caused by oxidative stress—which as we just saw is no small subdivision, as this oxidative stress to lipids, proteins and DNA eventually leads to a host of chronic illnesses, such as atherosclerosis and other cardiovascular diseases, cancer, diabetes, chronic inflammation and more.

A study from the *Biological Research Journal* (Gundogdu et al. 2014), in which the authors claim their Turkey to be the ancient homeland of many of the *Crataegus* species, analysed the fruits and leaves across species, observing the spectrum of phytonutrients was shared across them. They confirmed the leaves and fruit to have excellent flavonoid and procyanidin content, as well as glycosides, triterpene acids, citric acid, choline, acetylcholine, trimethylamine, caffeic and ascorbic acids, vitamins B1, B2, B6 and C, as well as 17 amino acids. As oft-repeated throughout this book, all these plants with their extremely rich nutrient profiles would have had preventative capabilities when consumed as part of the daily diet—today none of these are eaten and thus they are not performing their task as being the ideal preventative from the diseases of civilization. Subsequent analysis (Wen et al. 2017) from the College of Light Industry and Food Sciences, Guangzhou, China, confirmed the broad and beneficial spectrum of nutrients, notably phytonutrients, spanning the whole *Crataegus* genus, whether in China, Turkey or England, with only slight variations.

An extract from *Crataegus orientalis*, in a study from the School of Dentistry, Antalya, Turkey (Hatipoglu et al. 2015), showed an inhibitory effect on periodontal inflammation and alveolar bone loss (the bone where the teeth join the jaw). Admittedly again it's from rat experiments not human clinical trials, but the potential market and client base for pain caused by mouth inflammation is enormous, even if just used as an adjunct.

The heart disease benefits of hawthorn were further confirmed, along with the huge potential diabetic benefits, from an article in the *Journal of the Science of Food and Agriculture* (Aierken et al. 2017), where hawthorn extract lowered triglyceride, cholesterol and blood glucose levels in rats, as well as raising plasma insulin release from the pancreas. The conclusion pointed overwhelmingly to hawthorn extract representing a useful agent in the prevention and treatment of Type 2 diabetes.

* * *

From the Department of Food Science, Liaoning University, China (Zhu et al. 2015), the effects of *Crataegus pinnatifida*'s role in cholesterol metabolism were studied. Given haw pectin hydrolyzates (HPH) and haw pectin pentasaccharide (HPPS), lab mice exhibited higher cholesterol excretion, leading to the conclusion that HPPS could be a promising anti-atherogenic (atherogenesis is the formation of plaque in arteries) dietary ingredient to improve cholesterol absorption, thus a perfect nutraceutical when consumed as a whole food. Considering the revenues from statins alone far exceed £23 billion annually, with serious debate and concern that they do not even effectively lower cholesterol levels and have many unwanted side effects, an alternative is long overdue.

An Iranian epidemiological study (Nabavi et al. 2015) of the polyphenolic composition of *Crataegus monogyna* was overwhelmingly positive, confirming all the chemical profile analyses mentioned in the last two sections. The authors also expressed surprise about the lack of reliable human trials and understanding about which ailments may be aided by varying compounds, not least because so many parts of the plant are beneficial with many historical uses across a spectrum of ailments.

From *Natural Product Research* (Benabderrahmane et al. 2021), a study used a solid–liquid extraction method to identify for the first time a glucoside of quercetin and three other compounds, as well as demonstrating the bactericidal power of the fruit extracts from *Crataegus oxyacantha*. So far we've seen hawthorn as an anti-carcinogenic, anti-inflammatory, anti-diabetic—and now another arrow in hawthorn's quiver of health, antibacterial.

More antibacterial evidence came from the Traditional Chinese Medicine College of Ningxia, China (Niu et al. 2013), using water and ethanol extracts from the medlar (*Mespilus germanica*) and hawthorn (*Crataegus monogyna*). Together these extracts had good antibacterial effect against both *Staphylococcus aureus* and *Klebsiella pneumonia*, with better results against the latter, as there was from the ethanol extract.

Anti-carcinogenic ability was exhibited by another novel part of the hawthorn from Chinese research (Wu et al. 2017), with some interesting results from using the fiddly-to-obtain peel of the haws. The compounds in the skin of *Crataegus pinnatifada* were found to be a powerful antioxidant, also exhibiting cytotoxic activity against cancer cells, as well as inhibiting cell growth and acetylcholinesterase activity. This then prevents the cancer cells' abilities to proliferate and build its own network, furthering the argument of the holistic approach of eating as much of the available parts of plants as possible when in season, being the best prevention provided by Mother Nature for us after over half a billion years of harmonised symbiotic existence. It's been a mistake removing us from this symbiotic relationship, particularly when the alternative is derived from myopic reductionist, profit-driven methodology using man-made chemicals.

Looking at how the flavonoids and triterpenoids from the haws of *Crataegus pinnatifida* demonstrate significant anti-proliferative activity against all human cancer cell lines was work undertaken at the School of Food Science and Technology, Guangzhou, China (Wen et al. 2015). The research showed a potent action of causing mitochondrial death (also called cell apoptosis), preventing what is cancer's main danger, its uncontrollable growth, with hawthorn fruit extract stopping this. It beggars the question: why is no one else looking at this important anti-carcinogenic capacity?

An article in the journal *Chemico-Biological Interactions*, reporting a study by Shaanxi College of Food Engineering, China (Han et al. 2016) demonstrated again that the hawthorn fruit, in particular the peel of the fruit, is an excellent source of natural polyphenolic chemo-preventative agents and can also be used in the treatment of liver disorders. The School of Traditional Chinese Materia Medica, Shenyang, China (Huang et al. 2018), used another novel part of the tree, this time via an ethanolic extract from the seeds. They found not one, but seven new sesquineolignans, all of which had potent neuroprotective activity, inhibiting cellular apoptosis of the healthy human cells, while intensifying the apoptosis of the cancer cells. China's Shanxi Medical University (Cui et al. 2018) also showed that haw fruit extract was a candidate for further evaluation as a chemotherapeutic agent for human liver cancer, where it initiated the all-important cancer cell apoptosis, as seen from other laboratories across the world above.

Another article in the *Journal of the Science of Food and Agriculture* (Yang & Liu 2012), showed the phenolic compounds, which were highest in the leaves and flowers and lower in the haw berries, which were highest in procyanidins, were excellent for the circulatory system and therefore the heart. Procyanidins are present to a lesser degree in red wine, hence the confusing information often circulated about red wine being healthy. This informative study also showed the variations from differing stages of maturation of leaves and flowers, which changed the composition, quantity and quality of compounds. With recognised cardio-protective, hypolipidaemic (lowers the fat in the blood), hypotensive, antioxidant, free-radical scavenging and anti-inflammatory potentials from the whole plant, the article concluded by asking for better clarification and adding that 'more clinical studies are needed'.

These procyanidins are already recognised as being beneficial, which is why they can be found standardised in commercially available cardioprotective remedies, notably Dr. Schwabe Pharmaceuticals' WS 1442 special extract, Crataegutt novo 450 and CardioMax, all containing 18.75% oligomeric procyanidins. Research from the Medical University of Silesia, Zabrze, Poland (Zorniak et al. 2017), studying WS 1442, confirmed the vasorelaxant, anti-inflammatory and scavenging responses from the extract.

Finally, in a study of possible concern for meat eaters, with all the controversy about nitrates and nitrites being carcinogenic, another of the chemicals used for preservation, butylated hydroxyanisole (BHA—E320) and its close cousin butylated hydroxytoluene (BHT—E321) have also been linked to both tumours and endocrine disruption (that's messing with your DNA, 'screwing with your genes'). From the *Journal of the Science of Food and Agriculture* (Papuc et al. 2018), it was reported that an ethanol extract from the haw berries was a more effective preservative than the commonly used and seeped-in-controversy BHA, at least for the minced pork in their experimentations. Considering what BHA looks like and how it is made, it's flabbergasting that this stinky, waxy filth was ever considered to be a sensible addition to the food chain.

* * *

As already mentioned, it is somewhat of a surprise that these plants offer so much evidence for their possible use in the pharmaceutical world for a whole spectrum of ailments, but it appears they have been all but bypassed by researchers from America and Britain. Any research conducted, especially from the past fifteen years in America and ten years from the UK, nearly always has the opposite results to those being reported from elsewhere around the world, that is, the results are negative. Older research, with a positive outcome from Reading University, School of Food Biosciences, England (Walker et al. 2006), involved a very rare human trial (with claims to be the first randomised controlled trial), using current diabetic patients taking prescribed drugs; those taking a hawthorn extract from *Crataegus laevigata* showed a significant hypotensive difference, with a diastolic blood pressure reduction, demonstrating the positive effect is indeed cross-species from rats to humans (which is why animal experiments are conducted in the first place, because humans are also animals) and across the whole genus of *Crataegus*.

There was also research from the Department of Complementary Medicine, University of Exeter, England (Pittler et al. 2003). Using epidemiological methodology, the authors concluded that hawthorn extract had significant symptomatic control and physiologic outcomes and could be considered as an adjunctive treatment for chronic heart failure. Another older epidemiological study reported in the *American Journal of Health System Pharmacy* (Rigelsky & Sweet 2002) outlined the positive lipid, cholesterol and heart failure attributes exhibited by hawthorn, notably the flavonoids.

These two older epidemiological studies from the UK and US both corroborate the more recent research conducted worldwide. Why the switch to predominantly negative associations within the past decade from both the US and UK? Is it pharmaceutical lobbying? Because the rest of the world's more recent research merely confirms the earlier American and European studies, which were emphatically positive.

Fishy science with an industrial whiff

To briefly expand on the observation above, there have been well over 3,100 scientific trials studying the hawthorn, with the vast majority of these from outside of the Western pharmaceutical umbrella. Iran, Korea, China, Turkey and Poland are all out in front with the number of plant trials conducted, with predominantly very positive results. Conversely, with far fewer Western studies many tend to look at negative implications of hawthorn, with no suggestion of its heart or hypertension benefits and the new evidence of its carcinogenic benefits, or the antibiotic properties, especially not among those conducted in the past 15 years—almost nothing positive.

These diverse results from the two sides of the pharmaceutical world, US and Europe vs China, Korea, India and Iran. Who to believe? The Western pharmaceutical corps, who are hell-bent on profits and have lobbied so hard to only allow new member states to join the EU if they relinquish their traditional herbal lores and knowledge, or the rest of the world who conduct scientific research for science's sake and cures, not potential financial profit?

A general conclusion from the vast bulk of work carried out is very similar to that reported throughout this book—it heavily suggests that irrespective of the difficulty in establishing specifically which ailment may be benefited by which chemical present in the plant, there is something positive happening and something that does not manifest with any negative side effects. Also, there are preventative capabilities of regular ingestion of these plants in varying forms, be that as a tea, soup, small addition to a salad, tinctures and creams—all have a role to play in overall holistic health. Which is not good news for Western medicine, as the dogma of reductionism requires isolation of individual compounds, so they can be patented and monopolised to treat isolated ailments. Non-GM plants can't be patented, copyrighted or 'owned'—so when whole plants actually work holistically and preventatively, those business accountants and economists 'tut' as the books can't be balanced with less profits and cheaper treatment would see GDPs shrink.

* * *

The above scientific evidence has therefore demonstrated that all the aerial parts of the hawthorn possess potent antioxidant and free-radical scavenging activities, owing to the presence of different bioactive compounds. These compounds are reported to have many pharmacological effects, notably neuroprotective (nerves), hepatoprotective (liver), cardioprotective and nephroprotective (kidneys), and to be anti-carcinogenic, anti-diabetic and antibacterial.

Despite this growing body of evidence, to date there has been little attempt to formulate a coherent understanding of the potential health effects of hawthorn, which is not at all surprising considering the vast sums of money at stake, even if just half of the above turns out to be true.

Eating

Hopefully, by now you're wanting to know how you can get some of this preventative capability and curative potential from this miracle tree inside of you. Best of all, it's not only free, it comes with other advantages for physical and mental health, especially if you have to go out foraging for some.

With a regional name of bread and cheese, there is the indication hawthorn was once a common snack because the unopened flower bud has a distinctly buttery flavour, and these can definitely be eaten at the same time as the leaves in late April and May. It's such an easy foraging target, with definite health links to eating just a few flowers in a salad along with a couple of freshly chopped-up leaves in late spring, it's not going to be necessarily diet-changing, but it could well be life-changing. It could start a process that ultimately leads to a reharmonising with Mother Earth, opening the door to a reconnection we previously took for granted, as there once was a time when we knew we were a part of Nature, not apart from it.

* * *

Here's a very brief overview of some culinary uses of the hawthorn from around the world, beginning with China. *Crataegus pinnatifida* (Chinese hawthorn), nicknamed 'red pearl', possesses one of the best-tasting haws at 4cm, the size of a crab apple. These are used in various Chinese snacks and eaten as dried flakes—it was once common for a Chinese doctor to give a few to patients as they left. Most popular are Shandong haw flakes, which are the dried haw cake, resembling a small hotdog, thinly sliced, being made from the dried pulp remains from the fruit leather at the end of this section. Another style is *tanghulu*, the very popular fruit kebabs, skewered and dipped in sweet sauce, sometimes with sesame seeds and sold by street vendors. The haws are also used in jams, jellies, juices, booze and other drinks to make among others Qingsongling, the official beverage of the Chinese delegation to the Barcelona Olympics in 1992. Haws are also commonly dried and added to teas, and as wine becomes more fashionable in China a hawthorn wine has become popular, with health benefits, as haws are so rich in polyphenols. In China, the harvest of haws is surprisingly large, with official records stating the total to be about a million tons annually, which is 3% of the total of China's apple harvest of 35mt, now 50% of the world's total. The haw harvest is definitely significant and indicates its use has always been widespread. Archaeological evidence from the Jiahu site near to the Yellow River indicates an alcoholic beverage being made for at least 9,000 years using haws.

Over in Central America, the most common hawthorn is *Crataegus mexicana*, with many local customs using the 2.5cm-size golden fruit. One of these sees candied haws stuffed into piñatas and broken open during the pre-Christmas *Las Posadas* celebrations. It is also cooked into a Christmas punch, *Ponche Navideño*, using haws, guava, orange, apple, tamarind, cinnamon and sugar cane, with regional recipes and rum and other spirits added for the adults, deriving from the ancient Aztec *texocotl*, called *tejocote* today. Haws are also candied as *rielitos*, mixed with sugar and chillies for a popular sweet in Mexico.

Crataegus aestivalis is being grafted onto hawthorn rootstock by orchardists in Southern USA for commercial production of the haws; these are 2.5cm in diameter, high in pectin, and good for jams and jellies, fruit leather (great with rosehips), fruit pies and wine. In Iran the fruit of *Crataegus azarolus*, called *zalzalak*, is eaten as fruit or used in a jam.

* * *

Hawthorn leaves (before the haws emerge) go well in a beetroot and potato salad. In making wines, jelly and fruit leather, loads of the haws can be eaten from late September, along with the other autumn fruits, right through to early December, with the following recipe preserving them for a few months.

Fruit leather Depending on what is at hand, added to the mix will be some blackberries, elderberries, sloes, rosehips, crap apples or quince. Weigh about 600g haws and 600g other fruit—you don't have to add anything else apart from the haws: no rules, just clues. There's no need to de-seed but do remove stalks and other sharp bits as they might ruin your sieve. Place the fruit in a saucepan and cover with water (about 350ml) and bring to the boil for a few minutes, and allow to simmer quite briskly for a further 10 minutes. Then mash all the fruit together with a potato masher, squeeze through a fine-meshed sieve into a bowl. At this stage add a sweetener if desired such as coconut palm sugar, honey, apple concentrate etc., then pour the mix into a shallow baking tray lined with baking paper to no more than 5mm depth. Dry for a few hours on a very low oven setting, in the sun if warm or a food dehydrator. The consistency will become increasingly tacky, and when dry it will easily peel from the baking paper. Cut into strips, roll up and store in an airtight container for up to five months in a fridge … ours never lasts more than a week!

Ketchup Here's a great cross between tomato ketchup and brown sauce. Pick, destalk and clean 450g haws; prepare 280ml apple cider vinegar, 280ml water, 120g muscovado sugar (this can be replaced with coconut palm sugar or even done away with and adding more dates), 110g chopped dates, ½ teaspoon ground ginger, ½ teaspoon ground nutmeg, pinch of clove, allspice, salt and juice of ½ lemon. Add the haws, vinegar, water, dates and spices to a large saucepan, bring to the boil and simmer for half an hour. Squeeze the mixture through a jelly bag/muslin into a saucepan, add the sugar and lemon juice, and bring to the boil while constantly stirring for a further 4 minutes. There'll be about 400ml remaining; put this in sterilised bottles and seal. This recipe is an absolute cracker and is so versatile in various sauces or dips.

While digesting the ideal snack we're going to take a diversion to a different time and place, using the hawthorn for a very different purpose from its superlative health and culinary capabilities outlined above. It was once the plant of choice for hedgerows, and in darker times of less freedom and stolen common land it was a tool of subjugation—

Pleaching and a smeuse

As long as there has been agriculture in Britain (about 5,000 years), there have been field boundaries, made of either dry stonewalls or a living hedge. This original living hedge would never have been exclusively made from hawthorns, with the blackthorn and its sloes as fruit, a common companion; both are in the rose family, and look very similar at the end of winter when bare. By mid-March the blackthorn will start to blossom and continue throughout April, with the buds emerging from the spikes. The blackthorn leaves begin to emerge when the hawthorn leaves are already out and the hawthorn blossom begins to emerge as the blackthorn's blossom falls away. The leaves are very different, although the flowers are similar in size and shape, both with five petals. Blackthorn's flowers are more star-shaped, with the hawthorn's fuller and rounder, and both put on a most impressive display of flowers from March through to May. Other trees making up the hedgerow of old included oak, ash, elm, cherry, elder, hazel, dogwood, plum, crab apple and guelder rose. All of these also had important medicinal, nutritional, fire and hearth, building and cultural roles in the past and hopefully will do so again soon.

The Nervii of Flanders (a part of old Belgium) at the time of Caesar, and documented by him, had extensive pleached hedgerows of hawthorn and blackthorn to segregate their livestock, as well as making excellent defences, although evidently not good enough to keep the Romans out. Pleached hedges were also popular a little further south among the Gauls in present-day France, from the time of Asterix who did keep the Romans out, but with the help of a magic brew, not hawthorn, until the Normandy landings in 1944, when the British were also kept out—an area known as the Bocage made crossing the terrain impossible, even for 'friendly' tanks. An invention was required to battle the hedging before getting to the Nazis, coming in the form of a whirring giant mechanical Edward Scissorhands' attachment, called a Rhino tank, with 600 tanks having them attached to their fronts in order to clear a line to the front.

The fruit and the tree both derive their names from the Old English *haga* for hedge, from the Anglo-Saxon term *haguthorn*, 'a fence with thorns', indicating they have a considerable ancestry as being utilised as a hedge. It is thought that half the hedges in Britain predate the Industrial Revolution (from around 1750) and as many as a fifth may well date back to the Anglo-Saxons (roughly 410–1066), making the hedge an intrinsic part of the rural fabric of Britain. Give or take a few miles, there were a million miles' worth of hedges in the UK before WWII, a figure that has more than halved since then, with present rate of loss being from 4,000 to 8,000 miles per year—whichever it is, the loss continues. The government sadly encouraged farmers to remove hedges and open up farms after WWII in the false belief that bigger fields would lead to bigger yields and more national food security, when it actually introduced the final desecration to our Mother, monoculture.

More than just a fence, for millennia this living hedge has harboured a multitude of other plants and animals, from birds to butterflies, spiders, mice and ferrets, many of which act as deterrents and insect repellents, feeding on the weevils and beetles that have a tendency to eat crops if humans plant the crops near them. This hedgerow pattern has been largely replaced with massive open fields, barbed wire fences and other such fences, which are not only unsightly, they harbour nothing, except short-term economic viability and long-term unsustainability. In Dr Hooper's 1974 imaginatively titled *Hedges*, a formula is given to roughly work out the age of a hedge, done by counting the number of different woody species (mostly other trees) in 30 yards and multiplying by 110 years. It's not accurate, but it gives a rough idea, although it doesn't work for hedges more than a thousand years old—but basically, the more variety, the older the hedge.

For those who understand the importance of the hedge, they can appear quite fanatical, especially in their opposition to farmers' field expansions, or second home-owners from the cities ripping out whole lengths of hedge and replacing them with high fences and CCTV. The hedges in Devon alone are said to be home to about 600 species of plants, 1,500 species of insects, 65 species of birds and 20 small mammal species—making them a pretty special sanctuary requiring preservation, with many new fanatical supporters more than welcome to join the fray.

Hedges do have a small level of national protection from the Hedgerows Regulations 1997, with the local planning authority being responsible for applications for removal. This is because it's not just the protection of species living in and on the hedge, hedges also play a crucial role in preventing soil erosion and flooding, as well as reducing pollution.

* * *

Moving on to pleaching (aka plashing), not all hedges are pleached, just as not all stonewalls are dry stonewalls. It's the traditional way, but only a small percentage of hedges are actually pleached. So, what is pleaching? Pleaching is when a young tree, about 4 to 6 years old and roughly 6 to 8 feet tall, is bent over to about 30 degrees by making a deep cut near the base of the trunk, and as each tree is bent over it forms a much more solid structure. The unit of measurement for each sapling to be distanced apart when planting a hedge is a cubit, the distance from elbow to fingertip.

With the trunks bent, a little playing with the branches provides in a further couple of years a formidable hedge, with large impenetrable, near horizontal bars from the bent-over trunks, with an array of smaller curved branches full of ferocious thorns. This living fence obviously has immeasurable benefits over a barbed wire or wooden fence. The hedgerow (which is a term for the biosphere of the hedge and all that lives within it) is today essential for the continued survival of many British species, as this has become the last sanctuary from the endless destruction and shrinking biodiversity caused by modern monocultural agriculture, with its giant murderous machinery, sterile seeds and genocidal chemicals.

Which brings it on to the other anecdotal story for this chapter. In this instance I was extremely grateful for the lack of pleaching and the poor maintenance of this particular hedge.

I used to take an hour to go for a cycle each day, to get away from the desk, using a cycling app (Strava—Culinary Caveman). Sometimes I might be going at quite a pace to be competitive in a segment, usually the length or section of a road or hill to virtually (via the app) race against others. This day I had decided to use the uncommon northerly wind and go for a 3-mile segment, having started at pace before approaching a tight right hander I decided to not race and eased up … luckily, because immediately upon turning I'm suddenly looking straight at the radiator of quite a sizeable lorry, which thankfully was going slow at it was just about to take the 90-degree corner, left. Slamming on his brakes further gave me the hundredth of a second needed to pick a line straight into the hawthorn hedge, where there was a small vertical column of light at the bottom two or so feet. I aimed the front wheel at this and crashed through the hedge, coming to a stop about three bike lengths in. Taking a deep breath and feeling myself, arms, legs, all undamaged, no cuts, lacerations or holes and bike in one piece, check.

Dismounting and carefully going backwards to the road, I saw the lorry driver looking at me in utter disbelief, like he's just seen a ghost, or more pertinently, created one. His relief was palpable as he thought I had gone under his lorry, having only been in his view for such a short while and not realising I had managed to get across him and into the hedge. If I hadn't gone into the hedge I would have face-planted the radiator grill, wiping myself and the bike out, or I could have bounced back off the hedge if it had been maintained into the lorry and under a giant wheel. I reached up to shake the driver's hand, saying how close that was, perhaps even a Near Death Experience. Cycling away with a huge exhale, a smile spread across my face. Thankfully it was the winter and the hedge was bare, showing the gap to aim for, or perhaps the wee-folk thought me worthy of a magical helping hand?

I've since discovered that there is rather poetically in the old Sussex dialect (as this is where it happened) a word that could well apply to that gap—a *smeuse*—a hole in a hedge created by small animals frequently scampering through it. With more than 20 species of mammals criss-crossing our hedgerows, including hedgehogs, mice, stoats, rabbits and badgers, it shouldn't be hard to spot a smeuse or two. Whether it was a smeuse and not magically induced I'm not too concerned, being appreciative either way!

The reason for going on about pleaching, apart from it being fairly interesting (I hope), was in an attempt to set a scene, as it's another reason why the hawthorn has been included as one of the seven (not that another reason was necessary). This is because it teaches us a crucial lesson from history, where this hedge was used to divide far more than just fields—used originally to keep animals in, it was soon to be used to keep people out, segregating the 'haves' from the 'haves not'.

Enclosure

> I don't think people realise how the establishment became established. It simply stole the land and property off the poor, surrounded themselves with weak minded sycophants for protection, gave them titles and have been wielding power ever since.
>
> (Tony Benn, 1925–2014, proper Labour MP 1950–2001)

While researching this chapter, I asked many people I met what they knew about Enclosure. Only one person, a rather radical woman from a farming family, knew of the details; not even my Eton-educated friend (everyone needs a token posh friend!) knew any of the specifics, but being so well read he had heard of it and most helpfully suggested I read John Clare (which I did, and an excerpt appears in a few pages). Apart from him, an Irishman, a Scot, a Canadian and an Australian have managed to tell me little more than the very basics, but no other person from England (I must have quizzed well over a hundred) had any idea of what this was. Another example of cultural amnesia.

The fact we've lost touch with the stories and lores of our forefathers and foremothers and no longer venerate a thorny shrubby tree as we had done for anything from the past 2,000 to 20,000 years is kind of understandable. The fact no one has any idea that only a couple of centuries ago the landed gentry decided to increase their holdings in our country to the tune of 11.5% of the entire landmass of Britain, free of charge, bar the lawyers' fees and the thruppence ha'penny thrown at the dispossessed commoner if they were very lucky. This is when the land transferred from being common land to private land, in fact more a theft than a transfer. It should have started the revolution. Why not?

There were plenty of protests, only in those days it was almost impossible to centralise a protest, with many people rarely travelling more than a dozen miles from their hometown or village during their entire lifetime. The punishment for any poor starving workers daring to protest invariably led to a gruesome death, involving being pulled apart, entrails and head removed and put on a spike or hung from a tree. That barbarity appeared to act as a useful deterrent and kept others from complaining.

During these times wool merchants were among the richest of all, with landowners seeing the opportunity for vastly increasing their land and wealth, and sheep grazing seen as preferable to commoners roaming and subsisting. This was the cultural background when the first Act based on enclosure was passed in 1235, when Henry III gave landowners the right to convert common land to private land. Some landowners seized on this, violently evicting the poor farmers and replacing them with sheep, which for a few centuries thereafter was a major part of Britain's early GDP. The process also implanted the belief among the rich that the poor were literally of less use or value than sheep.

What has just managed to make it through the cultural filters of time is that there were rebellions against this theft of the land. It's a series of protests that has managed to be ongoing from no doubt long before William Fitz Osbert was hanged for leading a popular revolt against the rich in 1196; or the Peasants' Revolt of 1381 led by Wat Tyler and Jack Straw; 1549 gave us Kett's Rebellion, when a direct protest against enclosed land saw landowner Robert Kett's fences smashed, but instead of turning on the protestors he sympathised and joined them, becoming the leader of up to 16,000 men by the time they took Norwich, the country's second largest city at the time with about 12,000 residents; predictably Kett was eventually beaten and hung and vilified for centuries as a rebel, although now he is finally recognised as a leader of a just cause and represents exactly the type of landowner required when our imminent peaceful cultural revolution happens.

Kett's Rebellion occurred straight after Beltane (May Day) festivities, as did the 1607 Midland Revolt in Northamptonshire, and as have other uprisings, protests and riots of old, happening after mass congregations, where people get merry, share their troubles and sometimes get suitably roused by speakers to allow the mob to be enacted. The Midland Revolt, led by John Reynolds aka Captain Pouch—a pouch around his neck he claimed gave protection for all even if it was just mouldy cheese—was as a direct response to enclosed land becoming land for sheep. The Midland Revolt spilt into the Newton Rebellion, when one of Britain's most vile men, Sir Thomas Tresham, had enclosed as much as 27,000 acres, destroying 350 farms and displacing over 15,000 people, his fences resulting in possibly England's last battle directly between the rich and poor.

Then 1648 saw the rising of the Levellers and Diggers. The leader, Gerrard Winstanley, famously said: "Was the earth made to preserve a few covetous, proud men to live at ease; or was it made to preserve all her children?" In April 1649 the protestors levelled some land on St George's Hill, Weybridge in Surrey and planted vegetables, at a time when fresh vegetables were beyond the means of the peasants. The landowners were unhappy and wanted this to end before the peasants got ideas above their station, calling on the intimidating New Model Army to intervene. The Army leader Sir Thomas Fairfax saw no wrong in the protest and departed, refusing to take any action, but this provided only temporary peace. Today, St George's Hill is sadly and rather ironically a golf course and an exclusive gated-community for multimillionaires.

Another Hill, Octavia Hill (1838–1912) deserves a name check here as it was through her work with the Commons Preservation Society, founded in 1865, that Epping Forest was secured from enclosure; Hampstead Heath too could easily have been lost to a planned open gravel mine, and Octavia also helped preserve Regent's Park, Kew Gardens and Wimbledon Common, allowing Uncle Bulgaria and the other Wombles to be "Making good use of the things that we find, things that the everyday folk leave behind", still today. Hill also co-founded the National Trust in 1895—a person we have a lot to be grateful for but who is unknown by most people today.

One of the last mass protests against enclosure and specifically the right to ramble took place in the Peak District, where as many as 15,000 would take weekend walks in the fresh air as an escape from the hard week's work and smoggy air of Manchester and Liverpool. The landowner, the Duke of Devonshire, was unhappy with this and tried to stop it. As a result the 1932 Kinder Scout trespass was organised, with hundreds walking in defiance to the Kinder Scout moorland. Scuffles with the police and game-wardens saw prison terms for those arrested, including the organiser Benny Rothman—under a jury of eleven aristocrats including two brigadier-generals, three colonels, two majors and three captains! The ramblers' bravery did eventually lead to the formation of the National Parks, including much of the Peak District, as soon as a Labour government came to power in 1945—thanks to Benny Rothman, a hero. Surprisingly, still today, unbeknown to most, all the remaining public land in the cities to this very day is still being enclosed and given to a developer or an Earl or Duke.

A common denominator among all these revolts is that those wanting a little more equality in society, by highlighting the immorality of the rich, were often killed. Even those advocating peaceful protests, such as John Reynolds at the Midland Revolt calling for no bad words or threatening behaviour, often descended into provoked violence. History has vilified these brave souls who dared defy the cultural order of the day, and it is only more recently that they have finally been honoured for their resolution.

> They hang the man and flog the woman
> That steals the goose from off the common,
> But let the greater villain loose that steals
> The common from the goose.
> The law demands that we atone
> When we take things we do not own,
> But leaves the Lords and Ladies
> Who take things that are yours and mine.
>
> (English folk song, c. 1764)

A twenty-first-century version might read:

> They bang-up the poor couple in the penitentiary
> Who could neither afford the Community Charge or a cuppa tea

> Whilst leaving the ruinous bankers free
> For destroying the economy and causing austerity.

The practice of enclosing continued throughout the late medieval period, with the occasional protest always threatening to flare up into further protests across the country. This theft of land from the commoner became firmly embedded in law thanks to the Enclosure Act of 1773 and its countless amendments, until by the beginning of the twentieth century the rich had managed to turn over from commonly owned land more than one-ninth of the total area of Britain, some 6.8 million acres, into their personal possession. That's 21% of all of the landmass of England (11.1% of Britain), on top of what they had already, of course.

To get some perspective on the vastness of this theft, imagine a football pitch such as Wembley, the playing surface with the added bit around the edge of the pitch, which all comes in at just under two acres. Then times this by 3.4 million, which is admittedly difficult to imagine—that's far more than the entire South East of England. And we are supposed to just accept that it's fine as it's history now, they were perfectly entitled to do it anyway, making and changing the law at their will. Incredibly this happened some 5,200 times between 1604 and 1914, just to cement in law that this theft was absolutely nothing of the sort, in the same way the man in the sky can annul a sin, the lawyer can absolve a crime … "nothing to see here … move along, please".

* * *

Back to hawthorns and pleaching. At the height of this enclosure between 1750 and 1850, at least 200,000 miles of hedgerow were planted. The process culminated in the 1845 Enclosure Act, which basically gave landowners the rights to all the common land that was adjoining their land. From there once being large communal open fields and woodlands, land was partitioned off with hedgerows and the old boundaries disappeared. The hawthorn, along with the blackthorn, was used to make the distinction between ours and yours (with only a why between them!)—an impenetrable barrier, laid down to establish these divisions. The hawthorn certainly didn't complain, or have any views on the preposterous theft that was occurring—it was only too pleased to be given the chance to spread its roots across much of the country. Just as well, because now there is so much wild hawthorn, the fact it can be economically valuable and medically priceless means its time will come again soon.

It's hard to comprehend for urbanites today the utter cruelty and sense of theft associated with enclosing land (it's happening today in the cities with the rest of the common land, and hardly anyone is batting an eyelid as so few know anything has gone on at all). Beyond modern imagination, the closest analogy we may have today is if every day you play football on a field, or play in a local park, for it to be suddenly closed, fenced off and sold and a mansion and gardens built on it, with you put in prison for trying to go there. We have little idea of the land today or what common land meant in the past—when it meant survival, with land to graze your couple of sheep and goat on, or the river to fish in—whereas today for many it's just somewhere to take the dog for a shit! Which barely touches the surface of the scandal of having the woods used for firewood enclosed and the people excluded, with laws put in place whereby any taking of future firewood could land you in prison. It was an absolute disgrace, and I've failed to get even close to the abject ignominy and feeling of smallness and uselessness that must have pervaded as the rich took the land and paid the lawyers to make this theft perfectly legal and acceptable.

* * *

We, the people, have been marginalised in precisely the same way that the tribal populations around the world have seen laws and paper divide and take their inherited land away from them. This is exactly what happened to the tribes and people of Britain, much of it surprisingly just a few generations ago. The land can and did sustain anyone not wishing to enter servitude or to supplement diets when the harvests failed, with the Forest Charter of 1217 protecting this right (we'll come back to this Charter later because it might be old but it is very relevant). Then suddenly the woods actually belonged to someone, or the fields used for grazing were staked out ready for future pleaching, and all because a piece of paper the rich wrote some laws on makes it okay.

This slow incremental transition of common land, in dribs and drabs, week in, week out, soon began to add up. It was extremely difficult to gauge how this slow land grab was taking place, when even today it is extremely hard to establish who actually owns what in this country. Living in a small cottage in the countryside in poverty, as well as being a keen forager, I can imagine life in 1750 and appreciate the hurt and sense of loss of people at being outlawed from lands that would have provided much, making the forests with their firewood, nut and mushroom harvests suddenly out of bounds and the trapping of mammals or foraging for plants often resulting in transportation to the colonies. Foraging is a timeless pursuit, not an illegal menace.

Being an armchair judge today is definitely not the same as actually living and being present during enclosure, and so emotive is the subject that impartiality is not so easy. So in the defence of enclosure it has to be recognised that at various times, at various places throughout Britain the state of common land was, to say the least, varied, and it is these areas of ruination that gave a legitimate reason to make the land more productive. During the thirteenth to nineteenth centuries there had been repeated plagues, illnesses and hostility, resulting in there not being enough or even any villagers remaining to maintain common land, meaning the land would quickly become unruly, so if enclosed the estate could maintain it and a farmer utilise it and make it productive … or so it was claimed.

It can't be forgotten that the rich had an absolute and cocksure belief in their superiority over the commoners and the servitude expected from them. Of those not only noticing the change to the countryside but also documenting it for posterity from the people's perspective, John Clare (1793–1864) is perhaps the most famous poet of the time, lamenting on the evils of enclosure, which bless his soul, literally led him to madness—

> Unbounded freedom ruled the wandering scene
> Nor fence of ownership crept in between
> To hide the prospect of the following eye
> Its only bondage was the circling sky

One mighty flat undwarfed by bush and tree
Spread its faint shadow of immensity
And lost itself, which seemed to eke its bounds
In the blue mist the horizon's edge surrounds
Now this sweet vision of my boyish hours
Free as spring clouds and wild as summer flowers
Is faded all—a hope that blossomed free,
And hath been once, no more shall ever be
Inclosure came and trampled on the grave
Of labour's rights and left the poor a slave.

(excerpt from *The Mores*)

It is the Enclosure Act of 1773 that ironically opens the door to the future changes. It simply—not that the process of land redistribution would ever be simple—needs to be repealed and all enclosed land not deemed as being productive can be transferred back to common land, to be divided into thousands of permacultural smallholdings and forest garden projects. Fair reparations will be paid for all land reclaimed only after savings from non-payment of taxes are deducted from soon-to-be-illegal and redistributed offshore accounts. Although of course the law has not been written so that it can be replaced with fair and just egalitarian virtues. Heaven forbid, I'll be saying "socialism" next … oh! I just did …

Because this is where we can find the land to begin to implement a reversal of the destruction of the soil of this nation and to begin to use a new, sustainable and healthy agricultural system. Not a system that has poisoned and destroyed the biodiversity of not only the land it is grown on, it has poisoned those who eat it as well. We will be returning to land reform in the final chapter because the solution to food security cannot be fully implemented without the land and a regeneration of the soil.

It's not 'just' the land that got stolen/reappropriated

If nicking 20% of England, those 6.8 million acres, wasn't bad enough, consider the fact that Britain is one of the only countries in the world that doesn't actually own any of its essential utilities, such as the housing, electricity, gas, water, trains, postal, communications and even garbage—it's also the only feudal country in the world owing to the presence of 'freeholds'.

The 'Sale of Britain' had been discussed for a while during the Thatcher Tory government (1979–90), and under the economic guidance of Nigel Lawson it was seen as a convenient budget balancer and a vote winner to sell off the public utilities. There could even have been good intentions in the selling-off via shares to the public, giving everyone a truly vested interest in the running of the country. The services and utilities were split into millions of shares, which indeed the public did buy in their droves, although the plan probably didn't include the possibility that when all the shares came back on to the market, being sold as prices rose, they could be bought by just a few individuals, often in the guise of pension/hedge funds. When the Thatcher government took office in 1979, the global value of hedge and pension funds was somewhere in the region of £2 trillion, today it's £31tn, nearly double the GDP of America at £16tn. The largest pension fund is the American Federal Old-age and Survivors Insurance Trust Fund, which oversees £2 trillion in assets, a similar size to the GDP of the UK.

Here's a few of the bigger players in the arena of utilities and services, mercilessly sold off and who now call the shots, when it should actually be about 1/67,000,000th mine! Even if it appears to be a British company, with or without HQs in the UK, they are often a PLC with shares concentrated with a few larger pension funds and financial services.

The big six energy providers in Britain are British Gas (actually Centrica, a PLC although British-based), EDF (French), E-ON (German), Scottish Power (Spanish), npower (German) and SSE (a Scottish PLC).

Among the telecommunication companies, where once BT had it all, there is now Giff Gaff owned by O2, which is owned by Telefonica from Spain; Lycamobile is a private British firm; Talktalk is a British PLC; Three is owned by Hutchinson Whampoa out of Hong Kong; Virgin Mobile and Media has been split into several companies with many owners, although Richard Branson maintains a slice of each pie; EE was Orange and T-Mobile, now owned by BT but a PLC, not the people; Sky is owned by American Comcast (Disney and Fox), and Vodafone is a British PLC.

The water for England and Wales has been split into 25 or so separate companies, with some surprising owners. For insance, Anglian Water is owned by a Canadian Pension Plan. Northumberland Water is owned by a company registered in the Cayman Islands with a HQ in Hong Kong, CK Hutchison, which also owns Three. Hong Kong is also associated with Yorkshire Water, owned mainly by HSBC and Citigroup (USA). Wessex Water is owned by YTL, a Malaysian corporation. Thames Water is owned by a Canadian pension group along with Kuwaiti, Chinese and Abu Dhabi investment groups. Affinity Water, which supplies much of central, some east and south east counties, is owned by Morgan Stanley and a UK investment company. South East Water is owned by Australian Funds; Sutton and East Surrey gets its water via the Japanese Sumitomo Corporation, while the rest are PLCs trading out of Britain predominantly owned by foreign shareholders. Dividend payments to shareholders in these water companies have exceeded £57 billion since 1991, money that should have been added to the public coffers, instead of exclusively making the rich a little richer, while ramping up prices of all the public services. Only recently Thames Water hit financial trouble even though they've a guaranteed 10-million customer base—proving how efficient private companies really are.

Garbage and waste disposal companies are the same: Veolia is French; Biffa is a PLC with heavy American shareholding from several investment companies; Suez is French and FCC Environment is Spanish.

On top of this are all the fears about the Tories selling off the NHS today, which have to sound a little louder when one considers that the current CEO of NHS England, Simon Stevens's previous job was as CEO and Vice President of an American £22.5bn HealthCare. His character reads well, it must be admitted, but if anyone could broker such a scandalous sale he is perfectly positioned and would definitely be more than capable. Remember that NHS England has already sold millions of confidential patient files to American universities (under the guise of pharmaceutical companies) and insurance companies, for research purposes, apparently. Which looks like the US Big Pharma already know the names and addresses of their future customers. Meanwhile all services, equipment and provisions have already been outsourced to private companies, mostly foreign, meaning the NHS has already effectively been sold, being little more than an abbreviation.

* * *

Anyway, enough of all that, that's two pages that have nothing to do with hawthorn. So let us give the hawthorn some final consideration, because, if you remember, hawthorn had a connection to Jesus Christ, royalty and sacred turf forming the foundation of modern British society. That's in addition to the hawthorn being able to provide medicines for heart disease, and act as an anti-inflammatory, antibiotic, anti-diabetic, anti-carcinogen and all the other ailments mentioned throughout this chapter. I sincerely hope you will now look at the hawthorn in a new light, a light that is carried forward to the following chapter as this one has an important message as to how to reconnect us all back to this green and pleasant land.

As this chapter's been a bit more poetic than the others, let's end with some fine words, not about the hawthorn, but by a Hawthorn, Nathaniel Hawthorne, whom we met earlier with his May-Pole poem:

> Happiness is a butterfly which, when pursued, is always just beyond your grasp, but which if you sit down quietly, may alight upon you.

BLACKBERRY

(Rubus fruticosus agg.)

We end with a plant we should all be familiar with. Most of us have picked and eaten blackberries in August or September at least once, and all of us are guaranteed to have either tripped on one or been bloodied by one, getting a thorn embedded or snagged in our skin or an item of clothing. Also known as a bramble, the blackberry has to be the most foraged plant in the British Isles. The fact that it is so abundant, being yet another Human Plant, it found us, so it could hardly have been ignored.

It is the bane of many professional and amateur gardeners alike, as nothing beats the feeling of pulling up some plants while weeding and inadvertently including a bramble, which holds firm to the root when tugged, while the hand is pulled through the thorns ... ouch! It's a plant looked down upon with as much derision as the nettle, even by those who know there is a reward in the autumn from the fruit. What this plant actually offers, when we finally tune in, is significantly more than thorns and berries.

Botanically and plantality

It does clearly state Blackberry as the chapter heading, although, as with the other chapters this loosely includes all the other plants of the *Rubus* genus, which in most instances share nutritional and chemical compositions across all the species, as was seen with the *Urticas*, *Taraxacums* and others from the book apart from *Cannabis*, it being an only child with hops as its closest cousin. There was a temptation to widen this seventh chapter to also include the *Ribes* genus, as blackcurrants (*Ribes nigrum*) have a lot to give—plus '*Ribes* and *Rubus*' sounds uncannily like two mischievous characters from an Elizabethan play—but as blackcurrants only proliferate in England under agricultural conditions, the ever-present blackberry won out.

The *Rubus* genus is fairly broad, with over 300 species, possibly as many as 700, with botanists failing as usual to agree about membership. The most species held within one collection befalls Barry Clarke with the British National Collection at Hiller Gardens in Hampshire, with about 250 species. All the plants look comparable, with similar-looking fruit, with the main difference between a blackberry and a raspberry being that the hull/stem comes away with the fruit when a blackberry is picked, whereas the hull is left on the plant for a raspberry, leaving a hollow fruit.

Other fairly well-known species from the *Rubus* genus include *Rubus aboriginum* (garden dewberry), *Rubus arcticus* (Arctic raspberry), *Rubus armeniacus* (Himalayan blackberry), *Rubus caesius* (European dewberry)—the second most common wild blackberry in Britain, its main feature being that the fruit has fewer and larger drupes/segments, usually with a matt sheen—*Rubus chamaemorus* (cloudberry), *Rubus idaeus* (raspberry), *Rubus occidentalis* (black raspberry), *Rubus spectabilis* (salmonberry) and *Rubus ursinus* (trailing blackberry). There have been several hybrids developed as well, mixing two different species, such as the Tayberry, a cross between *Rubus idaeus* and *Rubus fruticosus*. If you were into the scientific study of brambles, you'd be known as a batologist. Only those batologists can identify and name all the similar-looking species, be that the *Rubus tuberculatus*, *Rubus nemoralis* or *Rubus bertramii*, even the rather dubiously named *Rubus schoolcraftianus* and *Rubus cockburnianus*!

Within a short while if not your first foray along the hedgerow, it will become evident there is a whole array of variations of blackberries on display—some will have a purple colouration to the thorns or stem, others will have simple arrangements of thorns, while yet others have layers of treacherous teeth; in others the leaves will be larger, heavily serrated, smooth or barbed, with different numbers of segments making up the fruit, also with variations in flavour. All of these differences will usually indicate a different species.

An important aspect for any food item is of course the flavour, with each species being unique, and wild varieties having different flavours from neighbouring plants. Most plants exhibit a desirable taste most years of fruiting, although some plant's fruits are often not worth the effort, being inconsequential and misshapen most years, yet every so often producing larger and tastier fruits, depending on the weather throughout the year (a warm, dry April is good for autumn fruit). Appreciate them all, pay them some respect and enjoy and share the harvest.

* * *

As with plantain and dandelions, brambles are one of the first plants to colonise newly disturbed, drained, spoiled, compacted or trampled lands. They also provide a far quicker boundary than the hawthorn and other tree-pleached hedges; planted in the spring, by the autumn they can grow into a barrier with impenetrable depth, whereas a hawthorn, blackthorn and plum hedge takes a couple of years to become established.

Brambles, with their foraging roots, are an ideal plant to prevent soil erosion as well as replenishing and phytoremediating the soil. Blackberry roots are a perennial although the canes (the fruit-bearing branches/stems) are biennial—this is because the first year's growth is of a new cane known as a primocane, which can grow up to 6 metres as it meanders across the floor or in tangled arches and clusters, producing only leaves and no flowers in the first year. This primocane in its second year swaps the primo prefix to become a floricane, where the cane grows no more and flowers emerge and, after pollination, fruit will appear, after which the cane dies. In America, all the *Rubus* are collectively called 'caneberries'.

The flowers are not too dissimilar to the individual blossom of the hawthorn, both being in the *Rosaceae* family. They serve as a valuable late spring and summer food for bees and other pollinators, when the availability of pollen from all fruiting plants and trees is significantly less than in the spring, with its abundance of tree blossoms.

Confusingly, it may be a berry in name, but botanically blackberry is known as an aggregate fruit, composed of a varying number of drupelets. By definition, a berry is a fruit produced from the ovary of a single flower in which the outer layer of the ovary wall develops into an edible fleshy portion (the pericarp). This all proves botany is bonkers, because under this definition the banana, chilli, courgette, cucumber, grape, hawthorn and tomatoes are all berries, while blackberries, raspberries and strawberries are actually berry imposters!

Rather worryingly, the *Oxford Junior Dictionary* defines a BlackBerry as a smartphone rather than a fruit, demonstrating short-sightedness by an editor who no doubt owned such a phone at the time, although does anyone own one today? The brambles with the fruit known as a blackberry, however, are very much still there in the real world.

* * *

As with any plant, once you actually start to recognise and pay attention to the plants that surround you daily, see what makes them stand out, and give them some recognition and respect, then others from the same or similar species seen elsewhere will show you that in the same way a person has a personality, a plant has a plantality. It's a buzz just knowing the names of the plants around us, but the thrill is seeing particular plants displaying unique traits that make them stand out among the crowd—the rainbow Mohican in a sea of commuter dark grey. When the plantality of the flora is drawn into focus, it becomes impossible thereafter not to see in this attuned harmonic visualisation of Gaia's resonance of togetherness. Everything on this beautiful planet, from the insects to the fish, fungus, trees and moss, are all just getting on with one another—and then there's *Homo destructus*, being forced down an alternative path, constantly being fed lies about this being the best path—let go of all that indoctrinated human nonsense and start to suck in the plantality all around us, everything trying quietly to get its head above the parapet of conventionality. Add the plantality to your personality and feel the sense of connectiveness thrive.

Challenging etymology

The etymology of the word bramble appears a done deal, as the Old English *braembel* meant a rough prickly shrub; this term can be traced further back to the Proto-Germanic *braemel* or *braemaz*, from whence we also get broom; and back even further in time to a language called Proto-Indo-European, a sort of original mother tongue from where many languages derive today, where the word *bh(e)rem* meant to project or point. It would appear a closed case, with a clear route far back into time.

Although, with a 'question everything' attitude, it's not such a done deal, as there is a perfectly viable alternative. *Bram* is a well-known and popular abbreviation for Abraham or Ibrahim, who just happens to be quite a significant figure from history across much of the world. Abraham is virtually unique in that he has the potential to unite the three main monotheistic religions of Christianity, Judaism and Islam. In Judaism he is recognised as the first Patriarch, the first Jew, and being a contemporary of Noah he was the first to believe there was only one god. In Islam he is seen as a link in the chain of prophets from Adam all the way to Muhammad; Ibrahim appears in the Qur'an in 35 chapters and is seen as the First Pioneer of Islam. In Christianity he is not as exalted, but still central, appearing the second most to Moses in the Bible, with various stories of the Canaanites and the famous one when he went up the mountain to sacrifice his own son, only for God to declare him the archetype of what a true believer should be by being willing to sacrifice that which was held dearest to him—some may call him a schizophrenic murderer!—God then humanely sacrificed a bush and a goat instead of Abraham's son Isaac.

So the origin and use of *bram* as the first syllable could be argued as venerating the Father Abraham and the second syllable, *ble* actually means in French *blet* (cereal or grain), while in Greek *ble* is blue. So why could the most prolific of forageable foods across a vast expanse of the world, and one that still unites us, not be actually originally known as *Abraham's blue grain*? Terms such as grain, fruits and corn were often interchangeable/mistranslated for crops with a different meaning today, as we tend to be more specific—just as meat often meant food in the past, even if referring to vegetables. It's just a suggestion, not to be taken too seriously.

In Gaelic the bramble goes by the name *dris-muine*, meaning 'prickle thorn', while a letter in the Ogham alphabet, *muin*, is often described as a vine, which is highly questionable as grapevines were absent from Celtic Ireland, so *muin* is more likely to be a bramble.

As with all languages and words worldwide, not everything is written down and as a consequence many words get lost (even if written down, books still get burnt or rot away, or were full of lies), but a few names still remaining for the berries around Britain include—Blackbide, Blackbutter, Black Heg, Blegs, Bly, Brameberry, Brambleberry, Bramble-kite, Brummel, Bumblekite, Dewberry, Fingerberry, Goutberry, Gouthead, Moocher, Scaldberry, Scaldhead and Thimbleberry. It is also known as Lawyers or Lawers as a direct comparison from the trouble in escaping their attention once caught in their clutches! The name scaldhead comes from an old belief that too many blackberries caused a scalp disease.

So little folklore, so much Devil's piss

Rather surprisingly for a plant so omnipresent, there is a shortage of folk stories. Before researching this chapter, the assumption was that there would be reams of folklore about this constant human companion, with even some reverence, particularly after the considerable amount bestowed on the previous six plants. It was a shock to find out there isn't even a recognised Blackberry Day, although if there was one, it'd probably be in late August to early September when many bushes are laden with ripe fruit.

This seems to have been, once again, a plant just hiding in plain sight, waiting patiently for us to refocus. It would be absurd to argue it wasn't an important dietary addition in the past, being not just abundant but with no poisonous lookalikes to catch out the unwary. But the most popular and timeless pursuit of gathering blackberries from the wayside and hedgerow seems to have fallen prey to our cultural amnesia. I've spoken to various older people when out and about, and only a diminishing few have any remembrance of days 70 or so years ago, when whole families went out together, whole villages or streets from rural towns, days out to the heath and beyond picking blackberries together throughout the day, and as is the case when humans congregate, some drank homebrew all evening and partied. The health imbued in community spirit and actions of mass common interest are the very things that bonds and unites the people together. So 70 years of enforced individualism, consumerism and egotism have created their own range of damaging mental health diseases of civilization. The common unity we all have to the world around us has been wrested away from us in only the past few generations. Nothing brings us together more than the common goal of gathering our food from this very generous planet, a community we were once a part of with every plant and animal.

* * *

The only significant blackberry folklore is about picking the berries late in the season, after Michaelmas Day (29 September). There is the story about St Michael casting Lucifer (the Devil) out of heaven, and Lucifer's fall was 'cushioned' by a bramble, causing his awfulness to curse the plant. So folklore determines St Michael's Day/Michaelmas as being the cut-off for when the fruit becomes unpalatable. This was a significant feast day of old, especially as it also marked when the crops should be harvested, as one of the four Quarter Days (Lady Day 25 March, sowing—the original New Year's Day; Midsummer Day 24 June; Michaelmas 29 September, harvesting; and Christmas Day 25 December). It is said, that on the evening of Michaelmas the Devil flies around and pisses on all the blackberry plants, with variations including him spitting on them, marking the leaves or treading on them. Irrespective of the devilry hocus-pocus, there is the definite fact that the fruit leaks a little when ripe, and as the fruit tends to ripen at different times on the same branch, the over-ripe berries attract bacteria so that by the end of the season stems are full of ripe fruit as well as plenty of mouldy individuals. Depending on rainfall and sunshine throughout the year, some are still to be found ripening until early November in sheltered enclaves. If spores are consumed of the most common *Botryotinia (Botrytis cinerea)*, a grey fluffy mould, sickness could prevail, or if it's a greenish white mould it's probably a type of penicillin.

Apart from Devil's piss, very little folklore could be found. There is the practice of graves sometimes having brambles planted on them, done to keep sheep and other grazers away, although conversely, it may also have been seen as helping to keep the dead in. Another countrywide custom was the passing of ill people (and cattle) through a naturally formed arch of a bramble branch, with local variations, some involving bread and butter, always crawling from east to west to mimic the passage of the sun. This was said to cure hernias, rheumatism and rickets; in Essex a patient would crawl seven times from side to side while saying "In bramble, out cough, Here I leave the whooping cough."

A cut blackberry stem (or other thorny specimens such as the briar rose) in the shape of a cross was also worn as a preventative against whooping cough, although eating it would probably have been more beneficial.

For a plant so constantly cursed, a reversal to being praised instead could well initiate a change for the human psyche towards one of positive mental awareness—to see the good in plants, to recognise their plantality, to listen to what they have to say, especially when it potentially benefits us.

History, ethnobotanic and traditional uses

Neolithic remains from the stomach contents of a man found on the Essex coast at Walton-on-the-Naze contained blackberry seeds, as reported in the most informative and compendious *Flora Britannica* by Richard Mabey (the godfather of British plants). This claim couldn't be corroborated but out of respect for Mr Mabey it's worthy of a mention, if only to name-check this very important person and all he has done to raise our understanding of the plant world, as well as schooling many new foragers with his book *Food for Free*.

Not that it is of any surprise to know our ancestors ate blackberries, it is always good to know for definite. Firm evidence for this comes from the Iron Age Haraldskær Woman, a remarkably well-preserved bog burial from Jutland, Denmark, dated to 490 BC. With the peat preserving her skin and internal organs, blackberry seeds and millet were identified in her stomach remains. Found in 1835, these were originally thought to be the remains of Queen Gunnhild of Norway, who died around AD 9; King Frederick VI of Denmark-Norway had the remains interred in a carved sarcophagus, thus preserving them. All told there are around 700 bog burials, with most assumed to be human sacrifice from this period and area. Unsurprisingly, what with the bramble's occupation of human spoil sites, banks, ditches, latrines and with abundant fruit for two to three months, the seeds are frequently found at archaeological sites across Europe and the British Isles. We've always eaten blackberries, even tens of millions of years ago when we were still a shrew.

We know that the ancient Greeks were cultivating raspberries by 370 BC. Pedanius Dioscorides (AD 40–90), he of the magnificent *Materia Medica*, used the berries as a gargle to treat sore throats, with the chewing of the fresh leaves to stop gums from bleeding and the eating of blackberries to treat gout. The ancient Chinese used the unripe berry to treat kidney disorders, urinary incontinence and impotence. Blackberry leaves were used in the Middle Ages to soothe burns, scalds and other skin problems.

Culpeper praised the plant as a remedy for almost all ailments, from wounds and ulcers to fevers and itching. On his walks across the Sussex Downs from his grandfather's parish at Isfield, where he was raised, across to his ancestral homelands near to the Long Man of Wilmington, brambles, along with nettles, plantain, cannabis, dandelions, hawthorn and yarrow, would have been a constant companion. Culpeper recommends the buds, leaves and branches as being good for mouth ulcers and other sores; he says the flowers and fruit if unripe are very 'binding', so good for diarrhoea or spitting blood; he regards the root as good for kidney stones, the dried brambles as effective for easing heavy periods, and the berries useful for piles and poisoning by snake venom. The leaves, branches, fruit and flowers, he goes on, make a pleasant-tasting drink, all of which can be distilled, being good for fevers and other ailments. As with the local name of scaldhead, the leaves when boiled up with lye will cure the dreaded itch (often used as a synonym for scabies) as well as dying hair black. Unsurprising, Culpeper considers brambles a store-cupboard essential, with the juice being able to be kept all winter if mixed with some sugar or alcohol. He also elucidated on how to make types of medications from the herbs, be that oils, ointments, pills, plasters, syrups and lohochs—this last is a lickable, paste-like medicine, also known as a lambative, or pectoral, usually for a chest disorder.

* * *

Continuing with ethnobotanical evidence among the indigenous First Nation tribes, *Rubus* comes out as being the most popular of all plants for utilisation and history, as expected with all the amazing health qualities and the fact it tastes so good. The primary use of the *Rubus* genus is as a food source, although there are also a wide range of nutraceutical and practical applications. With so many *Rubuses* used by the tribes across the entire continent, it would be fair to say that blackberry and yarrow, dandelion, hawthorn, plantain and nettle were all plants very close to their hearts.

Among the *Rubus* utilised, a selection would include *Rubus allegheniensis* (Allegheny blackberry), *Rubus arcticus* (Arctic blackberry), *Rubus arizonensis* (Arizona dewberry), *Rubus canadensis* (smooth blackberry), *Rubus chamaemorus* (cloudberry), *Rubus fruticosus*—which goes by the rather negative name of shrubby blackberry, indicating the superlative range of varieties available to them, *Rubus hawaiensis* (Hawaii blackberry), *Rubus idaeus* (American red raspberry), *Rubus laciniatus* (cutleaf blackberry), *Rubus leucodermis* (whitebark raspberry), *Rubus nivalis* (snow raspberry), *Rubus occidentalis* (black raspberry), *Rubus odoratus* (purple-flowering raspberry), *Rubus procumbens* (wild blackberry) and *Rubus pubescens* (dwarf red blackberry). All look remarkably similar, and just as with the European varieties all share phytochemical spectrums, with variations determined by soil, climate and altitude.

Every single Nation already mentioned throughout the book from the Abnaki to Zuni, and all the others who have ever lived for however many millennia on the continents, have remained vital and healthy from perpetual consumption of the *Rubuses*. The medicinal uses range from infusions of the leaves and roots for diarrhoea, dysentery, piles, venereal disease, influenza, measles, rheumatism, sore throats, coughs, childhood whooping cough, lung trouble, toothache, a general tonic, as a stimulant, for urinary ailments, headaches, blood remedy, fever, tuberculosis, a body wash, skin and eyes, boils, bowel complaints, stomach ache, as a women's medicine, to expel parasites, for low and high blood pressure and powdered stems for wounds and cuts; the Haisla and Hanaksiala used the flowers in their shamanic 'flower dance', while the Makah had sprout parties of dancing and singing in honour of the abundance of shoots and buds.

* * *

Throughout the seventeenth century, blackberries were drunk as a cordial, often added to brandy or wine with various spices, as well as a fashionable addition to gins.

Beers were popularly brewed in the eighteenth and nineteenth centuries using blackberries as a chief flavouring and nutritious addition. The nineteenth-century American Eclectic physicians recommended a blackberry syrup preparation for dysentery, gonorrhoea, vaginal discharges, childbirth recovery and 'cholera infantum'—not cholera, but infectious diarrhoea, a condition that is still one of, if not the biggest killer of children in countries with a low GDP.

Alfred Vogel (1902–1996) suggests blackberry juice as an effective stimulant for the kidneys (also bilberry and blueberry), as well as the fresh leaves of blackberry, hawthorn, raspberry and blackthorn for a spring cleansing tea. In Vogel's native Switzerland these leaves, along with those of the strawberry, coltsfoot, cowslip and plantain, were used as an infusion and drunk throughout the day. He was keen on all berries, not wishing to put one above another, stating that berries were good for those suffering from multiple sclerosis, cancer, rheumatism and arthritis. We will see in the scientific research later that Vogel's along with Culpeper's uses are all very much corroborated.

In Eastern Europe, the young shoots of raspberry (*Rubus idaeus*) have been used as herbal remedies for colds, flus and fever for some time. In Korea, *Rubus coreanus* was used as a traditional remedy for liver and kidney diseases, spermatorrhoea (involuntary ejaculation), prostate and urinary disease as well as allergic ailments.

In Chilean folk medicine, *Rubus ulmifolius* (looking exactly like a *Rubus fruticosus*, although thornless, going by the name of elmleaf in English) is used for its hypoglycaemic activity. Italian traditional medicines used the same *Rubus ulmifolius* for abscesses, diarrhoea, boils, haemorrhoids, intestinal inflammations, red eyes, treatment of ulcers and vaginal disorders.

Leaf extract of *Rubus fairholmianus* and *Rubus moluccanus* collected in the morning has been used by the *Koch-Rajbongshi* and *Rangia* tribes of north India, Bangladesh and Nepal to reduce headaches. The leaves have been reported to possess insecticidal properties as well.

In Irish folk medicine, bramble leaves, roots and fruits were a common remedy for ailments such as colds, coughs and flu, thanks to the plant's antiseptic and astringent properties. Bramble leaf tea is a common folk remedy for diarrhoea and dysentery the world over. So good was it as a cure for dysentery that there are reports during the American Civil War of both Union and Confederate soldiers calling for ceasefires so as to allow them time to pick blackberries in an attempt to abate their loose bowels, allegedly even from the same plant.

* * *

Preparations from teas to tinctures of blackberry leaves, fruit, root and bark have been used for digestion, mouth and eye ailments and relieving toothache. As with raspberry leaf tea, blackberry leaf is very similar and can be used for the same beneficial results, not just as a birth-aid in the final trimester of a pregnancy, but good for everyone else as well all the time, with a tea made from the roots being employed for labour pains.

Blackberry oxymel was once very popular, being made by covering the fruit for two days with white wine/cider vinegar, then crushing and sieving the fruit mass with half its weight in honey added. It is a perfect cold remedy/assistant, also used for gout and arthritis, which can also be frozen into useful ice cube-sized doses.

Herbalists have recommended blackberry jelly, cordial or wine for their potent restorative powers for centuries. There were remedies mixing blackberry juice with butter to treat various swellings, while the leaves have also been used for cleaning wounds, as a blood stauncher and chewed for toothache, with blackberry juice recommended for colitis. The oil from the seeds has some superlative health benefits and, as with nettle seeds, has often been overlooked. Blackberry seeds, as with many of the seed oils from berries, have a long association with skin conditions, especially eczema and psoriasis, having particularly healthy, beneficial and unique quantities of omega-oils.

Tinctures are concentrated plant power, readily available from health stores although making them yourself is a simple step to becoming empowered in controlling your own health. Three-quarters fill a Kilner jar (any jar with a lid is fine) with any aerial parts or the roots of any of the seven plants from this book, and of course thousands of others as well, then top up the jar with vodka or brandy and steep for two months, with a daily shaking. Decant into a small dropper bottle and you have medicine to help prevent and even remediate most of the ailments encountered throughout this book.

In folk medicine, *Rubus idaeus, Rubus coreanus, Rubus fruticosus* and many of the other *Rubuses* are all considered as being anti-allergenic, antibacterial, anti-carcinogenic, anti-diabetic, anti-diarrhoeal, anti-dysentery, anti-inflammatory, antimicrobial, antioxidant and an aphrodisiac. Traditionally *Rubus fruticosus* has also been used for treatment of skin diseases, itching, sore feet, cuts, burns, scabies, eczema, kidney and heart problems. It is also traditionally used in childbirth, with its high levels of vitamin K acting as a muscle relaxant and blood coagulator, along with the other inherent health-bestowing qualities of the leaves and berries. The other nutrients all contribute to improved digestive health, heart function, the immune system, eyes, bones, skin and memory as well as blood quality and weight management—it is pretty much another all-rounder.

In folk medicinal records it is often not possible to trace the actual species used in the past. Actually this is of little consequence, because as with all the other plants from the book, those sharing a first name more often than not share nutrients and health capabilities across not just the same field, even across different continents.

* * *

Fibres can be easily stripped from blackberry stems/canes and woven into hardy baskets and bags; it's also strong enough to be used for thatched roofs, even beehives, and with extra processing a soft yarn can be achieved. This could form a particularly lucrative cottage industry, as the processing of the stems is only a little more labourious than with nettles and hemp. The plant yields various dyes, ranging from orange from the root to bluish from the berries and black from the leaves—still used today in parts of Scotland and Ireland to dye wool.

Humans have been rather versatile in their pursuit of fibres, yarn and cloth, using a multitude of plants such as the willowherbs, yarrow, leek, horseradish, daffodil, honeysuckle, hops, wild garlic and willow to bamboo and coconut (coir), jute, flax, kenaf, pineapple leaf, raffia, eucalyptus, papyrus, sisal, cedar bark, straw and banana leaf, plus many other trees, mushrooms and seaweeds. All the modern plastic fabrics, such as acrylic, nylon, polyester and lycra, are all polluting petroleum products, and all can easily be made sustainably and environmentally from hemp and other plants in the very near future.

Collecting brambles was a way for children in the past to get a little money of their own—at the beginning of the 1900s a pound of blackberries could earn a child at least an old penny, called a copper, enough to get some sugared liquorice and/or a ride at the fair. When Cooper's Marmalade of Oxford won the contract to make marmalade and jams for the troops during WWI, many locals had a nice little side-earner from picking a wheelbarrow full of blackberries and taking it to the factory.

In Devon, bramble shoots were used as a type of fishing rod, to ensnare and pull young rabbits from their warrens. Brambles do also create a safe haven for many species of insects and small mammals, as well as offering protection for shoots of young trees away from the attention of grazing deer. Brambles are vital for certain species of caterpillars and moths, notably larvae of around 30 species of *Lepidoptera* from the garden tiger moth to the mottled beauty, skippers, the emperor moth, peppered moth and many more. The pollen is enjoyed by bumblebees, honey bees and solitary bees, butterflies, flies, hoverflies, moths, wasps and lacewings, as well as many nesting on the leaves and stems. Among the insects, the berries are feasted on by wasps, picnic beetles, Japanese beetles, stinkbugs, June bugs and crickets.

The leaves are eaten by deer, goats and other grazers, with the fruit enjoyed by birds and mammals alike from mice, voles and badgers, all allowing for transference of seeds across the countryside from fox scat to bird droppings. The dense bushes offer excellent protection for several species of birds, and crows, finches, starlings, thrushes, waxwings, wood pigeons, pheasants and nuthatches among others all also enjoy the fruit. Blackbirds, dunnock, finches, robins, thrushes, tits, warblers and wrens nest within bramble thickets and many more seek sanctuary when needed within them also.

Any problems?

The biggest problem I have with blackberries is those pesky flies, which disperse when going in to pick a fruit from the bush. What are they doing? What they actually do is probably best not known, as the more common fruit fly, the spotted wing drosophilia (*Drosophilia suzukii*), an invasive species from Asia, lays six to ten eggs in ripe fruit, and hatching in a day the larvae eat some of the fruit and fall to the ground—which is why I only eat the berries from the bush if they are perfect; if they are for jams or leathers it makes no difference as they will get boiled and actually provide some really important and unique nutrients. It's not a selling-point but it's the truth, if people are squeamish about such things they have to realise it pales into insignificance compared to much of what goes on in the food industry. Having never felt any ill effect, I advise that a blind-eye can be turned towards the flies.

There is the caution of tannins in large amounts being toxic and may cause stomach distress and vomiting, with the leaves and roots having the highest concentrations, so those with a chronic gastrointestinal condition might best swerve drinking and eating too much of blackberry leaves and roots. However, as with all foods, some people are more intolerant or allergic to different foods, even at different stages of their life. One of my daughters was often sick after eating a few blackberries when we went picking years ago, a problem she no longer suffers from today, although it wouldn't have been the tannins causing her sickness as that's mainly found in the leaves.

Sadly, the only caution would be to try to eat organic or unsprayed wild-foraged ones as the non-organic farmed ones have unsatisfactory levels of agricultural chemicals applied to them. There is basically next to nothing to say from a cautionary standpoint, apart from, perhaps, don't eat the thorns! As once again scientific evidence shows that the potential toxic intake level of any part of the plant is staggeringly high (if organic), outlining these plants as being safe, even when consumed in larger than normal quantities as a nutraceutical. Which most definitely is not the case with the following—

The biggest threat to the future of all Gaia's organisms

Being a plant found over most of the world, the bramble does have a multitude of beetles, aphids, flies, larvae, moths and weevils living among it, all of which are a part of the biosphere of the plant and animal world. These natural 'pests' have for far too long been exterminated by being drenched in highly toxic agricultural chemicals. There are plenty of safe ways in which to deal with these pests and competing plants—there always has been, although the allure of modern science and space exploration from the 1950s onwards blinded the public, with a faith in science being a part of the natural flow of evolution as we hurtled into the sci-fi world. Being well into the 2020s now, many sci-fi films and futuristic series from the twentieth century were actually set in what is the past today, be that Orwell's *1984*; *Mad Max*, set in the mid-1980s; *Space: 1999*; *Death Race 2000*; *2001: A Space Odyssey*; *Back to the Future 1* and 2 have 2015 as their future; *The Running Man* set in 2017; with *Blade Runner* and *Akira* both set in 2019 and *The Jetsons* were set in 2020.

Today doesn't feel so futuristic. We might have computers, satellites the Hadron Collider, iPhones and the James Webb Space Telescope, but almost everything else, from electricity generation to automobiles, has progressed only fractionally in the past two centuries. When the superficial layer of consumerism is removed, we as a species are little different from humans from the 1600s or 6000 BC—apart from the modern man-made chemicals, of course.

As we saw in the Nettle chapter, in 1930 before the world was given a repeated dosing of pesticides, about 30% of crops were lost to pests and diseases; today after 70 years of spraying deadly chemicals nearly 40% of crops are lost to pests and diseases. The loss to crops on organic and permacultural smallholdings is only about 2% to 5%—with no use of chemicals, just the tried and tested methods of companion planting, slug watches by catching them when they come out at dusk and dark or using ducks and chickens, ash and copper barriers, ponds for frogs and a healthy soil with its own bacterial self-defence. Spiders, dragonflies, toads, small mammals and birds all feast upon annoying pests and parasites—only these have all been tragically exterminated along with the weevils, beetles, butterflies and bees, as agriculture took a deadly chemical turn after WWII. It has been a period when governments and the public were blinded by the veil of truth drawn by modern science, as finally we reached what many commentators said at the time was the pinnacle of human evolution.

In actual fact it marked the near complete dislocation of the human from the Mother. The agricultural chemical scientists were never connected to Mother Nature, being deliberately dead-set against and opposed to Her in many instances, as their goal for decades after WWII was to seek out ways to replicate, banish, dominate, control, exterminate and violate all that was natural. The pernicious saturation of perilous poisons throughout the air, soil and water has rarely been considered a problem by the public, with the social engineering and programming of industry-led government steering a path to keep these real problems out of sight.

Consider the whole scandal of DuPont and Teflon. The dangers and cancer links have been publicly known about for a couple of decades now, with the scandal being that DuPont were fully aware of the dangers before going into production with PFOA (perfluorooctanoic acid), seeing only the profits from the marketability of this miracle non-stick pan. The science was experienced first hand from their workers' corroboration of the rat experiments, tragically with their babies suffering similar abnormal birth defects. There is now a film, *Dark Waters*, but just like Michael Moore's ground-breaking films and documentaries *Bowling for Columbine, Fahrenheit 9/11, Sicko, Capitalism: A Love Story, Where to Invade Next, TrumpLand* and *Fahrenheit 11/9*, all exposing the corrupt system, alas these become quietly accepted without stirring the hoped-for revolution. In America you can still buy a gun quicker than a car, and across the whole world over you can still buy a genome-mutating, cancer-causing, non-stick frying pan.

* * *

It is not just what we cook our food in that has deleterious chemicals, our very food has had millions of tons of the deadliest chemicals ever invented by man(not so)kind sprayed all over them, all of which have intimately joined with our organisms, while all science is proving is that these chemicals are persistent, mutate genes and cause cancer. They have been sprayed on our food for decades and now we are at the breaking point, but still the public believe there couldn't be any danger from the deadliest chemicals ever invented! The roots for most of these corporations making pharmaceuticals, plastics, pesticides and fertilisers is often from warfare and the mass killings and maiming of humans—they are the Devil's work, thoroughly morally depleted. In a caring, productive, sustainable culture these industries would never be allowed to prevail, let alone dominate decision-making of governments. Not for the last time it has to be said: "the greatest danger faced by biological organisms on this planet is from man-made chemicals".

Infuriatingly, the reason why these agricultural chemicals require another mention is because the entire journey from seed to plate for non-organic commercial blackberries is saturated in chemicals. Yep! Modern agriculture turns one of our healthiest plants into a definite risk factor for cancers, chromosomal mutations and neurological dysfunctions. It is the extensive spectrum of chemical applications to the commercial fruit that is the only real precaution in this whole chapter, if not the whole book.

Coming with the double-whammy that even the wild blackberries get a tough deal, with some of the heaviest-duty poisons being liberally applied to them, such as metsulfuron, which kills whatever it comes into contact with, going by the alluring brand names of Mustang, Associate and Escort. The Australian government recommends applying triclopyr (with the brand names Grazon, Victory and Renovate), and from late spring and when actively growing it suggests glyphosate (Roundup, Rainbow and Samurai), words deliberately used to confuse the farmer and home owner—they're not going to call them names closer to their true colours, such as Armageddon, Desecration, Four Horsemen or Satan, are they?

In California 128.5 tons of chemicals were sprayed to eradicate wild blackberries in 2018. This might not sound a lot, but these chemicals are lethal if just 1g is ingested. With 40 million residents in California, its entire population could be easily wiped out more than three times over, each and every year from just the chemicals used to eradicate wild blackberries. Just think of all those other plants being eradicated, on the golf courses, gardens, roadsides, agricultural fields and margins. Depressingly, this 128 tons is merely a deadly drop in the ocean of pestilence.

When it comes to commercial blackberries this chemical spraying regime is intensified. Caneblight is a serious worry and sometime problem for American blackberry growers, so a regular dosing in a variety of fungicides helps prevent this and unduly worrying the farmers. Orange rust, leaf and cane rust, fungus, bacteria and various viruses all warrant a dousing, as does the propensity to mould from the leaking juice, meaning a fungicide is also used for transport and storage to prevent mould formation. Strawberries came out worst in several American pesticide residue tests of all fruits, but the other berries with all their nooks and crannies to shelter some pesticide weren't too far behind.

A special mention needs to be given to Atrazine and its maker, Syngenta, for the inglorious title of the herbicide most commonly found in drinking water in the USA. It is used globally, with corn being a common target. In independent research (Hayes et al. 2010), 10% of African clawed frogs exposed to Atrazine become fully functioning females! This endocrine disruptor affects hormones, depletes semen, lowers testosterone, increases risk of miscarriage, lowers birth weight, is neurotoxic and a carcinogen. With over 32,000 tons sprayed worldwide (90% in USA), just imagine the chaos already caused to humans with this and BPA combined.

Oikonomia, oikonomia, oikonomia: spraying millions of tons of the deadliest chemicals ever invented on to our food; spraying millions of tons of the deadliest chemicals ever invented on to our food. These words need repeating again and again and again. We need to understand what a precarious position it is living under a false definition of economics, endangering not just the human race, but all planetary systems from the air, water and soil, with nothing more than lip service from the governments to reverse this.

So who is going to stop this then? A good place to begin would be checking out the wonderful work of the people at Pesticide Action Network, with massive respect to them, with a wealth of information on their sites—pan-uk.org and the international action site at pan-international.org. Get informed, then we can all stop this.

Unsurprisingly, the manufacturers of these agricultural chemicals are adamant that their products are perfectly safe, just as they fail to mention the countless lawsuits currently in progress against them. Meanwhile, independent research looking at the other end of the spectrum, where health might reside in blackberries, is once again dominated by countries our media definitely is prejudiced against, such as Iran, Taiwan, China, Pakistan and India, who have found out some rather interesting facts—

Last scientific studies to digest

Following similar lines to the previous chapters, where the scientific research not only substantiates traditional uses, other diseases our plants might be beneficial for are often uncovered, particularly with all these new diseases of civilization killing us all, at an ever-increasingly younger age.

We can start with an American researcher who has stuck with berries for some time now, Gary Stoner at the College of Medicine, Columbus. His many studies looking at cancer prevention from berries are well-documented in many notable journals, all using freeze-dried black raspberries (*Rubus occidentalis*). It was shown that raspberries not only inhibited the initiation of a tumour developing, they also prevented its proliferation, be that of the oesophagus or colon. Stoner's work from the mid-1980s had already demonstrated that the polyphenol, ellagic acid (abundant in seeds and pulp) inhibited oesophagus tumours in rats. The freeze-dried powder proved to retain much of its nutrient profile over a period of two years, with the fragile vitamin C suffering the worst; fresh fruit would have yielded even better results, but as these are not available year round, freeze-dried allowed for comparable results all year. Stoner's studies used up to 60g of dried powder, which is the equivalent of about 475g of raw fruit per day, indicating that small doses may act as preventative, but to be considered medicinal and enact a cure, doses need to be significantly increased. Stoner's faith had been in the powder for over 15 years of research and is a massive nudge in validating the almost certain nutraceutical benefits achievable from the *Rubuses*.

Another rare positive American study (Mace et al. 2014) showed the effectiveness of the black raspberry (*Rubus occidentalis*) on cancer cell proliferation, using cyanidin and quercitin phenols obtained via an ethanolic extraction. This extract proved to be an anti-proliferator, with the scientists concluding that black raspberries have a use for both therapeutic intervention and development of drugs.

Further anti-carcinogenic evidence came from the *Nutrition and Cancer* journal (Feng et al. 2004), where blackberry extract inhibited proliferation of human lung cancer cell lines and provided other antioxidant abilities.

A Brazilian study (Petreanu et al. 2015) used a flavonol (a polyphenol) obtained from the leaves of *Rubus rosifolius*, testing its effectiveness against a variety of human cancer cell lines; encouragingly, it did the best against the glioma cell lines (brain and spinal cord cancers, with only a 25% survival rate for patients), as well as being effective against breast cancer, a multi-drug-resistant ovary, kidney, lung, colon and leukaemia cancer cell lines.

A pan-European and university study (Giampieri et al. 2018) used an extract from the whole fruit, showing the berry extract (blackberry, raspberry, blackcurrant, blueberry and strawberry) to have an effect on human liver cancer cell lines. They reported that among all fruit it was blackcurrants that had the highest polyphenolic content.

A study from the University of Helsinki (Pajri et al. 2016) further corroborated the positive colon cancer association, using the cloudberry (*Rubus chamaemorus*); the authors attributed the chemo-preventive effects to the ellagitannins (a polyphenol), which covers quite a few of the major cancers.

A novel way of attempting to understand what actually causes the anti-proliferative activity was carried out by Iranian researchers from Shiraz University as published in the *Nutrition and Cancer* journal (Tatar et al. 2019). They focused on telomerase inhibition, and without getting drawn into a long-winded account, future studies on telomeres and their associated enzyme telomerase could very well open the door to cellular longevity and the prospect of organisms living for longer. It's potentially massive, and they used blackberry extract—which is all we really need to appreciate at this moment.

South African research (George & Abrahamse 2019) looked at the apoptotic (cell suicide) capabilities of blackberry against human breast cancer cells, using not the leaves or fruit, but this time an acetone extract from the roots (again, as with the other plants, every part of the plant is beneficial). Causing no harm to healthy human cells, the root was both an efficient anti-proliferative and induced cancer cell death, making it also a cytotoxin. Boom!

* * *

Young shoots from raspberry (*Rubus idaeus*) were analysed at the Medical University of Gdansk, Poland (Krauze-Baranowska et al. 2014). Using a methanolic extract the researchers determined it was efficient as an antibacterial agent against *Helicobacter pylori*, *Bacillus subtilis*, *Clostridium sporogenes*, *Staphylococcus epidermidis*, *Neisseria meningitidis* and *Moraxella catarrhalis*. They put the antioxidant activity down to the high ellagic acid and sanguiin H-6 content in the shoots. Sanguiin H-6 was also shown to have *in vitro* cytotoxicity against leukaemia and cervical cancer cell lines and dermal fibroblasts, further assisted by the shoots also being rich in ellagitannins.

More Polish studies, this time from Lodz University (Klewicks et al. 2016), again looked at the ellagitannins (they identified six separate ones) and the bacterial strain *Geotrichum candidum*. This is not a dangerous bacteria as it can be found as part of a healthy human biome, but it can get out of hand around spillages of effluence and is often found in the rotting process of meats, dairy and breads. The results demonstrated the effectiveness of ellagitannins, making them a suitable environmental cleaner when *Geotrichum candidum* was excessive, as opposed to using the highly toxic antibacterial and fungal chemicals currently available.

Research from the *Tropical Journal of Pharmaceutical Research* (Radovanovic et al. 2013) also looked at the polyphenols, this time for their antioxidant and antimicrobial properties. They performed admirably against *Escherichia coli*, *Pseudomonas aeruginosa* and *Salmonella enteritidis*. Polyphenols, this time obtained from boysenberries, were looked at by the Nigata University, Japan (Furuuchi et al. 2018) in suppressing age-related endothelial cell disorders. Endothelial cells line every blood cell, creating a barrier between blood and body tissues, allowing nutrients through and waste to be discharged, and are vital as real health can only be ascertained with good cell health. The researchers showed that these endothelial cells acted to maintain vascular homeostasis and inhibited endothelial dysfunction.

Antiviral activity was reported in the *Oral Surgery, Oral Medicine, Oral Pathology and Oral Radiology* journal (Danaher et al. 2011), specifically against the herpes simplex virus (HSV-1). A blackberry extract inhibited the early stages of replication of the virus, and also showed virucidal activity if the disease is already established. At least half of the human race is positive for HSV-1, and coupled with the over-use of drugs (the biggest brand in the UK is Zovirax) there has been an increase in drug-resistant strains of this extremely common virus. Which screams DANGER, as this is the same as the drug resistance to influenzas and antibiotics. The dangers of a serious pandemic on the scale of the 1918 Yankee/Spanish Flu or even the Plague or Black Death are definite, and the new mega-conurbations (Tokyo 37 million, Delhi 28m, Shanghai 25m and 27 others with over 10 million residents) have created the ideal breeding ground for an outbreak of a truly devastating future pandemic. The commercial market for herpes, fortunately a non-deadly virus, is currently worth in the region of £4.6 billion in revenues annually, with excellent future expectations as markets and sales are expanding.

An Italian study from Milan University (Sangiovanni et al. 2013), looking at the polyphenolic ellagitannins obtained from blackberries and raspberries, studied how they performed against *Helicobacter pylori* and the anti-inflammatory properties in relation to controlling gastric inflammation. For all of this the ellagitannins proved effective, as well as being an agent to assist with peptic ulcers. This was corroborated by Brazilian research (Nesello et al. 2017), using both the black mulberry (*Morus nigra*) and their local raspberry (*Rubus niveus*), with the raspberry decreasing ulcerated areas the fastest, and both being noted as providing gastro-protective value.

* * *

Looking again at oxidative-stress and the role berries may play in neurodegenerative diseases, the anthocyanins from several berries looked promising, notably with regards Alzheimer's. A report in the *International Journal of Molecular Sciences* (Ma et al. 2018) showed blackberry's free-radical scavenging abilities offered neuroprotection for both prevention and symptomatic relief from Alzheimer's—a disease worth £3.1 billion to global coffers and GDPs, equalling £3.1bn to protect. Shackled to the current economic system nothing will change, when surely it would benefit the people and health services not to spend so much money on drugs and start to look for preventative alternatives. Instead years of drug dependence are only ever suppressing symptoms but never actualising a cure, while causing other diseases from the side effects. It is time to actually focus on the people's health, not corporate wealth. The economic system is more at fault than the erroneous aspects of the health industry. It's therefore time to spend a fraction of that £3.1bn on proper scientific research involving plants, prevention and a holistic perspective, not a reductionist, patent and profit ethos.

Researchers from the School of Life Sciences, Tamil Nadu (George et al. 2014) using the local *Rubus fairholmianus* and acetone extracts of six compounds demonstrated that all six had anti-proliferative and anti-inflammatory capabilities. Across the border in Pakistan, where this region has the benefit of a double harvest from blackberries in the spring and the autumn, another extensive epidemiological study, published in *Molecules* (Zia-Ul-Haq et al. 2014), corroborated the wide range of uses the chemical constituents of the blackberry have been used for in health and medicine, being an important functional food item—because food's only real function is as a delivery system for nutrients and energy, with all that water, carbs, protein, minerals, vitamins and of course the newly discovered, yet always having been there, phytonutrients.

From the *Nutritional Neuroscience Journal* (Shukitt et al. 2009) rats whose diet had been supplemented with blackberry polyphenolic extract to the tune of 2% of total diet showed improvement in motor performance and memory tasks. Sticking with brain health, an Indian study published in the journal *Scientifica* (Keservani et al. 2016) also provides a good general summary of health, as well as informing where a great word originated—as used throughout this book, when a food item has the added bonus of being rich in a particular nutrient, this can then work as a 'nutraceutical'. The word was coined by the founder of the Foundation for Innovation in Medicine, Stephen L. DeFelice in 1989, by joining nutrition with pharmaceutical—a portmanteau term—I was stupid/arrogant enough to assume I had actually made this word up, just as with etymologicum, although I'm definitely claiming plantality.

* * *

Obesity looks to be continuing its impressive rise to becoming the new normal, with of course the associated rise in drugs attempting to do something with this fat, sadly often doing little to negate the obesity while having significant side effects. We've seen that there was evidence that nettles, dandelions, yarrow and hawthorns all helped with fat absorption, distribution and insulin production, which is true also for the blackberry. The entire weight-loss industry is worth somewhere in the region of £55.5 billion globally, with America spending nearly half of this at £27 billion. The fat-busting pills, with Orlistat as the biggest-seller among others including Naltrexone (an opioid inhibitor) and Bupropion (an antidepressant), have global sales worth at least £7.8 billion annually. Which makes the Korean results published in *Evidence-Based Complementary and Alternative Medicine* (Oh et al. 2016) particularly pertinent, as by using fractionation with unripe *Rubus coreanus* fruit, mice put on less fat, with the researchers concluding that the five bioactive compounds they used could be a useful agent for body weight control by downgrading adipogenesis (the formation of fat cells) and lipogenesis (the process of fat being made from glucose).

More potentially good news for the obese, this time using an extract of the whole fruit from chokeberry, raspberry, bilberry and cranberry at Poznan University of Life Sciences, Poland (Kowalska et al. 2017), concluded that the extract could be a valuable supplement for both obese and non-obese patients suffering from insulin resistance. The diabetic market is worth around £42 billion, probably £45 billion or more by the time you are reading this, with patient numbers rising while the drugs have yet to come close to a cure, not least because in many instances this can only come from the patient making lifestyle changes.

Also from the University of Life Sciences (Szymanowska et al. 2018), this time the Lublin campus, are results across a range of ailments, using predominantly anthocyanins (unsurprisingly, a polyphenol) from raspberries (*Rubus idaeus*). The findings were rather impressive, with an inhibitory effect comparable to the NSAIDs of Ibuprofen and Naproxen. Szymanowska's work on raspberries showed not only anti-inflammatory capabilities and leukaemia cell proliferation inhibition from purified fractionated anthocyanins, but interestingly the crude extract of the whole plant (holism) outperformed the chemical fractionated extracts (reductionism).

The higher medicinal value of the immature fruit was shown by a study from the National Institute of Horticultural and Herbal Science, Korea (Seo et al. 2019), where the researchers compared the immature and mature fruit from *Rubus coreanus* to show an anti-inflammatory response in both states of maturity, with a stronger response from the immature fruit. Another Korean study, using 15 healthy adults (Lee et al. 2011), showed that an extract decreased oxidative stress and provided health advantages, even in healthy individuals.

Mexican studies published in the *Journal of Agricultural and Food Chemistry* (Cuevas-Rodriguez et al. 2010), using their local wild blackberry, leaned towards the rich anthocyanin content actualising the antioxidant and anti-inflammatory responses, as well as advising on its nutraceutical implementation.

A Chilean study (Avila et al. 2017) of native berries, principally the Chilean raspberry (*Rubus geoides*), strawberries (*Fragaria chiloensis*) and currants (*Ribes magellanicum*), focused on the polyphenols offering positive intracellular antioxidant responses. This meant there was not only better communication between the cells but there was also better transit of information and nutrients. This is the most basic aspect of initial health, and it is at this level that we have adversely affected the health of the whole bunch of 3 trillion or so cells that makes us whole (some say there are as many as 30 trillion human cells and even more bacterial cells)—we have compromised ourselves from the cellular level up by allowing an unhealthy cultural existence to predominate. It's as if berries when eaten in high doses (about 350g+) act as a cellular servicing team, and by not having this service, the body as a whole begins to deteriorate—consider the analogy of a car not having any annual services, just poor fuel added: it would deteriorate quite quickly, stopping completely in just a few years—with any number of our intricate parts coming under relentless pressure from constant oxidative stress and the diseases this is a significant risk factor in (cancers, heart disease, dementia etc.). An organism can only be as healthy as the cells it is made from.

Polyphenols from berries appear to work from the cellular level up, scavenging free radicals and feeding and maintaining a healthy organism. Considering that the Chilean study was published in the very aptly named journal *Oxidative Medicine and Cellular Longevity* and the fact the berries are so tasty and readily consumable, it beggars the question: why aren't we already eating far more of these as a tasty prevention from illness? That'll be because the economic system is neither commonsense nor is it *oikonomic*.

Every part of these plants appears beneficial, as has been further demonstrated in a study looking at the polyphenols from the leaves of 26 wild blackberry species growing in south east Poland (Oszmianski et al. 2015). This study was excellent as it included a good cross-section of wild varieties, meaning the researchers could establish that it was the *Rubus perrobustus*, *Rubus wimmerianus*, *Rubus grabowski* and *Rubus pedemontanus* that contained the highest amounts of these polyphenols, with *Rubus perrobustus* having the most. All of these species look indistinguishable from the common blackberry, *Rubus fruticosus*—in England there would be comparable results from the many species growing here. The study also showed that the leaves have a healthy content of polyphenols and a good range of flavonoids, notably kaempferol, quercetin, luteolin and apigenin. The authors' conclusions were resoundingly for all parts of the blackberry to be part of functional foods, that is, nutraceutical.

Another extensive epidemiological study from Italy, published in the *Molecules* journal (Afrin et al. 2016), highlighted that berries not only decreased inflammation but also reduced polyps, a major risk factor in cancer as polyps have been shown to mutate to cancerous tumours, especially when negative environmental factors are thrown into the equation along with lack of exercise, poor diet, excessive alcohol, smoking, family history, stress, even poor posture. Considering the high costs of cancer drugs to patients and health services, a far cheaper preventative and what appears to be an inhibiting factor would be consuming a large quantity of berries as a nutraceutical over a course of several weeks—as humans would have always done seasonally even before we were *Homo*. Evidence from all the studies massively favours the positive aspect of berries and specifically those polyphenols, which berries are by far the richest in of all food sources, found across many families from either the *Rubus* (berries), *Ribes* (currants), *Aronia* (chokeberry), *Vaccinium* (bilberry, cranberry), *Fragaria* (strawberry) or *Sambucus* (elderberry)—they are all incredible, offering preventative and curative capabilities for a range of illnesses or maintaining optimum health. A punnet a day keeps a lot more than the doctor away.

* * *

Good news garnered from quite a few of these scientific studies is that the extracts of the whole fruit and blended juices appeared to function strongly nutraceutically. The possibility that a pint a day of a berry blend smoothie would be berry good for you appears to be a near-definite future health slogan. Pun(nets)s aside, when has the prescription of a pound of berries a day for a week ever been given? Although, as we've seen, there's loads of science suggesting this should be considered. Have a read again through the rather extensive list of ailments these plants benefit, then consider actually picking a few kilograms of blackberries to eat throughout the day and make some juice, vinegar and jelly from or even make enough nettle soup for a couple of days. No one is doing these ridiculously easy and surprisingly effective remedies any more, and even those who profess to know the powers of plants neglect the basic preventative procedures of eating these readily and abundantly available Human Plants.

Can't get away from those polyphenols

Hopefully, by this stage of the book, the regularity of that word 'polyphenol' is familiar and recognised as a very probable human health provider. It has been more prevalent throughout this chapter than the others and was the focus of much of the scientific research reported above, all because blackberries are very nearly the best source of these polyphenols. They are just beaten by the overall winner, the blackcurrant, although unless you have a local organic pick your own farm, you will have little to no chance of foraging several kilograms from the wild (perhaps in Croatia or Siberia, but never in England). Blackberries, on the other hand, can be foraged from August, sometimes even from July, right through to mid-October (playing with the curse of the Devil) when bucketloads can easily be harvested, tons if you really wanted.

Now, if poly is many, what's a phenol? A phenol is a family of organic compounds characterised by a hydroxyl group attached to a carbon atom that is part of an aromatic ring.

Which doesn't really help much. The whole terminology of many fields of study is a language of jargons all to themselves. All we really need to know is that these phenols, also known as phenolic acids, are found in all fresh food groups and they are abundant in cereals, legumes, oilseeds, fruits, herbs, vegetables and of course blackberries and blackcurrants.

A Costa Rican study (Azofeifa et al. 2013) on the antioxidant and anti-inflammatory activity of the phenolic compounds found in their local blackberry (*Rubus adenotrichos*) showed them to have a greater antioxidant effect than Trolox, concluding that the blackberry-derived phenols had beneficial effects against oxidative stress and inflammatory processes. Trolox is basically an analogue (a chemical copy) of vitamin E being looked at for its breast cancer potential, while blackberries and their phenolic acids have already shown themselves to be effective against this and many other human cancer cell lines. The global market in antioxidant drugs like Trolox is worth around £6.2 billion annually—where the phenols from blackberries were deemed better.

The European Journal of Nutrition (Tavares et al. 2013) reported on a Portuguese study with some very interesting considerations. Using extracts of wild blackberries (*Rubus brigantinus* and *Rubus vagabundus*) and a commercially bought one, the polyphenols present were held responsible for the neuroprotective effects observed. Interestingly, the commercially bought extract had no benefits, which has to be from the processing and pasteurisation, while the wild extract had both antioxidative and anti-inflammatory benefits—whoa! I'll rewind and come again with that because that's not the first time we've heard that—the commercially bought extract had no benefits, while the wild extract had both antioxidative and anti-inflammatory benefits. Says it all really.

Research and epidemiological study from the Czech Republic (Skrovankova et al. 2015) gave an excellent overview of all the phenolic compounds found in a range of berries, including strawberries, cranberries, redcurrants, blackcurrants, blueberries, raspberries and blackberries. With several flavonoids and the phenolic acids, these compounds appear to be among the leading contenders from studies looking at anti-inflammatory, anti-tumour, antioxidant and other nutraceutical pharmacological benefits. The authors did note that processing denatures these compounds, so until further research indicates the best way to preserve them, the advice has to be to eat as many as possible of the fresh fruits to experience an improvement in health.

Skrovankova's work was further corroborated by a later epidemiological survey published in the *Frontiers in Pharmacology* journal (Beata 2018). This established the health potential of berries, notably the phenolic compounds, which decrease by as much as 90% during the ripening process. The anti-oxidative capabilities are exciting as cardiovascular disease, cancers, inflammation and neuro-degenerative diseases are heavily linked to oxidative stress.

Research from the Khyber Medical University, Pakistan (Ali et al. 2017) provided a great summary of the polyphenols and uses from scientific evidence. The team used *Rubus ulmifolius* (elmleaf or thornless blackberry) to extract all they could, finding several phenols. They noted that these offer potent antipyretic activity (to prevent or reduce fever) as opposed to the toxic aspirin, paracetamol and nimesulide. They also highlighted the rich flavonoid content (all flavonoids are polyphenolic compounds) displaying biological activities that include antibacterial, anti-carcinogenic, anti-inflammatory, antioxidant, antiviral and direct cyto-protection for the coronary and vascular systems, the liver and the pancreas. Rutin, quercetin (flavonols) and hesperidin (flavanone) all inhibited inflammation, and rutin has already been used in the treatment of the chronic phase of inflammatory arthritis. Flavonoids, like rutin and apigenin, are very potent antioxidants against free radicals as well. They also gave toxicity safety levels of a staggering 6g/kg of the consumption of the isolated methanolated extracts—equivalent to a 70kg human eating quite safely 420g of pure extract, which is never going to happen. The Khyber researchers concluded that owing to the variety of pharmacological activities, flavonoids are best referred to as nutraceuticals—which is always so encouraging to read from scientists.

Without wishing to artificially champion just the polyphenols, there are so many plant chemicals that form an intricate web of potential health for humans—although polyphenols are dominating proceedings and have been mentioned since the Nettle chapter—berries are not only the best source of these polyphenols, they must be the tastiest also. Although it won't be just one or two of these polyphenols or other phytonutrients individually, it's always going to be the orchestra as a whole that sounds the best.

Polyphenols, along with all the other phytonutrients, are sadly the first to be destroyed by modern food processing, hence why it is imperative to include as many fresh vegetables and fruit in the diet as possible, with organic and locally grown being the best of all.

* * *

In summation of the health benefits implicated by scientific research, it can safely be said that the blackberry and others from the *Rubus* family all share chemical compounds found in the fruit, roots, young stems, seeds and leaves that can be said to be anti-carcinogenic and an anti-proliferator of human cancers of the breast, brain, colon, lung, ovary, liver and leukaemia; an antioxidant more effective than the market leader; anti-inflammatory; to improve memory, reaction times and assist in Alzheimer's; to be gastro-protective; antibacterial; antiviral; to help with obesity via fat metabolism as well as balancing insulin and blood cholesterol; and help get rid of polyps and ulcers while giving cells a service and much more. Phew!

The B17 story

This is a story that could have been lifted from a movie script. It all begins with something most people have no idea is even there, let alone the prospect that it might either be a poisoning threat or on the flipside a cure for cancer. The something in particular here is the pit found inside of an apricot stone, also called the kernel, within which the highest quantity of a certain chemical compound can be found. This inner seed, as it were, found throughout the fruits in the Rose family, such as peaches, apricots, cherries, almonds, apples, hawthorns and of course blackberries, contains a compound called amygdalin, aka laetrile. This is a compound mired in controversy, as amygdalin is a cyanogenic glycoside, which is converted in the body to the toxic hydrogen cyanide; scientific studies strongly suggest this form of cyanide eats cancer cells, notably lung ones. This laetrile caused quite a commotion in the 1980s, with actual armed raids by the police accompanied by Food and Drug Administration (FDA) officers, busting into health stores, threatening the owners at gunpoint while shouting "Take me to your apricots!" Apricots were focused on because these had been used in the following research and are the best source for this laetrile.

There are other sources of laetrile, as it is actually quite common. It is found in nuts such as almonds and macadamia, pulses including the butter and lima beans, vegetables such as carrots, celery, and the seeds of millet, flax and buckwheat—where it is consumed in a healthy diet without any mention of these containing a cyanide-forming compound. The apricot kernel may be the largest reciprocal for this laetrile but it never forms any part of a normal diet.

As early as 1924 amygdalin was promoted as a cancer treatment, where it caused no controversy being touted as such. So, picking up this story from what is the world's oldest private cancer hospital, the Memorial Sloan Kettering Cancer Center in New York, established in 1894, where coincidentally the poet Nathaniel Hawthorne's daughter Rose studied in 1896 before establishing her own order, the Dominican Sisters of Hawthorne. It was at this hospital that the highly respected Dr Kanematsu Sugiura worked in the 1950s and 1960s, and while studying the effectiveness of laetrile against lung cancer cell lines, recorded surprisingly positive results, where the laetrile appeared to miraculously only take out the cancer cells, not the human ones, causing apoptosis within the cancer. It was as if the cyanide specifically hunted down the cancer cells. The results demonstrated that laetrile was capable of stopping the spread of malignant lung tumours, so it was obviously massive news, as was the kerfuffle that followed.

What could have been seen as a breakthrough in cancer treatment was actually silenced. With the FDA alerted to this unregulated *Rosaceae* renegade, the health stores were raided at gunpoint for the unspeakable act of selling apricots—even stoned apricots were confiscated, or were the FDA stoned? Why go to such lengths? It brings to mind a Shakespeare quote, from *Hamlet*: "The Lady doth protest too much." Could there be some truth in the anti-carcinogenic claim? Certainly the rapidly expanding cancer industry would have been none too keen on this news affecting their staggering projected income growth, specifically from lung cancer with all those smokers. Those rosy growth rates for generations to come now suddenly looked in serious jeopardy, from an apricot of all things.

Unfortunately, problems further emerged for the Memorial Sloan Kettering Cancer Center when in 1974 one of their scientists, William Summerlin, made false claims of a successful skin transplant between white and black mice—he actually crudely used permanent marker to colour the white mice black! Sloan Kettering were put centrestage, and as laetrile was already controversial—with it also having associations with the ultra right-wing John Birch Society—Sloan wanted to distance themselves from any associations, turning their back on laetrile to avoid future embarrassment. The decision was no doubt assisted by the fact that among the board of directors of Sloan Kettering were major investors in petrochemicals, pharmaceuticals and tobacco, even CEOs from cancer drug manufacturers (such as Bristol-Myers Squibb, who made the chemo drugs)—the Sloan Kettering Center was therefore financed by some of the major causes of cancer and those directly involved in making a direct profit from it. This marker pen fiasco would be the least of their problems if a large wedge of their funding was stopped because of 'inconvenient' research results upsetting their paymasters.

It wasn't just Dr Sugiura's research either, because also in 1974 other scientists at Sloan Kettering returned positive results with laetrile and cancer, so this had to be buried. For the next three years deliberately falsified research and statements about the ineffectiveness of laetrile were made by Sloan Kettering. However, a young scientist in the PR department who'd been making those statements, Ralph Moss, realised a cover-up had been taking place and felt uncomfortable he had been used to unknowingly spread lies, so he came clean and exposed the cover-up in his book *Doctored Results*—he was predictably and promptly sacked. He has since been endlessly accused of lying, spreading fake news and mercilessly instilling false hope in millions of cancer sufferers worldwide.

But is Moss really the one lying? He was after all privy to Kanematsu Sugiura's results and the later ones, so it's unlikely. Moss himself is most adamant about the fact he is telling the truth, having written 12 books and made three documentaries on cancer and directing The Moss Reports (themossreports.com—a good place to start for more information if you're interested) to back up his claim.

Why would Moss be the one lying? Because there is of course those 150 billion reasons why the truth is not important. The fact that the FDA went to such extraordinary actions to protect the citizens of the USA from this deadly dietary disaster is admirable. When it comes to the cancer industry and the money involved, this instils as much trust as we have for your average politician or CEO of a polluting corporation. As the former Associate Chief Counsel for Drugs at the FDA said, "Nobody is going to pay $70,000 for a new cancer drug, if laetrile can be bought for 75 cents"!

Work has continued on laetrile, and results from 2003 to 2014 show positive results in relation to bladder, cervical and other cancerous tumours, as well as kidney disease. Since it was banned in 1963 in the US, many thousands of cancer patients have travelled to Mexico for laetrile treatment. Certainly not all successfully, not least because laetrile is seen as a last resort when all traditional methods have failed and the patient is often at Stage 4 already, having been given only weeks or months to live. So it's a last shot and it is unscientific and unrealistic to compare survival rates from this treatment course.

* * *

There is a warning on some cancer sites of people having died from cyanide poisoning while trying to administer kernels to themselves, although research only uncovers the unfortunate death of a three-year-old leukaemia patient in 1979 who died and had tried laetrile, and of a two-year-old from Turkey who died of accidental poisoning after eating 10 apricot kernels, not as a self-administered anti-carcinogenic remedy. Turkey, with its prolific output of apricots, appears to be a hot spot for infant sickness from eating the kernels, which is surprising in itself as the kernel is very hard to extricate from inside the stone. There is no evidence of other deaths, only patients suffering symptoms of non-lethal cyanide poisoning by trying to cure themselves from cancer—although, it must be said, chemotherapy probably kills more cancer patients than the cancer itself. The terrible death of a patient from cyanide poisoning via laetrile treatment, if it has ever occurred, is no different from a patient tragically dying from the chemo drug taxol, for instance. Taxol is derived from the yew tree, but there are no armed raids removing taxol from the apothecary shelves or unsanctioned yew trees being chopped down.

There is also confusion about laetrile being given a vitamin number, that of B17. This makes it sound an important part of the diet, which it might very well be, essential even, although it is not actually classified as a vitamin. This misinformation derives from the eminent cancer doctor Ernst T. Krebs from the early 1950s, who wanted laetrile to be classified as a vitamin so that he could sell it as a food supplement as opposed to a drug, which required a lot of trials, research, regulations, time and money to achieve. Calling laetrile a vitamin avoided all this, and it could be packaged and sold immediately. Dr Krebs, to his credit, had already discovered pangamic acid in 1951, which is legitimately known as vitamin B15, with the best source also being apricot kernels, so Krebs was definitely no dodgy tax-avoiding businessman, but definitely into apricots.

It does look as if laetrile is another of those compounds found within Human Plants that are essential for us to consume, in admittedly tiny amounts. There's far more laetrile in an apricot kernel, but you aren't going to be naturally eating those, unlike the much smaller amount of laetrile in a blackberry seed, because they hold preventative phytochemicals for diseases that only reared their head in significant numbers once the food with the prevention was either stopped being eaten or destroyed in the processed food chain. With nettles preventing neurological degenerative diseases via their ChAT, could it be that laetrile really might significantly assist in the prevention and cure of cancer? Even if it has nothing to do with laetrile, the polyphenols more than compensate for their rise to prominence as an essential preventative, once prevalent in the food chain, now deficient. It has been highlighted in thick marker pen, from the failure of the commercial extract to have any effect in the scientific research from above.

It is similar to the catching out of Ribena, who rightly claimed blackcurrants to be high in vitamin C, but failed to mention the vitamin had all disappeared by the time it's put in a bottle with water and an equal part white sugar, and sold for a nice profit. A school child's science study in 2007 proved to be more scientific and truthful than GlaxoSmithKline, one of the world's largest pharmaceutical corporations, who received a derisory £163,000 fine for falsely claiming Ribena to be rich in vitamin C, when there was actually none—Ribena is now owned by Suntory who have replaced most of the sugar with artificial sweeteners and fortify it with artificial vitamin C.

What I know is that every time I prise an annoying blackberry seed from my teeth, I have the instant gratification as I crunch through it with my incisors of some definite benefit and prevention each time.

Foraging and Shinrin-Yoku

Late August, given heavy rain and sun
For a full week, the blackberries would ripen.
At first, just one, a glossy purple clot.
Among others, red, green, hard as a knot.
You ate that first one and its flesh was sweet
Like thickened wine: summer's blood was in it
Leaving stains upon the tongue and lust for
Picking.

(Seamus Heaney, *Blackberry-Picking*, 1966)

Foraging is making a comeback, not out of necessity, not yet at least, but it might be a good idea to either learn or make sure a friend is familiar with it. We as a species have only not foraged for food on this bountiful beautiful planet for a tiny recent blip of our existence upon Her. Now we don't even notice Her, shepherded as we are into a narcissistic, thumb-skimming, brain-numbing, ego-shattering life of mediocrity and global homogenisation. One thing is for certain—as for sure as the three things we all share of being born, living and dying—all civilizations collapse.

In no shape or form is there a proposition here that we all need to start foraging for our food, particularly as with 67 million inhabitants in Britain this is simply not possible—the plan for solving the existing food crisis is going to be thoroughly laid out in the next chapter. The proposition here, and why blackberries have been so important, is that they signify loud and clear that food used to come from a different place, keeping alive a vision of a world before 'best before' dates, barcodes and petroleum-based plastic wrap.

Gathering wild foods from Nature's larder opens up a whole can-of-worms of philosophical conjecture, as we step back into a timeless realm, reuniting with our ancestors. Foraging also opens up a unique capability of reuniting us with our Mother. I've been taking people out foraging for a few years now, and I've noticed a change in those wanting to come along, especially at the wonderful Into the Wild festivals, a drink- and drug-free weekend with crafts, talks and world music. A decade ago a few would participate in the walks, most of whom it appeared knew more than I, but with each year more complete novices have showed an interest. At the last festival well over a hundred wanted to come along for each walk and talk.

This indicates a growing interest in foraging, which is great of course, but what has been most striking is what I have heard from many participating on these foraging walks, being bemused about not really knowing why they have come along although they sensed they needed to learn, guided by an inner voice telling them it would be a good idea to do so. Have these people been subconsciously harmonised with Mother Earth as She sings the siren song, calling Her children to perform an essential collective cultural cleanse? This is the only thing that can bring all the people together, as this planet is demonstrably under severe threat from the civilization we are living under, and all it takes to reconnect is to put this book down and just spend a little time in some woods, a meadow, the beach, a river, a sunrise or sunset, and revel in the fact we can appreciate it, as we are all part of a much larger, fantastical whole, all collectively on this incredible journey through space. Which might read like absolute bollocks, but if you were to spend time with me in the woods, or in the hedgerow and wayside, to actually feel the connection with Her is not something always transmittable through words—it can only be felt by being there with your soul. We are not really 'being' anywhere at the moment, are we? Not really being a Human Being, more a consumer, tax number and a part of the problem.

* * *

Being in woodland has a tangible effect on people. I've seen it lots, especially among kids. Coming from the frenetic electrical environment habited by most humans, whether at home, in school, the office, shopping centre or the automobile, every cell within the human body is affected by this, but when it comes into contact with the greater dominant frequency of the trees or the open countryside, the human frequency tunes into this more harmonious resonance. Whether such phenomena are quantifiable and measurable by modern science is another thing, because if modern science can't see it or measure it, by definition it isn't there. This is not the failing of Nature, where this phenomenon will always prevail, it's modern science's failure.

I began to notice quite a few years ago that when entering a woodland or forest environment, there appeared to be an invisible type of veil or boundary (not always), where upon crossing the wooded threshold there was a perceptible change in frequency, barely noticeable but sensational nonetheless. It also feels as if visual focal length shortens and auditory reception sharpens along with time slowing down. Is this the human frequency aligning to the frequency of the woodland? Sometimes while sitting with my back against a tree, birds have landed on me, heads to the side, looking me up and down—had I been tuned into the same frequency as the tree? With the same resonance as the tree, the birds see me as if not the tree itself, but of a similar harmonious, unthreatening frequency.

This ability to 'feel' the woodland and forest is not a special personal gift as it can be felt by everybody, and it actually has quite a weight of scientific evidence to back it up. The Japanese prescribe wood walking as a cure and as an aid to many diseases from depression to cancer—obviously it's not being hailed as a cure for cancer, but it can lift the spirits, giving an associated boost to the immune system and hence a useful adjunct to other therapy. This timeless, tranquil pursuit goes by the name of *Shinrin-yoku*, a word finding its origins surprisingly as recently as the 1980s, when the Forest Agency of Japan wanted to encourage more people to experience the benefits of the forest, especially as a counterpoise to the hectic and electronic lifestyle of the rapidly expanding megaconurbations of Tokyo and Osaka with 37 million and 19 million residents respectively.

Shinrin-yoku translates as 'forest bathing', which isn't taking a literal water bath in a forest—which if in an old enamel bath with brass taps and lion's feet would be a wonderfully surreal experience!—it's bathing in the air and ambience created by the forest. The British papers in 2023 headlined that woodland walking boosted the immune system by as much as 50%, no doubt quoting earlier Japanese research from the Nippon Medical School (Li et al. 2007), where there was also a 50% boost to creativity after four days' immersed in nature, as attested to by artists and non-artists alike (Atchley et al. 2012). A 50% boost, if it were a drug or supplement, would be hailed and marketed as a life-changing, ground-breaking miracle.

The Shinto Buddhist lifestyle may not be so actively followed in these days, but it still strongly reverberates through Japanese culture with its environmental outlook, so lacking in many religions. Japanese research demonstrates the boost to the immune system, with less of the stress hormone cortisol released (Park et al. 2010) and more of the pleasure ones (serotonin); mental health is enhanced and insulin production (also a hormone) and diabetes (Ohtsuka et al. 2008) are benefited, as also is heart disease as blood pressure can be reduced. *Shinrin-yoku* is not just a walk in the forest, although that is also beneficial; accompanying the walk would be breathing exercises, particularly close by trees with advantageous aromatic terpenes (such as pines and cedars, as demonstrated from work by Joung, Dayawansa, leading Japanese anime artist Miyazaki and many others), meditation, and appreciation and identification of some of the flowers and plants encountered on the walk—it's an immersive experience. It's also best to sometimes take your shoes and socks off and physically connect with the earth. Shoes break the connection, barefoot creates it.

* * *

Back to foraging, on a practical level concerning blackberry foraging, apart from definitely not doing it barefoot, the advice is to avoid wearing white or loose easily snagged clothing, because it's a near certainty it will stain and snag. There is of course the proviso, as there always is, that the area needs to be free from pesticides, but I hope you see the dark irony considering the millions of tons of chemicals sprayed on our food every year.

A definite health WARNING is that, regrettably, in studies from both America and Europe, supermarket berries (along with grapes) often come out as the worst offenders for pesticide residues—please be aware that these deadly, carcinogenic and genome-mutating chemicals are applied with an oily, clinging substance to prevent it from washing off in the rain, which makes it difficult to clean off at home as well. All the more reason to forage your own, although more and more farmers appear to be actively putting off 'townies' foraging blackberries by erecting signage that they have been sprayed, irrespective of whether or not they actually have been.

When picking, as the berries only ripen on the branch, usually from the tip of the branch backwards, pick those that give themselves up with little to no tugging; just assist them on their way to your container, and do not overfill as the bottom ones will get squashed. The great thing about them is that each bush has a distinct flavour, being a part of their plantality. It's definitely worth remembering where the sweetest, juiciest ones are found.

In Britain the blackberry generally flowers from June till September and bears fruit from August to October. The arching branch of toothed tangledness will begin to die back after two years, by which time it would have been replaced a dozen times over with equally gnarly clones, maintaining an impenetrable barrier. It makes them a perfect companion when planted within a living hedge, filling in the gaps and providing flavoursome, colourful and medicinal properties when utilised.

* * *

Croatian research using locally sourced wild varieties of raspberry (*Rubus idaeus*) and blackberries (*Rubus discolour*), from the University of Zagreb (Purgar et al. 2012) demonstrated that nutrient profiles were only slightly varied across the *Rubus* species, with wild blackberries always outperforming the cultivars. *Rubus* are particularly abundant in Croatia, where at least 33 species can be found, with it still being very much a part of their culture to forage for them.

A quick summary of the better-known nutrients available from 100g of blackberries includes 21mg Vitamin C, which is about 30% of Recommended Daily Allowance (RDA)—bear in mind the RDA is the minimum amount considered to avoid deficiency and subsequent deficiency illnesses, so a doubling of RDAs is generally a better idea of how much to aim for daily. Vitamin C is a powerful antioxidant and antibacterial as well as assisting in the production of hormones. The same 100g sample includes 0.6mg of manganese, about 25% of the RDA; this is a mineral assisting in blood sugar levels, insulin production and production of thyroid hormones. Then there is 20mcg of Vitamin K, about 20% RDA; this helps with bone health and blood clotting. Also present is 5.3g of fibre, about 20% of RDA, with blackberries being the most fibrous of all berries and most fruits—all those seeds and the husk do wonders for digestive and co-rectal health (88g of the fresh blackberries is water and the remaing 12g is fibre, minerals and vitamins etc.). There's also 0.2mg of copper, about 8% of RDA; this is essential for nerve and brain health as well as bone growth and about 6% of the RDA of Vitamin E, B9, potassium and magnesium. There is also about 3% of calcium, zinc, iron, Vitamin A, B2, B3 and B5. Finally, probably more importantly, blackberries are one of the richest sources for all those phytonutrients, particularly the polyphenols.

Considering that it is very easy to eat well over 100g of blackberries and the fact, as the science proves, that it's very hard to eat an excessive detrimental quantity, it can do no harm by gorging on the berries before the winter months, taking on board a massive amount of these essential phytonutrients to assist with everything from cellular health to seeing out the worst of the winter. One can easily imagine our distant ancestors feasting on seasonal foods when abundant—it's what we've always done, or rather it's what we used to do.

Blackberry cultivation

For a plant so ubiquitous, it is somewhat surprising that proper inroads into developing harvestable varieties didn't take place until well into the nineteenth century in America and Britain. Admittedly, the sight of unruly brambles doesn't lend itself to the idea of a neat, controllable cultivated crop. Also with so much available from the wild for at least 10 weeks a year, the need was maybe never felt—as with domesticating nettles, why bother when they are so prevalent? Far less prolific in the wild, with less threatening thorns, the raspberry was a much earlier choice for cultivation, recorded as grown for eating by 370 BC in ancient Greece.

One of the original popular cultivated blackberry species was the Himalayan blackberry (*Rubus armeniacus*), a confusing common name as it actually heralds from Armenia. Popular for its large if somewhat sour berries, these can still be found throughout Europe today and also much of America. They were originally imported by one of America's most famous plant and seed men, Luther Burbank, in 1885. He named the seeds he sold Himalayan Giant, which sadly today down much of the West Coast is deemed an invasive weed, keeping many gardeners in employ removing it. Another introduced Eurasian blackberry, *Rubus laciniatus*, has been grown since at least 1770 for its excellent fruit, and has subsequently undergone selective breeding with a thornless variety now available, often known as Cutleaf, whose leaf looks most dissimilar to the rest of the genus, being more like the Artemesias of wormwood and mugwort.

A breakthrough in the domestication and cultivation of blackberries was prickle-free erect canes, first developed in the UK at the John Innes Centre in the early twentieth century, albeit with a poor initial flavour. Fuller flavours appeared towards the end of the century with varieties such as Chester Thornless, Black Satin, Loch Ness,

Smoothstem and Triple Crown. Further work on cultivars from the University of Arkansas, USA, produced erect varieties able to be grown in abundance agriculturally. There are now varieties that are effectively an annual, where the primocane becomes a floricane in just one year.

Blackberries have had mixed popularity as market produce, the leaky berries being a major hurdle, although currently Mexico, Chile and Siberia are growing them in significant acreage. Mexico is now the third largest commercial grower of blackberries in the world after the US and Siberia, which is surprising considering that Mexico wasn't even growing blackberries before 1989. There has been an awkward development because of the nature of maturation of the berry, which attains its fullest flavour and sweetness at maturity, which is also the time it leaks. This wetness attracts mould, which can begin to affect the berries within six hours, and within 48 hours a whole consignment can be spoilt. This is not such a problem if the market is local, but with today's multinational distribution centres, mass deoxygenated storage rooms and frozen containers, it still proves extremely difficult. Sadly, the chosen answer is a saturation of antifungals and anti-mould agents for transport and storage. This is why it'll always make more sense to keep it local, and if cultivated keep it organic—with the future potential always waiting in the wings to utilise other revenue streams such as cordials, jams, dyes, sweets, tinctures, cough and cold syrups, etc., even diversify by processing the stems for use in fabrics with any waste being sold for bioplastic and biofuel.

Yum!

For millennia blackberries have been used for pies, jams, jellies, pickles, tinctures, cordials, wines, liquors, cheeses, ketchup and dyes. Recipes abound on the internet, so, to keep it short here, mention has to be made of a jelly that can lay claim to being one the best flavours to ever pass across my taste buds. It's a jelly hard to find in any store, and after making it you will realise that a price for that taste and effort makes money irrelevant. It is a perfect introduction to this plant we all need to get reacquainted with. The best bit about this one is of course all the benefits garnered from foraging, such as getting outside, sharing a timeless pastime with someone else and the sense of a completed goal.

The following proportions can be scaled up or down depending on how big your haul of deep purple beauties is—1.3kg of washed blackberries; 2 large cooking apples, washed, cored and diced; 450ml water; juice of 1 lemon (preferably a fairly juicy one, not one of those squeeze as hard as you can for a mere dribble variety); 1kg granulated or preserving sugar; sterilised jam jars and lids.

In a big saucepan place blackberries, apples and water, bring to the boil and simmer for 20 minutes, or until all fruit is soft; then add the lemon juice. Mash up into a mess with a potato masher. Place in a jelly bag or tea towel and allow a couple of hours for the contents to drip out. Place the drained juice into a saucepan. For every 600ml of this liquid add 450g of sugar and slowly raise heat while stirring the sugar in. When the sugar dissolves bring to the boil and reduce to simmer for a further 10 minutes. Skim off any scum—my youngest daughter always used to magically appear at this moment with a spoon in hand. Fill to the brim in jars washed and heated in the oven to 130°C. Store in a cool cupboard until needed. Consume in moderation. The surviving nutrients from the boiling outweigh the negativity of the sugar. CAUTION: gluttony brings forth many a deleterious disorder, opening the door to the diseases of civilization.

The berries can be marinated in sugar and white wine vinegar, then splashed onto grilled goat's cheese on a multi-seeded artisan bread, with a chopped dandelion and yarrow leaf salad. It may sound poncey but it tastes great. Young shoots can be eaten in omelettes, with lemon and olive oil, with fish or added to a vegetable broth, as can the young leaves and immature fruit buds.

When walking out and about, a fresh, soft, new leaf, or an unopened leaf or flower bud is a simple little snack, with a flavour best described as 'Nature's butter'. Raspberry leaves should not be considered the exclusive health domain for pregnancy; as shown in the scientific research above, the leaves from all the *Rubus* family contain a wealth of polyphenols and other beneficial compounds. Raspberry leaves are traditionally drunk for the claimed benefit of relaxing the uterus in preparation for childbirth, assisted by the superlative folic acid (Vitamin B9) also found in the leaves. B9 is a vitamin essential for cellular reproduction, which is what being pregnant is all about, and with the leaves being the best source, everyone, everywhere should be popping a couple of leaves in their fresh herbal brews, if not their mouths.

With literally tons available, the trick in the past would have been how best to preserve the berries, because there is no time to waste once the fruit has been picked and the race against the fluffy mould begins. This is why there are so many recipes and uses such as jams, ales, wines, liquors, brandies, pies, jellies, syrups, oxymels, vinegars, and (a Regency favourite) blackberry water ice—when it was considered to be at the height of affluence, fashion and gastronomy to serve a fruit ice between courses.

* * *

We end with a waste product from jam and food manufacture, the seeds. Seeds have been neglected somewhat throughout this book, being little powerhouses and storehouses of energy and nutrition, so to hopefully make up for this neglect the following paragraph should have gone into the scientific research section, but is added here so not to get lost within all the other positive scientific research.

An extra bonus for utilising berry seeds is that there is a considerable quantity available for free from jam, juice, cordial and food producers, as the seeds are often just separated and thrown away. That's almost criminal, because the seeds have in some instances even better nutrient profiles than the other already healthy parts of the blackberry plant. Seeds from all fruit have an already niche market, and, considering the excellent nutritional availability, the cheapness and the current lack of marketing or public awareness makes this a certain area for market growth. The oils are rich in linoleic acid/Omega6, alpha-linolenic acid/Omega3 and oleic acid/Omega9 in a healthy ratio. This is not to forget the tocopherols, phospholipids, phosphorus, potassium, calcium, magnesium and iron. It's no surprise to hear that the polyphenols are the richest component of the seed oil, including those with a multitude of health benefits such as egallic, chlorogenic, coumaric or caffeic acids and quercetin. The scientific evidence corroborates the range of traditional health benefits, spanning from anti-inflammatory to dealing with cholesterol, insulin, cancer cell lines, antiviral and antibacterial to skin ailments and split-ends in hair—the oils in many instances do an even better job than the other parts of this fine plant. An excellent overview of all seed oils was published in *Molecules* journal 2019 (Fidelis et al.), showing how they can create a circular economy.

Blackberries deserve a final shout-out as being the one wild plant that has kept our foot in the door with our connection to Mother Nature, a reminder of freer times long ago, as we've always foraged them, making it possible for us to kick the door wide open and expose the whole world of edible, healthy and utilisable plants. The fact they are among the healthiest and tastiest of forageable plants has undoubtedly helped to preserve their popularity.

* * *

Well, there they were—seven plants to save the world. There was a lot to take on board, so the next chapter condenses it all, outlines how much money is at stake, focusing on the pharmaceuticals and how we can overcome the greatest hurdle and threat to our future, the agriculture and food industries and their perverse and unnatural concept of providing food. The past seven chapters contained all the pieces needed to make a sustainable future; the next chapter will attempt to put them all together and provide connective answers and a solution.

SOLUTIONAL SUMMATION

(Respondere ad finem)

> You never change things by fighting the existing reality. To change something, build a new model that makes the existing model obsolete.
>
> (Buckminster Fuller)

The *Oxford English Dictionary* defines 'solution' as "Bringing to an end. The action or an act of solving a problem, difficulty, etc.; a means or method of doing this; the condition of being solved. Also, an explanation, an answer, a decision."

From the fifteenth to the seventeenth centuries, the definition of solution was a little different—it was more specific to the paying of a bill, or the action of releasing a person or thing, a form of deliverance, to be made free. You had reached your solution when freed from the shackles of servitude or slavery.

For this book, from its title, if the intention and claim is to save the world, an end and deliverance is required from the cultural systems destroying the essentials of our ultimate survival, from the very planet itself all the way down to our genome, our DNA. Most of the answers have already been touched upon throughout the book, with some briefly expanded upon. Being as there were so many uses, benefits and astronomical wads of wonger, this Solutional Summation is required to cram them all into their appropriate compartments—as well as offering a reformation of the agricultural system and land ownership, both vital necessities for when we finally have that change of mind and really understand that enough is enough, because it is not one facet of our culture needing a tweak to safeguard our future, since every system of our civilization is in need of reformation, from the economic, education, agricultural, food, health, industrial, legal, political and social.

* * *

What it all boils down to is the notion of change. Make no mistake, we are living in extraordinary times—the cultural system that has been in control for the past 70 years is just another brick in the Industrial Revolution's wall, which is crumbling rapidly into unsustainable dust. By being inherently unequal and essentially immoral our present civilization had no longevity, and with time it has merely become the proverbial ticking time bomb—but how can we accept this error, move on and make the required changes? As David Bowie would say "Ch-ch-ch-ch-changes, Turn and face the strange, Ch-ch-changes, don't wanna be a richer man, Ch-ch-ch-ch-changes, Turn and face the strange ... Ch-ch-changes, where's your shame? You've left us up to our necks in it."

With regard to changes, when I began researching and writing this book it could not have been envisaged how the world would pan out in the coming years. When writing the first draft of the first chapter in 2017 I had serious reservations about how the general public were going to embrace this notion of change, even if for the better—and then three years later the Covid-19 virus, a media and marketing campaign and lockdowns came along, changing people's lives more than most could have ever imagined. It also appears to have become a topic where divisiveness is rampant, and as the quest of this book is togetherness I've steered clear of the Covid-19 topic as much as possible, as C19 and others such as B17 appear to cause cognitive dissonance for some people, where the oppositional contention is mentally digested as if it were a food allergen, causing discomfort. However, the lockdowns did give us a taste of how easily change can be initiated—if you'd call being put under house arrest a change!

The vital change of this book is one focused on us having just a slight change of mind to accept that we all share a common unity in creating a sustainable culture to allow the forthcoming generations to enjoy this planet with more connection, community, respect and peace than the past few generations have managed.

The crazy thing is, this change of culture won't lead to more anxiety or austerity, it won't create mass unemployment, homelessness, poverty, civil wars, hike inflation, with no luxuries available and queues for fresh produce. These negative outcomes are actually manifesting right now globally owing to our current catastrophic culture. The opposite is actually more likely to occur when we finally remove the sociopathic rulers and leaders of governments, countries and corporations, and contentment and even abundance may very well befall us.

I've no claim to be a polymath, not least because I'm rubbish at maths, yet these plants, my only speciality, have opened the door to what will ultimately be a necessary reformation of our cultural systems. It may be considered a revolution, but it has nothing to do with stand-offs with the police, balaclavas and smashing the windows of a bank or MakkyDs: thinking violence is involved is merely playing the ruler's games, and they haven't even told us they're playing a game, let alone shown us the rules. The coming revolution will be permacultural, reconnective, self-sufficient and sustainable.

> Reality is created by the mind. We can change our reality by changing our mind.
>
> (Plato)

Systemic reformation

1. Health system

An end to the diseases of civilization

It would be perfectly understandable if after just the third plant it had become somewhat blurry as to which plant was best for which disease, with their potency proven against so many chronic diseases. After delving deeper into the worldwide folklore and science the seven plants were shown to be of benefit repeatedly in relation to the diseases of civilization—cancer, heart disease, diabetes, obesity, dementia, depression, iatrogeny, neurological dysfunctions and more. Upon reflection this should have been expected, as these are Human Plants, instrumental in our evolution, so why not our future?

* * *

Nettles with their unique acetylcholine enzyme and potential benefits for neurological conditions, from Parkinson's to general dementia, epilepsy and even autism, will be a huge medical breakthrough, but there was a lot more. Nettle's abundance of polyphenols and other phytonutrients when put under scientific scrutiny demonstrated anti-arthritic, antibacterial, anti-carcinogenic, anti-diabetic, antifungal, anti-histamine, anti-inflammatory, antioxidant, anti-rheumatic and antiviral properties, as well as being an astringent, decongestant, detoxifier, diuretic, galactagogue, nutritive, poison eliminator, phytoremediator for the soil and a perfect tonic for humans.

Dandelions absolutely bossed it in all medical research, firstly against human cancer cell lines of the blood, breast, colon, gastric system, pancreas and skin—not just killing the cancer cells but also eliminating them from the bodily systems. Scientific research has also demonstrated dandelions are anti-arthritic, antibiotic, antioxidant, antiviral, liver-protective and effective against hepatitis C; dandelions increase bile production and its subsequent release as it's no good just producing it, with bile playing a vital role in the absorption and digestion of fat as well as preventing the formation of gall, liver and kidney stones; dandelions assist cholesterol, heart disease, arterial sclerotic deposits and blood pressure; then, the chlorogenic acid found in dandelions has anthelmintic activity (kills intestinal parasites) as well as a novel way of metabolising lipids; dandelions also assist with constipation and act as a perfect probiotic, establishing healthy gut biota and reducing colon/rectal cancers; they improve bone health and blood sugar balance, while the latex sap acts as an effective fungicide and insecticide. Dandelions are a general tonic for all the organs, cleansing the blood, assisting anaemia and even reducing plaque build-up on teeth; and last but not least, dandelions are an anti-inflammatory.

Moving on to yarrow, the plant with both yin and yang and a higher vibration, traditionally it has been used for a wide range of illnesses. Iran leads the way in yarrow research, with these scientifically corroborated benefits—it decreases blood sugar levels and protects the pancreas; it decreases pain associated with dysmenorrhea, as well as oral mucositis caused by chemotherapy and general radiation protection from both chemo and radiotherapy and electro-magnetic radiation (phones, wi-fi, electricity etc.); it benefits epilepsy, Alzheimer's and Parkinson's, stroke, multiple sclerosis and leishmaniasis (skin-eating parasites). The rest of the world contributed to the corpus of yarrow medical knowledge by showing it can also act as an anxiolytic (reducing anxiety); an emmenagogue; to prevent tumour proliferation; an antibacterial, including the super-nasties of MRSA, *Shigella*, *Staphylococcus* and *Salmonella*; it has hypertensive, cardio-depressant, vasodilator and bronchodilatory capabilities and is yet another anti-inflammatory. It also effectively kills the larvae of the Asian Tiger mosquito, a seriously annoying, frequently biting, disease-transferring pest, buzzing near your head soon.

The fourth plant, plantain, brought with it a somewhat surprising smorgasbord of health benefits, including, among those scientifically verifiable—prevented carcinogenic synthesis, decreased tumour size; offered protection to the cardiovascular and gastric systems and the liver; brought relief from bronchitis; it is another anti-inflammatory; an antispasmodic; it reduced LDL cholesterol levels; is antibacterial; antiviral; antiseptic; anti-carcinogenic (stomach, bladder, bone, lung, kidney and cervix cancer cell lines); anti-diarrhoeal; has anti-plasmodial (malaria) action; prevents calcium oxalate stones forming; improves kidney function; reduces oxidative stress and diabetic foot ulcers, and is also effective in controlling and curing oral mucositis; it speeds up and improves wound healing and the immune system; it has phytoremediation abilities and helped our lovely Bella back to health.

Hemp, CBD and the inherent confusion caused by marijuana doesn't detract from cannabis having many health benefits, even under the cautionary note that the marijuana industry will be worth a trillion pounds globally when legalised. The epilepsy benefits have already been capitalised on by GW Pharmaceuticals with Epidiolex, suggesting that the traditional uses from centuries ago were not unfounded. It shows positive results for patients suffering from rheumatoid arthritis, fibromyalgia and juvenile arthritis, and also has anti-inflammatory benefits. A big bonus is that marijuana has proven to be a safer alternative to the dangerous opiated painkillers so over-prescribed, causing severe addictions and hundreds of thousands of deaths. Its appetite-enhancing attributes are vital and will become more so, for an ever-widening array of very sick patients. Its action as a muscle relaxant benefits symptoms associated with diseases such as multiple sclerosis and epilepsy, with proven spasmodic benefits and acting as a useful seizure treatment. The cannabinoids can relieve PMS and anxiety, with differing blends of THC and CBD offering help for a host of ailments. With so much money at stake, the benefits of marijuana may have been a little over-hyped, with the truth getting lost in a cloud of its own haze. The main concern with marijuana is the THC, as being 'stoned' as a side effect might not always be desirable or even safe for many people, causing either clumsiness or adversely affecting mental health and daily life. Cannabis's overlooked uniqueness lies in the freshly pressed seed oil and the outstanding contribution this can make in preventing inflammation, particularly as so many diseases have inflammation as direct associated symptoms.

The sixth plant, hawthorn, covered the current biggest killer from our civilization, heart disease, which would be more than enough, but not for hawthorn, as it's full of surprises. Not least are the connections to Jesus, royalty and fairies. From a health perspective it's worth noting that Traditional Chinese Medicine has employed hawthorn in remedies for digestive ailments for many centuries, just as throughout Europe and America as recently as the early twentieth century. It seems that the deterioration in the Western diet, with processed carbohydrates and demineralised foods becoming the norm, has become a major risk factor in the rise of heart disease, especially cloggy arteries and the surge in obesity. Hawthorn appears to be able to remedy various aspects of the heart disease arena, from angina to arrhythmia, high blood pressure, cholesterol and atherosclerosis. Digestion and heart disease aside, hawthorn also weighed in with positive results from research for skin cancer and cancer cell proliferation; it is another anti-inflammatory and assists oxidative stress and diabetes, offers liver protection and is an antibacterial.

The final plant, blackberry, demonstrated omnipotence scientifically, as well as boosting the social aspect, exercise and mental health benefits from foraging and forest bathing. There is positive scientific evidence for blackberries (and their close relatives in the *Rubus* genus) across a vast range of ailments, including various cancers (skin, lung, oesophagus, breast, brain, colon, liver and leukaemia), and also inhibiting the proliferation of tumours, as well as polyps. The whole B17 affair might have been a fiasco, but the anti-cancer properties of blackberries are far too clear to ignore, especially when combined with the efficacy of dandelion root. Another plant with antibacterial abilities, it has proven effective against *Helicobacter pylori*, which affects 60% of humans, *E. coli*, *Staphylococcus* and *Salmonella*. Antiviral capabilities were demonstrated against the herpes simplex virus (also affecting 60% of humans), and no doubt the polyphenols so common throughout the seven plants also have potent antiviral activities, be they preventative or curative, especially when combined. Blackberries were also beneficial against oxidative stress and could play a role in neurodegenerative diseases such as Alzheimer's, while also displaying anti-inflammatory benefits and assisting with obesity by positively affecting fat production and deposition. Blackberries appeared to benefit us from the fundamental cellular level, the very basis of health. The fact they can do all of this and have held the tenuous link humans have with Nature makes them very special indeed.

Medicine

With all the folklore and scientific corroboration there should be no delay in making medications, tinctures, balms and pills, while also incorporating these plants into the diet as effective nutraceuticals. This is not to forget all the shampoos, toiletries, creams and beauty products, because when the effectiveness of these is more widely known, they will become more extensively used.

The almost complete absence of human trials to ascertain the veracity of these health claims from these omnipresent plants is perplexing, as there has to be a series of in-depth trials conducted to fully establish the health potential of so many plants, not just the seven from this book. Having personally experienced how the health industry and science is interested in only economics, when Covid-19 hit, a customer of mine happened to be the doctor in charge of the local Covid ward, as well as being a professor in contagious diseases, so I asked her to read up about yarrow, which she did, and it was very enthusiastically received. However, it was made clear there would be no trials on yarrow's preventative or curative potential, even if proven, as there was no economic benefit at the end of the research by using the whole plant as grown in Nature—a patentable phytochemical to trial would be required.

Shockingly, pharmaceuticals have quietly become dominated by petroleum, where analgesics, antibacterials, antimicrobials, anti-histamines, cough syrups, creams, gels, lubricants, ointments, rectal suppositories and salves all have petroleum as their backbone. Take, for example, aspirin: the bit that works is salicylic acid, originally identified from the willow tree and also found abundantly in meadowsweet—not that you're getting anything from a plant with an aspirin today as it's now a synthetic version called acetylsalicylic acid, which is made from chemical reactions involving petroleum. It's the same for coatings of pills and capsules, all petroleum-based, making mass production of pharmaceuticals possible—as mentioned before, the third biggest individual killer of industrialised inhabitants is one of an iatrogenic nature, predominantly prescriptions from the doctor, taking out about 750,000 humans every year (and uncounted animals and other living organisms from water pollution etc.).

Vaseline is basically petroleum oil, as are most skin creams, usually with some acetylsalicylic acid added; this also applies to most perfumes and anything with butyl, ethyl, propyl or methyl in the ingredients list. By artificially implanting petroleum into the health industry, cash registers rang to the tune of several billion pounds for the petroleum people annually. This was all planned for by a person I'm reluctant to mention too much about for fear of being labelled anything from a conspiracy theorist to a leftie or alt-right, but it is the truth to state it was John D. Rockefeller, the oil baron with his monopolistic mentality about any business he was involved in, who set about a deliberate vilification of plants and herbs to confuse the public and help sell his pills, of which aspirin in 1917 was his first step into the pharmaceutical monopoly he established for these pills and creams made from petrol. This man (or his business managers) not only created the pharmaceutical industry we know of today, he is also responsible for decades of anti-herbal rhetoric as well as creating numerous educational facilities and publications espousing the glory of the science of the pharmaceutical. Rockefeller may have seen oil as a godlike substance, having given him so much wealth and power, as well as driving America to industrial dominance, so why shouldn't it give health as well? After all the pharmaceutical industry is just that, an *industry*, created as a business opportunity.

Emphasis needs to be put on the fact that in no shape or form is the entire cachet of Western medicine being dismissed—modern surgery is unarguably amazing but not necessarily all the pharmaceuticals, especially as already stated, health is a combination of physical, mental and gut health. As Western medicine only deals with the first and damages the other two, why should such public faith be bestowed upon this clearly detrimental and erroneous approach to health without it being questioned? Particularly when a known deleterious ingredient such as petroleum forms the foundation of this approach today. Pharmaceuticals have taken over because they have a long shelf life, can be mass-produced for the human flock and monopolised through legal patent channels with exorbitant prices. Nutraceuticals and herbalism has a completely different approach, and after presenting the case for the seven plants of this book, they should definitely play a central role in future health, by preventing ill-health and maintaining a balanced organism, allowing cellular, gut, mental and physical health to flourish.

There is no health sense by taking crude oil or natural gas apart and putting it back together again to make medicine, only economic sense. Common sense would have ensured a healthy soil with healthy plants and a healthy food chain feeding a healthy nation. Common sense finds plants with the constituents to prevent, ease and cure ailments and then to consume these. Our culture has deliberately prised us away from Nature and immersed us into the industrial world run on economic sense.

2. Economic system

> If we wish to diminish the love of money which, we are told, is the root of all evil, the first step must be the creation of a system in which everyone has enough and no one has too much.
>
> (Bertrand Russell)

What is money?

The basic function of money is to facilitate the buying and selling of items or services with a widely recognised and accepted form and value, be that cowrie shells, barley bushels, cows, coins or banknotes. Credit notes, effectively IOUs, were and have been effective to perform trade for millennia, although this requires a certain level of knowledge about the debtor, while a widely accepted form of money negates this buyer/seller familiarity.

The earliest use of metal for money is tenuously argued as being used in Babylonia around 2000 BC, with firmer evidence for minted coins occurring almost simultaneously in China and Anatolia (modern-day Turkey), taking the shape of mini bronze shovel-heads in the Guanzhuang mint from 640 BC, just beating the electrum (a naturally occurring mixture of gold and silver) coin/token from the Anatolian Lydian empire from 630 BC. The Ganges civilizations also had metal coins, with current evidence from about 600 BC, and undoubtedly future global excavations and finds will push the dates further back. Paper money wasn't invented until the seventh century AD in China and only becomes a fixture in European currencies from the eighteenth century, because before then bills of exchange (credit notes) sufficed.

We get the word money via the French *monnaie*; the French in turn got it from the Latin and the Roman goddess Juno Moneta whose temple was also the mint (a word also derived from *Moneta*) where the coins were made in Rome.

The Greeks knew full well that when money first reared its head the lust for more would always be greater than the universal value it could potentially bestow, which is why Aristotle gave us his Midas story in about 350 BC. It is indeed this universal value that has yet to be expressed within the current economic system. This universal value should be one of the pillars of a civilization as it is indisputably one of the keys in establishing an equitable system.

With the root of the word economics coming from the ancient Greek for 'good household management', *oikonomia*, it truly is time our cultural systems become underpinned with this as an ideology in action, not the unsustainable ideal of continual growth. Most economists do concur that our current economic system is in its death throes and a new economic system will inevitably replace it.

Along with economics, we see another word that has undergone a change of definition and our understanding of, the word capitalism. When Marx was writing *Das Kapital* in 1867 and for long before, it was explicit that capital was the workforce, yet as the financial institutes levered more power this capital morphed into being money, not people. Not a minimal switcheroo.

* * *

Today, somewhat confusingly, there are three systems to consider when talking of money: monetary, economic and financial.

The monetary system has three levels. First is the money itself, which is held by the public, businesses and certain government holdings such as the Bank of England and its £200bn worth of gold (which can also be considered money). The second level is commercial banks, institutions that safely hold the public's money via deposits with which they make loans to individuals, businesses and governments—buying bonds and bills etc. The third level is the central bank with its monopoly on the production and quantity of money in the system.

The economic system is all the production from industrial to financial, the allocation and distribution of all goods and services within a country or society. As with the monetary system it is a type of social system, a reflection of social enterprise and productivity, although catastrophically the economic system has become dominated more recently by a facet of the financial system, a system that allows the transfer of money between savers and borrowers, providing financial services encompassing financial management, accountancy, credit-card and insurance companies, asset management, credit unions, investment funds and stock brokerage.

The biggest problems come from the financial services. Those crazy debts are little more than gambling gone wrong, but here's the rub—in 2021 these services provided a positive accounting entry to Britain's GDP of £174bn, nearly 8.5% of the total, while employing over a million people within the sector, which equates to 3% of the total workforce making 8.5% of the money, which gives the sector a cloak of near invincibility.

* * *

The economic system was always on shaky ground, having experienced a crisis every decade or so somewhere in the world, with the occasional one having global ramifications. With a conspiratorial cabal of money alchemists, making money from debts and gambling, this economic system became progressively and excessively vulnerable, even though the public have been paying the cost of bail-outs, which amounted to a staggering 23% of global GDPs in 2008 (£11.5 trillion) after the sub-prime economic collapse of that year, when many banks globally fell from unadulterated gambling debts (unsecured and overpriced mortgages), and the public paid the price. Here is when all banks should have been nationalised, but instead the costly bail-outs were balanced by printing more cash, and with it all economic 'rules' suddenly got jettisoned and a hush-up ensued.

Incredibly, the Bank of England has printed three times the cash that was in the system since 2008, with an extra £895bn, irrespective of the fact that printing extra cash is inflationary; unsurprisingly, all the talk today is about soaring inflation! Calling it quantitative easing (with a definition not as straightforward as printing more cash!), it also proved to be a convenient way for the wealthy to massively increase their wealth (often via shares, land and holdings), syphoning much of the extra cash printed with price rises across every sector of life; whether petrol, electricity, gas, food, building supplies, insurance or bicycles, all prices have risen by an average of 33% in the past three years. Brexit initially gave a good excuse, as did Corona, then the Russian invasion of Ukraine, now it's Israel in Gaza and it'll be climate change next—BUT prices at wholesale for the original producers only experienced a slight increase, while blatantly showing this to be the case as unsurprisingly all global corporations posted record profits from 2021, be that energy suppliers, food producers or pharmaceutical makers, many even being so bold as to clearly state prices were increased for no other reason than to take the opportunity to make more money!

Insidiously even wars are ultimately over money, trade and business, draped in a flag of foreign policy. Wars are sold to us over spurious human rights abuses, acts of war and aggressive foreign policies, when actually they are about control of resources, which is of course money by way of future profits. The US invasion of Iraq in 2003, with some help from the EU—spawning much of the terrorism problems thereafter—was sold to us with the fear of weapons of mass destruction, when what really happened was we were all media-bombed with psychological weapons of mass distraction, while Haliburton via Vice President Dick Cheney cleared up the restructuring and oil wells. The same with Russia and Ukraine, BlackRock are basically contracted to rebuild Ukraine after a proxy war, while conveniently curtailing Russia's gas and energy exports.

* * *

Religions such as Christianity and Islam have strict rules condemning the making of profit from any money via loans, seen as immoral. Was Jesus himself (although the actual documentary evidence is to say the least somewhat flimsy) actually crucified for his constant calling out of the evils of the moneylenders as much as for any heretical claims of kingship? What we do know is that as early as AD 33 there was a financial crisis brought about by excessive lending and overpriced and unregulated mortgages (sound familiar?), causing Emperor Tiberius to create a three-year 100 million sesterces programme to restore credit. Talk about not learning from history.

There is definitely a move to eradicate money in the form of banknotes at the moment, and the rationale for doing so is not so easy to comprehend with so many conflicting opinions vying for attention. The French have taken the initiative to confound this rule by going to restaurants with a 'no cash' policy en masse and ordering multiple meals and then refusing to pay in any other form but money; the same has happened at petrol stations and it has made an impact as the policy is being reversed, as it should being legal tender. But money such as this is not what causes all the problems with economics, as Sir Mervyn King (ex-head of the Bank of England) said: "most people assume economics is the study of money, when in actual fact economists rarely if ever actually talk of money *per se*", as money has nothing to do with modern economics, just as economics has nothing to do with *oikonomia*.

Our lives are undisputedly governed by economics. To allegedly help to understand or assist this governance, occasionally as a dissenting voice, we have economic theory, hundreds of them—from the Ability-to-Pay principle to the Capital Asset Pricing model to Competitive Advantage with some Dual Economy theory thrown in along with Equilibrium theory or the Fundamental Disequilibrium theory, Keynesian or the La Chatelier or Marshall-Lerner principles, even Marxism and Privatisation, the theorems of Rybczynski and Slutsky, or Syndicalism, Thatcherism or Trickle-Down theory, the laws of Verdoorn and Wagner or Wicksell's theory of capital—all are just that, theories. Whereas the reality today is somewhat different, as it appears the economic system has been tailored to funnel money (profits) to the highest echelons while keeping the masses on the brink of poverty.

An erroneous backbone of many theories has been the notion of a continual growth in production, sales and profits, ever since Adam Smith's *Wealth of Nations* in 1776, laying the foundations for our 'free' market economy, structured and governing society around this underlying principle of growth. Principles or theories based around sustainability, longevity and equality never got heard above the greedy clatter. Smith has to take some of the blame for our present economic crisis, although he did also state that "Landlord's right has its origin in robbery. The landlords, like all other men, love to reap where they never sowed, and demand a rent for even the natural produce of the earth." His earlier work of 1759, *The Theory of Moral Sentiments*, shows he was a well-rounded philosopher and perhaps a little unfairly bears the brunt of anti-capitalist derision, not least because at the time he was particularly unhappy with mercantilism and slavery, recognising both as thoroughly immoral ways of doing business.

Financial historian Charles Kindleberger has highlighted how there has been a crisis every decade for the past 400 years, stating it was the system at fault more than the ineptitude and corruption of the banksters. The economic problem has always had any holes plastered over instead of finding the root of the problem, so perpetual failing is the only possible outcome. Exactly as Felix Martin concludes in his *Money: The Unauthorised Biography*, after outlining money's history and continual incongruence, money is indeed found to be corrupt and erroneous, calling for the reformation that never took place and will only ever take place with a realistic understanding of money.

After WWII there was a significant shift in the apparatus controlling our industrial cultures, transitioning a global military complex into a global consumerist society. With money being given an even louder voice as the financial services began to dominate, the potential dangers became so damningly obvious that many famous people commented upon it. The following quotes show that wealth not only trumps health, it also beats democracy. Starting with someone who knew all sides to the money story right at the beginning of the transitional period directly after WWII, Britain's Prime Minister for the years 1945–51, Clement Attlee, said:

> Over and over again we have seen that there is another power than that which has its seat at Westminster. The City of London, a convenient term for a collection of financial interests able to assert itself against the government of the country. Those who control money can pursue a policy at home and abroad contrary to that which is being decided by the people.

It's worth noting that the City of London, also called the Square Mile, is indeed a separate entity from the rest of London and Britain, with its own government, police force and Lord Mayor, with no subservience to British parliament, laws or constitution—even the King needs to be invited by the Lord Mayor to visit. The only thing it shares with the rest of London is a name and geography.

Clement Attlee's quote is supported and even bettered by a person of undisputable great intellect, Albert Einstein, who in a 1949 essay entitled "Why Socialism?" wrote:

> Private capital tends to become concentrated in a few hands, partly because of competition among the capitalists, and partly because technological developments and the increasing division of labour encourages the formation of larger units of production at the expense of the smaller ones. The result of these developments is an oligarchy of private capital, the enormous power of which cannot be effectively checked even by a democratically organized political society ... The consequence is that the representatives of the people do not in fact sufficiently protect the interests of the underprivileged sections of the population. Moreover, under existing conditions, private capitalists inevitably control, directly or indirectly, the main sources of information (press, radio, education). It is thus extremely difficult, and indeed in most cases quite impossible, for the individual citizen to come to objective conclusions and to make intelligent use of his political rights.

Since these public disclosures, those in the spotlight have successfully deflected all the lights away from themselves and through psychological weaponry (social engineering) have managed to intensify the public's bickering among themselves, continually using their control of the media to drive ever more wedges in (divide and rule), preventing any unification of the people against the true enemy of the people, their own sociopathic leaders.

* * *

Three reasons that may well contribute to our current method of economics being a doomed system include, first, as we saw above, it being restricted to not only the generation of profit but a continual increase in profits, where smaller than expected profits are considered a loss, even if not a literal one, with perpetual growth being only theoretical, not realistic.

Secondly, by a quirk of mathematics the rich predominantly just get richer, creating an ever-widening chasm in wealth, as highlighted in Thomas Piketty's book *Capital in the Twenty-First Century*—a surprising bestseller in 2014, being little more than dry economics—which does corroborate what Einstein says above (those five sentences are worth a reread) regarding the owners of capital usually only ever getting progressively richer with a straightforward equation $r > g$ (where r is rate of return on private capital and g is economic growth rate). Basically, instead of everyone benefiting from their work and seeing an increase in returns as they get older, the money only gets channelled to a few individuals ... as it has done for centuries. It was this and other moral contradictions that convinced Marx capitalism couldn't last too long, as well as leading world-renowned economist John Maynard Keynes to say in 1936 in his *The General Theory of Employment, Interest and Money*: "the outstanding fault of the economic society in which we live is its arbitrary and inequitable distribution of wealth and incomes".

Thirdly, modern economics excludes all free work, housework, child-care and charity and has serious problems accommodating for any un-profit-led environmental considerations, such as carbon emissions, which soon got corrupted and led to an illicit trade in carbon credits such that larger polluters could pay to continue polluting as before, which is exactly why the environment has been desecrated—it was all economically justified from the beginning. Talk about a false economy, as now we face a genuine man-made catastrophe from chemicals to climate.

Crypto or Green Currency?

Cryptocurrencies such as Bitcoin aren't the alternative to a moneyless society, at least not in their current guise. Underlining the unsustainability and ridiculousness of what is a commodity linked to profits and growth, not a currency, get this crazy fact—that somewhere in the self-justifying blockchain of computers, equations and algorithms, the vast amount of hard drives whirring to the Bitcoin beat consume allegedly as much energy as the entire power usage of Sweden. Add this to the other cryptocurrencies such as Ethereum (who have admittedly recently radically reduced their power requirement), Tether and Binance coin, and the power consumed is nearer to that of Argentina! Tether is a little different in that it is tethered to the price of the dollar, and is known as a stablecoin, along with USD Coin, which is centralised and backed by cash and US Treasury bills. So this electricity cost has to be factored in and reflected in a rising price, and if it keeps going up in value it's a dangerously volatile inflationary item—it's not a currency, it's a non-tangible, pretty much unusable commodity, useless in a store; it's a Ponzi scheme and it will prove to be a relatively short-lived once some rich people have gotten richer from it. Incredibly, Bitcoin had a market cap in 2023 of £370 billion, Ethereum £175bn, Tether £53bn, USD Coin £44bn and Binance Coin £42bn. Which is enough to make quite a lot of loud noise irrespective of any truths.

At the heart of so much revolving around cryptocurrencies is the drive to make a profit. That's the prime reason people get involved with Bitcoin, to make a profit—it's not to join a currency revolution and undermine the financial system. People didn't trade in cryptokitties because they wanted to breed or love cats, it was to make money. NFTs are ridiculous, and even the name, non-fungible token, essentially means you have nothing. There was recently a NFT artwork by Beeple, selling, among much press attention, for £56m; but it didn't, it was £56m worth of Ethereum tokens, swapped for some other token to artificially inflate their price, and you definitely can't be spending any of these in Tesco. It's all a scam.

* * *

That said, there is a change in store and with potentially many billions, if not trillions worth of goods that can be taken to the marketplace from just these seven plants alone, an agreed and trusted monetary system is imperative. We'll see in the next section some very rough guesstimates from sales of nutraceuticals replacing a minuscule portion of pharmaceuticals. If these revenues were banked with a new Green Currency, along with all the future other industrial uses from petroleum-free plastic to fabrics, hempcrete, rubber, paper, bathroom and beauty products etc., this will amount to a significant amount of money. So long as the faith is there and of course the actual clout to use the Green Currency freely, be that in the cinema, shop, utility payment or bar bill, with wide acceptance its success is almost guaranteed.

If the value of the total Green Currency in circulation never exceeds the value of the produce being grown and manufactured, the proclamation of the currency—"I promise to pay the bearer"—is actually guaranteed (albeit in produce), which surely makes it as real, if not more practical, than being on the gold standard. The UK left this standard in 1931 and the USA in 1933 (the world of finance and economics has moved on in the last 90 years, although not as much as perhaps imagined, as we rapidly move into a new economic crisis in its cyclical flow through time). Although, sadly, if it is left to the present political system to enact any societal changes it will only happen when it truly is too late. What do the politicians actually do anyway?

So, it's not necessarily the monetary system causing austerity, it's the broader financial services burning gigantic holes in the economic system. Money without doubt is a catalyst to greed and violence, from individual to international conflict, but with a better legal and moral code, transgressions will be minimised. The ultimate caveat to take from this section is that any reformation will require universal value being attached to the value of money—by universal value I mean a re-evaluation of the values attached to different professions, especially ironing-out gender and race inequalities, as well as the general genuine levelling of all wages.

Show us the money

Running with the theory that money is a social invention and function, meaning its existence is somehow intrinsic to society, there has been as of late the very real threat to banknotes and coins themselves. Money is and should always have been considered as the realisation of the product of a nation, which is an approximation of the worth of its production. In such a realm money is a self-regulated system, which in this book's case is the Green Currency and the new industrial applications and revenues opened up by the seven plants. For this section we'll be adding the revenue potential of these plants in just one sector of industry, pharmaceuticals, and only those mentioned in this book.

Taking as a completely theoretical and minuscule fraction, let's assume 0.0001, a 10,000th, or 0.01% (all are 1 in 10,000) of the current revenue from a drug can be captured by nutraceuticals and deposited in our fictional Green Currency Bank, and only if science has already shown one of the seven plants to be as effective if not more so than the conventional drug for that ailment.

Going through the chapters in approximate order, we first saw the dangers of Diclofenac and the possible role nettles can take as an effective replacement for NSAIDs (all of the seven plants are anti-inflammatories), which already weigh in with yearly global revenues of £20 billion (a rolling total will be in the brackets after the annual revenue, so a 10,000th of this £20 billion is a £2 million deposit for the Green Currency).

The acetylcholine and neurological market nettles could open up is enormous, within a sector of drugs where those for dementia alone total £20bn (£2m from this + £2m from NSAIDs = £4 million in the kitty). Dementia and what is referred to as the rare neurological diseases (including Alzheimer's, narcolepsy and multiple sclerosis) are all rising in incidence, with stocks in companies making the drugs also set to rise, and according to market analysts they are sure-fire buyers in the stocks trade with at least 10% returns annually for investments—which makes it look like they are betting on continuing and broadening the equation of sickness = the health *industry*.

Then the extremely positive results from a lectin found in nettles and its antiviral effectiveness against SARS (that's the Coronavirus family)—what price can be put on this if it proves to be true? Not that there would be any advice to go foraging for some nettles, instead there would be scare stories about people selfishly putting themselves in danger foraging or that the amount of the lectin cannot be guaranteed by picking it fresh, and it's always best to take a standardised pill ... say the pill-sellers. The potential market is not included in this tally as it's just speculative, even if priceless.

The global fertiliser industry is worth annually in revenues £150 billion, with nettles and some comfrey being a common and potent mix, good enough for my PermaLife plot and for many others that I know—the fact that it has been banned in France irrespective of it being a centuries-old sophisticated discipline stinks of the nefarious tactics employed by those pestilent lobbying people working for the fertiliser industry. This crazy £150bn will be vastly reduced, as most smallholders of the future will be taught how to make their own soil and fertiliser, as permaculture will finally oust monoculture—not that the loss of this amount of money is going to be swallowed easily by the world's governments to begin with, but we'll not count this in this calculation, even though green plant feeds and animal excrement, including humanure fertilisers, will be a significant element in the regeneration of the soil and the healing of the earth.

We will leave it here and not even consider adding any other potential earners from nettles via the food industry, beauty and hair sectors, textiles or building. Obviously nettles aren't going to completely monopolise the markets of the NSAIDs and other illnesses they have been scientifically proven to be effective against, but they will become a fixture and their inclusion as a nutraceutical as a part of the diet will act as a preventative negating the necessity of dishing out so many drugs to begin with, generating less short-term wealth but more long-term health.

* * *

Moving on to dandelions, taking on the conventional cancer industry, whose revenues of £150bn (£15m + £4m = £19m) every year could be seriously put under threat, by what many people might consider a bit of a joke, dandelions. Although it is nothing in comparison to the sick joke that it is civilization, our lifestyle, that is the biggest risk factor in getting cancer. The cancer industry has already managed to legislate against any competition to their monopoly of treatments, with cancer laws preventing doctors from even considering any other avenue apart from the mainstream chemotherapy or radiotherapy. Add to this the global diuretics market worth in the region of £1.2bn (£120,000 + £19m = £19.12m), with dandelions outperforming the commonly prescribed Furosemide.

Dandelion leaf extract also outperformed the hepatitis C drug Sofosbuvir, with around £5.3bn in revenues (£530,000 + £19.12m = £19.65m). Dandelions proved to be effective in many medical fields, being anti-inflammatory, whether that be in the blood vessels, stomach lining, lungs or liver, with no attempt to monetise these here. However, being better than Sulfasalazine in dealing with inflammatory bowel disease and its £5bn (£500,000 + £19.65m = £20.15m) in revenues, that's easier to quantify. We'll leave it here and not include the potential huge windfall available when and if taraxagum (tyres and other rubber items) takes off, or the phenomenally huge ethylene market worth currently a quarter of a trillion pounds sterling every year.

* * *

Yarrow certainly lived up to being named after the hero Achilles from Greek legend, potentially filling the Green Currency coffers across a range of ailments. Easing the pain associated with dysmenorrhea brings drug revenues in excess of £2bn (£200,000 + £20.15m = £20.35m), with the rather more common and growing market for diabetes drugs reaping about £50bn in revenues (£5m + £20.35m = £25.35 million). Oral mucositis caused by chemotherapy has drugs with revenues worth about £1bn (£100,000 + £25.35m = £25.45m). The fact that yarrow has been shown to help with the damage caused by all types of radiation, from chemotherapy and radiotherapy, nuclear and electrical, will be worth a huge amount of money, although only speculative at the moment, so it'll have to be ignored. What won't be ignored, as it was skimmed over in the nettles calculation, is multiple sclerosis (MS), which individually or when combined with nettles and blackberries appears to offer significant benefits, though with MS drugs having revenues of £20bn (£2m + £25.45m = £27.45m), the pharmaceutical companies might be a little more willing to ignore this fact. What is only small fry from sandfly, the skin-eating leishmaniasis, where yarrow, thyme and propolis were more effective than Sanofi-Aventis' drug Glucantime, with 'only' £100m in revenues (£10,000 + £27.45m = £27.46m).

Yarrow is better than diazepam in reducing anxiety, not to mention the benefits of avoiding all the side effects; diazepam has global revenues of over £2bn (£200,000 + £27.46m = £27.66m), which of course doesn't include all the other drugs prescribed for anxiety, tension and stress. Yarrow is also a broad-spectrum antibacterial agent against some of the worst killers, such as *Shigella*, MRSA, *Staphylococcus* and *Salmonella*; only finding figures for MRSA, the revenues are around £3bn (£300,000 + £27.66m = £27.96m), which, as with the anti-anxiety/stress drugs, is only adding a portion of the real total to the imaginary and fractional 1/10,000th. Yarrow could prove to be the SARS herb, with its respiratory benefits and blood-moving ability, as many of the deaths were blood clots on the lungs. Yarrow has yet another trick up its sleeve as it has been shown in research to be a remarkable defence against any cytokine storm, where the immune system goes haywire, often causing death, associated with organ transplants, AIDS and SARS.

With yarrow's efficacy in the asthma and COPD arena, the revenues here exceed £35bn (£3.5m + £27.96m = £31.46m). Not to double-count the NSAIDs, but compounds in yarrow were found to be more efficient than indomethacin, prescribed for similar problems as the vulture-destroying Diclofenac—so a formulation with both nettles and yarrow would probably be even more effective and still carry none of the awful side effects, because they are nutritive nutraceuticals, not toxic pharmaceuticals. Also these nutraceuticals would all be sold as over-the-counter (OTC) remedies, and this huge market worth globally somewhere in the region of £90bn (£9m + £31.46m = £40.46m) is definitely under threat from the sharp rise we will soon see in cottage industry salves, balms, tinctures, syrups and ointments being produced as tertiary industries within the smallholding revolution (initial processing of the plant being the secondary industry, and growing the first). Let's leave it there without trying to monetise what may be a substantial market in the future for mosquito repellent and larvicide, or the radiation cleansing soon required, the anti-aging market or the food and herb flavouring potential.

* * *

Bringing us over the halfway line, plantain is surprisingly omnipresent and incredibly healthy. Plantain appears promising with regards dementia, which was already accounted for with nettles, and when combined with nettles in a formulation could produce a host of safe and effective nutraceutical options. Plantain also benefited cancer (accounted for in dandelions) and heart disease (which will be accounted for in hawthorns) and was an effective anti-inflammatory on various parts of the body, and along with yarrow was a larvicide, killing mosquito larvae, and had the ability to clean up industrial and chemical pollutants from the soil (phytoremediation).

Plantain did outperform the drug given for internal calcified stones, allopurinol, with revenues for this ailment of £500m (£50,000 + £40.46m = £40.51m). Plantain also outperformed another drug for oral mucositis, chlorhexidine, accounted for in yarrow, underlining the incredibly broad range of ailments, diseases and conditions all these seven plants can potentially deal with. Plantain is one of, if not the best plant antibacterial and antifungal, with global revenues in the antifungal market somewhere in the region of £8bn (£800,000 + £40.51m = £41.31m), as with all the other illness sectors, rising year-on-year in revenues, territories and customers.

The cholesterol market is huge, including the biggest-selling drug of all time, Lipitor, with total revenues from 1996 to 2012 of £125bn, but whose efficiency is highly disputable and its side effects many; plantain acts positively for both cholesterol production and deposition, accounted for later in hawthorn. Individually plantain may have added little to the kitty, but this is not to downplay it, as without a doubt it will become priceless once its use becomes universal, solely and in combination with the other six plants.

* * *

Not many figures were mentioned in the Cannabis chapter, although with anything from 25,000 to 50,000 industrial uses, the potential is huge, but for this calculation none will be included. In 2020 there was an estimated £5 billion worth of revenues from hemp and hemp products, a figure set to rise by as much as 27% per annum. This increase is eclipsed by the medical cannabis market, which is seeing rises of well over 1,000%. Cannabis itself, on the other hand, is already worth significantly more than that, and from an averaging of future estimates the annual net worth of marijuana will be at least

£200bn by 2025. However, there should be little emphasis on numbers, because the US CBD market is already worth over £10bn and the cannabis market has been valued currently at £110bn, although with no consensus on the figures there is no reason why cannabis cannot become as big as beer, worth worldwide a guzzling £500bn, with a similar amount for alcoholic spirits. However, the mood is confident that 1/10,000th of those turning a profit from this plant will see it prudent to invest in some Green Currency, so without even thinking of the huge textile or paper potential, or the £400bn spent on concrete every year or the £500bn spent on petroleum plastics, or the already insane growth rates of cannabis pharmaceuticals, being ultra-conservative we can add just the current global market value of £110bn (£11m + £41.31m = £52.31m) to the running total, which will be eclipsed when there are coffee-shops à la Amsterdam in cities and towns around the world.

* * *

Moving swiftly and briefly on to the magical hawthorn, this had some of the same bases covered, notably those to do with heart disease and with hawthorn proving to be beneficial against angina, with revenues for this sector of heart disease topping out at £7.8bn, or the statins market at £23.4bn, similar in size to the high blood pressure drugs, the entire heart disease market comes in at £115bn (£11.5m + £52.31m = £63.81m). Hawthorn also shows cancer and liver benefits, diabetic and anti-inflammatory uses, all accounted for already, so no double-counting, but no reason not to double up on the nutraceutical ingredients, which provide increased nutrition as opposed to the dangerous side effects of doubling up on pharmaceuticals. There will be revenues to come from foods, snacks and drinks, with industrial uses being found for the seed oil—all excluded from the tally.

* * *

Completing the accounts for the Green Currency is blackberries, which surprisingly had a limited amount of folklore but had attracted some hard-hitting scientific research across the broadest spectrum of illnesses seen for the seven plants. Having already accounted for cancer in dandelions, this does not discount blackberries and other fruits from the rose family taking a more central role in cancer prevention and curing. Its effectiveness against herpes (HSV-1), which had revenues of £4.6bn (£460,000 + £63.81m = £64.27m), is a safe deposit, although its benefits to Alzheimer's have already been calculated with nettles, so probably another formulation in the waiting, as the variations are huge among these seven, let alone the rest of the plant kingdom. The benefits of blackberries in the diabetes arena is a sector with revenues of about £50 billion a year was already accounted for with yarrow, a figure eclipsed by the weight-loss industry with revenues of £55.5 billion (£5.55m + £64.27m = £69.82m), where blackberry's phytonutrients did excellent things with fat, fat absorption and distribution. Finally, as it beat the antioxidant Trolox in trials, which has revenues of £6.2 billion per year (£620,000 + £69.82m = £70.44m), we get our final figure—over £70 million, a significant amount of money to consider when establishing a new method of exchange and barter, if not a currency.

A Green-backed economy

By only adding a few of the potential pharmaceutical revenue streams and only adding a one ten-thousandth of the current revenues and ignoring all potential incomes from foods, fertilisers, textiles, building, plastics, paper and energy, this £70m is very conservative. With more of these plants being grown, new revenue streams will be opened up from the processing of the plants grown across a countrywide network of permacultural smallholdings, extracting and separating usable parts, then being in control of the next stage of production—the thousands of products from yarn, or weaving the fabric into clothing, blankets, bags, twines and ropes, as well as making a variety of health and food produce from tinctures, balms, creams, syrups, pills, oils, breads, seasonings and soups, even hemp and nettlecrete bricks and blocks. It's imperative that cultivation, processing and production become more local and with more people leaving the cities to fulfil this repopulating of the countryside, local becomes a much larger market, assisting the longevity of local enterprises, all using the trusted, stable and strong Green Currency.

Once this 1 in 10,000 is achieved, with greater customer awareness, education and word of mouth, the market share could easily be ten times as much. With proper law and regulations in place, the biological poisons should no longer be produced, meaning the future market share will potentially become 100% for truly sustainable and safe products. Putting all the previous figures in some context, the entire annual global revenues from pharmaceuticals is currently about £1.2tn, up from £317bn in 2001—and £1.2tn buys a lot of ammunition in the war against herbalism.

The actual sums are immaterial, particularly as the revolution of our cultural systems will direct emphasis away from money, losing its importance as business focus becomes 'small, independent and local', as opposed to the brutality of 'immense, dependent and global'. The intention is to merely show that the figures involved are very large and all of this is potentially available to be redirected towards more ethical and sustainable small businesses.

Here's a shocking fact underlining how inherently sexist the world of business and money is—it's been estimated that the wonderful women of the world put in about 12.5 billion hours of unpaid care work every day, which amounts to a contribution to the global economy of at least £8.3 trillion every year, which is equal to the second largest GDP, China or three times as large as the entire global tech industry. It's worth a mention at least.

3. Industrial system

Clearly the industrial system is phenomenally complicated and convoluted. An essential component of this reformation in the industrial system once again revolves around not just the changes in machinery and raw materials, it also involves this change of mind, one focused on sustainability, small and local and spreading the wealth, not concentrating it. Why do we continue giving profits to polluting industries with a clear deficit in morality, as opposed to sustainable, local and family-run businesses with environmental conviction?

Here's a brief rundown and recap on how the seven plants may be of assistance in just a few of the industrial sectors.

Building

The stalling in the building industry to implement any sustainable or environmental practices beggars belief, considering that back in 2005 new houses were supposed to be carbon neutral by 2012. Environmentalism and any greening have been ignored in the main, flying in the face of all the existing hempcrete developments, which exceeded predicted energy savings, yet still most new housing developments have zero attempt to incorporate any new or sustainable technologies within them.

Hempcrete was looked at in some detail in the Cannabis chapter, with its possible role in replacing a portion of concrete's 17 billion tons produced every year (that's the total of concrete's ingredients, mainly sand, stones, clay and lime). Concrete is not even the best building material, it's just the cheapest, and that's all that matters in our skewered economic world. What hasn't been factored in as a cost of concrete is the annual 2 to 4 billion tons of carbon dioxide (from a global total of about 40bnt) and other greenhouse gases emitted as a consequence of its production. Hempcrete and nettlecrete would not eliminate this figure entirely as limestone is still used, and it is the crushing and heating of the limestone that uses as much as half the energy requirements—although the temperatures are significantly lower for the plant cretes, and when used with recycled crushed concrete overall CO_2 emissions should be halved.

It also needs to be understood that a significant proportion of the concrete poured is not essential, and when the new sustainable/ecological mindset is employed (not profit-driven), the requirements for concrete will also be at least halved. It's worth reflecting on the fact that many massive building projects serve little more than massive pay packets for those constructing them, who also pay 'backhanders' to fellow tie-wearers, hand-shakers and 'party donors', the high-speed rail network HS2 being a perfect example.

The lack of raw materials has been once again the stumbling block here, as technology has brought us not just blocks, bricks and cretes, but also now 3D printable hemp houses, with hemp insulation and all carpets, curtains and furniture coverings also being made from hemp.

Fabrics

Fabrics, as we have seen, are used for a whole range of purposes, from bags, bed sheets, blankets, carpets, chair coverings, clothing, curtains, interiors of cars, boats and trains to toys and dolls, towels, tents, parachutes, ropes and nets. Hemp and nettles provide a perfect source for fabric, with brambles and yarrow also good; all grow in a broad range of environmental habitats, doing so with limited water and pest control required, with the capability of maturing in dense swathes and the added bonus of the textile utilisation being just one of the revenue streams available from a small field of nettles, hemp, yarrow and brambles.

There is no call to replace the destructive monoculture of one crop with another. The answer is not going to be enormous fields full of these seven plants (or any plants), but lies more along the lines of an intricate patchwork of smallholdings collectively producing a huge amount of raw material. Producing longer-lasting clothing and fabrics ultimately means that less needs to be made, which will become tied in with the change in public perception to consumables. A £35 t-shirt that doesn't lose its shape or colour for years, while enabling all in its chain of production to be paid a fair wage, encompassing a multitude of environmental and social benefits is way better than the cheap rubbish available at the (not so)supermarket.

For some reason there has been, as with all the other sectors of industry, an initial high interest and PR storm about hemp products from clothing to houses and then little follow-up. Again it's due to a lack of raw materials, but when hemp gets fully legalised there will be an upsurge in 'waste' from the high-value initial recreational and medicinal uses, creating many tons of usable plant waste for other industrial sectors, notably the building, plastic, fabric and ethanol industries.

Transport

With recent high-end sports car innovations in carbon fibre and similar composites such as graphene, it is now possible for components from the seats, chassis and even parts of the engines and gearboxes to be made from plant sources, with the tyres made from dandelions, the panels and seat coverings etc. made from vegetable plastics and fabrics, and the engines fuelled by bioethanol or vegetable oil, although the future of such engines will change also. There is much to do in the public transport arena.

The stagnation of innovation in the entire transport industry is staggering, with little to no advancements made in over a century until the heavily publicised electric cars were introduced—with electric cars being as unenvironmental and unsustainable as a regular car. Those batteries are no more ethical than the early sugar and cotton industries, using rare resources and requiring huge amounts of energy to be made, although of course the biased and erroneous reportage of their planet-saving capability is what the public are told and would probably rather hear. Not the fact that the 500kg of batteries for an electric car contain around 12kg of lithium, 30kg of manganese, 15kg of cobalt, 100kg of copper and 200kg of aluminium (batteries currently are usually nickel-manganese-cobalt oxide or lithium-ion). To counter the unsustainability of electric cars, there is a new utilitarian solar car (Sono Motor's Sion), which, although still partly electric, can charge fully from solar alone and has a range of about 100km over a week or 300km+ if electrically charged. It's a start at least. Or a restart, considering that Japan's Tama Company produced a battery-powered 2-seater half-ton truck with 4.5bhp and 500kg load and a 4-seater 2-door passenger car capable of 35kmh and a range of 65km as long ago as 1947.

Of course, electric cars need to be plugged in to charge, and charging often comes from a power station run on anything from wood chip to gas and coal or nuclear, completely negating any environmental advancements. There are some alternatives now for energy providers who don't use oil, coal, wood chip, nuclear or fracked gas, the number one among these being Ecotricity, who provide both electricity for home and on the move at service stations. Ecotricity's owner Dale Vince is a definite character, setting a very good example—he also runs Forest Green Rovers FC, the country's and probably world's only vegan football team, whose home ground is recycled, wooden and zero-carbon. Other notable companies include Octopus and Good Energy (many others have been squeezed out of business during the current collapse) who are worth switching to.

This is not a long-term solution but it is the best way in the interim to wrest control away from the unsustainable petroleum polluters. Be warned of the current fashion by power generators of claiming their output to be 100% regenerative, which is classic greenwashing as they are burning mature trees for electricity, including virgin forests from the Americas, the very same electricity charging all those electric cars also being sold on a false pretence of being environmental—why on earth do we need pylons stretching across the countryside, sending electricity hither and thither when we have ample technology to locate energy sources where required? It's the whole mindset requiring that change, because energy will be cheaper in the long run if individually and community owned, with little to no changes needed for the conversion to switch off from the grid and guarantee a local, secure supply.

* * *

There is another development, heavily financed by Shell, namely hydrogen fuel cells, with pressurised hydrogen available from (selected) petrol stations. This, though, is definitely not a long-term solution, because the hydrogen has to come from somewhere, currently from natural gas—a finite resource and obviously controlled by the petroleum industry. It could also be fractionated from water, although again this is a finite resource with none being replenished, and just as with petroleum, there will be a point in the near future where wars are fought over water. Even though there are innovative energy-free ways to obtain the hydrogen from salt water, there might well even be more innovative machinations in the not too distant future that are capable of extracting hydrogen from some form of methane and fungus exchange. Currently there are already three manufacturers making hydrogen-powered cars—Honda, Hyundai and Toyota—with some rather sporty models to be launched soon.

Some of the biggest changes to the transport system have already been discussed. It is more than just swapping the fuel and oil to run on plants—there will be a new generation of batteries made from hemp; vehicles will be charged from the propulsion and braking of the vehicle; solar panels will be placed across the roof, bonnet and windows; and tiny wind turbines and fans across the front and side funnels and efficient dynamos attached to the wheels/axles. The only input in the very near future will be the manufacture of the vehicle itself and any subsequent servicing and tyres. The carbon footprint of a standard-produced car (figures are more for electric cars) is in the region of 10 tons, and with 75 million cars made every year this equals about 750 million tons of CO_2 emitted just for car production. Cars are only the best in an economic system, but as part of an environmental system they are the worst. There needs to be more than a switch from petrol to vegetable oil, as future transportation will use neither.

The main innovation we have yet to see will be when there is a radical change of how propulsion works, where a totally different system emerges, such as magnetism or sonic. Such technology may well have already been invented, being kept under strict wraps by the industrialists.

Petroleum

The petroleum industry has done a magnificent job in making and selling a whole world of stuff from their raw material, with congratulations due for a job well done in the business world. In many respects petroleum is one of the main components of modern capitalism, refined and fractionated into a multitude of other chemicals and often being the main ingredient in products as diverse as adhesives to agrochemicals, condoms, vinyl records and surfboards—indeed, quite a few of the estimated 45.72 billion different 'things' on our planet will be made from petroleum plastic. The average capitalist citizen consumes products made from petroleum at the astonishing rate of 3.5 gallons of oil and 250 cubic feet of gas every day!

Irrespective of the fact that for most people petroleum is only considered as the stuff that makes the car run, which is itself an unfortunate connection as motorcars weren't initially intended to be fuelled by polluting petroleum. The first engine designed by Rudolf Diesel (patented in 1893) was to be fuelled with peanut oil, although his first design made in 1897, owing to various unfortunate complications, ran on petrol, with the peanut oil model presented at the Paris World Show of 1900. Another surprising fact about the manufacture of cars was the unveiling in 1941 of a car made from plant matter (apart from the metal chassis, engine and gearbox), mainly pine fibre, ramie and hemp, a product of the Soybean Institute funded by Henry Ford. Taking twelve years to develop, it was launched in 1941, with a dramatic publicity shot of Ford himself hitting the bonnet with a large hammer. This produced no dent, demonstrating the strength of a plant car, which was 25% lighter than the rest and fuelled of course by hemp oil.

This threat to the motorcar industry and its source of fuel was undoubtedly another nail in the coffin for hemp, because hemp and cannabis were of course criminalised in 1937 in America (and soon thereafter in much of the world), making a hemp car a complete non-starter. As seen in the Cannabis chapter, its versatility made it a threat to the paper, textile, medicine and food industries as well, so the possibility of cars being made and run on hemp was just never going to happen with so many industries having their toes trodden on. The fact that plastic can be made from hemp was not an issue in 1937, although the extent of DuPont's involvement in the debacle is not known, never being far from controversy, as it was they who emerged after WWII with many of the patents for various plastics derived from petroleum.

Over ninety years ago, the largest car manufacturer of the time came up with an environmentally sustainable solution that we are supposed to believe can't be equalled or even bettered today, although some manufacturers such as Mercedes and Porsche have incorporated plant plastic panels on a small scale into some models. It's not a technical hurdle that is stalling uptake and development, as plant plastics were used in the iconic East German Trabant car with its panels of duroplast (cotton and phenol), with 3.1 million cars being made from 1957 to 1991. A genuine hurdle is and has always been the lack of any guarantees of a constant supply of the raw materials, whether that be seeds for oil, stalks for ethanol, both nettle and hemp for fabrics and plastic composites, all still severely hampered by the law today even in legal States and countries worldwide, with licences to grow hemp extremely thin on the ground—with clear nepotism and shady manoeuvres, as seen among the Tory party in the UK—even if, and possibly because, this plant can generate billions of environmentally sustainable £€$s.

* * *

The fact that the petroleum industry is basically the same players as petrochemicals and plastic is not so shocking, but throw in the mainstay to what we erroneously thought was a health industry devoted to improving our health, whose only concern, it appears, is the legal concern to make profits to pay dividends to shareholders, then there has to be concern. Not least the whistle-blowers showing that cures and health have been absent from the 'business' agenda since the 1950s.

Somewhat worryingly petroleum can also be found in the food we eat, the agricultural system, cleaning, bathroom and beauty products. Food products with preservatives in (which is all processed foods) will undoubtedly have originally been crude oil, as also would many of the colourings. In the agricultural system, petroleum-derived products include not just the plastic in the polytunnels, twine, tubing and wrapping, it's also the backbone of fertilisers and pesticides, and the fuel to drive the tractors, crop sprayers and delivery trucks. Then there is the (not so beautiful) beauty products and other bathroom products all derived from petroleum—with sodium lauryl sulphate, found in most bathroom products and sodium saccharine in toothpastes, chewing gum, food and medicines, both derived from petrochemicals. It's everywhere.

The mere fact that petroleum has managed to ensnare so many other industries in its unsustainable, unenvironmental and unhealthy fake rainbow sheen is bad enough, there are barrels full of other problems with this dodgy devil. One is the uncomfortable fact that for the past 40 years and no doubt for decades before, Britain and of course other nations have fought over secret contracts with oil-producing nations to secure uninterrupted supplies, as oil not only lubricates the cogs of industrialism, most of the cogs are made of oil as well. Clearly it's important for a society dependent on petrol and petrochemicals to secure its supply. Only it turns out we've been swapping it with the Saudis for weapons, to the tune of a whole army's worth of equipment and knowledge—far more equipment than could ever actually be used by such a small nation's army. Billions of pounds swapped for oil, and all done on the false promise that these most blatant weapons of war would only be used for 'defence'. I wonder how the parents of innocent Yemeni children killed by this defence feel? These countries with some of the worst human rights, we are supposed to believe are led by close friends of King Charles III. To them it's just good business, proving it's all down to individual perception, and I personally have no idea how the CEO of Exxon or BP views the world and their business practices.

What is the view of the CEOs of Chevron and Texaco about decimating the Amazon, or ExxonMobil in Indonesia, Shell in Africa? Do they not see the ravaging of the indigenous peoples and animals with their oil wells, creating access roads, drilling and allowing constant oil leakages into all rivers, becoming so heavily polluted as to be unusable for potable water and unliveable for wildlife? Or do they just see the money? What is often a previously pristine setting becomes wasteland after the oil corporation moves in, makes the money and leaves, without cleaning up the barren, toxic environment. What can the impoverished locals do? Whose health coincidentally has also taken a nosedive, with cancers and birth defects and crops devastated, while the lawyers, a world away, stall, lie, divert, appeal, violate and mitigate. All the oil companies have done this globally as it's their economic business model.

The above is just a tiny speck of the flotsam on a wave of scandal that is the oil industry. It goes without saying that the sensibility of putting petroleum in our food, medicine, crops, bodies and plastic is not common sense, just economic sense. With the intention of this book being a return to *oikonomia*, the whole perception of what makes economic sense will have to be viewed in the future along the lines of environmental sense and moral sensibilities.

* * *

The oil industry is currently worth somewhere in the region of £1.32 trillion every year. This is of course just the beginning of the web of other industries, such as the automobile industry, with much of their output still dependent on petroleum, bringing in revenues worth about £2.3 trillion every year, while the plastic industry is worth about £500 billion every year. Plastic is taking a lot of deserved flak recently, although the problem is not with plastic *per se*, the problem is with polluting petroleum-based plastic, and the ridiculous overuse of it. The good news here is that there are already many compostable and safe plant, seaweed and fungal plastics for use in packaging and wrapping, and with further developments and incentives the range will expand to cover all requirements, putting a large dent in the petroleum requirements.

At the current petroleum usage levels there is no way bioethanol or vegetable oils could provide a substitute, being beyond the agricultural capacity. To continue with this monocultural model—enormous fields of a single crop, involving huge machinery compacting the earth, spraying deadly pesticides and artificial fertilisers, and using sterile seeds—while erroneously assuming a change of the crop, such as hemp being grown will magically provide a solution, is by its very definition, madness.

Obviously if no oil was available there would be a massive hit to the global economies and the obsession with GDPs, hence why the Western countries have been falling over themselves to guarantee continual supplies instead of rolling out alternatives. As countries are run on economics, with numbers being more important than people or their health and welfare, these enormous revenues are seen as 'untouchable' and the very foundations of national security. This has allowed the oil industry to have an iron grip on legislation and use governments as puppets to prevent alternatives, while even establishing educational facilities and forums to maliciously spread false science to ameliorate their polluting guilt and greed. It is after all the petroleum industry that is responsible for the vast majority of greenhouse gases in the environment, notably electricity production and transport, since other sectors have remained roughly equal since 1990. Far more important than swapping one oil for another oil is swapping the machines requiring the oil for machines requiring no oil, be that the next generation of batteries and solar and wind turbines, and we haven't even started with sonic technologies.

With a change in consumer spending and habits, with associated taxes and surcharges placed on unsustainable manufacturing products, such as making short-lived plastic tat from China more expensive, while relaxing business rates and taxes and giving subsidies for sustainable and local manufacture, there would be a massive swing away from cheap superfluous consumables. We must hold manufacturers to task over planned obsolescence, where products have to be designed to last longer, repair more easily and with more extensive warranties. There will be a reformation of global conglomerates, with mass downscaling, while at the same time allowing more people to be employed within the sectors. It's not a Luddite uprising, but it will be an uprising of artisans and quality, with all sustainable innovations and technologies being incorporated, producing a revolution of quality in build, use and design. Perhaps most importantly, the changes will genuinely be better for the planet and health, as well as re-establishing not just rural and urban communities but also human's connection with the entire planet.

4. Agricultural system

The great agricultural myth

Being an advocate of wild plants and a keen forager and botaniser—walking slowly and observing the plants—it is perhaps only in keeping that mine and Mother Earth's (if I may be so bold as to speak for Her) natural foe would be monoculture. Vast fields of wheat or rape represent a green desert of denatured soil with little to no natural biodiversity, and are nothing short of a despicable despoliation of the land. At the same time, of course, the very same fields are seen by many as they drive through from the cities as being bucolic, a sure sign of fecundity and productivity from our green and pleasant land.

It can't be both though, can it? It's either a desecration or a consecration, although of course reality is no more than an individual's perception. After 20 years of direct observation, 35 years of walking and cycling in the countryside and 16 years of research, reading and writing, on top of studies when at university long ago, here's my opinion—

The agricultural system is not the saviour it is portrayed as, being ultimately more devolutionary for *Homo sapiens* than evolutionary. The first revolution undergone by humans was the Agricultural Revolution, when we switched from being a gatherer of the abundance of wild foods to growing just a few varieties. This played out at different times across the globe, starting about 10,000 BC in some parts of Western Asia, South East Asia and possibly South America, and was a revolution not felt in Britain until about 4000 BC. This also meant a narrowing of the nutrient intake, as a varied wild diet includes a vast spectrum of phytonutrients, whereas with wheat and meat it is far more limited. Only relatively recently have hunter-gathering peoples been so marginalised in all aspects of their existence, so any modern comparisons are unrealistic with the scenario millennia ago when agriculture began to push these peoples further away. No doubt agriculture was initiated as climatic conditions compromised the natural abundance, as food security is paramount, although as it spread across Europe the dearth of wild food may not have been the main consideration for the introduction of agriculture. The study of archaeo-anthropometrics (the study of old skeletons) clearly shows humans became smaller and sicker after the Agricultural Revolution had swept through.

The domestication of plants and animals has been sold to us as being one of the most important events in our social evolution, on a straight trajectory to civilization and ourselves today. Without agriculture—a culture based around agrarian endeavours—we would have none of the luxuries available to us today. Although the truth is, when we domesticated plants and animals, we also domesticated ourselves, and this is when power and prestige were ceded to a tiny minority. Thereafter, those who controlled the grain controlled the people, with towns and cities being the perfect place to breed as many humans as possible to work all their lives to make the few hugely powerful and rich. Thus began millennia of social engineering—just because there isn't a government office with "Ministry of Social Engineering" on a brass plaque, doesn't mean there isn't one—preventing the people from seeing the truth, about how they have managed to legalise their immorality, wealth and land.

* * *

How can we survive and feed a population of eight billion without agriculture? The current agrarian culture of monoculture, as we know from modern reports, is not only unsustainable, it causes more harm than the food it delivers, as confirmed by the 2019 International Food Policy Research Institute, which alarmingly stated: "the cost of the damage caused by agriculture is greater than the value of the food produced".

This just screams for a new culture of food procurement, and there is one, with a positive-sounding name and a system of growing food that is in harmony with the land it is grown on, replacing unsustainable monoculture with sustainable permaculture. Simple as that, and later in this chapter we will see how this can be achieved, because living in a world governed by an unsustainable erroneous agricultural system makes the problem a little worse every year and a little more irreparable, just as with all of the other cultural systems, making that reformation more urgent each and every day.

A lesson from history has to be that all those great civilizations from the Sumerian to Babylonian, Egyptian, Minoan, Greek, Roman, Shang or Mayan, are all from history, no longer, finished, dead, kaput! For those living in the middle period of these great civilizations, collapse was unimaginable, so rich, great and better than the rest did each one see itself. We today are no different. Although all these civilizations were brought to their knees when the food sources from the outlying regions of the civilization no longer arrived, be they intercepted or not successfully harvested owing to climatic disruption and then, in the blink of history's eye, civilization gone. Which means the most important thing is to have food security and that security, particularly for an island, cannot consist of a reliance on imported food. The Security Service, MI5, think we are four meals away from a revolution, Lenin said chaos was three meals away and many others put it between two and nine, which means after a couple of days with no food, anything could happen.

We currently have all the knowledge and technology to create a completely sustainable, detrimental chemical-free cultural system supporting 10 billion people. As education improves and becomes relevant, with this sharing and ethical culture taking root, birth rates will plummet, as will population, and alongside the depopulation of cities back to the land, world population could easily fall to 5 billion in just a couple of generations (50 or so years), in direct contrast to the many models that predict at least 10 billion by 2050. Whether it's 5 or 10 billion, they'd all be better fed from a massive reduction in the oppressive international food trade, a surge in permacultural smallholdings, consuming more vegetables, less meat and processed food, which would also vastly improve general health.

Keep it small, keep it local and keep it sustainable—this simple mantra will guarantee continuance, abundance and a fairer, more caring world. The human race has gotten so out of balance, we neither care for ourselves (look at the amount of obesity, depression, alcoholism and drug dependence, either prescribed or illegal, hardly suggesting a balanced disposition, more like someone who is at dis-ease) nor about our neighbours, the environment or anything else unless it feeds the ego, often through a rush of dopamine released by fats/sugars/drugs/likes on social media. When we start to care about ourselves, our health and where this comes from, then the connection is made that it is the environment that brings us wellbeing, it is the food from the soil that nourishes us and it is other people that make us happy. That's all we need.

* * *

It was the Green Revolution that introduced/sold this monocultural system, and in so doing it indebted the world's farmers with an unnecessary short-sighted money-making scheme, which will in only a generation or two leave the world's soil near to barren (many argue it is this already). In just the three generations since 1950 when the expensive machinery, irrigation systems, agro-chemicals and sterile seeds enslaved the world's farmers, the precious soil experienced an unprecedented barrage of what has become another mantra of the book—spraying the deadliest chemicals ever invented by man(not so)kind directly onto our food for 70 years, while illnesses once rare have become the norm, the diseases of civilization.

The Green Revolution was a multi-pronged attack on squeezing all riches from the less greedy world. With fewer people needed to grow the mechanised crops, the infrastructure of countries imploded, making poorly paid factory workers of millions, producing the consumer products for the West. The Green Revolution was never about feeding the poorer nations and avoiding starvation … duh! It was all about feeding the rich nations cheaply and providing biofuels, while enslaving the poorer nations to make our useless tat, emptying our pockets while making a few almost bottomless. The results of the Green Revolution (beginning in 1950 with the pre-war science reaching a peak in the late 1960s) wasn't excess food for the world from these acclaimed high-yielding grains and chemicals, paradoxically it appears to have only brought starvation to the poorer nations.

The Green Revolution was a classic smokescreen for an ulterior motive of economic dominance, by converting self-sufficient countries into economically enslaved ones with a debt-trap. Selling monoculture included the sales of massive machinery, tractors, combine harvesters etc., as well as yearly seeds, fertilisers and pesticides by way of massive government loans. This was all meticulously planned for at the Bretton Woods meeting in New Hampshire, USA in 1944, when 'austerity' was fictioned-up and slowly spread like a plague to enslave humanity; this meeting also saw the establishment of the International Monetary Fund (IMF), World Bank and, in 1948, the General Agreement on Trade and Tariffs (GATT, becoming the World Trade Organization in 1995). These seemingly benign organisations, advertised as a form of charity, where the rich lend to the poor to enable the poorer nations to enter into trade agreements to free them from impoverishment, is a load of bollocks—all it ever was intended for was economic slavery, where the IMF, sold as a lender of the last resort, established Structural Adjustment Programs (SAPs) for poor nations to either pay for previous loans or future ones. As soon as a country signed up for a SAP (over 70 nations did), this is what the IMF enforces the debtor nation to do—sell off all national industries and resources (usually to a foreign corporation and government), prioritise debt repayments and corporate welfare over infrastructure investment and human welfare, demand suppression of wages and restrict unions, and dictate the level of public spending. All of which are diametrically opposed to what may be considered a charitable action. Even less so when one considers that for every £1 given as financial aid, £5 comes back as interest from those SAPs.

* * *

The dramatic increases in crop yields experienced from the first years of the Green Revolution soon slowed up. With the introduction of new crop varieties and new pesticides the bugs had no chance for the first few years, giving a few unadulterated increased harvests. Soon enough the unnatural monoculture experienced problems, with depleted soils, denser stands, multiple cropping and genetic uniformity being an invitation to pests and fungus, leading to more pesticide usage. Just 10 years after the much-hailed and loudly trumpeted success of the Green Revolution, its collapse was not quite so loudly reported.

The Green Revolution ripped out and replaced a timeless, sustainable food procurement system (smallholdings) with one ruled by overseas profits, politics, trade deals and debt, a system uninterested if the locals starve, making it actually illegal to feed themselves before loan repayment! Switching from sustainable slash and burn and crop rotation to monoculture suddenly left the soil exposed to a great rate of soil loss, further compounding the drop in yields after just a few years—a fact known about for decades if not centuries, with studies from the 1970s in Iowa showing that for every acre under monoculture a staggering 10 tons of top soil was lost!

In Australia topsoil is being lost at about a tonne per acre. With thin soils to begin with, this equates to a 1mm loss per year, which soon adds up, or rather is lost forever, as it takes roughly 100 years to replace this 1mm naturally. Soil loss is just one of the many side effects of monoculture, alongside loss of biodiversity, extinctions, human diseases, algal blooms and deoxygenated oceans. The Global Soil Partnership estimates that a staggering 75 billion tonnes of topsoil is lost from arable land worldwide every year. Using monoculture and other non-regenerative practices, soil, just as with water and petroleum, is a non-renewable, which will only make the inevitable crash of unsustainability harder and faster. Simply keeping soil covered with plants prevents a large proportion of wind and water soil erosion, significantly compounded by agriculture as merely ploughing the soil destroys most microbial life and the glomalin glue, exposing the soil to water and wind erosion, with just exposure to the air being sufficient to kill the microbes, releasing carbon and nitrogen back into the air.

It was ignorance about soil, its accumulation, deposition and erosion that seriously confused me at university. Now it's no longer a mystery and from understanding the soil another great mystery has possibly been answered. I was fortunate to study at UCL, which is conveniently next to the British Museum (BM) and the vast library at Senate House. These places provided me a refuge from the mostly erroneous dogma I was being force-fed at the Institute of Archaeology. Within the BM there are Sumerian friezes depicting muscular winged gods holding bags in a specific manner (a yogic mudra); they also appear to be holding pinecones and wearing watches, but the last two weren't the source of the intrigue, even if they were extremely peculiar. It was those bags, and near identical ones can be found as far afield as South America, Greece and Egypt and extending back into time to about 10,000 BC at Gobekli Tepe in Turkey. What is it with these bags? Understanding soil now and its microbes provided a possible answer as to what could have been in the bags, and the answer to a real fundamental question: what is required to start a civilization, a culture and remembered peoples? Besides water, soil is the essential prerequisite, particularly a healthy soil with all the microbes and mycorrhiza, where just a cubic centimetre would in only a few years improve the fertility of vast swathes of land, making it possible to provide for a growing population (with some suitable seeds also stashed in the bag).

Feeling somewhat smug about having a possible answer for those bags, it also proved to be true that it was the soil that provided the answer for sustaining a sizeable Mayan population in South and Central America. I was told on my course that the Amazon rainforest was (a) vastly old and (b) had a low carrying capacity (it could support only a few people from its limited calorific load). It was not strictly true, because the ancient inhabitants had significantly improved the soil by chopping the trees down and adding biochar to the soil. This was *terra preta*, Portuguese for black earth, made from charcoal, bones and other organic matter, which centuries later still provides added fertility. Biochar is currently being championed by Craig Sams of Green & Blacks fame with his Carbon Gold; he also opened Britain's first macrobiotic restaurant in 1968 and founded Whole Earth Foods, and is a truly lovely and wise fella. Investigations have also uncovered a similar biochar practice in West Africa, proving that the answer to food security is, as one would have assumed, the soil. We'll return to the soil in more detail in a couple of sections.

* * *

Back to the demise of the Green Revolution. Throughout the 1970s it became obvious that trade agreements, immoral contracts and pest-infected new crop varieties were the reason for starvation, with droughts being a convenient excuse, when in fact even during drought years some of the poorer agricultural nations had exported *more* during famines. By growing only a narrow variety of crops solely for export and leaving less and less for the locals, with such low wages, starvation is the only outcome. Droughts and floods are commonplace, yet starvation only took hold after the Green Revolution, with Westerners being sold the story that impoverished Third World peasants and farmers had found themselves once again on the brink of ruin and starvation owing to their uneconomic backward ways …

Of all the more visible and obvious polluters desecrating the soil, air and water, it is modern agriculture that is surprisingly the single most destructive human activity. There is not one aspect of monoculture that can be said to have been sustainable, with no proof we are better off today thanks to it. The world's most valuable and essential commodities for our survival and optimum health are the soil, water and air; today we've got shit soil, shit air and shit water, and we've also got shit health—it's not a coincidence.

It's all well and good writing about the agricultural system being dead and in need of being replaced, but we all know *actions speak louder than words*, so for the past two years I've had my hands in the soil, even made my own soil to prove this point—that monoculture is definitely not the best way to provide food for the people.

PermaLife 1:0

I took it upon myself to learn the real way, by finding a little bit of land to grow some vegetables on. I was recommended to check out permaculture before getting stuck into the land, and immediately recognised it as a modern remix of many of the principles laid down by the Henry Doubleday Research Association and the Soil Association in the late 1960s and early 1970s. They produced some excellent publications as a reaction to the damage being caused by the agro-chemicals and monoculture, with much on good composting, pamphlets on which I had been given years before by my mother-in-law (who also painted the cover of this book—thanks, Kari, you're a star). PermaLife 1:0 had a planting and growing regime that allowed for some of the plants already there to remain, and after the second year, the growing felt as if some kind of a symbiotic relationship with the plants and the soil had been formed, allowing what wanted to grow to grow within reason, increasing the biodiversity, encouraging companion plants and with it the range of predators for the pests. Although not particularly green-fingered, I wanted to empower myself with the knowledge and practical skills of being able to grow my own food. I needed also to demonstrate that not only was monoculture the bane of the countryside, but permaculture, or good allotment/organic sense, was a viable, sustainable, more productive and healthier option, accessible to all skill levels, and which can easily feed the world while replenishing the soil and our health.

I had a total harvest of just under 200kg each year, on what amounted to a twentieth of an acre (50m^2), where the following chemical-free, vegetables were grown—a range of potatoes, butternut squash, cucumber, tomatoes, marrow, beetroot, courgette, runner beans, broccoli, onions, lettuces, quinoa, radishes, chard, sorrel, spring onions, leeks, garlic, chilli, peppers, curly cress, land cress, chickweed, buckshorn plantain, broad beans, aubergine, physalis, raspberry, blackberry, basil, mint, rosemary and oregano.

The last of the potatoes had been eaten by December and by the end of January the butternut squashes and beetroots had all gone; in total our land provided less than 20% of our yearly vegetable intake for the five in the house, which I considered a failure, but it was at least 200kg of healthy food—and the aim was not to grow as much as possible but to learn as much as possible.

This 200kg would equate to four tons per acre if scaled-up. For the figures used in a later section of this chapter this four tons has been doubled to eight tons per acre for three hopefully justifiable reasons—firstly, fruit and nut trees were absent; secondly, help from someone with horticultural knowledge would have increased yields significantly; and thirdly, there was absolutely no attempt to maximise coverage and yields, the focus being on learning and observing. Judging by some of the local allotments and smallholdings, 20 tons per acre is easily achievable from skilled growers, so eight tons per acre represents a very achievable harvest by any person willing to get their hands dirty.

There is a permacultural very small smallholding of only a tenth of an acre just outside Los Angeles (urbanhomestead.org) where the growers produced 3.175 tons, were 97% self-sufficient and turned a revenue of $20,000 from the excess produce. This is an extreme intensive case, but it would be a whopping 31.75 tons, plus $200,000 revenue if scaled-up to an acre. This represents the upper benchmark for potential yield, and it would require full-time devotion and probably more than two people. So if my 200kg from 50m^2 represents the lower benchmark—which is actually already on a par with mainstream agricultural productivity—the range in yield is huge, depending on soil fertility, competence, time and climate.

PermaLife was a closed system, with nothing brought in (apart from some seedlings raised at home). All the soil was made from composts of kitchen scraps, grass cuttings and leaves, with fertilisers made from nettles, comfrey and yarrow as a green liquid feed, and some local seaweed added to the soil also. The water was collected rainwater, with pest control being companion plants and ash around the bases of plants—a pond for frogs and toads, a few ducks and a family of hedgehogs would have been ideal, as slugs were the biggest problem.

PermaLife enforced the first precept to good plant health and that's soil health, as without this anything grown in it will be deficient. This was demonstrated by the chain of problems caused by demineralised soil, culminating in our own ill-health, as explained in my last book, *The Seven Deadly Whites*, which ended asking the reader to reflect upon how soil is a crucial element to human health because—

A nation is only as healthy as the soil it stands upon

Seven billion microbes, really small living things, microorganisms, the major groups being bacteria, archaea, fungi, algae, protozoa and viruses, can all be found in a single heaped teaspoon of a healthy soil. This is nearly the same number as the entire human population, except these microbes live in harmony with one another—when not confronted with man-made chemicals, that is.

The fungal part of that teaspoon consists of the mycelium filaments, the below-ground bit of what can be considered as the fungal body, with the observable mushroom above ground being considered as the fruit. The body can be truly enormous, with the biggest single organism on Earth being a type of honey fungus in Oregon, USA, covering 4 square miles and being at least 2,500 years old, maybe even nearer to 8,000. If the mycelium forms a symbiotic relationship with an individual plant's roots this is known as a mycorrhizal fungus (where *myco* is fungus and *rhiza* is root in Greek), distributing water and nutrients to the roots in exchange for sugars created via phytosynthesis. The fly agaric, *Amanita muscaria*, is a mycorrhizal fungus found symbiotically around the roots of several species of trees, with a preference for silver birch, acting as an intermediary between the plant world, the fungal world and the world of the soil. It is this symbiotic relationship that likely allowed for the initial forays onto land by the plant kingdom from the water some 600 million or so years ago.

The multitude of microbes found in healthy soil play roles within and without humans beyond what is considered conceivable; for instance *Mycobacterium vaccae* acts as an antidepressant and may assist with PTSD, but is extremely common within healthy soil. What is also often forgotten about is that it is the gut flora, bacteria in the main with a few fungal friends, that are essential in the production of many nutrients, such as the vital vitamin B12, or L-ergothioneine, an anti-inflammatory that may be crucial in unlocking doors to longevity by slowing the aging process. That peculiarly satisfying reaction to the smell of the land just after rainfall, petrichor, is actually a soil microbe being released, geosmin. And not only do microbes make rain smell lush, they might well be even responsible for the rainfall in the first place with their ability to 'seed' rain within clouds, forming the nucleus of a raindrop more effectively than inert dust can do—another damning effect of monoculture and its deficit of microbes in the soils, exacerbating drought conditions. Which gives some credence to the idea of a raindance, stirring up the dust, at least enough to seed a raincloud.

Microbes have been thoroughly vilified, with the public being fed by adverts and scary news stories of bacterial infections, deaths and wipe-outs, with antibacterial and antibiotic products sold to us as saviours, guaranteeing 99.9% death to all bacteria ... to all bacteria, oh really? When so many are so useful and so few are actually terrible, do we always need to kill all of them? It seems very likely that it was actually the soil that gave us the bacterium found within our human organism, and therefore all organisms, with their decision to coalesce ultimately leading to us today, so their exclusion today is a reckless and ominous omission.

Mycorrhizal fungus can alter/boost the flavour profile of foods. This is because the fungus has allowed for more nutrients to be absorbed and created, which means mineralised food tastes better, fuller and more complex—it's those phytonutrients imparting much of the flavour. Ever had a tomato in February from Spain? That lack of flavour is sadly not just absence of flavour, it's lacking what food should be all about—an efficient delivery system for nutrients. Experiments have already found mycorrhizal fungus improve bread wheats, tomatoes, lettuce, basil and undoubtedly whatever food plant is being tested—all will benefit from what was there for near eternity before man's interference. Different mycorrhizae can alter the flavour of the crop, producing distinct differences in taste, shape, size and appearance within the same strawberry species, for instance, while wheat grown in rich mycorrhizal soils produces bread with a fuller flavour (flavour is usually synonymous with nutrients). It's a vital network just beginning to be studied by modern science—being just as neglected as the study of the human gut and its bacterial and 50 or so species of fungal tenants.

* * *

Albert Howard (1873–1947), a pioneer of organic farming in the 1940s, saw it as paramount that it was on "the efficiency of this mycorrhizal association [that] the health and well-being of mankind must depend". Howard brought his ideas straight back from the agricultural practices he observed in India, writing such seminal books as *An Agricultural*

Testament in 1940 and *Farming and Gardening for Health and Disease* in 1945, stressing the danger of modern agriculture destroying the mycorrhiza. The success of agriculture is determined by this mycorrhiza, or at least it was …

Our understanding of this kingdom inhabited by fungi is, to say the least, limited. Those seven billion living things found in one teaspoon of soil could actually be an underestimate, and there could actually be more microbes in a teaspoon of healthy soil than all the humans who have ever lived, somewhere in the region of 150 billion. Either 7 or 150 billion, both are more than impressive. Or, as mushroom expert Paul Stamets puts it, "There are more species of fungi, bacteria and protozoa in a single scoop of soil than there are species of plants and vertebrate animals in all of North America (if not the world)."

Mycelium, the fungal filaments and strands, sometimes tough and cord-like and other times thin, fibrous and all but invisible, spreads out across the top organic layer of the soil (where abundant there could be more than a mile of mycelium in that teaspoon of soil), interconnected and composed of thousands of species all living in harmony. Spanning across the globe, it is this mycelium that has been referred to as the internet of the planet or the WoodWideWeb, although I prefer Gaia's nervous system, with rivers Her arteries.

The important part to remember is that there is a class of fungi whose role is symbiotic with plants and when left to flourish results in healthier plants, while fertilisers and pesticides either kill or make redundant these mycorrhiza. This breaking of a timeless symbiotic relationship has been an error by agriculturalists, almost certainly an economic decision as replacing the knowledge of this free source of nutrients and immune system with man-made sellable chemicals guaranteed continual future profits—running in tandem with the suppression of the knowledge of plants being key to human health, making human's dependent on sellable pharmaceuticals.

* * *

A thoroughly neglected aspect of the desecration caused by monoculture has been this mycorrhizal loss and the general loss of microbes in the soil, an unrecognised critical component. It's a double-whammy of woe because not only has the soil less minerals in it today than ever before, the loss of the fungal organisms that made the minerals in the soil biologically available to the roots of the plants have gone. Mycorrhiza perform the essential role of feeding the plants as well as optimising the immunity, fertility and health of the soil, just as Albert Howard said last century.

Only a century or two ago, rich soils on flood plains would have held about 10 times the carbon (1000%) of today, with it being the mycorrhiza sequestering the carbon to the soil, making all the trillions of little microbes, worms, roots and nematodes (being made from carbon). This represents a healthy soil, one in balance, maintaining its health with a full complement of minerals, which have also depreciated by a similar 1000%.

There have been stories floating about for quite a few years stating there are only sixty, twenty, twelve or five harvests left from agricultural soils. They all probably derive from the report in 2014 by Maria Helena Semedo of the Food and Agriculture Organization, stating we had no more than 60 harvests left. Sadly, it is far worse, because with no fertiliser added to any of the monocultural fields, the soil would struggle with just one more harvest! Yet for those regenerative farms, some already have 10,000 years of annual harvests to look forward to.

It's a broken system, with the soil merely being an inert medium to hold and grow a plant that is a shadow of its once former nutritious self. It is of course the soil alone that determines the potential health of any plants grown in it; there is no magic involved where all plants grown wherever in the world have the same nutritious content. If the soil is depleted of minerals, then so will be any plants grown in it and subsequently the food made from them; they will be less tasty as well—plants do of course also get sustenance from the air.

A couple of fifty-year-long studies from Australia and England clearly showed this depletion. The Australian research revealed that apples had lost 80% of their vitamin C from 1948 to 1991, and a staggering 99% of the vitamin A precursor carotene had disappeared from what it was named after, carrots. Mineral depletion in England showed a 76% reduction in copper, 46% decrease in calcium, 27% in iron and 24% in magnesium across samples of many vegetables from 1940 to 1990, with this loss having continued undeterred for the thirty years since these studies. The reality today is that supermarket vegetables contain at best a quarter, more realistically a sixth to an eighth of the nutritious content they did when our grandparents ate the same vegetables in 1950. This means we have to eat up to eight times the quantity to consume the same amount of essential nutrients. Sadly, today less fresh food is being consumed to begin with, a loss compounded further by denatured, processed, carbohydrate-dense food being on average 80% of the daily calorific intake—not forgetting the disruptive man-made chemicals. The food chain is thoroughly demineralised because of the poor soils and processing, and this depletion has opened the door to the diseases of civilization.

This demineralisation is either coincidental or causal in regards the past 50 years and the huge escalation in the diseases of civilization, coupled with a devolution of the human genome. In this period we have undergone a dramatic change in general health and number of illnesses suffered. Since the intensive and aggressive implementation of global monocultural practices the food chain has not only become dangerously demineralised, the chemical load of the food from soil to plate has risen significantly.

It's a complex problem, because as the soil gets more depleted, the food has less minerals, which would prevent this deficiency manifesting as a disease; this is made worse because the soil has become toxic from the agro-chemicals, and the best defence against these chemicals is the very same minerals now depleted. A notable and hopefully soon to be vilified and banned chemical is Monsanto's glyphosate—worryingly, the initial patent for glyphosate was as a boiler cleaner and descaler. A later one, US Pat. No. 7771736 B2, was as an antibiotic … an antibiotic! Give that some reflection, please, especially after all we've learnt about the association of the soil and human health. The comparisons between the health of the soil and the health of the human is stark, not least because human health is dependent on the health of the soil, bacteria in particular, and man-made chemicals have killed all of these and damaged the immune systems of both the soil and human and all biological systems in the process.

The unarguable consequence is that agricultural soils have little to no mycelium, and a teaspoonful of it wouldn't contain 7 million microbes let alone 7 billion, resulting in less mineralised plants dependent on the NPK (nitrogen, phosphorus and potassium) fertilisers. This in turn means there are far less phytonutrients available from the plants, and without these phytonutrients the human becomes compromised, as these phytonutrients are the essential co-factors enabling all the better-known nutrients to function properly. This deficit of phytonutrients compromises the ability of any organism's functionality, plant or animal, be that cell reproduction, hormone production and every other aspect of health and life. It's a catastrophic spiral of barrenness, involving the minute mycorrhizal hyphae to the phytonutrients and human cells—these all but invisible phenomena have a huge future role to play if humans are ever to get back on the road of true health, homeostasis.

Respect requires a mention of Paul Stamets and the fact that he has already mapped out a fungal future, with their ability (and his many patents) to be used in industry for packaging, plastics, clothing and furniture. There's also mycoremediation, where various species of fungus have been shown to consume and chemically alter (usually to an inert result in regards human toxicity) pollutants as diverse as pesticides, biological weapons, lignin, TNT, pharmaceuticals, nuclear radiation, polyurethane plastics and probably anything else thrown at it. Mycelium also filters water, the process of mycofiltration, being able to remove heavy metals and even bacteria such as *E. coli*.

Mycorrhizal fungus is most definitely one of the keys to a prosperous and sustainable future. If you have any interest in fungus, read or watch anything by Paul Stamets, or the modern pioneer in mycorrhizal studies Suzanne Simard's *Finding the Mother Tree*, or Merlin Sheldrake's *Entangled Life*. There has been limited mention of fungus for two reasons: one, they are not plants and two, there was not mushroom in this book to squeeze them in—"sorry … I'll get my coat".

It would not be an exaggeration to state that science knows more of distant space and planets millions of light-years away than it does the first 30cm within our own planet's skin. Dark matter, forming the bulk of what is out there in the universe but can't be seen, was first proposed by the appropriately out-there-named astronomers Jan Oort in 1932 and Fritz Zwicky in 1933. As for soil's own dark matter, it was not discovered until 1996, with the glue holding the soil together, a glycoprotein, glomalin; this also just happens to hold between one third and one quarter of the world's soil carbon, yet it remains particularly elusive and, for the most part, invisible.

As for carbon and its role as a greenhouse gas assisting global warming, of course the prime culprit again is fossil fuels, yet the soil plays a significant role for a solution as it is the top metre of the Earth below the surface, which holds nearly 500% more carbon than the rest of the biosphere, all the living flora and fauna, in all the air and water, five times as much in just over 3 feet—with erroneous agricultural practices (and that includes ploughing) being a significant contributor to releasing this soil-stored carbon back into the atmosphere. The soil is still the best solution to carbon capture, even when marginalised, ripped and torn apart by agriculture—with regenerative agriculture not only is the soil health improved, but a feature of this increase in health is the increase in carbon within the soil. It then becomes its own positive feedback loop—more carbon, more plants, more water retention, more nutrients, more carbon, more plants, and around and around—it occurs naturally and automatically once the optimum conditions are met, conditions that are never going to be met with monoculture.

If the problem is too much atmospheric CO_2, the answer is easy: it's shifting the carbon from the air to the soil, which is done by regenerating the soil and increasing plant coverage, eliminating bare soil, stopping ploughing, pesticides and fertilisers, and the rest is done for us.

* * *

Modern agriculture just takes and gives nothing back. Agricultural lands are compacted by very heavy machinery, stripped bare of their crop, exposing the soil to erosion from rain and run-off or blown away in the wind when dry. Run-off is exacerbated as the soil is impermeable owing to compaction, which also discourages worms and the other aerators, all further discouraged by the agro-chemicals actively killing them and all the microbes, decreasing the yield and healthiness of the soil and plants.

Smallholdings could be a significant part of the revolution we are waiting for regarding soil regeneration using permacultural practices. A web of smallholdings can not just feed the country, they can act as a preventative to many of the diseases of civilization by growing nutraceuticals in fully mineralised soils. Smallholdings can also considerably reduce carbon emissions, with healthier soils sequestering carbon, as well as creating drought-resistant soils along with increasing biodiversity, revitalising local economies and preserving traditional knowledge.

Massive fields of single crops profit no one, except those supplying the equipment for the monstrous monoculture. Modern monoculture is firmly entrenched in warfare, literally at war with nature—after WWII it all smoothly shifted from a military production complex to the agricultural system, with the nitrates left over from bomb-making becoming the backbone of fertilisers and the biological warfare agents becoming the pesticides, and all using the extensive production facilities and global distribution networks and logistics established during the war for global consumerable domination.

Chemicals that are known to be carcinogenic and endocrine-disrupting (real-time mutation of genes, the DNA, the blueprint to life and health altered) are directly sprayed on our food, mixed with oils so the rain doesn't wash them off—and nor does the cursory splash of water in the sink at home. Self-regulation in the industry allowed the manufacturers of these chemicals to conduct their own experiments and tell us if there is a danger or not, conveniently always finding no danger in their money-making schemes, while the conclusion of independent research only ever finds danger. Who to believe?

* * *

Having personal relations with the local farmers around me, it'll be convenient to blame the farmers for this severing of the Earth's biosecurity—although that would be unjust as the farmers have been brainwashed by the marketing and representatives from the agro-chemical corporations. Farmers have just been following guidelines and regulations, theirs was not to reason why, although many of those with the temerity to do so now probably follow organic and permacultural principles. Most of the farmers, organic or not, can see that the current model is destroying our most prized asset—the soil.

The public and the farmers have been fooled into believing that chemicals are essential to grow our food—which by making the soils barren and unhealthy has actually become true, the soil becoming like a drug addict, dependent on a chemical fix to survive. This underlines the mess we have gotten into—if the soil was healthy to begin with, it is its own remediator and no chemicals would be needed. Monoculture has stretched out this drug dependence to the nth degree and with vast swathes of the same crop the pests and diseases literally have a 'field day'. Sadly, the world's crops have fallen prey to Satan (money), while Demeter (*oikonomia*) has long been banished.

The rise of organic farming in the late 1960s and 1970s and that of permaculture a couple of decades later were at their core a reaction to the deadly chemicals used in agriculture, as well as the desire to provide more nutritious food, because the proof of a functioning food system performing to its potential has to be the healthiness of the nation it feeds. It's a sentiment I no doubt inadvertently plagiarised from the hugely influential, timeless and inspirational *The One-Straw Revolution* by Masanobu Fukuoka, with his take on natural farming, being one seriously tuned-in Japanese farmer and philosopher, who wrote: "The ultimate goal of farming is not the growing of crops, but the cultivation and perfection of human beings." The book is full of gems, so read it to see what the future could still be, as envisaged in 1975.

* * *

The association of good plant health and the health of those eating it is an uncommon thought for most. The health of the soil and its association with plant and human health is given even less thought, as is the more personal thought of the health of our own gut biota and its association with our overall health. The microbes of the soil that maintain soil health are one of the first victims from agro-chemicals, just as the biota of our guts maintaining our health is the first victim from Western medicine (notably antibiotics). Chemicals have knocked out the natural health systems of both the planet and humanity, and this has to be a major risk factor to our overall health today. The destruction of the immune systems of the soil and the human is the first business step of the agricultural and health industries these days, as this guarantees a dependence on the agrochemicals and pharmaceuticals thereafter. It's not personal, it's simply business and erroneous economics.

If every decision-maker and politician had actually made their own soil, grown their own food, was tuned into the seasons, understood the effects of the weather and pests and the positives of adding a vital home-made compost and fertiliser, they would also be able to clearly see the damage being caused by the modern agricultural system. But they don't and they're not; they are only tuned into the money. Hence this book, with all those astronomical figures from industries as diverse as health, clothing, food, plastic and building, all of which the seven plants could reform; these are real. These indicate a real economic worth to the environmental objectives, creating a mass of new revenue streams and associated employment. The growing of enough food, full to the brim of its potential mineral goodness, the knock-on effects this has on health and wellbeing, savings to the NHS, as well as reprioritising what we actually need—and what we actually need is good food and water, and a sense of belonging with some companionship.

Is there any surprise that the food chain is a significant risk factor for obesity, diabetes and widespread ill-health? This is clear to see from a look at the customers and their trolleys in the supermarket. Food should only ever have been allowed to compete on healthiness and quality, not on price and quantity. Industrial food, the product of monoculture and processing, is making us sick and it is also making the Earth sick. In flashing twenty-foot neon lights it says—A NATION IS ONLY AS HEALTHY AS THE SOIL IT STANDS UPON.

Permaculture

We've come across this word quite a few times already in this book, being coined in 1978 by Bill Mollison, as in *perma*nent agri*culture*, another portmanteau term. It is Bill Mollison and David Holmgren who are considered the fathers of this regenerative, sustainable, abundant and efficient method of smallholding farming, adapting and cataloguing their natural farming in Tasmania, fronted now by the charismatic Geoff Lawton, an early student of Mollison. Holmgren introduced the 12 permacultural design principles that make permaculture a definable entity, laying down a viable model for anyone lucky enough to own their home with some land.

Mollison may have been the first to document the word permaculture, but he was definitely not the first to consider the best form of crop growing as being permanent agriculture. Decades earlier, Franklin Hiram King's 1911 *Farmers of Forty Centuries*, written after touring China, Korea and Japan, described the impressive system of a multitude of smallholdings as being one of "permanent agriculture". King's tome was a valuable resource book for what is essentially organic farming, and it also further encouraged many more to expand their forty centuries' old techniques, being loosely defined also as 'Ecological Farming', which amounts to any of the sustainable practices, working in harmony with nature, boosting soil health, sequestering carbon as a natural consequence of enriching and encouraging the microbes, and leaving the land in far better condition after conversion.

My own inspiration came from the Henry Doubleday Research Association (HDRA). Founded by Lawrence Hills in 1954, and writing many of the pamphlets and small books himself, the association was named after Henry Doubleday (1810–1902), who was big on comfrey, which is how I stumbled upon HDRA, as comfrey and I had a thing a while ago also! Considering there was an HDRA in Australia, it's quite feasible Mollison himself got inspiration from these fantastic publications, although Tasmania has a rich history of regenerative agriculture, what with the Living Soil Association of Tasmania, founded only months after the English Soil Association in 1946, which in turn grew into the Organic Gardening and Farming Society, which undoubtedly was also a source of some inspiration for Bill Mollison and David Holmgren. Permaculture not only restores the fertility, it adds to the total soil depth, a complete reversal of the modern agricultural system.

* * *

It's worth a very brief overview of a few other inspirational characters influential in the general organic and permacultural ethos. We can start a hundred years ago with Rudolf Steiner, who in 1924 gave a series of lectures and published his highly influential *Agriculture Course*, which became the handbook for what became known as Biodynamic farming, similar to permaculture but also incorporating moon phases for planting with added ethics and philosophy (not a million miles away from Culpeper's astrological perspective).

Another twentieth-century term is Forest Farming, where trees are deliberately planted to protect the soil, provide shade, food and medicine for humans and animals, while encouraging birds and bees, providing future wood, with cash crops being grown among the undergrowth. This forest farming stems from an appropriately entitled 1929 book by J. Russell Smith, *Tree Crops: A Permanent Agriculture*, which was the main influence of this forest farming technique taken up and made famous in Japan by Toyohiko Kagawa in the late 1930s and 1940s, and subsequently brought to the UK by the pioneering Robert Hart, who co-authored *Forest Farming* in 1976.

It would be amiss if no mention is given to another pillar of sustainability and organic farming, Lady Eve Balfour, who published her magnificent *The Living Soil* in 1943, founding the Soil Association along with Jorian Jenks and Friend Sykes in 1946. Hoorah! Lady Eve Balfour was the first to actually conduct an experiment comparing the difference between modern agriculture with the new chemicals and fertilisers or traditional, organic with an emphasis on composting and soil health. Their science may not have been too solid but the Haughley Experiment, being where it took place, did show positive results in yields, nutrients and worm numbers—easily seen, earthworms will be noticeably depleted from agricultural soils, which miss out on the humic acid provided by the worms; they facilitate phosphorus absorption as well as the bioturbation (mixing up the soil and nutrients), as well as provide aeration and water storage; in short, healthy soil will have plenty of worms. What's not so easily seen is the 90% of the world's nematodes living in the top 15cm of soil, eating harmful bacteria, making nutrients bioavailable; they act as food themselves, now pretty much gone, along with all the other microbes from mycelium to bacteria, all depleted from the soil we grow our food in. When all this is added to more recent understandings of the role of microbes in the soil, it has to be said that the founder of the Soil Association was absolutely correct to question the *modus operandi* of growing food the modern man way, with behemoth, brutish machinery and science's deadly chemicals.

* * *

In France they have a new term 'Agroecology', for a farming system that has been accepted as being potentially more profitable than conventional agriculture and is predominantly organic and sustainable, and as well as reintroducing native plants and animals the natural biodiversity is increased. With an impressive 41,600 farms making the switch from conventional to agroecology, all posting a healthy 25% average increase in incomes, it is a major step forward in France to directly challenge the mainstream agricultural practices and more than makes up for their nettle interference—*vive la France*!

Let's get this straight, our perspective on agriculture is not what feeds the world. Monoculture feeds some of the Western world, and that is usually as far as we look, while the rest of the world already lives from what are predominantly organic smallholdings. In fact 70% of the world's population survive on these smallholdings for their food, using just 25% of the total energy expended in food production. Which means that the 30% of the world's population fed from industrialised agriculture (monoculture) uses a staggering 75% of the total agricultural energy to do so. The disparity is clear and the path forward is no more complicated than observing how most of the world already feeds itself, without the corrupting and malign influence of Western man's short-term profit and long-term desecration—the monocultural complex of sterile seeds, deadly poisons, huge expensive machinery and lifelong debt has no future.

The UN Secretary-General, António Guterres, has said that in order to rebuild the global food systems after the impact of the ramifications of the Corona Crisis fully hit home, sustainable food systems must be prioritised. Of course, any implementation of his words is done based entirely upon how they are interpreted. It could be seen as investing in permacultural practices globally, giving each country food security, an undeniable sustainable food system.

Or, as it is the money that shouts the loudest, genetically modified organisms (GMOs) will be presented as the saviour safeguarding global food security. But consider that the monocultural system is completely unsustainable and using GMOs is a continuation of this system that has just managed to survive for no more than 70 years, while agriculture has at least managed well over 7,000 years, and the gathering of wild foods and hunting kept us perfectly alive and vital for hundreds of thousands of years before then. The best answer today has to be the options least damaging to the planet, which would be permaculture and forest farming, where the soils are actually improved. Without the revolution talking place beforehand, the path will inevitably be the GMO one as it involves all the same players from the same teams as those already playing the money game (look no further than the investment mammoths Vanguard and BlackRock and individuals such as Gates and Buffett). GMO will just take us further away from where we should be going, perhaps irretrievably—what is the reality once unsustainability can no longer be sustained? Are the diseases of civilization an early symptom of the manifestation of this unsustainability?

Ten acres with planning permission

Joseph Chamberlain (not related to Neville) in the 1880s gave the rousing offer that three acres and a cow should be available to anyone wishing to take up farming. This sadly never materialised, although this radical Liberal MP did manage to get through the Small Holdings Act of 1892 and the Small Holdings and Allotments Acts of 1908; another was passed in 1925, after Chamberlain's death. These Acts enshrined the provision of allotments in law and also allowed for the creation of County Farms (council-owned farms rented to aspiring young farmers whose main obstacle was initial finance). The ethos of Three Acres and a Cow is kept alive by the admirable group of troubadours, mingling folk, history, land ownership and politics into amusing and thought-provoking shows (threeacresandacow.co.uk).

The twenty-first-century version needs an update to 10 acres with planning permission. Considering that there are some amazing prefabricated houses out there already, it would be an ideal opportunity to begin production of sustainable, low-impact, carbon-neutral houses, utilising hemp and nettles—immediately creating a demand for these plants, enabling many initial smallholders to start growing the raw materials for other sustainable houses to be built—combined with sufficient solar panels, windmills and water storage to achieve self-sufficiency for each and every smallholding as they roll out across the country.

Complete self-sufficiency has to be the way forward and the goal, with an intertwined community of smallholders all collectively creating an intricate web of supply, demand and distribution. The task in many respects is to relearn how it used to be done and still is done in many parts of the world and supplement this with appropriate modern innovations.

* * *

A plot of 10 acres allows for the growing and the processing by hand of most crops manageable by two adults, with any required machinery co-operatively used and owned. A small co-operative of all local smallholdings makes for a bigger entity, especially if labour is easily movable when required, which means co-operation within the co-operative, with days spent together to jointly process crops, such as pressing fruit and oils or decorticating hemp, nettles and flax etc. This smallholding co-operation is a definite way forward if the longevity and diversification of the land is to be guaranteed.

Agreed, 10 acres is a lot of space, a bloody big field in fact, but when broken down into sections, from 10m squared to an acre, there becomes a multitude of growing areas. With approximately two acres devoted to a few of the seven from the book, an acre or two for grain, an acre or two (with or without polytunnels/greenhouses) for vegetables and salad crops, an acre or two for some tree farming, half an acre left as a meadow with a pond, an acre left fallow, and with crop rotation and diversification there are a million variables on what 10 acres can provide—and that's even before animals such as goats and chickens get considered.

It just takes a few smallholdings to garner the media attention for more to take up this call-to-farm, and the more that grow an acre of nettles or hemp, the more there is for secondary and tertiary industrial processing, increasing the chances of sustainability for those endeavours. The crazy bottom line is as we have already seen: (a) the agricultural system has been declared by those who have studied it to be creating more harm than benefit, and (b) 10 acres of monoculture creates just £300 profit, or 20 or so tons of grain. At best that is, for those declaring-a-loss taxpayers fund welfare and subsidies. If you can grow 60 heads of broccoli permaculturally, and if sold for as little as £1 each, you still accrue £60, of which £30 would be seen as a reasonable profit margin. Those 60 broccoli plants, requiring just a few square metres, can earn as much as an acre of corn or wheat. Do the maths. Stand in a 10-acre field and contemplate the countless positives of regenerating it permaculturally and feeding the locals healthy vegetables. The more who do it, the greater the likelihood of success and permanent change.

Smaller farms allow for more individual attention, maximising yields, with sensible and soil-enriching crop rotations, meaning that with time productivity actually increases as the soil becomes increasingly more fecund and healthy—the exact opposite of what has happened to the monocultural fields, providing decreasing yields from increasingly unhealthy soils. The countryside should be a vibrant mix of communities across the entire country, all interconnected with a dense tapestry of business and social interactions, selling, distributing and bartering local produce to all corners of the country.

There is not only a change of life possible from taking on a 10-acre plot, there is also the potential to be self-sufficient, learn real-life skills, meaningfully meet more people, be completely fulfilled and be part of a revolution. These 10-acre smallholdings will feed the people mineralised food, support the immune system and nourish the soul of the nation back to health.

* * *

The need for food security is essential, especially when at the mercy of shaky international politics, as seen with wheat and the Russian invasion of Ukraine. Improvements in efficiency and longevity of food provision will become paramount as climate change alters crop and land availability, as flood plains do as their name suggests and flood. With the sea level rising the biggest problem will be flooded fertile agricultural river floodplains no longer being able to be used for agriculture—meaning food security will become the top priority; securing this is surprisingly simple, 10 acres at a time.

I have it on extremely good personal and private authority from an advisor on climate change to world leaders that we have already gone way beyond the tipping point and climate change is a foregone event, hence the stagnation of many governments. The climatologists admit the focus was erroneously put on CO_2 and that they themselves are caught in this trap of conceding that a change of mind is required before moving forward and rectifying, yet this admittance that a mistake ever took place is one of the greatest challenges. The emphasis needs to move away from CO_2, particularly as a greater danger is posed by other more potent greenhouse gases, notably methane, not just from cows farting as fossil fuels are by far the current top emitter, but the imminent release from previously frozen tundra across Siberia, Alaska and elsewhere of billions of tons of a greenhouse gas with twenty-one times the global warming potential (GWP) of CO_2 and a lifespan of 12 years (CO_2 has a five-year lifespan).

The next biggest human polluting greenhouse gas after CO_2 is nitrous oxide (N_2O), with 275 times the GWP of CO_2 and a worrying lifetime of 121 years. But this is nothing compared to the worst-polluting greenhouse gas, sulphur hexafluoride, SF_6 (used in electricity production/transmission), which has a stupendous lifetime of 3,200 years and a GWP of at least 23,500 times that of CO_2! (these figures are direct from the horse's mouth at nationalgrid.com). This is clearly the one to rectify yet it's the one least talked about, no doubt because there's so much money in energy.

* * *

The 10-acre plan is not short-term, it would take a few years before each 10-acre plot works efficiently. Preparing the land is a huge task, planting trees for borders and connected forests, shelter and shade will take several years, but once the 10-acre system is up and running to nearly full capacity the soil will have been fully regenerated with plants super-rich in nutrients bringing increased health and taste to all those who eat it, as well as supplying an ever-increasing amount of other 'waste' materials for various small-scale industrial sectors. Many factors determine this, not least the implementation of the secondary processing of the seven plants to be used for not just nutraceuticals, also textiles, plastics, fuel, colourants, vegetablecretes etc. As the industrial sector localises and downsizes, the viability of plant alternatives will soon drive demand and as demand is increased more land can be turned over to these plants.

It will all start with government grants, legislation, planning consent and a framework of action as this 10-acre solution actually intends to be the food source of Britain, as well as providing an income for millions, improving the health of the human and soil, and all that other hyperbole jazz you've been reading about. This 10-acre revolution needs to be done from both the private sector, that is, people who have the finances to buy their own 10-acre plot and by those who have funding, either through private channels or government and council means—we'll look at how this land can be obtained now, as it's not the insurmountable hurdle it might at first appear to be.

Imperatively, by creating tens of thousands of smallholdings, forest gardens and farms, permacultural plots or whatever the terminology, Nature is being allowed to cohabit all of these plots and the soil gets regenerated. This will provide sanctuaries for what is an ever-diminishing population of insects, small birds, mammals, flora and the like. It will repeople the countryside, restoring communities and rolling out urban food gardens in cities across the country and reconnecting people with other people and the land. All that as well as improving the food and health systems. Can't be a bad thing, can it?

How to get the land ... the peasants are revolting

There are a few ways the required land can be acquired to guarantee our future food security. The most appealing is repealing the Enclosure Acts, performing an unclosing of the land, possibly requiring some form of a miracle, although where there is a will there is a way, especially if that will is amplified by the populace. It only takes one well-publicised 10-acre plot to get the ball rolling.

Of course, it is far easier said than done, and this proposition is merely one suggestion as to how Britain can feed itself. This proposition for many other countries around the world would be far easier to fulfil because Britain is virtually alone in its apparent air of mystery about who actually owns much of the land beyond it all simply being the King's, which surprisingly it is—thanks to our shady law. The world's primary feudal landowner is King Charles III, king of 15 countries and head of a Commonwealth consisting of 54 countries, including a quarter of the world's population; he is the legal owner of a staggering 6.6 billion acres of land, one-sixth of the Earth's land surface (this figure is hugely contested on so many levels from size to validity). Irrespective of it being corrupt, sinful or shrewd, it is undeniably impressive.

Notwithstanding the arrogance and the ability of the rich to write their own laws, making it 'legal' in the past to take the common land away from the people on the pretext they were 'to make it more productive'—it is this exact phrase that could prove to be pivotal, allowing a return of at least some of the land back to the people. Initially, when faced with food shortages in the seventeenth and eighteenth centuries it could have been seen as common sense to desire the land be more productive than it was. However, there doesn't appear to have ever been any later checks or confirmation that the enclosed land had indeed been put to more productive use—so that'll be a good starting point.

* * *

It was exciting to see a land reform document outlining what is needed to be done to start the return of the land to not just the people, but to make it profitable and sustainable. This was written by, among others, George Monbiot (Earth's voice on many issues) and Guy Shrubsole, the author of the excellent recent book *Who Owns England?*. This land reform document was brought out with the Labour Manifesto for the 2019 election, titled *Land for the Many*, but with Labour losing the election this ground-breaking paper shamefully gets binned. How frustrating that the right thing to do to move forward into the future is tied to a political party. Its focus was on urban land as most of the people live in cities and towns, and it was not necessarily attempting to answer the same questions as this book, although it's part of the same solution. The paper addressed issues involving community rights to buy and compulsory purchase orders on land and buildings left vacant and/or run down for a certain period, as well as easier access to mortgages, extensions to rights to roam and more public outdoor spaces and allotments. The proposal of this book is that much of the urban housing crisis can be eased by the creation and restoration of houses and communities in the countryside, with the food being produced via 10-acre smallholdings. An economic shift is achieved by the rejuvenation of rural communities, and this relocation would therefore provide ample vacant urban accommodation, at a lower, more reasonable cost.

* * *

So, how much land roughly are we talking about to become potentially food-self-sufficient as a nation?

Firstly, it needs to be known how much is eaten in weight of food every day from different countries to get some idea of how much we should be producing. Is there a set weight or calorific intake? Often it is said we need around 2,500 calories a day, which is meaningless as there are a million ways to get 2,500 calories, and calories are measurements of energy created by candles! Your body requires air, water, minerals, carbohydrates, protein, vitamins and phytonutrients, and it demands a minimum of these to perform basic functionality—the RDAs. There's no intrinsic value attached to a calorie, it's the food it's attached to, and we all know that unhealthy food (many would say nice) is high in calories and low in nutrients, while fresh food is low in calories and high in nutrients. Which makes an average food weight per day requirement just as meaningless as calories, as 1kg of cola is not the same as 1kg of salad. If you ate 1kg of varied salad every day you'd thrive; if you drank only 1kg of cola a day you'd not survive.

There has to be a way to equate harvest to number of possible people who can enjoy it, and for this a standard of 1kg per person is used, whether a baby, 90-year-old, athlete, builder or couch potato. The average weight of the modern diet—currently 80% derived from highly processed carbohydrates—is 2.5kg/person/day, which is quite literally nearly a ton of shit a year, especially when compared to the nutritional content of 1kg of fresh organic vegetables. By comparison an Al Jezeera documentary has mentioned it took 6,000 tons of food to feed a city of 10 million every day (600g/person), making 2,190,000 tons in a year. Two millennia ago, Alexander the Great's army of 65,000 required 88,500kg of grain and 136,000kg of water, which if scaled up to 10 million would be 15,000 tons of food and 25,000 tons of water every day—1.4kg grain a day each for a soldier. Which makes 1kg a day a fairly good average.

By using the results of a total novice at PermaLife, which with tree harvests on the 10-acre plots, helped with advice from someone with expertise and knowledge and an attempt to actually maximise yields, the expected harvest can be conservatively set at 8 tons/acre. With food crops grown on six of the 10 acres, 48 tons of fresh nutritious produce can be harvested per plot. If there were 10,000 of these 10-acre plots (100,000 acres used), 480,000 tons of produce would be grown, enough to feed 1.315 million people. Some 68 of these 100,000 acres of smallholding networks are enough to feed 90 million people (Britain has 67 million); 68 is used as this totals 6.8 million acres, the exact amount of land stolen from the common people by the rich up to the twentieth century via enclosure. The answer therefore to the above question is—all the enclosed land would be more than sufficient to feed the entire country, with a 23% harvest surplus to trade with.

Here's an interesting coincidence to put what might appear a ridiculously large land requirement of 6.8 million acres. During WWII, given food shortages, a 'Dig For Victory' campaign emerged, which successfully fed the nation of about 45 million from about 6.5 million acres of additional land that was put to food growth. This extra land came from gardens, parks and any green urban spaces, with very few country estates contributing—it was down to the people ... surprise, surprise! It wasn't even the nicked land used, so imagine how incredibly productive we could become if all the best land was legally reclaimed. It proves there is easily enough land in Britain to feed itself, twice over if necessary.

This 10-acre revolution would also employ many others not directly involved in agriculture, such as the 680,000 super-efficient, low-impact, stylish, hemp pre-fabs to house 2.7 million people on the 10-acre plots. These will be the new houses required in the future, not those awful shoebox, characterless housing estates being built everywhere today outside of towns. These are soulless sociological failures, encouraging zero community atmosphere and are only being built because it's one of the last ideas of a floundering economic system that has already bankrupted itself at least twice before over house prices.

* * *

It's not going to be practical or possible to unclose all the enclosed land, especially as some of it will be either coastline, cliffs and mountains, marginal agricultural land, or already under good cultivation methods such as organic. We have already seen other ways of obtaining the land, the County Farms among them; these may have just 200,000 acres left, or 20,000 10-acre plots, which could grow nearly a million tons of fresh vegetables, enough to feed about 2.75 million. It should be in the interests of national security that council-owned land at least doubles in the next 20 years to put 10% of the nation's food requirements directly under council care.

Land either in foreign ownership or land-banked waiting for the time the Local Authorities allow for planning permission should be taken back, as left vacant it still increases in value every year regardless, de-incentivising any other use. Compulsory Purchase Orders (CPOs) need to be placed on such land and a fair price paid for it, especially land that can be repurposed as allotments and Public Cultivation Centres (PCCs). It has to be the councils who are actively stockpiling land to grow food on, not the rich speculating.

Any enclosed land never put to any other use than forming a larger country estate should also be unenclosed, as these were certainly fraudulently enclosed to begin with. Voluntary hand-backs and charity from those landowners who consider themselves to have excess would hopefully account for a few thousand acres—this could be incentivised by government tax relief on, say, inheritance tax being part-offset by the donations of land given back. Donations from the rich would be very welcome—how good would it feel to know that by giving away just 100 acres from an inherited 10,000-acre estate, 10 families can start a new life, improve the soil while creating enough food (480 tons) to feed over 1,300 people, as well as all the secondary and tertiary processes and manufacturing possibilities.

Then there is the Crown Estate, which controls 2 million acres, including half of the foreshore around the nation—so not all the 2 million acres is applicable to farming but a marine harvest is still possible. There has to be quite a bit that is superfluous to their needs, which is a very subjective quantification of what are 'needs' and what are not. The public already pay for a minimum of four palaces (Buckingham, Kensington, St James's and Holyroodhouse), four rather massive houses (Sandringham, Clarence, Highgrove and Bagshot), three castles (Balmoral, Hillsborough and Windsor) and loads of other 'cottages', houses, manors and parks, all exclusively for one family.

Obscenity laws could be rolled out, with a brief section on this later, as the land can always be claimed back as being the right of the nation and its population; it certainly shouldn't be owned by non-domiciles and foreign pension and hedge funds, be vast grouse shooting estates, or the publicly funded country estates, some of which were themselves initially built on the monies immorally earned from slavery or payments from slave abolishment.

* * *

If we are looking at the prospect of the UK ever being self-sufficient in food, there is definitely a lot more to be done to come even close to this. Today we are about 80% self-sufficient in dairy and beef, and that's about it. For the rest of the fruit and vegetables consumed we are in the region of supplying 20% to 50% of requirements. Obviously, bananas and pineapples would be difficult to produce—although somewhat surprisingly Iceland is currently the largest grower of bananas in Europe, in greenhouses, of course—but do we really need apples from New Zealand? Now we have Brexited there needs to be some serious consideration given to feeding the people, and the only logical answer would be to step up production, educate the public about what should be eaten and to significantly reduce meat consumption, as this puts an unreasonable, uneconomic and environmental strain on the grain and land.

When looking for the first steps for England to take, it might be an idea to follow the lead of Scotland where their Land Reform Acts (2003 and 2016) could eventually make the compulsory purchase of land obligatory for sustainable development and for local communities, as well as making it much easier for rural communities to buy their land.

Scotland also has much better right-to-roam regulations, with far more of the country accessible than England; all of Scotland can be camped on whereas only Dartmoor in England allows this (recent law changes saw this flip from allowed to prohibition, back to allowed, for now). Scotland might also be a better country for the 10-acre proposition as their government is far more forward-thinking and progressive than England and its shady secrets. Scotland, via its first Land Reform Act of 2003, has enabled over 500 community-led organisations to take control of over half a million acres already, which could be 50,000 10-acre plots, producing in excess of 2.4 million tons, enough to feed 6.6 million humans.

Anders Holch Povlsen appears to be a billionaire with the right attitude required for the future. He's just bought Topshop from someone who many would call a rogue (many would use a harsher word), Philip Green, as well as owning the fashion retail brand ASOS among others. He also owns 221,000 acres spanning 12 estates in Scotland, making him the country's biggest private landowner. His plan is to rewild as much as possible, then provide for walking and access and preserve a natural state. Only knowing about him from research, not personally, he and his wife Anne appear honourable and very decent people, so nice in fact they might be open to giving up some suitable south-facing, well-drained land for a few 10-acre plots? It'll be interesting to get a couple like this involved, as it would be a minuscule dent in their land but could make a huge difference to the traction of the 10-acre proposition nationwide.

Forest charter, allotments and PCCs

> Great historical shifts are sometimes symbolized by minute changes to everyday behaviour.
>
> (Alvin Toffler, *The Third Wave*)

Not everyone wants to throw in the towel of their present life and take a roll at Life Roulette to see if they might enjoy some hard work outdoors as a farmer, although undoubtedly there are many thousands willing and capable to hear the call. There are plenty of other much smaller and practical options, all helping with reconnecting the human to the planet and bringing us all back into unison. Whether that be from just a houseplant and pot or two on a table by the window in some sun, to a few larger pots on a patio, to those fortunate to have a garden, or raised beds and what should be and once was the right of every person to feed themselves by growing their own food on an allotment.

Some of the Enclosure Acts in the eighteenth century rather ironically took the land away with one hand while the other allowed for a provision of small parts to be used as allotments for workers to grow some food and feed their families, possibly done so as not to contravene the Forest Charter of 1217. Only the Magna Carter of 1215 is remembered from this period, but for centuries it was this Forest Charter that held sway as it involved the common people, and it was on the statute books longer than any other legislation until the Tories repealed it in 1974.

The Charter was groundbreaking in many respects, being the first to make provision for the environment, and also asserted the rights of the individual and secured rights for widowed women. As well as guaranteeing free access to forage for food (from fish to honey), water, firewood and clay for utensils, it also crucially retained the common land already under threat from enclosure as long ago as the thirteenth century. This Forest Charter with pertinent modern additions should be reinstated on to the statute books forthwith, not least because all the subsequent Enclosure Acts were in contravention of this Forest Charter and should never have been allowed to be passed as law (not that the commoners could afford to pay for a lawyer to prevent land becoming enclosed). No doubt this is why the Tories repealed it, to stop any embarrassing questions or even legal public action about where their land came from historically—slightly tangentially, this was actually a very contentious point of law in America, where it refused to accept genocide as a human rights violation from when it was proposed by the UN in 1948 until 1988, because if it did it could lay itself open to nationwide First Nation litigation as it is well-documented that the white folk committed genocide upon each and every tribal nation.

* * *

As urbanisation increased, Local Authorities towards the end of the nineteenth century began to accommodate for allotments around towns and cities, helped by Joseph Chamberlain's Small Holdings Acts of 1892 and 1908, also the 1925 Allotments Act, where it became part of our statutory rights of living in England that an allotment be provided if asked for (excluding London). An amendment is needed as there is no time allowance for those placed on a waiting list—with well over 70,000 around the country waiting. This should be changed to one or two years' wait at maximum, after which councils can use a Compulsory Purchase Order (CPO) on unused land for community use.

Allotments are a wonderful piece of almost forgotten recent history, keeping alive this thin thread of millennia-old natural law, allowing people to be personably empowered to grow their own food. Today, allotments aren't seen as particularly revolutionary or militant, but they are a brilliant font of priceless horticultural wisdom, knowing what grows best on the local soil. Expanding allotment availability would be a fantastic way to increase the urban food supply and make it possible for more urbanites to get involved with horticulture to see if it is for them.

Local Authorities have been selling off allotments, public land and County Farms for decades now, usually only with the short-term vision of balancing accounting books. Allotments were at a peak during WWII, with as many as 3.5 million; today there are about 265,000 across the UK, a similar figure to 1872 when the population was only 28 million as opposed to 67 million today. The acreage of County Farms has recently more than halved, from 426,000 acres in 1980 and for much of the twentieth century, to only 208,000 acres today. These sales suggest the Secretary of State has either not been paying attention or has been swayed to sell off land to interests and building corporations who line the party coffers or have other allegiances. The Enclosure Acts underline how the law works progressively in favour of those in power; boldly underlined and run through with a highlighter pen are those 5,200 Enclosure Bills over a period of 200 years (1670 to 1870), which averages out as a land parcel theft occurring twice a month for two centuries, with each theft of each parcel of land being perfectly 'legal' as a Bill for it was written on paper and got passed by Parliament.

This 10-acre proposition will take place in the countryside, but it's only going to happen if people from the cities want to farm. A worthy question to ask the public would be: who wants to move to the countryside, grow the food for the country, get a brand-new house and life with full self-sufficiency, connectivity and purpose?

It would only be democratic for the *demos* to participate in something apart from a little over half voting for which monkey sits on a leather chair in the House of Commons—even its name is a slander upon the people it fails to represent. A proportion of inner-city residents would not only be willing, they would make a marvellous success of the transition to the land; and as more succeed more will want to swap their constant poverty and troubles of urban sufferance for some rural graft. This change of life, losing no comforts or technology, only gaining a sense of fulfilment, happiness, contentment and community, will work for not just the poor in the council estates but also the rich who need to reconnect, because many are just as unfulfilled irrespective of their wealth. We are all just people and we all need the same things, and it's definitely not the things being sold to us that will bring fulfilment and connection.

* * *

Another way to reconnect urbanites with one another and the earth is to transform areas of public parks or abandoned industrial sites and establish a network of Public Cultivation Centres (PCCs), getting the public involved in community gardens, learning about the soil, plants, nature, food and health. There are plenty of local people who want to participate in some voluntary community action if offered the chance to do so. It needs just a fraction of council expenditure on public areas to be spent on the initial infrastructure of these public food and health centres, helping to pay for printing, raised beds, fruit nets etc. As the local population becomes more knowledgeable and involved about where their food comes from and can see the connections between food and health—as by growing some a connection with nature evolves—this will be a perfect stepping-stone for those wanting to take up the 10-acre transition.

This self-sufficient ethos takes over automatically, as demonstrated in many places across the world with guerrilla allotments. America was hit hard after the economic collapse in 2008, causing a distinct whiff of anarchy to take over Detroit and other economically deprived cities. Detroit has experienced constant economic decline since the 1970s, not to mention racist divides, and being dependent on car manufacture it was severely hit by cheap foreign imports. The situation was so dire that in 1975 Mayor Coleman A. Young (Detroit's first African American mayor) allowed for 3,000 vacant plots to be turned to productive gardens using the catchy 'Farm-a-Lot' slogan to provide food for the residents. Council support waned during the 1980s but has been reinvigorated with the more recent economic crisis and the sad fact that the local population simply do not have any fresh food in their diets. Urban Farming became a thing again, with over 2,000 vacant houses and gardens now being utilised. It is technically illegal, but to hell with the law when the people need to eat, which shouldn't mean losing freedom or a diet of cheap processed food. It may not be gaining much mainstream media coverage, but there is without doubt a movement that has happened in Detroit, and notable among these urban farms is D-Town Farm, Earthworks, Georgia Street Community Collective, while Keep

Growing Detroit coordinates and advises over 1,600 urban food gardens and farms; meanwhile the Michigan Urban Farming Initiative is continually expanding and boasts of giving away 23 tonnes of food donated since 2012. Let's all raise our hoes in support of these good people.

This fresh food anarchy is not just confined to the economic destitute of Detroit, it's a nationwide movement and has been brewing for decades. In New York Hattie Carthan was planting trees in underprivileged areas in the 1960s, while the Green Guerillas in the 1970s and 1980s took over vacant lots for just a $1 and grew food on them—there are now thousands of plots growing food for the community. Similarly, in Atlanta, Georgia, 7 acres was bought by the Local Authority and turned into a forest garden, allowing residents to freely pick the produce, volunteer and learn about the usefulness of plants for food, health and medicine first-hand, with guest speakers and special weekend features. So successful and popular has this been that other potential sites are being considered around the city as Atlanta, with a population of just over half a million, definitely has social-economic reasons preventing many residents having access to vegetables. These community food gardens are proving to be a vital step in many residents' path back to health and companionship, and are definitely a vital step in reconnecting us all to our true Mother. Our Mother provides while our Culture divides.

Someone who has been on YouTube for over a decade now showcasing what has been going on within the smallholding, sustainable, environmental and alternative food and farming worlds has been Kirsten Dirksen. It's definitely worth giving her positive channel a watch.

All this local growing has another bonus, and that is getting close and personal with all the local microbes, allowing for a free transfer from soil to gut, exactly as has occurred for hundreds of millennia. It's only in the past few generations that there has been a significant dislocation from Nature. Spending time in Nature and children playing outdoors has significantly decreased over the past few decades, while conversely there has been a significant increase in the prevalence and incidence of auto-immune diseases, mental health, other diseases of civilization and general ill-health. The complex of microbes within our gut does far more than just digestive services, as science is showing repeatedly recently; it also controls the immune system, genes, hormones and enzymatic activity, more so than the brain can be held responsible for. So this 10-acre scheme will hold many unseen and unintended positive side effects, notably this re-establishing of the symbiotic relationship between soil microbes and human microbes, boosting the immune systems and health of both the soil, human and planet.

Land prices and subsidies

As frustrating as it is that a considerable amount of the land has been wrongfully acquired/stolen in the past few centuries, made all the worse with the recent shortsightedness and ineptitude of having sold off all the nation's utilities in the past few decades, it's not irredeemable. If the land could be taken originally, it can just as easily be taken back, because it has to be understood that no human being 'owns' the land—what a peculiar state of mind to find such a concept acceptable in the first place. Humans are one of the most recent species on this planet, so when did we begin to think it was all ours to do as we please with? Humans make up approximately 0.01% of the living biomass of our planet, but regrettably in many areas we are 100% responsible for the ecocide adversely affecting, even creating extinctions, among the other 99.99%.

Even if as a last resort the land has to be bought back, the cost of the 10-acre project is not exorbitant. The cost of 10 acres at the current market rate is about £100,000; add in the eco-dwelling and some equipment probably gives it a start-up price of about £250,000 ready-to-go, made available with a government mortgage given at 0%, with the government paying the landowner a small amount a month with 0% interest off the £100,000. The fabricated ecological building company producing all the hemp and nettlecrete houses for all the new farmers moving to the country from the city will of course be publicly owned, being an instant market for the hemp and nettles being grown on the new 10-acre plots.

Surely that's quite impressive, only taking £250,000 (and that is probably an overestimate) for a farm of 10 acres and up to four people to be self-sufficient and carbon neutral for the rest of their lives, while contributing to the new superlative diet of the nation for this and future generations. This price includes all the online tutorials and preparation, a mentor from a local permacultural smallholding who can assist in the initial years' farming, with every farm having a member who will in turn be a mentor at another farm. As a comparison, my sister recently got a very small one-bedroom basement flat in Brighton, between a main road and a rail-line, for more than this amount.

* * *

Moving on to subsidies. These hark back to WWII, when farmers were encouraged to grow more crops as the torpedoes of the Nazi U-boats were sinking so many supply ships that national food security was compromised. A cash incentive was given as more work and hands were required to increase productivity, so the more land a farmer had to be worked the greater the subsidy. It proved to be a very effective method to increase agricultural production after the war, when the hedges were ripped out and fields expanded. So successful it was there became a surplus, creating butter and grain mountains in the late 1980s, causing agricultural prices to drop with the poorer nations feeling the consequences, so the farmers were encouraged to not grow quite so much but still get the subsidy, for doing nothing, continuing to this day. Which doesn't sound very reasonable, does it?

With about £3.5bn given out by the UK government in 2020 to landowners and farmers by way of subsidies, part of this will have prevented some farmers from falling through the net of bankruptcy but the bulk just makes the rich richer, allowing them to invest the subsidy to purchase more land or properties abroad, as well as paying for workers. The largest payments of subsidies included £1.6m to James Dyson (from 2016 figures); the Dukes of Norfolk and Northumberland were both given £475,000; Prince Khalid Abdullah al Saud £400,000; the King's Sandringham Farms got £560,000; while the Duke of Westminster's Grosvenor's Farms received £440,000; Richard Drax (he of the mahoosive wall on enclosed land in Dorset just before Dorchester—you'd know the one if you've driven besides it) got £411,000; Viscount Matt Ridley (who argues against subsidies but gratefully pockets them) £315,000; Richard Benyon, another beneficiary from slavery and enclosure who just also happened to be Rural Affairs Minister between 2010 and 2013, pocketed £278,000; then there were 24 non-royal dukes who own between them 4 million acres, 17 of whom received £8.4 million in farm subsidies; 14 marquises pocketed £3.5 million for their 100,000 or so acres, and a depressingly large number of other rich individuals were paid out as well. There is no way these people needed or deserved any money, especially if one considers that most smallholdings regenerating the soil, even if poverty-stricken, received £0. Those Dukes and the King were given a lot more for other lands and farms, and these are just some of the biggest individual payments, to the richest people in the country ... or am I just being picky?

There is within subsidies a vast gulf between on the one hand the hardworking farmer who genuinely struggles with his 200 acres, managing to just survive thanks to the subsidies; and on the other the landowner who has seen his family's land increase significantly throughout enclosure and who now reaps hundreds of thousands of pounds for literally doing nothing and being paid for this stolen land. Subsidies need to be phased out, and if farmers are not able to earn enough from their vast acreage they can get Universal Credit and start to grow an acre or two of permacultural organic vegetables to supplement their diet and income. Two hundred acres is enough for twenty 10-acre smallholdings, not just making a living and surviving but positively thriving and supplying all the local food and more.

The fact that many farmers actually earn more from subsidies than farming says a lot, with the average farmer receiving up to £30,000 in subsidies in the UK in 2019, while the average cereal farmer lost £10,000 from produce grown. In no way do we wish to disrespect the farmers and say they are doing a poor job, when they are merely pawns caught within a broken system. How much clearer could the problem be—when growing the crops to make the food that every one requires for survival can't even be done to cover costs? What does this say about the state of the agricultural industry in the UK? Quite loudly it says that it is not economically viable and, as we have seen, modern agriculture isn't environmentally viable either—so why are we continuing with an agricultural system not fit for purpose? In a system when many farms earn as little as £10/acre after costs, it seems ridiculous to continue when the same acreage could be earning dozens if not hundreds of times as much from being a network of 10-acre, soil-improving smallholdings. Permaculture proves beyond doubt that there is loads of potential in growing crops; the blame for any current failure has to fall on the agricultural system.

Subsidies aren't essential, as proven in 1984 when the New Zealand government ended all subsidies for farmers. Instead of bankruptcies and a national food shortage, the farmers restructured their agricultural models, resulting in a 40% increase in productivity and value of produce. Today it isn't just Europe and America propping up their unsustainable agriculture: of the total $700 billion given out worldwide as subsidies, Indonesia spent about $30bn, much of which directly contributes to destroying primary forest to grow palm or rubber in futile attempts to raise the national GDP, erroneously assuming this to be a magic formula. This acronym GDP is held as sacrosanct, when we should really be taking a leaf out of Bhutan's book where they have what is known as GDH, Gross Domestic Happiness, a genuine philosophy guiding government decision-making since 2008.

The Agricultural Bill, which replaced the EU's Common Agricultural Policy (CAP), does specify that it is looking at improving the soil for future generations, as well as the air, water and better practices—so immediately the mainstream agricultural system will try to fix its broken frame around these new money incentives and make all their endeavours deliberately appear to be done solely to improve the water, livestock welfare, wildlife diversity and soil, while still compacting the soil with heavy machinery, spraying deadly poisons (just a little less) and perhaps incorporating better fertilising methods to give the PR some spiel to wax lyrical about. Of course, what should be done is to split up all the farms into 10-acre smallholdings and the fertility of the soil would not only go through the roof, so would the food production and the health potential of the food made from the crops grown in these newly rejuvenated, mineralised, man-made chemical-free soils. This is where the subsidies should be going—to the smallholders who genuinely have the betterment of Mother Earth and the food produced as their goals, not profits, new 4x4s and holidays.

Owing to Brexit there has been the Direct Payments to Farmers Bill aimed at covering the shortfall from the end of EU subsidies. Considering that small organic farms have never received a penny, while the large desecrating monocultural farms have been given the bulk, it makes no sense. The new subsidy is known as the Basic Payment Scheme, which only begins at 5 hectares, about 12.5 acres, so all 10-acre smallholders would not receive a penny, while those above 5 hectares would. If 5 hectares continues to be the cut-off, it might be necessary to reconsider the 10 acres and upgrade it to 5 hectares with planning permission, because such a cash incentive would ultimately mean more crops grown sustainably, while improving the soil and consumer health.

Brexit is an ideal potential opportunity to get some priorities in order, to start to display proper economic sense as determined by *oikonomia* and invest in some 'good household management' in not just the agricultural system but all cultural systems. Especially as we could do away with subsidies, reclaim the necessary land to guarantee national food security, regenerate the countryside, employ millions while preventing a vast proportion of future illness at the same time. Boom!

Pharmas to farmers

We saw how each chapter on the seven plants had plenty of scientific corroboration for their health benefits, meaning the farmers have the very real capacity and potential via improved fertility and education to maximise the nutraceutical potentiality of the nutrient-dense plants growing in newly mineralised soils. This would make a portion of the current pharmaceutical output redundant by growing nutraceuticals, literally switching from pharmas to farmers.

An early hurdle will be ending the spraying of agro-chemicals, which can be done easily enough with a little tutoring, because we can never get to real health if we keep ingesting deleterious chemicals. The farmers are already responsible for never questioning the rationale in following the regime set by the seed and chemical sellers, even when they can see yields dropping and pests increasing—the farmers need to lose their blind faith and help in this revolution by listening to what has been going on in France, with their recent 40,000+ converts to organic farming and their increased profits and more importantly, the cessation of spraying chemicals, proving there is and has always been a better alternative. As we saw before, the percentage of lost crops is more today (37%) after decades of a deadly chemical regime than before in 1930 (30%), while organic and permacultural smallholdings suffer losses in the region of only 2% to 5%.

It's going to be much easier to change the agricultural system with the aid of the farmers, at least until there are enough 10-acre permacultural plots, proving that less compaction, more hedgerows and trees, better soil and no man-made chemicals leads to increased profits. It'll be great if all farmers not only 'knew' the marketing spiel from the agro-chemicals' representatives, but were also made aware of books such as Maria Rodale's *Organic Manifesto*, or at least the key facts, to make them aware of what they have been really doing to the countryside and offering a better way forward. Many farmers might be stuck in the agricultural dogma and dependent on subsidies and chemicals, but they are unarguably experts on the land they farm. With some education and proof that any changes will not cause economic disadvantage, it might even provide some pride for the farmer, as well as the undying respect and appreciation of those consuming the better crops grown.

Why would any farmer want to knowingly continue potentially giving someone cancer or inheritable genetic dysfunctions from the broccoli they grew? Broccoli has been proven to have one of the highest onboard levels of deleterious agricultural chemicals of all fruit and vegetables, making one of the healthiest vegetables dangerous. It's a dark irony when one thinks of all the children forced to eat broccoli on the insistence of it being good for them! Of course, it is excellent if organic, just like grapes and berries—but, imagine the horror in the realisation that this determination on getting their loved ones to eat their greens could actually have been the cause of a cancer later in life? It is an absolute disgrace that food we assume to be healthy is actually harmful, while the real healthy food is too expensive for the majority.

Our personal body-load of pollutants has gotten to the point where we are saturated, from our blood, skin, hair, organs and lipid deposits (many of these chemicals are fat-soluble and stored in body fat). There is no abatement to the production and use of these chemicals, whether sprayed on our food, as fire retardants on our furniture, leaching from plastic food coverings or 'beauty' products with their myriad toxins. With this chemical overload it is incredible we have struggled on for the past few decades, because when this toxic bioaccumulation was first publicised in the late 1960s and the dire warnings from the first Earth Day conference in 1969 (every 22 April is Earth Day, although of course every day should be), with immunologists, scientists, academics and researchers all coming to the conclusion we were soon to be doomed, the advice which was taken on board by many governments and measures have been taken to try to avoid this impending predicted disaster, which may well have extended the period of our survival. The fact we have gotten to 2025 would appear miraculous to many experts 40 years ago, although of course this disaster has only been delayed, not avoided.

5. Cultural system

Obscenity law

This heading is not intended in the sense of legislating against exposing your genitalia in public or calling someone a really rude word, it's about having an amount of wealth considered obscene, vulgar. There is the old argument that the wealth created privately is merely a reflection of the wealth created for society, but this is outdated and wrong; just as with any notion of a trickle-down, it clearly only funnels up. A cursory glance around the average high street suggests there isn't much wealth to be seen, yet the rich have themselves doubled and more their wealth this century, if not in just the last decade. As the general public continue to suffer austerity, the rich have actually seen their wealth grow. Clearly, the vast wealth of a few individuals is not benefiting society as a whole, so a wealth reform is long overdue to create an effective redistribution.

To properly account for everyone's wealth is going to be fun to enforce, but first off all money has to come out of offshore accounts and be redistributed equally to every human on the planet, except to those who had the accounts, who themselves will feel no ill-effect from the loss of money deemed surplus. With an estimated £20 trillion in these offshore accounts (probably more) this would mean a Robin Hood Redistribution of £2,500 per capita, which when used collectively could easily make millions of villages around the world self-sufficient in energy, water and food *ad infinitum*.

There's always the tax answer to consider. Everyone complains about the corporations not paying enough, or any taxes, but it never changes and the super-rich pay virtually zero personal income tax, while the corporations get tax relief. So we could start with the simple rule that all taxes are paid, with all corporation accounts being completely transparent. Depending on the severity of any transgressions, as corporations are anything but ethical, there is the possibility of companies becoming a public entity. This very matter has been raised with Purdue (the makers of OxyContin), who may receive a court order for a reorganisation to transform it into a 'public benefit company', run on behalf of the communities devastated by the opioid crisis. This is exactly what should happen to DuPont (Teflon), Ineos (plastics), Monsanto (glyphosates) and other mass desecrators who should be up for crimes against humanity and the environment, instead of getting tax breaks! What these mass polluters have committed is known as Ecocide, which incredibly is not even against the law, so since the 1970s and more recently brought to the fore by environmental lawyer-activist Polly Higgins (RIP), these corporations may soon have to pay for the environmental catastrophes they have caused.

Thomas Piketty, who we met a little earlier with his $r > g$ equation, calls for a general wealth tax, as have many in America such as Bernie Sanders and other Democrats, where the total wealth of an individual is taxed every year. This will always fail unless they learn from the fiasco in Europe when trying to implement similar taxes, with all profits suddenly shifted elsewhere with the establishment of offshore accounts in the 1980s. In the US and elsewhere there are currently huge failings in the actual tax paid, with proper taxation and no evasion, increases in corporate tax on larger companies and proper adherence to inheritance tax, which will surely even out the playing field a little.

* * *

If indeed the $r > g$ equation is true, how is it not possible for a worker-owned industry to not experience the same growth in returns? The only way to create equality is to have the means of production and the capital in the control of the workers, and this means every stage from the initial crops and manufacture to distribution and market outlets.

There are of course arguments against Piketty, as there are for all economic theories—it's a theory after all—but without being finicky, in essence Piketty is correct, just as Marx was, and the rich get richer as the wealth gets siphoned to a minute number of people.

Because $r > g$ does appear to be the case for the rich, it is imperative to implement taxation mechanisms and asset stripping, otherwise the problem of the super-rich will always continue to arise—it's almost as if a side effect of being rich is getting richer—and its very unavoidability makes it necessary to cap wealth for those unlucky enough to be experiencing uncontrollable wealth accumulation. If we say for argument's sake this upper limit of total wealth is set at a £1 billion, it would fall year-on-year until nearer to £10 million. Being poor and frugal I have no conception of what it's like to be wealthy and lavish, although a figure will have to eventually be agreed upon.

It is truly morally depleted and highly offensive that people find it necessary to flaunt their wealth and with people adulating them for doing exactly that—so there is an immediate problem as opinions differ as to what may represent obscene. It has to strongly offend the prevalent morality for it to be deemed obscene. The etymology of the word is uncertain, although the Latin *obscaenus* means 'boding ill, disgusting, indecent', fits the bill. Obviously each and every one of us has a different moral barometer with conflicting calibrations, but there are some things the people all agree on, and in times of austerity, rising prices, an energy crisis, refugees, homelessness, wars and social unrest, how can immense and increasing wealth ever be justified? This was exactly the question that resulted in the following pamphlet becoming rather fashionable in the nineteenth century. The time for the pamphlet is surely coming around again, the perfect antidote to the global digital media would be an 8-page pamphlet, possibly sparking the revolution.

The Co-operator

In yet another cycle of time, proving so much has been done before, a fellow Brightonian, Dr William King (1786–1865), was upset in the 1820s to see that with all the money being lavished by the rich, there should be no poor or criminals; he blamed the existence of both on society, not the individual, and primarily a lack of food. In May 1828 he began to publish *The Co-operator*, a 4-page monthly pamphlet beginning every month with the mantra "Knowledge and Union are Power: Power directed by knowledge is happiness: Happiness is the end of Creation". King's plan was to grow some produce and fund a produce shop, sell the produce and buy more land with the profits to make more produce and buy more land. The idea spread fast nationally, with 300 other co-operatives established, although within a decade most had failed, often from not garnering enough local support and trust. Here's an excerpt from one of the first editions:

> If the workman does not grow the food, nobody else will; and all the world must starve—not the workman only, but everybody else. How is it then, that when the workman has grown the food, or made a valuable piece of goods, he can scarcely get enough food to support life, or a tolerable stock of domestic comforts? Everybody else has plenty. But the man who grew the food, or cloth, or builds the houses, can scarcely get any food or clothing, or a house to live in! This is certainly a most extraordinary fact.

Through their sequential publications the pamphlets outlined a viable and reasonable means as to how the working classes could become their own masters, promoting a fair capitalist system whereby shops, stocked by co-operatives of workers, would collectively grow richer and empowered together, and as they sold more they could save more and purchase their own land and houses to achieve true freedom, repeating the process throughout the nation. *The Co-operator* was influential in the initial move towards co-operatives (even the stores, banks etc. of the same name today) and improving pay and conditions, undoubtedly encouraging the development of the trade unions emerging in the 1830s, along with the Chartists, whose intention was in upholding the People's Charter of 1838.

Dr King could not understand how any caring Christian society could allow such wealth injustices to continue for so long. Religious affinity aside, the injustices continued, although with more attention given to sweeping this ethical problem under the carpet. By being told we are much better off than those of the past, we believe it and accept our lot, and the system continues. But we know the system is unsustainable and that at some point unsustainability has to manifest as a phenomenon which signifies this end of unsustainability—in our case this appears to be the diseases of civilization and the corrosion and imminent failure of the life systems of the planet Herself, namely the water, air and soil. People are living on the streets and full-time workers needing food banks, while at the same time corrupt politicians sell £300 million mansions (Berlusconi) or businessmen like Philip Green, who, while 13,000 workers at Arcadia (Topshop, Topman Miss Selfridge and Dorothy Perkins) lost their jobs for Christmas 2020 along with 12,000 more at Debenhams, was sipping champagne on his £100 million yacht in Monaco.

Wetiko and follow the money

I had initially ended this book with subsections on karma, *wetiko*, collective consciousness, unity and the brain, in a fruitless attempt to stitch together disparate topics, hoping these would somehow create a utopian ideal where a sustainable and equal future will naturally emerge, like a beautiful butterfly from a chrysalis. Yeah right! So, out they went, being all the better for the change, leaving just this summary—

The veritable mountain of bad karma white man may well have stockpiled is not the reason we are getting forever closer to a catastrophe. It is our very culture that is causing the impending doom, a culture infected with the *wetiko* virus, even if unbeknown by most. *Wetiko* is a Cree word from Canada, who when cultures collided from the sixteenth century saw the embodiment of their tales of a cannibalistic entity famed for eating everything necessary for survival, living in these white people. The First Nation Peoples knew instinctively that the victory of the white man was temporary, the country would still be there when the white man, or at least his ways, have long gone, so clear was the unsustainability of our cultural existence.

Language itself is a part of this *wetiko*. Taoists believe the human race first strayed from the path of Tao with the emergence of language, getting in the way of *wu-wei* (a non-interfering, natural, intuitive and connective life). Language makes the *wetiko* virus more contagious, as well as providing the ultimate justification for the entire culture, by putting us under a spell with words—which is why etymological examples were prevalent throughout our text, to show that words themselves have hidden meanings and have often in the past been deliberately corrupted to hide the truth.

Yet another phenomenon outside of the Western realm of understanding alongside chi, chakras and karma, is collective consciousness, which sounds like a 1970s Funk band—"Please welcome to the stage Chi, Chakra, Karma and the Collective Consciousness". Understanding the concepts of karma and *wetiko* is useful, but they are not the change required for humanity to move forwards. Unity and collective consciousness definitely play a role, as it is the collective consciousness that can be tapped into to begin the changes outlined throughout this book. It us understanding we are all a part of the same whole. Because the undeniable truth is that it is She, Gaia, the Earth who nourishes us, whether white, black, brown, yellow, pink, small, big, fat, skinny, hairy, smelly, rich, poor, good, bad, Buddhist, Muslim, Tory or Anarchist. All are the Mother's family, and it is plants by their nourishment that allows us to live. Man didn't make these plants—if anything they made us, or the bacteria may have played a role, but that's a road too far for this book. What gives us life and health besides food has to be the common unity, and in this we share a collective consciousness that will ultimately empower us. I'll leave this here again—Our Mother provides while our Culture divides.

Why we have not tapped into this collective consciousness was summarised well by Andy Thomas in his book *The Truth Agenda*:

> Apathy, however, has been the biggest neutraliser of our own potency. The icy focus of the highly coordinated intent gives the elite the upper hand over a collective mind that has been weakened by distraction and deception, diverted socially and chemically … Our own collective power of intent has been eroded down to almost nothingness in a culture that seems no longer to know what it wants, nor is able to agree on what it thinks matters. The intent from the opposition, meanwhile, has grown ever sharper. It is time to wake up and learn how to use both the weight of numbers *and* intention to change this unacceptable situation. We must become as attentive and focused as those who are shaping the world for us to our detriment.

Yet the elephant in the room regarding what the problem is and what has to change (along with our reconnection with nature) began to take up so much room it was impossible to skirt around it. That elephant is of course money. We've seen all those huge numbers, all those revenues, profits and GDPs; we already know it to be the Devil, would sell our soul for it … whatever the problem, you can be sure most of the time it's the money. From the wars, murders, crime, fallings out, we all suffer stress directly from money.

One of, if not the biggest changes of mind is a shift in the value placed on money. Obviously we need to have a way to be paid for our time at work, pay for goods etc. That isn't going to change, and the Green Currency can easily accommodate for this. What needs to change is our value attached to money, the value attached to jobs and people who earn vast sums. Why is the cleaner, farmer or nurse worth a 100th of a banker or a 1000th of a footballer? This is a part of the universal value that also requires a change of mind.

Those astronomical revenues have been mentioned because it is these sums of money that actually run the economic world, so these biggest corporations, employing tens of thousands and reaping revenues of $100bn and more, are deemed untouchable. In a utopian world these massive corporations would produce and sell for just a fraction more than cost, and all necessities, with the bulk of consumables being superfluous, not to mention unenvironmental. What's missing from the life of most is not consumables, it is connection, community and other such ethereal free phenomena such as love and friendship.

Publicly owned corporations could function like this—profit is necessary to pay for all wages and overheads—but being profit-driven as opposed to people- or quality-driven is erroneous, considering it's hard to find evidence to show that private corporations are more driven and better led than public ones. There isn't any evidence, because it doesn't work like that in reality: the media may portray an image of private business being on a higher level than that obtainable from government/public-led ones, but it's just not true. Trains don't run any better when privately owned, television programmes don't get better, water doesn't run cleaner and factories aren't more productive. The only difference is that only a few benefit (massively) when private, whereas a public-run corporation would benefit everyone.

If a corporation employs globally 100,000 people, if that sector of industry was fragmented and restructured locally, the sector would probably actually need to employ more and the overall revenues would be undiminished. The truth is, the dissolution and restructuring of the global corporations would create more smaller companies, local jobs and increased local income, with profits staying put and not flitting away to foreign owners, centralised multinational factories, stockbrokers and hedge funds. If supermarkets ended it would mean those working there could shift to smaller specialist stores, with the high street revived—there is no unemployment and it doesn't stop industry being serviced or a product made. If Tesco closed down, the high street could gain a greengrocer, a butcher, fishmonger and a general store, creating local demand to supply them. If Amazon were to be outlawed, the high street would be fully restored, because for every pound gone the Bezos way is a pound lost from the high street. Wealth has become concentrated while poverty has spread, and make no mistake, money isn't just miraculously made, it's quite a strict one-in-one-out policy.

That straightforward equation from Piketty, $r > g$, is not exclusively only available to a rich investor putting in their capital (money) since capital is also the workforce itself, the original capital as from Marx's *Das Kapital*. Capital, if it now means money, can be accumulated collectively by the workforce or investors and borrowed against the Green Currency.

This immoral and unethical civilization and the nepotistic, rich-serving, corrupt economic system has to end, and end they are guaranteed to do. What else happens to something that is unsustainable? Our culture can either end when it finally extinguishes the planetary systems, or it can end simply by the people rejecting it, walking away and starting something better. As the German philosopher Arthur Schopenhauer (1788–1860) said: "All truth passes through three stages. First, it is ridiculed. Second, it is violently opposed. Third, it is accepted as being self-evident."

Socialism?

There's also no point in trying to skirt around this, and just as with money, it's more than deserving of its own section. Not wishing to be drawn into any political talk, all I will say is Socialism is not Communism—we are living in the twenty-first century. The title of this section is in homage of Einstein's essay, a tiny segment of which appeared earlier in this chapter. Socialism is most simply defined as a society led by egalitarian principles, all working for the common benefits of that society.

Socialism is the ultimate threat to capitalism, with the social engineering departments in the capitalist world fully focused in demonising this word. In America, the only ever true enemy for the past century and a half has been whatever threatens the capitalist monopoly, thus making socialism enemy number one (communism is seen as the same thing there), as all its principles are directly opposed to the ideals of capitalism. For this reason alone it would be fair to say that the powers-that-be in America have decided that socialism is indeed the last thing the public can sympathise with, it being such a direct threat—and for this reason alone we should probably embrace it.

Socialism is people power, and it's an irresistible force happening right now that will continue until the reformation is complete, but it is difficult to see it for what it is when placed in the thick of it. H.G. Wells (1866–1946), father of science fiction with his *War of the Worlds*, *The Time Machine*, *The Island of Doctor Moreau* and *The Invisible Man*, ruminated on this. A selection of his essays collected as *First and Last Things* from 1929 stated: "Socialism for me is a common step we are all taking in the realization of God's purpose of human organization and unity."

What some authors have to say at the end of their careers and lives can be extremely revealing as to their inner spiritualism and belief. For example, D.H. Lawrence (1885–1930), author (most notoriously, *Lady's Chatterley's Lover*), playwright, poet, painter, traveller and modernist, in his posthumous *Apocalypse*, saw it as fitting to end his last ever book thus:

> So that my individualism is really an illusion. I am a part of the great whole, and I can never escape. But I can deny my connections, break them, and become a fragment. Then I am wretched. What we want is to destroy our false, inorganic connections, especially those related to money, and re-establish the living organic connections, with the cosmos, the sun and earth, with mankind and nation and family.

Such a sentiment spans all of written history, as shown by some of the oldest texts in the world from India. The Sanskrit phrase *Vasudhaiva Kutumbakam* is found in all texts from the *Rig Veda* to the *Upanishads*; with no religious overtones, it means 'the world is one family'. Bob Marley and the Rastafari say "One Love", while the Sufis say there is a divinity within every person that can be realised if love, faith and optimism can be released, a belief also shared by the Catholics. To be as controversial as possible, let's conclude with the words Karl Marx used to finish *The Communist Manifesto*: "The proletarians have nothing to lose but their chains. They have a world to win. Working men of all countries, unite!"

Take a bow, the Seven Plants

Focus for this last half chapter had to be on how to secure food security for the long term, for not just the islands of Britain but the world as well. The agricultural system is unsustainable, needing to be replaced with a network of 10-acre permacultural smallholdings, providing mineral-rich, agricultural chemical-free foods—which is how as many as 70% of the world already feed themselves, if often from marginalised circumstances. Some 680,000 smallholdings, all feeding the nation the healthiest, tastiest and broadest phytonutrient-diverse diet ever, with food so good it acts as a nutraceutical, nourishing the immune system and acting as a preventative from most diseases. With the smallholdings bringing life and community back to the countryside, the regeneration of the soil also brings back the life and community of the soil, with the trillions of microbes restoring its immune system—an imperative step in restoring the human's immune system and health in general.

The seven plants will be central to the initial success of the permacultural smallholdings, and as soon as any of these plants begin to gain traction, immediately the potential for secondary and tertiary revenue streams from the various processing avenues becomes real. This opens the door to a vast surplus of vegetative matter for all manner of industrial applications from buildings to paper, bioplastic, clothing, transport and nutraceuticals. Cannabis will soon be grown in significantly greater amounts once antiquated and unjust laws finally fall away, providing an initial surge in raw materials for a diversity of uses, as will happen once nettles begin to be accepted as a valuable food source. Then the systemic reformation will have begun as it will happen internally, as by merely eating a mineral-rich source of food from real soil the entire human organism can be reharmonised from the cellular level up, creating the necessary environment for that change of mind to happen naturally.

With new revenue streams opened, further plants can be used to fulfil demand, whether dandelion leaves for the chlorophyll and supplements or its roots for rubber and nutraceuticals. As the veracity of the nutraceutical route is verified with further scientific and real-life effectiveness, more plants can be used for the same ends. It is actually happening right now; the seeds have already been sown for a sustainable future to blossom.

If seven plants really could initiate all of this, just imagine what we could achieve with a better understanding of all the plants on our beautiful planet ... food, medicine, homes, energy and water can all be individually and community-provided for, a scenario we could have accomplished a century ago, but unfortunately money, profit and greed got to rule instead, and have become definite causative factors in the imminent environmental and economic catastrophe we now face.

The seven plants are a good start, particularly as six are so easy to see everywhere. It's definitely worth getting to know them—as hard as it is to put into words, the difference in 'knowing' them is huge. If you identify a plant when out, see the nettle or bramble poking through the fence, the hawthorn in the hedge, the dandelion in the grass or the plantain by the edge of the path, there's a connection, and by doing so the entire plant kingdom becomes open to this same communion. All you need to do is recognise a plant, and this equates to saying 'hello'. You pay it a little respect, appreciate its plantality and make it a part of our world, because in reality we are of course all from One World, and it most definitely does not belong to humans.

* * *

I have saved the biggest quote till the end. Coming direct from the heart of the old ways, the ways of the hunter-gatherer, the leader of the Waorani, an Amazonian tribe for whom the destruction of the entire planet by our civilization is clear to see, Nemonte Nenquimo wrote:

> The Amazon rainforest is my home. I am writing you this letter because the fires are raging still. Because the corporations are spilling oil in our rivers. Because the miners are stealing gold (as they have done for 500 years), and leaving behind open pits and toxins. Because the land grabbers are cutting down primary forest so the cattle can graze, plantations can be grown and the white man can eat. Because our elders are dying from coronavirus, while you are planning your next moves to cut up our lands to stimulate an economy that has never benefited us ... You forced your civilization upon us and now look where we are: global pandemic, climate crisis, species extinction and driving it all, widespread spiritual poverty. In all the years of taking, taking, taking from our lands, you have not had the courage, or the curiosity, or the respect to get to know us ... It took us thousands of years to get to know the Amazon rainforest. To understand her ways, her secrets, to learn how to survive and thrive with her. And for my people, the Waorani, we have only known you for 70 years (we were 'contacted' in the 1950s by American evangelical missionaries), but we are fast learners, and you are not as complex as the rainforest ... When you say that you are urgently looking for climate solutions, yet continue to build a world economy based on extraction and pollution, we know you are lying because we are the closest to the land, and the first to hear her cries ... This forest has taught us how to walk lightly, and because we have listened, learned and defended her, she has given us everything: water, clean air, nourishment, shelter, medicines, happiness, meaning. And you are taking all this away, not just from us, but from everyone on the planet, and from future generations ... I say to all of you ... the Earth does not expect you to save her, she expects you to respect her. And we, as Indigenous peoples, expect the same.

APPENDIX

1. Nutritional profiles
2. Pesticide sprayed since 1950
3. Pollution
4. GDPs vs corporation revenues
5. Humans
6. Historical pandemics

1. Nutritional profiles

For most of the seven plants nutritional analysis carried out on the familiar dietary nutrients of vitamins and minerals has been minimal, except for blackberry and nettle, with only the former being considered a part of a 'normal' diet. The focus of this book has been the phytochemicals from scientific study. Useful for this appendix has been the superb resource of Jim Duke's amazing phytochemical and ethnobotanical database hosted by the USDA, with an overwhelmingly huge amount of information available, so a big thank you to the late Jim Duke, a folk hero in America with his legacy of the Green Farmacy Garden educating many future generations to come—

Nettles

(% of RDAs in brackets—there are no RDAs established for phytochemicals) 100g blanched *Urtica dioica* from Alaska provides (believe me if I could have afforded it on my frayed shoestring budget I would rather have done a full spectrum analysis of all seven plants picked locally)—2011IU Vitamin A (40% of RDA, composed of 101mcg Vitamin A RAE, 114mcg Carotene alpha, 1150mcg Carotene beta and 4178mcg Lutein + Zeaxanthin), 21mg Betaine, 481mg Calcium (37%), 17mg Choline, 18mcg Chromium, 0.076mg Copper (8%),

497mcg Vitamin K (416%), 57mg Magnesium (14%), 0.8mg Manganese (34%), 0.388mg Niacine (B3—2%), 71mg Phosphorus (10%), 334mg Potassium (7%), 0.103mg Pyridoxine (B6—8%), 0.16mg Riboflavin (B2—12%), 0.008mg Thiamine (B1—1%), 1.64mg Iron (9%), 0.3mcg Selenium (1%), 4mg Sodium, 0.34mg Zinc (3%), 42 calories, 87.5% Water, 7.5g carbohydrate (3%), 6.9g Fibre (25%), 2.71g Protein (6%), when dried 18% protein with a great balance and range across the amino acid spectrum, from the essential ones—Isoleucine, Leucine, Lysine, Methionine, Tyrosine, Phenylalanine, Threonine, Valine and Histidine, as well as a good amount of the non-essential Alanine, Arginine, Asparagine, Glutamine, Glycine, Proline and Serine.

Among the phytochemicals and compounds nettles have—acetic acid, acetophenone, acetylcholine, acetyl-glucoside, adenosine, aesculetin, agglutinin, alanine, alkaloids, alpha-bisabolol, alpha-pinene, alpha-tocopherol, angelic acid, apigenine, arginine, aluminium, arsenic (aluminium is shown to be part of biological activity with c. 80ppm, and although 0.06ppm of arsenic is a concern, with all other detrimental metals given as ppm, the intake of some in minuscule quantities may be useful but lead or radioactive materials have no positive biological functionality—these are American results, and British soil will have different impurities), ascorbic acid (Vitamin C), aspartic acid, astragalin, azulene, beta-carotene, beta-sitosterol, bisacetyl-neo-olivil, boron, bromine, butyric acid, cadmium (0.03ppm), caffeic acid, caffeoylmalic acid, calcium, campesterol, carbonic acid, carvacrol, catechin, cellulose, chamazulene, chlorine, chlorogenic acid, chlorophylls, choline, choline acetyltransferase, chromium, cinnamic acid, citral, citric acid, cobalt, coproporphyrin, coumarin, daucosterol, dehydroconferyl-alcohol, dienoic acid, ellagic acid, enoic acid, ferulic acid, fibre, flavonoids, flavonol, fructose, fluorine, folacin, formic acid, fructose, fumaric acid, gaba, galactose, gallic acid, glucosamine, glucose, glucuronic acid, glutamic acid, glycerol acid, glycerol, glycolic acid, herniarin, hesperetin, histamine, homovanllyl, hydroxycoumarine, 5 hydroxytryptamine, iron, isobutyl angelate, isocitric acid, isoleucine, isoquercitrin, isorhamnetin, kaempferol, lead (4ppm), lecithin, lectins, leucine, lignans, linoleic acid, linolenic acid, lutein, luteolin, lycopene, lysine, magnesium, malic acid, maltose, manganese, mannose, mercury (0.015ppm), methionine, molybdenum, mucilage, myricetin, naringenin, naringin, neochlorogenic acid, neoolivil, niacin, nickel, nitrates, nitrogen, nobilin, oleanolic acid, oleic acid, olivil, oxalic acid, p-coumaric acid, patuletin, palmitic acid, pantothenic acid, pectin, phenylalanine, phosphoric acid, phosphorus, pinoresinols, populnetin, porphyrins, quercetin, quinic acid, riboflavin, rosmarinic acid, rutin, scopoletin, secoisolariciresinol, selenium, serotonin, sililic acid, silicon, sitosterol, sodium, strearic acid, stigmasterol, succinic acid, sucrose, sulphur, syringic acid, tanninnic acid, tannin, tartaric acid, threonine, thymol, tin, titanium, tyrosine, ursolic acid, urtica terpenes, valine, vanillic acid, vanillin, violaxanthin, xanthophyll, xylose, zeatin and zinc, among more than 250 others in its chemical structure.

Dandelions (% of RDAs in brackets)

100g raw leaves provides—10,160IU Vitamin A (203%), 0.2mg Thiamine (B1—13%), 0.3mg Riboflavin (B2—15%), 0.8mg Niacine (B3—4%), 0.1mg Pantothenic acid (B5–1%), 0.3mg Pyridoxine (B6—13%), 27mcg Folate (B9—7%), 35mg Vitamin C (58%), 3.4mg Vitamin E

(17%), 778mcg Vitamin K (973%), 187mg Calcium (19%), 35mg Choline, 0.2mg Copper (9%), 3.1mg Iron (17%), 36mg Magnesium (9%), 0.3mg Manganese (17%), 66mg Phosphorus (7%), 397mg Potassium (11%), 0.5mcg Selenium (1%), 76mg Sodium (3%), 0.4mg Zinc (3%), 45 calories, 9.2g carbohydrates (3%), 3.5g Fibre (13%), 85.6g water, 2.7g Protein (6%), 44mg Omega3, and 261mg Omega6.

Jim Duke lists 280 separate chemicals found in the dandelion, and some of these chemicals include—2-4 arnidiol, 3,4-dihydroxycinnamic acid, aesculetin, alpha amylase, alpha tocopherol, aluminium (320ppm), arnidiol, arsenic, ascorbic acid, asparaginic acid, barium (65ppm), benzyl glucoside, beta-amyrin, beta-carotene, beta-sisterol, beta-tetrahydroridentin, boron, bromine, cadmium (2.6ppm), caffeic acid, campesterol, capric acid, carboxyarabinitol, cerotic acid, chicoric acid, chlorine, chlorogenic acid, choline, chromium, chrysanthemaxanthin, cobalt, coumaric acids, coumestrol, cryptoxanthin, cycloartenol, cynaroside, daucosterol, faradiol, ferulic acid, flavoxanthin, fructose, glucose, glutamic acid, glycerol, hydroxycinnamic acids, inositol, inulin, iodine, isoquercitrin, isorhamnetin, jasmonic acid, lauric acid, lead (6ppm), lecithin, levulin, levulose, lutein, luteolin, mannitol, molybdenum, mucilage, myristic acid, nickel, nicotinic acid, oleic acid, oxalic acid, p-coumaric acid, palmitic acid, pectins, phytosterols, pseudotaraxasterol, quercetin, quercimeritrin, rubber, rubidium, saccharose, scopoletin, selenium, sesquiterpene lactones, silicon, stearic acid, stigmasterol, strontium, sucrose, sugars, sulphur, tannin, taraxacin, taraxacolide, taraxasterol, taraxerol, taraxinic acid, tartaric acid, tin, titanium, tyrosinase, vanillic acid, violaxanthin, xanthophyll, xylose and at least 150 others.

Yarrow

Dr Duke recorded 149 chemicals including—1,8-cineole, acetylbalchanolide, acetylagelolide, achilleine, achillin, achimillic acid, adenine, alanine, alpha-bisabolol, alpha-cadinol, alpha-curcumene, alpha-phellandrene, alpha-pinene, alpha terpinene, alpha-terpineol, alpha-thujone, aluminium (20ppm), anacyclin, apigenin, arabinose, artemesia–ketone, artemetin, ascaridole, ascorbic acid, asparagine, azulene, beta-carotene, beta-elemene, beta-pinene, beta-sisterol, beta-thujone, betaine, borneol, bornyl-acetate, butyric acid, caffeic acid, calcium, camphene, camphor, caprylic acid, carvacrol, caryophyllene, casticin, chamazulene, chlorogenic acid, choline, chromium, cineole, cis-carveol, cobalt, cosmosin, coumarins, cuminaldehyde, delta-cadinene, dextrose, diazaphenoxanthin, dulcitol, eo, eucalyptol, eugenol, ferulic acid, fibre, folacin, folic acid, formic acid, furfural, galactose, gallic acid, gamma-terpinene, geranial, geranyl-acetate, glucose, glutamic acid, glycine, guaiazulene, heptadecane, histidine, humulene, hydroquinone, inositol, inulin, iron, isoborneol, isohamnetin, isovaleric acid, leucine, limonene, linalool, linoleic acid, linolenic acid, luteolin, luteolin glucoside, lysine, magnesium, maltose, mandelic acid, manganese, mannitol, menthol, millefin, myrcene, myristic acid, neryl acetate, niacin, oleic acid, p-cymene, palmitic acid, phenol, phloroglucinol, phosphorus, potassium, protocatechuic acid, pyrocatechol, quercetin, quercitrin, riboflavin, rutin, sabinene, salicylic acid, saponins, selenium, sesquiterpenes, silicon, sodium, stachydrine, stearic acid, stigmasterol, succinic acid, sucrose, swertisin, tannic acid, tannin, terpinen-4-ol, terpineol, terpinolene, tetramethoxyflavone, thiamin, thujone, thymol, tin, tricosane, trigonelline, vanillic acid, yomogo alcohol, zinc and quite a few others.

Plantain

Psyllium seed (*Plantago psyllium*)—100g provides 235 calories, 4.5g protein, 3.7g fat, 0.3g fibre, 334mg calcium, 20mg iron, 51mg magnesium, 1.6mg manganese, 63mg phosphorus, 811mg potassium, 54mg sodium, 2.1mg zinc, beta-carotene (Vit A) 4023IU.

Dr Duke lists just 102 chemicals found within *Plantago major* (there are many more chemicals than this as on average most plants have in excess of 550), including—acetoside, adenine, allantoin, amylfuran, anethole, apigenin, ascorbic acid, asperuloside, aucubin, baicalein, baicalin, benzaldehyd, benzoicacid, beta-carotene, boschniakinic acid, bourbonene, caffeic acid, calcium, camphor, capric acid, capronaldehyde, carvacrol, carvone, caryophyllene, catalpol, chlorogenic acid, choline, chromium, cineole, cinnamic acid, citric acid, cobalt, cyclocitral, cymene, damascenone, decadienal, dibutyl phthalate, dimethylfulvene, eo, fenchone, ferulic acid, fibre, fokienol, fumaric acid, geniposidic acid, gentisic acid, geraniol, heptanal, hispidulin, ionone, iron, isomenthon, isophytol, isospathulenol, lignoceric acid, limonene, linalool, linoleic acid, linolenic acid, luteolin, magnesium, manganese, margaric acid, megastigmatrienon, methol, menthol, menthone, methyl linoleate, mucilage, myristic acid, neo-chlorogenic acid, neomenthol, nepetin, nerol, nerolidol, niacin, nonanal, ocimene, octanol, octen, octenal, oleanolic acid, oleic acid, p-coumaric acid, p-hydroxy-benzoic acid, palmitic acid, pentadecanoic acid, pentyl vinyl ketone, phosphorus, phytol, phytone, plantagonine, plantamajoside, plantease, potassium, propyl vinyl ketone, pseudoionone, pulegone, riboflavin, salicylic acid, selenium, silicon, sodium, sorbitol, spathulenol, stearic acid, succinic acid, syringic acid, syringin, tannin, terpinene, terpineol, terpinolene, tetradecanol, thujone, thymol, toluene, tyrosine, tyrosol, ursolic acid, valeranone, vanillic acid, viridifloral and zinc.

Cannabis

Hemp seed oil (approx. values)–Linoleic acid/Omega6 (50–70%), Linolenic acid/Omega3 (15–25%), Oleic acid/Omega9 (10–16%), Palmitic acid/Omega7 (6–9%).

Dr Duke identifies 253 chemicals, whereas 545 have been identified from other sources among which are—3,4,5-trihydroxybibenzyl, acetaldehyde, acetone, alanine, alpha-bisabolol, amyrin, apigenin, arabitol, arginine, ascorbic acid, aspartic acid, azelaic acid, benzyl acetate, beta-bisabolene, beta-carotene, bibenzyls, borneol, cadaverine, caffeic acid, calcium, campesterol, camphene, camphor, cannabamine A, B, C & D—among 113 identified cannabinoids there is cannabichromene (CBC—with 8 distinct compounds), cannabichromenic acid (CBCA), cannabichromevarin (CBCV), cannabicyclol (CBL), cannabidiol (CBD—8 compounds), cannabidiolic acid (CBDA—3 compounds), cannabielsoin (CBE), cannabigerol (CBG—3 compounds), cannabigerolic acid (CBGA—2 compounds), cannabigerol monomethylether (CBGM), cannabigerovarin (CBGV), cannabinodiol (CBDL), cannabinodiolic acid (CBNDA), cannabinol (CBN—3 compounds), cannabisin D, cannabistilbene I & II, cannabitriol (CBT), cannflavin A & B, delta 1, 8, 9 &10 tetrahydrocannabinol (THC) tetrahydrocannabinolic acid (THCA), tetrahydrocannabivarin (THCV) and tetrahydrocannabivarinic acid (THCVA)—caryophyllene, cis-anethole, coniferic acid, coumaric acid, courmaroyl, cycloartenol, dammaradienol, delta-cadinene, fibre, geranyl diphosphate, glucosamine, hexanoate, histidine, humulene, iron, kaempferol, lariciresinol, limonene, linalool, linoleic acid, linolenic acid, luteolin, medioresinol, mevalonic acid, monoterpenes, muscarine, myrcene, niacin, oleic acid, olivetolic acid, orientin, phenanthrenes, phenylalanine, phosphorus, pinene, pinoresinol, quercetin, riboflavin, secolsolariciresinol, sesquiterpenes, spirans, stilbenes, syringaresinol, terpinolene, thiamin, trans-anethole, trigonelline, triterpenes and tryptophan.

Hawthorn

100g of fresh haws provides—7.6mg Vitamin C, 18mg calcium, 187mg potassium and 95 calories.

Results from Dr Duke's database for the hawthorn appear to be a mix of the Midland or Woodland hawthorn (*Crataegus laevigata*) and the English hawthorn (*Crataegus monogyna*). The Woodland hawthorn has 237 chemicals listed, while the *Crataegus monogyna* has only 153 chemicals documented including, among many others—5-hydroxytryptamine, acantolic acid, acetylcholine, acetyl vitexin, adenine, adenosine, aesculin, aluminium (19ppm), amygdalin, apigenin, ascorbic acid, beta-carotene, beta-sitosterol, cadmium, caffeic acid, calcium, catechin, chlorogenic acid, choline, chromium, citric acid, cobalt, copper, crataequinones, cynaroside, dimethyl-amine, dopamine, epicatechin, epinephrine, esculin, ethylamine, fibre, glucose, guanine, horhammer, hyperoside, iron, isoorientin, isoquercitrin, isovitexin, kaempferol, lauric acid, linoleic acid, linolenic acid, lupeol, luteolin, magnesium, malonylglucoside, manganese, maslinic acid, niacin, oleanolic acid, opc, orientin, oxalic acid, palmitic acid, pectin, pentacyclic triterpenes, phosphorus, potassium, proanthocyanidins, procyanidin, procyanidin B to G, pyonogenol, quercetin, quercitrin, riboflavin, rutin, saponins, schaftoside, selenium, silicon, sorbitol, spiraeoside, starch, stearic acid, sucrose, tartaric acid, thiamin, tin, trimethylamine, tyramine, ursolic acid, vicenin-2, vitexin, xanthine and zinc.

Blackberry (RDAs in brackets)

100g serving provides—50 calories, 0.3mg betaine, 8.5mg choline, 214IU Vitamin A (4%), 0.02mg thiamin (B1 2%), 0.026mg riboflavin (B2 2%), 0.646mg niacin (4%), 0.276mg pantothenic acid (B5 6%), 0.030mg pyridoxine (B6 2%), 25mcg folate (B9), 21mg Vitamin C (23%), 1.17mg Vitamin E (alpha tocopherol 5%), 19.8mcg Vitamin K (16%), 29mg calcium (2%), 0.165mg copper (18%), 0.62mg iron (3%), 20mg magnesium (5%), 0.646mg manganese (28%), 22mg phosphorus (3%), 162mg potassium (3%), 0.4mcg selenium (1%), 1mg sodium, (0.53mg zinc (5%), 1.39g protein, 5.3g fibre (19%), 9.6g carbohydrate (3%), 94mg Omega3, 186mg Omega6 and 88.15g water.

There has been limited chemical analysis on the blackberry (*Rubus fruticosus*) on Dr Duke's database, with only 21 chemicals itemised—apigenin, arbutin, ascorbic acid, beta-amyrin, boron, chlorogenic acid, ferulic acid, hesperidin, hydroquinone, inositol, lactic acid, malic acid, oxalic acid, resveratrol, sitosterol, stigmasterol, succinic acid, tannin, tiliroside and ursolic acid—with over 500 more, many of which are shared among all the *Rubus* genus.

Brazilian studies in 2018 on their common blackberry, *Rubus ulmifolius*, uncovered among the phenolic compounds—aminobenzoic acid, benzoic acid, caffeic acid, catechin, chlorogenic acid, chrysin, coniferaldehyde, coumaric acid, dihydroxybenzoic acid, epigallocatechin galiate, epicatechin, galangin, gallic acid, isoquercitrin, isohamnetin, kaempferol, luteolin, naringenin, naringin, quercetin, pinobanksin, pinkcembrin, salicylic acid, sinapaldehyde, sinapic acid, syringaldehyde, syringic acid, rutin and vanillic acid.

The raspberry's (*Rubus idaeus*) profile and probably the blackberry also includes—acetic acid, alpha-carotene, alpha-tocopherol, aluminium (220ppm), benzaldehyde, beta-carotene, beta-ionone, beta-sitosterol, boron, butyric acid, caprylic acid, chromium, cinnamyl-alcohol, citric acid, cobalt, cyanidin-3-0-glucoside, cyanin, dextrose, diacetyl, ellagic acid, epicatechin, ethanol, ethyl-acetate, farnesol, fibre, formic acid, fructose, furfural, gentisic acid, geraniol, glucose, lactic acid, levulose, linoleic acid, linolenic acid, lutein, malic acid, malitol, manganese, mannitol, mufa, oleic acid, p-coumaric acid, p-hydroxy-benzoic acid, palmitic acid, pantothenic acid, pectin, pedunculagin, procyanidin-B-4, propionioc acid, protocatechuic acid, pufa, riboflavin, selenium, silicon, sorbitol, stearic acid, succinic acid, sucrose, tannin, thiamin, tin, tocopherol, valerianic acid, pyridoxine, xylitol and xylose among others.

2. Pesticide sprayed since 1950

The year 1950 is a useful starting point for the saturation of the land by agricultural chemicals, although the use of pesticides dates back a lot further—the Sumerians (from ancient Mesopotamia) used sulphur compounds to control pests as long as 5,000 years ago. There was little development, or probably need for pesticides until around the fifteenth century when, no less stupidly than today, compounds of arsenic, lead and mercury were applied to crops, with no reports of adverse effects, which is definitely not proof of there being none.

The neonicotinoids are getting deserved press currently as being a leading contender for continuing the insect and bird Armageddon—but this has been caused by all the chemicals in the past 70 years, it's not just one that can be blamed. Neonicotinoids originate from the tobacco plant, which in the seventeenth century had nicotine sulphate extracted from the leaves to be used as an insecticide. More natural highly toxic plant chemicals were extracted for use as insecticides during the nineteenth century, notably pyrethrum from chrysanthemums and rotenone from the roots of various tropical plants. Arsenic was still being used as the base in most of these formulations until the 1950s when the white-coat brigade firmly took over, making arsenic look inert as the toxicity went through the roof with the introduction of the organochlorines such as DDT. The effects of this were documented in Rachel Carson's seminal 1962 book *Silent Spring*, with the observation that the birds were no longer singing, having been poisoned by eating the insects that had eaten from the sprayed plants.

The fuss caused by this and the clear evidence of the danger of these organochlorines caused a shift in the 1970s to the equally dangerous organophosphates and carbamates—these work by affecting the uptake of the acetylcholine enzymes and, as we saw in in the Nettle chapter, the current neurological dysfunctions are remarkably similar to the effects of these toxins specifically targeted to prevent acetylcholine functionality. Sprayed in extraordinary quantities every year, with only the industrial-led research telling us the chemicals are safe, while independent research strongly disputes this claim. One of the deadliest chemicals literally being forced down our throats is Monsanto's genome-disrupting and cancer-causing glyphosate, which, at exposure levels lower than 1 part per billion, destroys gut bacteria and nerve cells, and damages the kidneys and liver. With thousands of legal trials over the coming decades, the dangers of glyphosate will be only too clear to see, and hopefully production will be stopped and Monsanto and owner Bayer turned over as public concerns.

We met the nettle tea anarchist when we discovered that nettles are probably more effective than Diclofenac, without wiping out the vultures—well, this is a crochet evening with the Women's Institute compared to the teaspoon terrorist administering neat agricultural chemicals! The figures about to be revealed are beyond outrageous. Remember, these are by some distance the deadliest chemicals ever invented, and they have been willingly sprayed on our food for 70 years. Only now is evidence surfacing that these corporations knew full well of the potential devastation they were reaping, getting lost in economics and the profits, not any population control conspiracy. However, the outcome has been the same, a deadly saturation of carcinogenic and mutagenic chemicals, affecting not just the past few generations, also the ones yet to be conceived.

Pesticides (literally meaning 'the death of pests') are not just applicable to pests such as weevils, beetles and flies, but also refer to any or all of these—Algicides (algae), Avicides (birds), Bactericides (bacteria), Fungicides (fungi), Herbicides (plants), Insecticides (insects), Miticides (mites), Molluscicides (snails), Nematicides (nematodes), Rodenticides (rodents), Slimicides (algae, bacteria, fungi and slime moulds) and Virucides (viruses) = Planetary Genocide.

Be seated for this, as the numbers are staggering—there are different estimates as to the actual quantity of pesticides used every year, although we're not after exact figures here. It's just an example, and ten per cent here or there makes little difference. Estimates range from 2.4 to 4.1 million tons (mt) of pesticides used every year across the globe for the past few years, so taking 3 million tons as an average for usage today and a fairly confident 60,000 tons being used worldwide in 1950 as a starting date, there was a 50-fold increase in consumption from then till 2000.

Although not accurate for every year, this tally gives an indication as to the spread—60,000 tons were produced in 1950, 120,000 in 1951, 180,000 in 1952 and so on until 660,000 tons in 1960; the total continued to rise to approximately 1.32mt being made in 1970, 1.98mt in 1980, 2.64mt in 1990, 3mt every year by 2000, then remaining constant till 2020. In reality some years would have remained the same, while others would have much larger increases as new products are brought to the market. The total of all pesticides sprayed across our beautiful planet in 70 years therefore is a depressingly stupendous 140 million tons.

* * *

So how deadly has all this been? Taking a ridiculously conservative lethal dose of pesticide as being 5g to kill one adult – parathion, for instance, is deadly at 4mg/kg, so for a large 100kg adult a dose of 400mg (0.4g) would be lethal, and a teaspoon (5g) spilled in a gallon of water would kill more than 12 adults. Considering that 5g of cocaine or 5g of paracetamol are deemed a deadly amount, 5g of pesticide is way too conservative, but it'll do to be on the safe side—although this is definitely the dark side.

At 5g lethal dose, 1 ton of pesticide would be enough for 200,000 deaths from our teaspoon assassin if he/she were able to administer a teaspoon's worth individually to that many people. It would be enough to wipe out my home town/city of Brighton with just 1 ton, and there are 3 *million* tons being sprayed annually across the globe. This means our very busy and lightning fast teaspoon assassin could take out 600,000,000,000 people if they gave individual teaspoon doses out—that's six hundred billion deaths at the hands of the impossible assassin, because there's only 8 billion humans. In other words, the teaspoon assassin could wipe out 75 times the population of the world every year from the pesticides being sprayed on our food and gardens, roadsides and verges. But this conservative six hundred billion dead a year is still not quite the final figure: the total 140,000,000 tons of pesticide produced over 70 years could potentially kill twenty-eight trillion people (28,000,000,000,000)—most of it sprayed on our food! Have these numbers started to sink in?

Not wishing to add to this unbelievable potential carnage, 5g is a stupid amount considering that a millionth of a gram has been clearly shown to cause mutagenic damage, messing with chromosomes and DNA replication at this and even lower levels—a millionth of a gram—and there has been ONE HUNDRED AND FORTY MILLION TONS of the deadliest chemicals ever made sprayed—a millionth of a gram! I don't know about you, but I'm raging writing these words, and to make it even worse we've paid for quite an amount of this pestilence to be sprayed via subsidies from income tax and VAT. Apologies for being such a downer, but it was never going to be good news, and sadly it just gets worse—

3. Pollution

We saw what potential deaths and destruction those 3mt of pesticides could reap annually. Well, this pales into insignificance when one looks at the quantity of pollution produced from all the industries, from manufacturing to construction, transport and energy production. This figure is somewhere in the region of 500mt (which includes the 3mt from pesticides) every year and rising year-on-year, despite some countries reducing output in the past few years. This 500mt equates to a total somewhere in the region of 25 billion tons of highly toxic pollution created since 1950—with industry being polluting for many centuries, but nothing like on the scale of the past 70 years.

Toxicity of these industrial pollutants is for this example on a par with the pesticides, with a lethal dose from our teaspoon terrorist being one teaspoonful. The poison produced to continue with our consumerist capitalist culture produces enough deadly waste (the majority of which is simply pumped into the air or water) to potentially kill 100,000,000,000,000 people—a hundred trillion—every year!

These are completely insane numbers, but please stop for a minute and think about what you've just read. Our culture produces enough pollution to easily kill the human race 12,820 times over every year! The vast majority of this is deposited in the planetary systems our survival and wellbeing is dependent on. Remember the Parsi and the Towers of Silence? Whose way of life was to never desecrate fire, air, water or soil—can you imagine within their cultural consciousness allowing five hundred million tons of a pollutant? Yet we assume it's perfectly okay for it to be put in our air, water and soil. This chemical catastrophe has not been a product of real intelligence. If we were really clever, we would never have allowed any poisons to get anywhere near us or our food.

4. GDPs vs corporation revenues

These figures are an IMF 2022 estimate, and they get it pretty spot on most years. It clearly demonstrates how big these corporations are and who it is calling the shots. There have been dozens of omissions of companies in the following list worth in the tens and hundreds of billion pounds, so it is a tiny proportion of the better-known companies with some of the largest and quite a few countries' GDPs for comparison. This is revenue, not capital, or stockmarket worth as this includes futures and is speculative. The global lockdowns have made massive differences to the vast majority of the population, causing most small and medium businesses to suffer, while on the other side of this destitution coin, the biggest companies have made significant profits, posting their biggest profits in 2022. It's also worth considering that for the majority of the companies listed a majority of the shares will be held by just a few investment and hedge fund companies, namely Vanguard, BlackRock, Northern Trust, JP Morgan, Geode, Bank of America, State Street, Berkshire Hathaway; these in turn are owned by a few families, including the UKs royal one, Rothschild, Orsinin, Rockefeller, DuPont, Vanderbilt and Morgan; they all figure prominently, if rather secretly. It is with these few that the real power sways.

The list is definitely illuminating when the changes in the past decade are considered, as the last list I compiled was for the appendix of *The Seven Deadly Whites* using 2014 figures. The biggest change in the intervening years has been the huge rise in value of the health sector (service providing, insurance etc., not just pharmaceuticals), outperforming even the petroleum giants, with retail continuing to rise also (the list is illustrative and is far from definitive, with many massive companies missing).

GDPs WORLD £85 trillion

1. USA £21 trillion
2. China £15 trillion
3. Japan £3.6 trillion
4. Germany £3.3 trillion
5. India £2.9 trillion
6. UK £2.7 trillion
7. France £2.3 trillion
8. Canada £1.8 trillion
9. Russia £1.7 trillion
10. Italy £1.7 trillion
11. Iran £1.7 trillion
12. Brazil £1.6 trillion
13. South Korea £1.4 trillion
14. Australia £1.4 trillion
15. Mexico £1.2 trillion

16. Spain £1.2 trillion
17. Indonesia £1 trillion
18. Saudi Arabia £820 billion
19. Netherlands £800 billion
20. Turkey £705 billion
27. Thailand £444 billion
31. Nigeria £419 billion
    AMAZON £390 billion
33. Egypt £390 billion
    SAUDI ARAMCO £332 billion
41. Denmark £321 billion
    STATE GRID CHINA £320 billion
    CHINA PETROLEUM & CHEMICAL £319 billion
    APPLE £314 billion
42. Pakistan £312 billion
45. Chile £257 billion
    WALMART £247 billion
    UNITED HEALTH GROUP £247 billion
47. Czech Republic £246 billion
    VOLKSWAGEN £245 billion
    CVS HEALTH £242 billion
48. Iraq £235 billion
    TOYOTA £234 billion
    EXXONMOBIL £233 billion
49. Finland £233 billion
    BERKSHIRE HATHAWAY £229 billion
    SHELL £217 billion
    ALPHABET £214 billion
    McKESSON £213 billion
50. Portugal £212 billion
    SAMSUNG £186 billion
54. Greece £186 billion
    AMERISOURCEBERGEN £183 billion
55. Qatar £183 billion
    COSTCO WHOLESALE £174 billion
    IND & COM BANK of CHINA £173 billion
    CHINA STATE CONSTRUCTION £171 billion
    GLENCORE £169 billion
    CHINA CONSTRUCTION BANK £168 billion
56. Ukraine £166 billion
    MICROSOFT £154 billion
58. Hungary £153 billion
    TOTAL £153 billion
    PING AN INSURANCE £150 billion
    AGRICULTURAL BANK of CHINA £150 billion

MERCEDES-BENZ £148 billion
STELLANTIS £147 billion
AT&T £135 billion
BP £131 billion
THE HOME DEPOT £125 billion
AXA £120 billion
60. Morocco £119 billion
ALLIANZ £115 billion
FORD £113 billion
BMW £109 billion
ALIBABA £108 billion
DEUTSCHE TELEKOM £107 billion
61. Angola £104 billion
JP MORGAN CHASE £103 billion
META PLATFORMS £98 billion
COMCAST £96 billion
63. Ecuador £95 billion
ÉLECTRICITÉ de FRANCE £83 billion
NESTLÉ £79 billion
70. Guatemala £76 billion
E.ON £75 billion
SONY £75 billion
TESCO £71 billion
71. Bulgaria £70 billion
PFIZER £67 billion
74. Belarus £66 billion
PEPSICO £65 billion
76. Tanzania £64 billion
WALT DISNEY £61 billion
84. Ivory Coast £57 billion
UNILEVER £51 billion
90. Myanmar £50 billion
TESLA £45 billion
CATERPILLAR £42 billion
100. Latvia £34 billion
SAINSBURY £34 billion
COCA-COLA £32 billion
111. Senegal £23 billion
ADIDAS £21 billion
125. Albania £15 billion
ROYAL MAIL £14 billion
134. Jamaica £13 billion
ROLLS-ROYCE £12 billion
148. Rwanda £10 billion
eBAY £9 billion

151. Mauritania £8 billion
 JD SPORTS £8 billion
 FOOT LOCKER £7 billion
162. Montenegro £5 billion
 NEXT £5 billion
167. Fiji £4 billion
 INVESTEC £3 billion
178. Bhutan £2 billion
216. Tuvalu £53 million

5. Humans

POPULATION (000,000s)

	WORLD	UK	FRANCE	GERMANY	CHINA	NIGERIA
10,000BC	10 (probably nearer to 50)					
5,000BC	20 (probably nearer to 100)					
Year Zero	300	0.8	5	3	59.6	
1000AD	400	2	6.5	3.5	59	
1500	500	4	15	12	103	
1700	600	8.5	21.4	15	138	
1820	1,000	21.2	31.2	24.9	381	
1913	1,800	45	41.4	65	437	20
1973	3,900	56.2	52.1	78.9	881	79.8
1998	5,900	59.2	58.8	82	1,200	116
2021	7,800	67	65.4	83.9	1,444	220

Notes on table above: Estimates for historic world population are varied, to say the least—many exclude any at all from America before Christ, when there was probably 20 to 50 million (which is why the population of America was omitted) and before 1900 Nigeria was a series of ill-defined tribal regions. Archaeologists estimate ancient populations by extrapolating out actual evidence, which often leads to a considerable underestimate, but even with a 50 million global population in 10,000 BC, population density would have been minuscule compared to today.

Billion population milestones—c. 1810 1 billion, 1927 2 billion, 1960 3 billion, 1974 4 billion, 1987 5 billion, 1999 6 billion, 2012 7 billion, 2022 8 billion.

* * *

As with the pandemic deaths next, it is important to appreciate that the population of the past was only a fraction of what it is today. Farming families are generally stereotyped as being large, although the explosion in human population only became possible with the growth of cities. The city is an essential component of capitalism, and indeed the growth of the city is a perfect reflection of this.

As cities began to emerge, with their awful stench and unsanitary conditions, cheap labour and further divisions emerged. This is when enclosure begun in earnest (mid-1700s), when there was a displacement from the country with landowners evicting and demolishing villages, and at the same time severe vagrancy laws further handicapping the recently homeless.

POPULATION (000s)

year	1650	1750	1850	1900	1950	2020
London	500	675	2,700	6,500	8,300	9,300
Birmingham	7.5	24	232	505	2,220	2,600
Glasgow	10	35	300	750	1,750	1,670
Liverpool	2.5	22	250	685	1,380	901
Cardiff	1.5	2	45	160	310	477
Sheffield	2.5	10	175	450	645	730
Bristol	15	45	160	330	548	686
Norwich	20	25	65	100	110	145

Notes on table above: Norwich was surprisingly the second largest city in the UK for some time before 1650. The table shows the effects of enclosure emptying the countryside from 1750 to 1900, with cities clearly being a perfect breeding ground for humans.

Another observation from the above figures is that the cities reached their peak in terms of growth by the 1930s to 1950s, with only slight growth since then (usually with expansion and agglomeration of outlying areas), if not decline as for Liverpool and Glasgow. This indicates that the trend is for the decrease to continue, which will be as a result of less immigration, coupled with the move to the countryside with the roll-out of the 10 acres with planning permission, as quality food becomes a priority on the national agenda. In fact, the city is a huge part of the problem and 'it', as an entity, is well passed its sell-by date. The future needs many smaller self-sufficient communities, not massive, singular, depleting, dependent, cancerous cities.

6. Historical pandemics

Just to put the recent Covid-19 pandemic into some historical perspective, bear in mind that world population in 1900 was 2bn, in 1800 it was 1bn and in 1500 it was 500m, in AD 1000 400m and in AD 1 it was only about 300m, a 26th of what it is today. Thus a million deaths in AD 0 is the equivalent of 26m today, and a million in AD 1000 is the equivalent of 19.5 million today when comparing death rates as percentage of population (figures in **bold** are total deaths for length of plague/illness)—

AD 165–180—Antonine Plague—Roman Empire—smallpox?—**5m–10m**
AD 250–266—Plague of Cyprian—Europe—smallpox?—**1m+**

AD 541–542—Plague of Justinian—Eur/W Asia—plague—**25m–100m**
AD 735–737—Japanese smallpox—Japan—**2m**
1346–1353—Black Death—Europe/Asia/N Africa—plague—**75m–200m**
1520–1800—Smallpox—N and S America—**20m–100m**
1545–1576—*Salmonella enterica*—Mexico—**15m+**
1563—London Plague—London—**20,100**
1592–1593—London Plague—London (mainly)—**19,900**
1596–1602—Spanish Plague—Spain—**600,000–700,000**
1603—Plague—England (mainly London)—**30,000**
1609—Egyptian Plague—Egypt—**1m**
1625—Plague—England (mainly London)—**35,000**
1629–1631—Italian Plague—Italy—**280,000**
1636—Plague—England—**10,000**
1647–1652—Great Plague of Seville—Spain—**600,000–700,000**
1656—Naples Plague—Italy—**1.25m**
1663–1664—Netherlands Plague—**24,148**
1665–1666—Great Plague of London—**100,000**
1668—French Plague—**40,000**
1679—Great Plague of Vienna—Austria—**76,000**
1710–1712—Great Northern War Plague—Den/Swe/Lith—**164,000**
1738—Great Plague of '38—Balkans—**50,000**
1770–1772—Russian Plague—Russia—**50,000**
1772—Persian Plague—Persia—**2m+**
1775–1776—Plague—England—unknown deaths, possibly **20,000**
1812–1819—Ottoman Plague—Ottoman Empire—**300,000+**
1813—Caragea's Plague—Romania—**60,000**
1816–1826—First Cholera Pandemic—Asia/Europe—**100,000+**
1829–1851—Second Cholera Pandemic—Asia/Eur/N America—**100,000+**
1846–1860—Third Cholera Pandemic—Russia—**1m+**
1855—Norfolk's Yellow Fever—Norfolk/Portsmouth, England—**3,000**
1855–1860—Plague Pandemic—worldwide—**12m +** in India/China alone
1863–1875—Fourth Cholera Pandemic—Middle East—**600,000**
1875—Fiji Measles Outbreak—Fiji—**40,000**
1881–1896—Fifth Cholera Pandemic—India/Germany—**9,000+**
1889–1890—Flu Pandemic—worldwide—influenza—**1m**
1899–1923—Sixth Cholera Pandemic—Eur/Asia/Africa—**800,000+**
1910–1912—Pneumonic and Bubonic Plagues—China—**100,000**
1915–1926—*Encephalitis lethargica* Pandemic—worldwide—**1.5m**
1918–1920—Spanish Flu—worldwide—Influenza H1N1—**17m–100m**
1918–1922—Typhus—Russia—**2.5m** est.
1957–1958—Asian Flu—worldwide—Influenza H2N2—**1m–4m**
1977–1977—Smallpox—worldwide—**500m**
1968–1970—Hong Kong Flu—worldwide—Influenza H3N2—**1m–4m**
1972–1973—London Flu—US—Influenza H3N2—**1,027**
1981–present—HIV/AIDS Pandemic—worldwide—**32m–44m**

1996–2001—UK BSE Outbreak—worldwide—vCJD—**231**
2006—Plague—Congo–**61**
2009–2010—Swine Flu Pandemic—worldwide—Influenza H1N1/09—**151,000–575,000**
2013–2016—Ebola Virus Epidemic—West Africa—**11,300**
2014–2017—Bubonic Plague—Madagascar—**292**
2015—Indian Swine Flu—India—Influenza H1N1—**2,035**
2017–2018—Seasonal Flu—US—'Seasonal' Influenza—**60,000–80,000+**
2019–2022—Covid-19 Pandemic—worldwide—**5m?**

Add to this **650,000** from flu every year and **750,000** from iatrogeny, while heart disease and cancer each account for a whopping **10 million deaths** every year.

BIBLIOGRAPHY

My cataloguing and note-taking are horrendous, coupled with a broken hard drive and an ignorance of saving to desktop, which resulted in being faced with a succession of blank pages when trying to compile this list. Scientific journals make up the first part of each chapter's bibliography, followed by the websites and finally the books. All in all, the following corroborates the general thesis of this book, making it less a figment of my own imagination, more a trail of threads collaborating to form a comfy cardigan.

Introduction

Barker *Flower Fairies of the Wayside* (1985 Penguin).

Nettle

Adamski & Bieganska Herba Polonica 1984;30:17–26 *Studies on substances present in* Urtica dioica *L. leaves …*

Alford Alternative Therapies in Health and Medicine 2007;13(6):58 *The use of stinging nettles for pain.*

AlKhatib et al. European Scientific Journal 2014 July;10(21):302–309 *Leaf extract of* U. pilulifera *down regulates …*

AlKhatib et al. European Scientific Journal 2014 Aug;10(24):333–340 U. pilufera *mimics the actions of metaformim …*

Anderson et al. Phytochemistry 1978;17:1875–1877 *Water-soluble glycoprotein from* Urtica.

Badirzadeh et al. PLOS Neglected Tropical Diseases 2020 Jan;14(1):e0007843 *Antileishmanial activity of* Urtica dioica …

Barlow & Dixon Biochemical Journal 1973;132:15–18 *Choline acetyltransferase in the nettle* Urtica dioica *L.*

Behzadi et al. Avicenna Journal of Phytomedicine 2016 Nov–Dec;6(6):686–695 *Effects of* Urtica dioica *supplementation on blood lipids …*

Bhaskar et al. Food Science & Nutrition 2016 Jan;4(1):119–124 *Comparison of nutritional properties of stinging …*

Broer & Behnke Journal of Rheumatology 2002;29(4):659–666 *Immunosuppressant effect of IDS 30, a stinging …*

Chrubasik et al. Phytomedicine 1997;4:105–108 *Evidence of antirheumatic effectiveness of* herba Urtica …

Cummings & Olsen Wilderness & Environmental Medicine Journal 2011;22(2): 136–139 *Mechanism of action of stinging nettles.*

D'Abrosca et al. Scientific Reports 2019;9:4986 Urtica dioica *L. inhibits proliferation and enhances …*

Damme et al. Plant Physiology 1988;86:598–601 *The* Urtica dioica *agglutinin is a …*

Dar et al. Pharmaceutical Biology Journal 2013 Feb;51(2):170–180 *Pharmacological and toxicological evaluation of* Urtica dioica.

Durak et al. Cancer Biology & Therapy 2004;3(9):855–857 *Aqueous extract of* Urtica dioica *makes a significant …*

Esposito et al. Molecules 2019 Aug;24(15):2753 *Therapeutic perspectives of molecules from* Urtica dioica *extracts for cancer treatment.*

Fan et al. Nutrients 2020 Apr;12(4):1059 Urtica dioica *whole vegetable as a functional food targeting fast accumulation …*

Fattahi et al. Asian Pacific Journal of Cancer Prevention 2013;14(9):5317–5323 *Antioxidant and apoptotic effects …*

Fattahi et al. International Journal of Molecular and Cellular Medicine 2014 Spring;3(2):102–107 *Total phenolic and flavonoid contents of …*

Ghorbanibirgani et al. Iranian Red Crescent Medical Journal 2013 Jan;15(1):9–10 *The efficacy of stinging nettle …*

Gulcin et al. Journal of Ethnopharmacology 2004;90(2–3):205–215 *Antioxidant, antimicrobial, anti-inflammatory …*

Hadizadeh et al. Pakistan Journal of Biological Sciences 2009 Jan 1;12(1):58–63 *Antifungal activity of nettle …*

Hodroj et al. Nutrients 2020 Sep;12(9):2629 *Nettle tea inhibits growth of acute myeloid leukaemia …*

Hossein Motamedi et al. Jundishapur Journal of Natural Pharmaceutical Products 2014 Nov;9(4):e15904 *Introducing* Urtica dioica, *a native plant of Khuzestan, as an antibacterial medicinal plant.*

Irshaid & Mansi American Journal of Pharmacology and Toxicology 2009;4(2):22–28 *The effects of leaf extract of* Urtica pilulifera *L. on male …*

Jankauskiene & Gruzdeviene Zemdirbyste-Agriculture 2015;102(1):31–40 *Changes in the productivity of wild and cultivated stinging …*

Jiang et al. International Journal of Oncology 2011 Jun;38(6):1675–1682 *Suppression of growth and invasive behaviour …*

Johnson et al. Phytomedicine 2013 Jan 15;20(2):143–147 *Lipophilic stinging nettle extracts possess potent …*

Kan et al. Turkish Journal of Pharmaceutical Sciences 2009;6(1):21–30 *Fatty acid profile and antimicrobial effect …*

Kasouni et al. Molecules 2021 Oct;26(20):6248 *The unexplored wound healing activity of* Urtica dioica …

Khan & Joergensen Chemosphere Journal 2006 Nov;65(6):981–987 *Decomposition of heavy metal contaminated …*

Kianbakht et al. Clinical Laboratory Journal 2013;59(9–10):1071–1076 *Improved glycemic control in patients with …*

Klingelhoefer et al. Journal of Rheumatology 1999;26:2517–2522 *Antirheumatic effect of IDS 23, a stinging …*

Konrad et al. Planta Medica Journal 2000 Feb;66(1):44–47 *Antiproliferative effect on human prostate …*

Krakowka Current Archaeology 2016;322 *The cist on Whitehorse Hill.*

Kumaki et al. Antiviral Research 2011 Apr;90(1):22–32 *Inhibition of severe respiratory syndrome coronavirus …*

Levy et al. Environmental Research 2004 Jan;1:102–111 *Bisphenol A induces feminization in* Xenopus laevis *tadpoles.*

Obertreis et al. Arzneimittelforschung 1996 Apr;46(4):389–394 *Ex-vivo in-vitro inhibition of lipopolysaccharide stimulated …*

Mahlangeni et al. Journal of Environmental Science and Health 2016;51(3):160–169 *The distribution of macronutrients, anti-nutrients …*

Mansoub Global Veterinaria 2011;6(3):247–250 *Comparison of effects of using nettle* (Urtica dioica) *and probiotic …*

Marrassini et al. Phytotherapy Research 2010;12:1807–1812 *Evaluation of antiociceptive, anti-inflammatory activities …*

Mohammadi et al. Journal of Cellular and Molecular Biology 2016 Feb 29;62(2):62–67 *Effects of* Urtica dioica *dichloromethane extract on …*

Mohammadi et al. Journal of Gastrointestinal Cancer 2016 Jun;47(2):187–195 *The herbal medicine* Urtica dioica *inhibits proliferation …*

Namazi et al. Avicenna Journal of Phytomedicine 2018 May–Jun;8(3):254–262 *Protective effect of* Urtica dioica *leaf hydro …*

Neef et al. Phytotherapy Research 1995;9:45–48 *Hypoglycaemic activity of selected European plants.*

Oguz et al. Toxicology and Industrial Health Journal 2015 May;31(5):475–484 *Effects of* Urtica dioica *on oxidative stress …*

Otles & Yalcin The Scientific World Journal 2012:564367 *Phenolic compounds analysis of root, stalk …*

Paulauskiene et al. Plants 2021 Apr;10(4):686 *Influence of harvesting time on the chemical composition …*

Qujea et al. International Journal of Molecular and Cellular Medicine 2013 Winter; 2(1):21–26 *Effect of* Urtica dioica *leaf alcoholic and aqueous …*

Randall et al. Journal of the Royal Society of Medicine 2000;93(6):305–309 *Randomised controlled trial of stinging nettle for …*

Randall et al. Complementary Therapies in Medicine 2008 Apr;16(2):66–72 *Nettle sting for chronic knee pain …*

Rayburn et al. Alternative Therapies in Health and Medicine 2009;15(4):60–61 *Stinging nettle cream for osteoarthritis.*

Roschek et al. Phytotherapy Research 2009 Jul;23(7):920–926 *Nettle extract* (Urtica dioica) *affects …*

Rutto et al. International Journal of Food Science 2013 Article ID 857120 *Mineral properties and dietary value of raw and …*

Safarinejad Journal of Herbal Pharmacotherapy 2005;5(4):1–11 Urtica dioica *for treatment of benign …*

Salehzadeh et al. African Journal of Traditional, Complementary and Alternative Medicines 2014;11(5):38–40 *Antimicrobial activity of methanolic extracts of …*

Siouda & Abdennour Veterinary World Journal 2015 Dec;8(12):1458–1465 *Can* Urtica dioica *supplementation …*

Smallman & Maneckjee Biochemical Journal 1981;194:361–364 *The synthesis of acetylcholine by plants.*

Suomela et al. Studies in Conservation 2018;63;7:412–422 *Seeking nettle textiles …*

Teucher et al. Arzneimittelforschung 1996 Sep;46(9):906–910 *Cytokine secretion in whole blood of healthy subjects …*

Viktorova et al. PLOS One 2016;11(12):e0167927 *Native phytoremediation potential of* Urtica dioica *for removal of PCBs …*

Virgilio National Research Council of Italy, Institute of Biometeorology PhD paper 2013 *Stinging nettle: A neglected species with a high potential as multi-purpose crop.*

Vogl & Hartl American Journal of Alternative Agriculture 2003 DOI:10. 1079/AJAA200242 *Production and processing of organically grown fibre nettle …*

Yildirim et al. Journal of Cellular and Molecular Biology 2013 Nov 3:Suppl 59:1882–1888 *Evaluation of antioxidant properties …*

https://clinicaltrials.gov/ct2/show/NCT00422357 2007 *Effect of* Urtica dioica *on glycemic control in patients …*

https://www.coltivazionebiologica.it/nettle-mix/ *Nettle mix, organic fertilizer and organic remedy against parasites.*

https://www.nutritionvalue.org/Stinging_Nettles,_blanched_(Northern_Plains_Indian)_nutritional_value.html *Stinging nettles, blanched …*

https://www.terriwindling.com/blog/2019/06/the-folklore-of-nettles.html *The folklore of nettles.*

https://botanical.com/botanical/mgmh/n/nettle03.html *Nettles.*

https://ctr.hum.ku.dk/research-programmes-and-projects/the-margrethe-hald-archive-digitalization-and-dissemination/The_Nettle_as_a_Culture_Plant- *The nettle as a culture plant.*

https://www.c82.net/twining/plants/?id=125 *Urticacea.*

http://naeb.brit.org/uses/search/?page=1&string=urtica+dioica *Native American ethnobotany.*

https://www.healthyhildegard.com/stinging-nettle-health-benefits/ *Stinging nettle a health panacea.*

https://www.woodland-ways.co.uk/blog/primitive-crafts/paper-making/ *Paper making.*

https://en.wikipedia.org/wiki/Urtica_dioica Urtica dioica.

https://draxe.com/nutrition/stinging-nettle/ 5 *proven, remarkable stinging nettle benefits.*

https://www.cherylsherbs.com/home/herbal-profile-nettle-leaf-root-seed/ 2014 *Profile: Nettle, leaf, root & seed.*

https://ejfoundation.org/resources/downloads/the_deadly_chemicals_in_cotton.pdf *The deadly chemicals in cotton.*

https://www.drugs.com/sfx/Diclofenac-side effects.html

https://www.drugs.com/sfx/naproxen-side effects.html *Naproxen side effects.*

https://www.iceman.it/en/bowstring/# *Otzi the Iceman.*

https://www.nettlesfortextiles.co.uk [check out the whole site and Facebook page].

British General Practice 1995 Mar:162 letter by **Adrian White**.
Grand View Research 2017 Mar *Colorant market size worth $37.49 billion.*

Candlin *Tales of Old Sussex* (CountrysideBooks 1985).

Dandelion

Chatterjee et al. Evidence-Based Complementary and Alternative Medicine 2011;2011: 129045 *The efficacy of dandelion root extract in ...*

Cho et al. Clinica Chimica Acta 2002;317(1–2):109–117 *Alternation of hepatic antioxidant enzyme activities ...*

Choi et al. International Journal of Molecular Science 2010 Jan 6;11(1):67–78 *Hypolipodemic and antioxidant effects of dandelion ...*

Clare et al. Journal of Alternative and Complementary Medicine 2009 Aug;15(8):929–934 *The diuretic effect in human subjects of an ...*

Colle et al. Journal of Medicinal Food 2012 Jun;15(6):549–556 *Antioxidant properties of* Taraxacum ...

Davaatseren et al. Journal of Medicinal Food 2013 Jan;16(1):26–33 *Dandelion leaf extract protects against liver injury ...*

Davies & Kersey Contact Dermatitis 1986;14(ISS4):256–257 *Contact allergy to yarrow and dandelion.*

Diaz et al. Microbial Pathogenesis 2017 Aug;109:162–168 *In vitro assessment of* Argemone mexicana, Taraxacum ...

Diaz et al. Evidence-Based Complementary and Alternative Medicine 2018;2018: 2706417 *Isolation and identification of compounds from ...*

Flores-Ocelotl et al. BMC Complementary Medicine and Therapies 2018;18:95 Taraxacum officinale *and* Urtica dioica *extracts inhibit dengue ...*

Frohlichova et al. Environmental Monitoring Assessment Journal 2018 Feb 19;190(3):150 *An assessment of the risk of element contamination of ...*

Gargouri et al. Journal of Cellular and Molecular Biology 2017 Feb 28;63(2):67–75 *Dandelion-enriched diet of mothers ...*

Gomez-Arroyo et al. Environmental Science and Pollution Research International 2018 Feb;25(5):4739–4749 *Indicators of environmental contamination by heavy ...*

Gonzalez-Castejon et al. Phytotherapy Research 2014 May;28(5):745–752 *Reduction of adipogenesis and lipid accumulation ...*

Guo et al. Zhong Yao Cai 2015 Oct;38(10):2129–2133 *Anti-proliferation effect of* Taraxacum mongolicum ...

Hfaiedh et al. Environmental Toxicology 2016 Mar;31(3):339–349 *Hepatoprotective effect of* Taraxacum officinale *leaf ...*

Hu & Kitts Journal of Agricultural and Food Chemistry 2003;51(1):301–310 *Antioxidant, prooxidant, and cytotoxic activites of solvent fractionated ...*

Hudec et al. Journal of Agricultural and Food Chemistry 2007;55(14):5689–5696 *Antioxidant capacity changes and phenolic profile of* Echinacea purpurea, *nettle ...*

Jedrejek et al. Chemico Biological Interactions 2017 Jan 25;262:29–37 *Evaluation of antioxidant activity of phenolic fractions ...*

Jeon et al. Journal of Ethnopharmacology 2008 Jan 4;115(1):82–88 *Anti-inflammatory activity of* Taraxacum officinale.

Jeon et al. BMC Complementary & Alternative Medicine 2017;17:508–520 *Anti-inflammatory evaluation of the methanolic extract of …*

Kim et al. Immunopharmacology and Immunotoxicology 2000;22(3):519–530 Taraxacum officinale *inhibits tumor necrosis factor-alpha …*

Kisiel & Barszcz Fitoterapia 2000;71(3):269–273 *Further sesquiterpenoids and phenolics from* Taraxacum officinale.

Koh et al. Journal of Medicinal Food 2010 Aug;13(4):870–878 *Anti-inflammatory effect of* Taraxacum …

Kyu-Hyun Han et al. Gut and Liver 2017 Sep;11(5):655–666 *Heme oxygenase-1 induction and anti-inflammatory actions …*

Lee et al. Molecules 2012 Nov 7;17(11):13253–13265 *Effects of* Taraxacum officinale …

Lee et al. Journal of Ethnopharmacology 2016 Dec 24;194:1022–1031 *Anticancer effects of* Ixeris dentate *(Thunb. Ex …*

Liangliang et al. Molecules 2017;22(9):1409–1434 *Purification, preliminary characterization and hepatoprotective …*

Madcolo et al. Phytotherapy Research 1987:28–29 *Biological screening of Italian medicinal plants …*

Miller Archives of Internal Medicine 1998;158:2200–2211 *Herbal medicinals: Selected clinical considerations …*

Munkhtugs et al. Food and Chemical Toxicology Journal 2013 Aug;58:30–36 Taraxacum officiale *(dandelion) leaf …*

Ovadje et al. Pancreas 2012 Oct;41(7):1039–1047 *Selective induction of apoptosis and autophagy …*

Ovadje et al. PLOS One 2012;7(2):e30604 *Efficient induction of extrinsic cell death by dandelion …*

Ovadje et al. Oncotarget 2016 Nov 8;7(45):73080–73100 *Dandelion root extract affects colorectal cancer …*

Petlevski et al. Journal of Ethnopharmacology 2001;75(2–3):181–184 *Effect of 'antidiabetis' herbal preparation on …*

Piao et al. European Journal of Pharmacology 2015 Jun 5;756:38–42 *Taraxasterol inhibits IL-1 …*

Qian et al. International Journal of Biological Macromolecules 2014;64:392–394 *Preparation and antibacterial activity of oligosaccharides …*

Qureshi et al. Veterinary World 2016 Mar;9(3):269–275 *Histomorphological studies of broiler chicken fed …*

Rasheed et al. Current Pharmaceutical Biotechnology 2017;18(14):1116–1123 *Biomedical potentialities of* Taraxacum …

Rehman et al. Biomedicine & Pharmacotherapy 2016 Oct;83:881–891 *Therapeutic potential of* Taraxacum officinale *against …*

Rehman et al. Frontiers in Pharmacology 2017;8:875–882 *Effect of methanolic extract of dandelion roots on cancer …*

Saratale et al. Environmental Science and Pollution Research International 2018 Apr;25(11):10392–10406 *Bio-fabrication of silver nanoparticles using the leaf …*

Schutz et al. Ethnopharmacology 2006;107(3):313–323 Taraxacum *– a review on its phyto-chemical and …*

Sigstedt et al. International Journal of Oncology 2008 May;32(5):1085–1090 *Evaluation of aqueous extracts of* Taraxacum …

Swanston-Flatt et al. Diabetes Research 1989;10(2):69–73 *Glycaemic effects of tyraditional European plant …*

Sweeney et al. Journal of Herbal Pharmacotherapy 2005;5(1):79–93 *Evidence-based systematic review of dandelion …*

Takasaki et al. Biological Pharmaceutical Bulletin 1999 Jun;22(6):602–605 *Anti-carcinogenic activity of* Taraxacum *plant*. I. and II. 22(6):606–10.

Trojanova et al. Fitoterapia 2004;75(7–8):760–763 *The bifidogenic effect of* Taraxacum officinale *root.*

Wimgo et al. Journal of the Society for Biomedical Diabetes Research 2016 Summer–Fall;13(2–3):113–131 *The physiological effects of dandelion …*

Xiaohong et al. Fish and Shellfish Immunology 2018 Feb;73:197–206 *Effects of dietary dandelion extract on intestinal …*

Xueshibojie et al. European Journal of Pharmacology 2016 Oct 15;789:301–307 *Taraxasterol inhibits cigarette smoke-induced lung …*

Yan et al. Asian-Australasian Journal of Animal Sciences 2012 Oct;25(10):1439–1444 *Evaluation of* Houttuynia cordata *and* Taraxacum officinale *on …*

Yousefi Ghale-Salimi Urolithiasis 2018 Oct;46(5):419–428 *Antiurolithiatic effect of the taraxasterol on …*

Yuan-Yuan et al. Molecular Medicine Reports 2014 Apr;9(4):1381–1387 Taraxacum mongolicum *extract exhibits a …*

Yu-Cheng Li et al. Pharmaceutical Biology 2014 Aug;52(8):1028–1032 *Antidepressant effects of the water extract from …*

Zhao & Jian Heze Municple Hospital 2016 Jul 20;34(7):535–537 *Clinical effect of compound monoammonium …*

Zhi et al. International Journal of Molecular Medicine 2007;20(3):287–292 *Dandelion T-1 extract up-regulates reproductive hormone …*

Zhihao et al. International Immunopharmacology 2014 Apr;19(2):342–350 *Protective effect of taraxasterol on acute lung …*

Zhu et al. Biomedicine & Pharmacotherapy 2017;93:1010–1017 *Dandelion root extract suppressed gastric cancer …*

https://www.healthyhildegard.com/health-benefits-of-dandelion/ *The health benefits of dandelion.*

https://en.wikipedia.org/w/index.php?title=Lawn&oldid=900660192 *Lawn.*

https://en.wikipedia.org/wiki/Vaux-le-Vicomte *Vaux le Vicomte.*

https://en.wikipedia.org/wiki/Gardens_of_Versailles *Gardens of Versailles.*

https://www.planetnatural.com/organic-lawn-care-101/history/ *Lawn history.*

https://www.yourgreenpal.com/blog/every-wonder-about-the-history-of-lawns *The complete history of lawns.*

https://www.farmshow.com/a_article.php?aid=20546 *Growing dandelions for dollars.*

https://www.americanewshour.com/2020/01/03/polyolefins-market-size-growth-forecast-worth-414-58-billion-by-2026/172822/ *Polyolefins market size, growth forecast …*

https://owlcation.com/stem/Natural-Latex-and-Rubber-From-Common-and-Russian-Dandelions *Common and Russian dandelions: Nutrition, latex …*

https://www.scribd.com/document/38394329/Composition-of-Tyre *Composition of a tyre.*

https://www.nationalgeographic.co.uk/environment-and-conservation/2019/09/tyres-plastic-polluter-you-never-thought-about *Tyres: The plastic polluter you …*

https://www.gardeningknowhow.com/edible/herbs/dandelion/dandelion-seed-growing.htm *Dandelion seed growing: How to grow dandelion seeds.*

http://www.ourherbgarden.com/herb-history/dandelion-history.html *Dandelion history.*

https://www.webmd.com/vitamins/ai/ingredientmono-706/dandelion *Dandelion.*

https://www.preventdisease.com/news/12/042312_Canadian-Researchers-Studying-Anti-Cancer-Potential-of-Dandelion-Tea-and-Extract.shtml *Canadian Researchers …*

https://www.statista.com/statistics/288538/top-cancer-drugs-based-on-revenue/ *Leading cancer drugs worldwide by revenue 2021.*

https://www.biospace.com/article/global-top-10-cancer-drugs-by-sales-2021-/ *Global top 10 cancer drugs by sales 2021.*

https://khkeeler.blogspot.com/2013/04/ *A wandering botanist.*

https://news.psu.edu/photo/618976/2020/05/06/dandelion-team *Dandelion.*

https://www.drugs.com/sfx/keytruda-side effects.html# *Keytruda side effects.*

https://nutritiondata.self.com/facts/vegetables-and-vegetable-products/2441/2 *Dandelions greens, raw.*

www.thehealthiestalternative.com *More studies on dandelion root …*

https://www.verywellhealth.com/the-benefits-of-dandelion-root-89103 *Health benefits of dandelion root.*

https://extension.umd.edu/hgic/topics/dandelion *Dandelion.*

https://foodfacts.mercola.com/dandelion-greens.html *What are dandelion greens good for.*

https://www.thecandidadiet.com/healthy-benefits-dandelions *The healthy benefits of dandelions.*

https://globalnews.ca/news/212574/can-dandelions-kill-cancer/ *Can dandelions …*

https://changinglifestyleblog.wordpress.com/tag/dandelions/ *Dandelions.*

https://en.wikipedia.org/wiki/Taraxacum_officinale *Taraxacum officinale.*

https://www.cancertutor.com/dandelionroot/ *Is dandelion root effective in treating cancer?*

https://thehill.com/opinion/energy-environment/449282-hydroponic-dandelions-hold-the-key-for-a-critical-resource *Hydroponic dandelions for a …*

https://globalhealing.com/natural-health/what-are-benefits-of-organic-dandelion-leaf/ *What are the benefits of …*

https://continentaltire.com/learn/continental-constructing-tires-dandelions-0 *Continental constructing …*

https://www.organicfacts.net/health-benefits/herbs-and-spices *13 surprising benefits of dandelion.*

http://www.efloras.org/florataxon.aspx?flora_id=1&taxon_id=220013281 *Taraxacum officinale.*

http://www.botanical.com/botanical/mgmh/d/dandel08.html *Dandelion.*

Wright *The River Cottage Hedgerow Handbook* (2010 Bloomsbury).

Yarrow

Ayoobi et al. Iranian Journal of Basic Medical Sciences 2017;20:604–612 *Bio-effectiveness of the main flavonoids of* Achillea …

Ayoobi et al. Phytomedicine 2019 Jan;52:89–97 Achillea millefolium *is beneficial as an add-on therapy …*

Bansal et al. Acta Pharmaceutica Sinica B-Journal 2011;1(4):226–235 *Phenolic compounds isolated from* Pilea …

Baretta et al. Journal of Ethnopharmacology March 2012;140(1):46–54 *Anxiolytic-like effects of acute and …*

Begum et al. Biomedicine and Preventive Nutrition 2012;2(1):16–24 *Apigenin, a dietary antioxidant, modulates gamma radiation …*

Benedek et al. Phytomedicine 2006;13:702–706 *Choleretic effects of yarrow (*Achillea millefolium *s.l.) …*

Benedek et al. Ethnopharmacology 2007;113(2):312–317 Achillea millefolium *L. s.l. is the anti-inflammatory …*

Benedek & Kopp Weiner Medizinische Wochenschrift 2007;157(13–14):312–314 Achillea millefolium *L. …*

Boglarka et al. Phytotherapy Research 2008 Dec 23.PMID:19107850 *Flavonoids found in yarrow have antiproliferative …*

Cheng et al. British Medical Journal 2007 Oct 6;335(7622):699–702 *Adding fluoride to water supplies.*

Choudhary et al. Natural Product Research 2007 Sep;21(11):1032–1036 *Inhibitory effect of lactone fractions …*

Cinkilic et al. Chemical Toxicology 2013;53:359–363 *Radioprotection by two phenolic compounds …*

Conti et al. Parasitology Research 2010;107(6):1455–1461 *Essential oil composition and larvicidal activity of six …*

Csupor et al. Phytotherapy Research 2009 May;23(5):672–676 *Antiproliferative effect of flavonoids and sesquiterpenoids …*

Dalsenter et al. Reproductive Toxicology 2004;18(6):819–823 *Reproductive evaluation of aqueous crude extract …*

De Souza et al. Phytomedicine 2011;18(10):819–825 *Hypotensive mechanism of the extracts of artemetin …*

Diasa et al. Food Chemistry 2013;141(4):4152–4160 *Chemical composition of wild and commercial* Achillea …

Dogan et al. Journal of Academic Emergency Medicine 2013;4:89–91 *An unexpected anticholinergic effect …*

Fenech Mutation Research 2000;455(1–2):81–95 *The in vitro micronucleus technique.*

Giorgi et al. Phytotherapy Research 2009 Apr;23(4):540–545 *Antioxidant and cytoprotective properties …*

Guin & Skidmore Dermatology (Archive) 1987;123(4):500–502 *Compositae dermatitis in childhood.*

Hasson RAF Journal of Science 2011;22(3):11–20 *Antibacterial activity of water and …*

Hausen Dermatologica 1979;159:1–11 *The sensitizing capacity of* Compositae *plants …*

Hausen et al. Contact Dermatitis 1991;24(4):274–280 *Contact allergy to* Compositae *plants in patient …*

Hosseinimehr et al. Journal of Radiation Research (Tokyo) 2002;43(3):293–300 *Radioprotective effects of 2-imino-3- …*

Hosseinimehr Drug Discovery Today 2007;12(19–20):794–805 *Trends in development of radioactive agents.*

Hosseinimehr et al. Pakistan Journal of Biological Science 2007;10(4):637–640 *In vitro antioxidant activity of* Polygonium …

Hosseinimehr et al. Radiation and Environmental Biophysics 2009;48(1):95–98 *Radioprotective effects of hawthorn against genotoxicity induced by …*

Hosseinimehr et al. Cancer Biotherapy and Radiopharmaceuticals 2011;26(3):325–329 *The radioprotective effect of* Zataria multiflora *against …*

Innocenti et al. International Journal of Phytotherapy and Phytopharmacology 2007 Feb;14(2–3):147–152 *In vitro estrogenic activity …*

Jenabi Journal of Pediatric and Adolescent Gynecology 2014 Oct;28(5):402–404 *Primary dysmenorrhea: A double-blind randomized clinical trial.*

Jenabi & Fereidoony The Journal of Pediatric and Adolescent Gynecology 2015 Oct; 28(5):402–404 A. millefolium *is effective in minimizing the pain …*

Khan & Gilani Phytotherapy Research 2011 Apr;25(4):577–583 *Blood pressure lowering, cardiovascular …*

Kim et al. Mutagenesis 2006;21(6):361–367 *Mitochondrial dysfunction, persistently …*

Lakshmi et al. International Journal of Pharmaceutical Sciences Review and Research July–Aug 2011;9(2);Article-022 *Yarrow (*Achillea millefolium *Linn.) …*

Majnooni et al. Research in Pharmaceutical Sciences 2013;8(4):269–275 *Chemical composition and anxioltic …*

Miranzadeh et al. Nursing and Midwifery Studies 2014 Sep;3(3):e20249 *A new mouthwash for chemotherapy …*

Miranzadeh et al. European Journal of Oncology Nursing 2015 Jun;19(3):207–213 *Effect of adding the herb* Achillea …

Molochko et al. Vestnik Dermatologii I Verologii 1990;8:54–56 *(The antistaphylococcal properties of plant extracts in relation …*

Moradi et al. African Journal of Traditional Complementary and Alternative Medicine 2013;10(6):499–503 *Antispasmodic effects of yarrow …*

Nilforoushzadeh et al. Journal of Vector Borne Diseases 2008 Dec;45(4):301–306 *Comparison of* Thymus vulgaris *(thyme),* Achillea …

Orav et al. Natural Product Research 2006 Oct;20(12):1082–1088 *Phytochemical analysis of the essential oil of* Achillea …

Pain et al. International Journal of Cosmetic Science 2011;33:535–542 *Surface rejuvenating effect of* Achillea millefolium *extract.*

Potrich et al. Journal of Ethnopharmacology 2010;130(1):85–92 *Antiulcerogenic activity of hydroalcoholic extract …*

Reza et al. Indian Journal of Pharmacology 2014 *Effect of aqueous extract of* Achillea …

Rezatofighi et al. Jundishapur Journal of Microbiology 2014;7(2):e9016 *Evaluating the efficacy of* Achillea millefolium *and* Thymus …

Rithidech et al. Mutation Research 2005;585(1–2):96–104 *Protective effect of apigenin on radiation-induced …*

Saeidnia et al. PMC free article 2011;19(3):173–186 *A review of phytochemistry …*

Schempp et al. Hautarzt 2002;53(2):93–97 *Plant-induced toxic and allergic dermatitis …*

Sedelnikova et al. Mutation Research 2010;704(1–3):152–159 *Role of oxidatively induced DNA lesions in human …*

Shahani et al. Dose Response 2015 Jan–Mar;13(1):1559325815583761 *Radioprotective effect of* Achillea millefolium …

Shalizar et al. Chinese Journal of Natural Medicine 2012;10(4):247–254 Achillea millefolium *inflorescence aqueous extract …*

Shutler & Campbell Journal of Avian Biology 2007;38(1):7–12 *Experimental addition of greenery reduces flea loads …*

Stojanovic et al. Journal of Ethnopharmacology 2005 Oct 3;101(1–3):185–190 *In vitro antimicrobial activity of …*

Trumbeckaitea et al. Food Chemistry 2011;127(4):1540–1548 *Radioprotective effect of* Achillea millefolium *L. against …*

Uter et al. American Journal of Contact Dermatitis 2001;12(3):182–184 *Occupational contact urticarial and late-phase bronchial …*

Van der Weijden et al. Journal of Clinical Periodontology 1998;25(5):399–403 *The effect of herbal extracts in an experimental mouth rinse …*

Vitalini et al. Acta Biochimica Polonica 2011;58(2):203–209 *Phenolic compounds from* Achillea millefolium …

Xu et al. Journal of Agricultural Food Chemistry 2012;60(46):11625–11630 *Antioxidant and DNA-protectiveactivities of chlorogenic …*

Yaeesh et al. Phytotherapy Research 2006;20(7):546–551 *Studies on hepatoprotective, antispasmodic and calcium …*

Yamamori et al. Free Radical Biology & Medicine 2012;53(2):260–270 *Ionizing radiation induces mitochondrial reactive oxygen …*

Yassa et al. DARU Journal of Pharmaceutical Sciences 2007;15(1):49–52 *Three glycosides and immunological properties of* Achillea …

Zolghadri et al. International Journal of Molecular and Cellular Medicine 2014 Autumn;3(4):255–262 Achillea millefolium *L. Hydro-alcoholic extract …*

https://www.henriettes-herb.com/eclectic/madaus/achillea.html *Yarrow.*

https://www.botanical-online.com/en/medicinal-plants/yarrow-herb-properties *Yarrow herb properties.*

https://www.healthyhildegard.com/yarrow-health-benefits/ *7 Yarrow health benefits.*

https://www.healthyhildegard.com/yarrow-plant-uses/ *Hildegard's yarrow plant uses.*

https://whisperingearth.co.uk/category/herbs/yarrow/ *The multiple benefits and …*

https://thepracticalherbalist.com/advanced-herbalism/yarrow-the-wound-healer/ *Yarrow the wound healer.*

http://naeb.brit.org/uses/search/?string=Achillea+millefolium *Achillea millefolium.*

https://en.wikipedia.org/wiki/Aedes_albopictus *Aedes albopictus.*

https://www.ecdc.europa.eu/en/disease-vectors/facts/mosquito-factsheets/aedes-albopictus *Aedes albopictus – Factsheet for experts.*

https://www.statista.com/statistics/664868/global-top-energy-companies-by-revenue/ *Leading energy companies worldwide in FY 2020/21, based on revenue.*

https://en.wikipedia.org/wiki/List_of_countries_by_GDP_(nominal) *List of countries by GDP (nominal).*

https://www.forbes.com/sites/forbesstaff/2022/05/12/forbes-global-2000-list-2022-the-top-200/ *Forbes Global 2000 list 2022: The top 200.*

http://issg.org/database/species/ecology.asp?si=109&lang=EN *Aedes albopictus …*

https://en.wikipedia.org/wiki/Leishmaniasis *Leishmaniasis.*

https://en.wikipedia.org/wiki/Achilleamillefolium *Achillea millefolium.*

https://www.webmd.com/vitamins/ai/ingredientmono-151/yarrow *Yarrow.*

https://www.drugs.com/sfx/verapamil-side effects.html

https://www.drugs.com/sfx/indomethacin-side effects.html

https://phytochem.nal.usda.gov/phytochem/plants/show/22?qlookup=yarrow&offset=0&max=20&et= *Achillea millefolium (Asteraceae).*

http://matthewwoodherbs.com/Yarrow.html *Achillea millefolium. Yarrow.*

http://wildfoodsandmedicines.com/yarrow/ *Yarrow.*

https://www.electricsense.com - information on EMF.

https://oilhealthbenefits.com/yarrow-essential-oil/ *Yarrow essential oil.*

http://botanical.com/botanical/mgmh/y/yarrow02.html *Yarrow.*

https://witchipedia.com/book-of-shadows/herblore/yarrow/ *Yarrow.*

https://www.satyacenter.com/blog/category/health?page=3 *Yarrow environmental solution …*

https://www.bioethika.net/pdf/yarrow.pdf *Yarrow.*

https://www.prohealth.com/library/yarrow-herb-heals-wounds-much-84968 *Yarrow.*

https://draxe.com/nutrition/yarrow/ *Yarrow uses for wounds, inflammation …*
http://plants.usda.gov/java/profile?symbol=ACMIC *Profile for Achillea millefolium …*
http://www.chinatownconnection.com/chinese-superstitions.htm *Chinese superstitions.*
http://www.iging.com/intro/introduc.htm *Introduction to the I Ching.*

Baker *Discovering the Folklore of Plants* (1971 Shire Publications).
Bernays *Propaganda* (1928 H. Liveright).
Hayward *Dartmoor 365* (2020 Curlew Publications).
Hutchens Indian *Herbology of North America* (1973 Shambala Publications).
Milham *Dirty Electricity* (iUniverse Inc. 2012).

Plantain

Amini et al. Journal of Applied Animal Research 2010 March;37(1):53–56 *Effect of* Plantago major *on burn …*
Aziz et al. The Malaysian Journal of Medical Sciences 2005 Jul;12(2):22–26 *In vitro effects of* Plantago major …
Aziz et al. Journal of Ethnobiology and Ethnomedicine 2018 Jan 9;14(1):2 *Traditional uses of medicinal plants practiced by …*
Bajer et al. Journal of Food Science and Technology 2016 Mar;53(3):1576–1584 *Chemical composition of essential oils from …*
Beara et al. Journal of Agricultural and Food Chemistry 2009 Oct 14;57(19):9268–9273 Plantain *(*Plantago *L.) species as novel sources of …*
Beara et al. Journal of Pharmaceutical and Biomedical Analysis 2010 Sep 5;52(5): 701–706 *Liquid chromatography/tandem mass …*
Beara et al. Lebensmittel-Wissenschaft und-Technologie 2012 June;47(1):64–70 *Comparative analysis of phenolic profile …*
Cabrera-Jaime et al. European Journal of Oncological Nursing 2018 Feb;32:40–47 *Efficacy of* Plantago major*, chlorhexidine …*
Choi et al. Molecular Medicine Reports 2012 Oct;6(4):843–847 Althaea rosea *Cavanil and* Plantago major …
Chiang et al. Antiviral Research 2002 Jul;55(1):53–62 *Antiviral activity of* Plantago major …
Chiang et al. The American Journal of Chinese Medicine 2003;31(2):225–234 *In vitro cytotoxic, antiviral and …*
Chung et al. The British Journal of Nutrition 2008 Jan;99(1):67–75 *Asian* plantain (Plantago asiatica) *essential …*
Fernandez-Banares et al. American Journal of Gastroenterology 1999;94(2):427–433 *Randomized clinical trial of* Plantago *ovate seeds (dietary fibre) as compared …*
Ferrazzano et al. BioMed Research International 2015;2015:286817 *Determination of the in vitro and in vivo …*
Fleer & Verspohl Phytomedicine 2007 July;14(6):409–415 *Antispasmodic activity of an extract from* Plantago …
Galvez Peralta et al. Journal of Ethnopharmacology 2003 Oct;88(2–3):125–130 *Cytotoxic effect of* Plantago *spp. on cancer …*
Gómez-Estrada et al. Journal of Ethnobiology and Ethnomedicine 2011;7:article 27 *Folk medicine in the northern coast of Colombia …*

Gomez-Flores et al. Phytotherapy Research 2000 Dec;14(8):617–622 *Immunoenhancing properties of* Plantago major *leaf extract.*

Gonda et al. Acta Biologica Hungarica 2010;61 Suppl:25–34 *Screening of common* Plantago *species in Hungary …*

Hetland et al. Scandinavian Journal of Immunology 2000 Oct;52(4):348–355 *Protective effect of* Plantago major …

Holetz et al. Memorias do Instituto Oswaldo Cruz 2002 Oct;97(7):1027–1031 *Screening of some plants used in …*

Hriscu et al. Revista Medico-Chirurgicala a Societatil de Medici si Naturalisti din Lasi Jan–Mar 1990;94(1):165–170 *A pharmacodynamic investigation of the effect …*

Huang et al. Journal of Ethnopharmacology 2009 Jul 30;124(3):493–498 *Immunodulatory activity of the seeds of …*

Hussan et al. Evidence-Based Complementary and Alternative Medicine 2015; 2015:347861 *Anti-inflammatory property of* Plantago major *leaf …*

Jana et al. Journal of Environmental Management 2012 Nov 15;110:188–193 *Analysis of arsenic and antimony distribution …*

Jaric et al. Journal of Ethnopharmacology 2018 Jan;211:311–328 *Traditional wound-healing plants used in the Balkan …*

Karpilovskaia et al. Farmakol Toksikol 1989 Jul–Aug;52(4):64–67 *Inhibiting effect of the polyphenolic …*

Kartini et al. Pharmacogosy Journal 2019 Nov;11(6):1483–1489 *Antioxidant activity, HPTLC fingerprint and discriminant analysis …*

Ma & Ma Inflammation 2018 Jun;41(3):795–802 *Plantamajoside inhibits lipopolysac …*

Malinowska & Jankowski Environmental Monitoring and Assessment 2015;187:775 *Pesticide residues in some herbs growing in …*

Malizia et al. Chemistry Central Journal 2012;6(Suppl 2):S6 *Common plants as alternative analytical tools …*

Markvicheva et al. Biomeditsinskia Khimilia 2009 Jul–Aug;55(4):479–488 *Biodegradable polymer microparticles with …*

Matev et al. Vutreshni Bolesti 1982;21(2):13–7 *Clinical trial of a* Plantago major …

Mello et al. Journal of Pharmaceutical Pharmacology 2012 Aug;64(8):1177–1187 *Hydroxyl scavenging activity accounts for …*

Mello et al. Molecules 2015;20(10):17747–17759 *Protective effect of* Plantago major *extract …*

Najafian et al. Current Drug Discoveries Technology 2019;16(2):223–231 *Efficacy of* Aloe vera/Plantago …

Nostro et al. Letters in Applied Microbiology 2000;30:379–384 *Extraction methods and bioautography for evaluation …*

Ozaslan et al. The American Journal of Chinese Medicine 2007;35(5):841–851 *In vivo antitumoral effect of* Plantago …

Parhizgar et al. Iranian Journal of Kidney Diseases 2016 Jul;10(4):182–188 *Renoprotective effect of* Plantago major *against …*

Park et al. Phytotherapy Research 2007 Dec;21(12):1118–1123 *A 90 day repeated oral toxicity …*

Piyaviriyakul et al. Pharmacognosy Magazine 2017 Jul–Sep;13(51):393–399 *Effects of* Plantago major *extracts and its chemical …*

Romeh International Journal of Phytoremediation 2010 Feb;12(2):188–199 *Phytoremediation of water and soil contaminated …*

Romeh Journal of Environmental Health Science & Engineering 2014;12:38 *Phytoremediation of cyanophos insecticide …*

Saeedi et al. Acta Pharmaceutica 2013 Mar;63(1):99–114 *Evaluation of* Plantago major *L. seed …*

Sangian et al. Parasitology Research 2013 Nov;112(11):3697–3701 *Antiplasmodial activity of ethanolic …*

Sharma et al. Indian Journal of Dental Research 2016;27(4):433–436 *Antifungal efficacy of three medicinal plants …*

Shirley et al. Journal of Prosthodontics 2017 Aug;26(6):508–515 *In vitro effects of* Plantago major *axtract, Aucubin …*

Thome et al. Experimental Biology and Medicine 2012 Dec;237(12):1379–86 *Evaluation of healing wound and …*

Turel et al. Indian Journal of Pharmacology 2009 Jun;41(3):120–124 *Hepatoprotective and anti-inflammatory …*

Velasco-Lezama et al. Journal of Ethnopharmacology 2006 Jan 3;103(1):36–42 *Effect of* Plantago major *on cell …*

Wu et al. International Immunopharmacology 2016 Jun;35:315–322 *Plantamajoside ameliorates lipopolysaccharide …*

Yazd et al. The Iranian Journal of Kidney Disease 2018 Mar;12(2):99–106 *Protection against Doxorubicin-induced …*

Zubair et al. Journal of Ethnopharmacology 2012 Jun 14;141(3):825–830 *Effects of* Plantago major *L. leaf extracts …*

Zubair et al. Natural Product Research 2016;30(5):622–624 *Promotion of wound healing by* Plantago …

(no authors named) Eksperimental Klinicheskaia Gastroenterologia 2005;113(4):73–76 *Experimental study of the effect of the phytomixture …*

https://www.cloverleaffarmherbs.com/plantain/ *Plantain.*

https://www.herbalpedia.com/blog/?p=180 *Plantain.*

https://herbs.org.nz/herbs/plantain/ *Plantain.*

https://en.wikipedia.org/wiki/Plantago_asiatica *Plantago asiatica.*

https://en.wikipedia.org/wiki/Plantago *Plantago.*

https://www.botanical.com/botanical/mgmh/p/placom43.html *Plantain, common.*

https://www.drugs.com/sfx/chlorhexidine-topical-side effects.html *Chlorhexidine.*

https://www.drugs.com/npc/plantain.html *Plantain.*

http://www.armeniapedia.org/wiki/Plantago_major *Plantago major.*

http://tollundmanden.one/nye-analyser-af-det-sidste-maaltidsidereng/?lang=en *New analysis of Tollund Man's stomach contents.*

Orwell *Nineteen Eighty-Four* (Penguin Modern Classics, 2014).

Cannabis

Abrams & Guzmán Clinical Pharmacology and Therapeutics 2015 Jun;97(6):575–586 *Cannabis in cancer care.*

Aguirre-Velázquez Neurology Research International 2017;2017:2985729 *Report from a survey of parents regarding the use …*

Aldington et al. Otolaryngology: Head and Neck Surgery 2008 Mar;138(3):374–380 *Cannabis use and cancer of the head and neck …*

Andre et al. Frontiers in Plant Science 2016;7:9 *Cannabis sativa: The plant of the thousand and one molecules.*

Barrie et al. International Journal of Rheumatic Diseases 2017 Jul;20(7):789–797 *Endocannabinoids in arthritis …*

Bifulco et al. Oncology Reports 2007 Apr;17(4):813–816 *Endocannabinoids as emerging suppressors of angiogenesis …*

Biro et al. Trends in Pharmacological Science 2009 Aug;30(8):411–420 *The Endocannabinoid system of the skin in …*

Blazquez et al. FASEB Journal 2006 Dec;20(14):2633–2635 *Cannabinoid receptors as novel targets for …*

Blazquez et al. Neuropharmacology 2008 Jan;54(1):235–243 *Down-regulation of tissue inhibitor of …*

Boland University of Michigan 2014 MSc. Thesis *Life cycle energy and greenhouse gas emissions of natural fiber composites for automotive applications: Impacts of renewable material content and lightweighting.*

Borgelt et al. Pharmacotherapy 2013 Feb;33(2):195–209 *The pharmacologic and clinical effects …*

Bryant et al. Oncology Hematology 2012 July;83(1):1–10 *The intersection between cannabis and cancer …*

Caffarel et al. Oncogene 2008 Aug 28;27(37):5033–5044 *JunD is involved in the antiproliferative effect …*

Capasso The Open Neurology Journal 2017;11:61–73 *Do cannabinoids confer neuroprotection against epilepsy?*

Caro et al. Expert Review of Neurotherapeutics 2017 Nov;17(11):1069–1079 *The potential of cannabinoids in epilepsy treatment.*

Carracedo et al. Cancer Cell 2006 Apr;9(4):301–312 *The stress-regulated protein p8 mediates …*

Carracedo et al. Cancer Research 2006 Jul 1;66(13):6748–6755 *Cannabinoids induce apoptosis of pancreatic tumor …*

Casanova et al. The Journal of Clinical Investigation 2003 Jan;111(1):43–50 *Inhibition of skin tumor growth and angiogenesis …*

Cianchi et al. Clinical Cancer Research 2008 Dec 1;14(23):7691–7700 *Cannabinoid receptor activation induces apoptosis …*

Corroon et al. Journal of Pain Research 2017;10:989–998 *Cannabis as a substitute for prescription drugs …*

Cudaback et al. PLOS One 2010 Jan 14;5(1):e8702 *The expression level of CB1 and …*

Cunha et al. Pharmacology 1980;21(3):175–185 *Chronic administration of cannabidiol to healthy …*

Darkovska-Serafimovska et al. Journal of Pain Research 2018;11:837–842 *Pharmacotherapeutic considerations for use …*

De Jesus et al. Neurochemistry International May–Jun 2010;56(6–7):829–833 *Oppposite changes in cannabinoid CB1 and …*

Fayaz et al. BMJ OPEN 2016 Vol. 6, Issue 6 *Prevalence of chronic pain in the UK …*

Fernandez-Ruiz et al. British Journal of Clinical Pharmacology 2012;75(2):323–333 *Cannabidiol for neurodegenerative disorders …*

Fitzcharles et al. BMC Musculoskeletal Disorders 2014;15:258 *Rheumatologists lack confidence in their knowledge of cannabinoids …*

Fitzcharles et al. Arthritis Care & Research 2015 Nov;68(5) *Efficacy, tolerability, and safety of …*

Fraguas-Sanchez et al. International Journal of Pharmaceutics 2020 Jan 25;574:118916 *CBD loaded microparticles as a potential …*

Friedman Cancer Biochemistry Biophysics 1977;2(2):51–54 *In vivo effects of cannabinoids on macromolecular …*

Gleba & Harris Archaeological and Anthropological Sciences 2018 Print ISSN 1866-9557. https://doi.org/10.1007/s12520-018-0677-8 *The first plant bast fibre …*

Guzmán et al. British Journal of Cancer 2006;95:197–203 *A pilot clinical study of delta9-tetrahydrocannabinol in …*

Guzmán Cannabis Cannabinoid Research 2018;3(1):17–119 *Cannabis for the management of cancer …*

Haj et al. The Journal of Pharmacology and Experimental Therapeutics 2015 Oct;355(1): 66–75 *HU-444, a novel, potent anti-inflammatory …*

Hampson et al. Annals New York Academy of Sciences 2000:274–282 *Neuroprotective anti-oxidants from marijuana.*

Hart et al. Cancer Research 2004 Mar 15;64(6):1943–1950 *Cannabinoids induce cancer cell proliferation via …*

Hausman-Kedem et al. Brain and Development 2018 Aug;40(7):544–551 *Efficacy of CBD-enriched medical cannabis …*

Haustein et al. Biochemical Pharmacology 2014 Nov 15;92(2):312–325 *Cannabinoids increase lung cancer cell lysis …*

Held-Feindt et al. Journal of Neurochemistry 2006 Aug;98(3):886–893 *Cannabinoid receptors in human astroglial tumors.*

Herrara et al. Experimental Cell Research 2006 Jul 1;312(11):2121–2131 *The CB2 cannabinoid receptor signal apoptosis …*

Jia et al. Molecular Cancer Research 2006 Aug;4(8):549–562 *Delta9-tetrahydrocannabinol-induced apoptosis …*

Jones & Howl Expert Opinion on the Therapeutic Targets 2003 Dec;7(6):749–758 *Cannabinoid receptor systems: therapeutic …*

Katz-Talmor et al. Nature Reviews Rheumatology 2018 Aug;14(8):488–498 *Cannabinoids for the treatment of rheumatic …*

Kogan & Mechoulam Dialogues in Clinical Neuroscience 2007;9(4):413–430 *Cannabinoids in health and disease.*

Lastres-Becker et al. Neurobiology of Disease 2005;19(1–2):96–107 *Cannabinoids provide neuroprotection against …*

Leelawat et al. Cancer Investigation 2010 May;28(4):357–363 *The dual effects of delta (9)-tetrahydrocannabinol on …*

Leizer et al. Journal of Nutraceuticals, Functional & Medical Foods 2000;2(4):35–53 *The composition of hemp seed oil …*

Linares et al. Frontiers in Pharmacology 2018;9:315 *No acute effects of cannabidiol on the sleep-wake …*

Lopez-Rodríguez et al. Mini-Reviews in Medicinal Chemistry 2005 Jan;5(1):97–106 *Involvement of cannabinoids in cellular proliferation.*

Liu et al. Current Clinical Pharmacology 2010 Nov;5(4):281–287 *Cannabis-derived substances in cancer therapy …*

Maroon & Bost Surgical Neurology International 2018;9:91 *Review of the neurological benefits of phytocannabinoids.*

Massi et al. The Journal of Pharmacology and Experimental Therapeutics 2004 Mar; 308(3):838–845 *Antitumor effects of cannabidiol, a …*

McDougle et al. PNAS 2017 Jul;114(30):e6034–e6043 *Anti-inflammatory omega 3 endocannabinoid epoxides.*

McKallip et al. Blood 2002 Jul 15;100(2):627–634 *Targeting CB2 cannabinoid receptors as a novel …*

Mimeault et al. Prostate 2003 Jun 15;56(1):1–12 *Anti-proliferative and apoptotic effects …*

Mitchell The Asia Pacific Journal 2014;12(49):1 *The secret history of cannabis in Japan.*

Murillo-Rodríguez et al. Neurochemical Research 2018 Aug;43(8):1511–1518 *Systemic injections of cannabidiol enhance acetylcholine …*

Neale Nursing Children and Young People 2017 Sep 11;29(7):32–37 *Efficacy and safety of cannabis for treating …*

Noreen et al. Critical Reviews in Eukaryotic Gene Expression 2018;28(1):73–86 *Is cannabidiol (CBD) a promising substance for …*

Partland & Hegman Vegetation History and Archaeobotany 2018;27(4):627–634 *Cannabis utilization and diffusion patterns in prehistoric …*

Petrocellis et al. British Journal of Pharmacology 2013 Jan;168(1):79–102 *Non-THC cannabinoids inhibit prostate carcinoma growth …*

Piper et al. Journal of Psychopharmacology 2017 May;31(5):569–575 *Substitution of medical cannabis for pharmaceutical agents …*

Ramer et al. Biochemical Pharmacology 2010 Apr 1;79(7):955–966 *Cannabidiol inhibits cancer cell invasion via …*

Ramer et al. FASEB Journal 2012 Apr;26(4):1535–1548 *Cannabidiol inhibits lung cancer cell invasion …*

Rasche et al. Der Urologe 2018 May;57(5):558–562 *(Cannabinoid therapy in practice).*

Rezapour-Firouzi et al. Complementary Therapies in Medicine 2013 Oct;21(5): 473–480 *Immunomodulatory and therapeutic effects of …*

Rezapour-Firouzi et al. Complementary Therapies in Medicine 2014 Dec;2(6):986–993 *Activity of liver enzymes in multiple sclerosis …*

Roperh et al. Natural Medicine 2000 Mar;6(3):313–319 *Anti-tumoral action of cannabinoids …*

Russo et al. Chemistry and Biodiversity 2007 Aug;4(8):1729–1743 *Cannabis, pain, and sleep: lessons from …*

Sales et al. Progress in Neuropsychopharmacology and Biological Psychiatry 2018 Aug 30;86:255–261 *Antidepressant-like effect induced by …*

Sanchez et al. Cancer Research 2001 Aug 1;61(15):5784–5789 *Inhibition of glioma growth in vivo by …*

Sarfaraz et al. Cancer Research 2005 Mar 1;65(5):1635–1641 *Cannabinoid receptor as a novel target for …*

Satam et al. ACS Sustainable Chemical Engineering 2018 Jul 23;6,8:10637–10644 *Spray-coated multilayer cellulose nanocrystal-chitin nanofiber …*

Schleider et al. European Journal of Internal Medicine 2018 Mar;49:37–43 *Prospective analysis of safety and efficacy of …*

Schley et al. Brain Research Bulletin 2009 Jun 30;79(5):333–337 *Predominant CB2 receptor expression in …*

Sledzinski et al. Cancer Medicine 2018 Mar;7(3):765–775 *The current state and future perspectives of cannabinoids …*

Solis The Pharmaceutical Journal 2017 10 Jan *Cannabis for epilepsy: Is there enough evidence of efficacy?*

Stockings et al. Journal of Neurology, Neurosurgery and Psychiatry 2018 Jul;89(7): 741–753 *Evidence for cannabis and cannabinoids for epilepsy …*

Stockings et al. Pain 2018 May;159(10):1 *Cannabis and cannabinoids for the …*

Turna et al. Depression and Anxiety 2017 Nov;34(11):1006–1017 *Is cannabis treatment for anxiety, mood …*

Vaccani et al. British Journal of Pharmacology 2005 Apr;144(8):1032–1036 *Cannabidiol inhibits human glioma cell migration …*

Vara et al. Cell Death Differentiation 2011 Jul;18(7):1099–1111 *Anti-tumoral action of cannabinoids …*

Velasco et al. Neuropharmacology 2004 Sep;47(3):315–323 *Hypothesis: Cannabinoid therapy for the treatment of gliomas?*

Velasco et al. Life Sciences 2005 Aug 19;77(14):1823–1831 *Cannabinoids and ceramide: Two lipids …*

Velasco et al. Molecular Neurobiology 2007 Aug;36(1):60–67 *Cannabinoids and gliomas.*

Watson et al. Archives of General Psychiatry 2000 Jun;57(6):547–552 *Marijuana and medicine: Assessing the science …*

Whyte et al. Pharmacology 2010;85(6):328–335 *Cannabinoids inhibit cellular respiration …*

Yasmin-Karim et al. Frontiers in Oncology 2018;8:114 *Enhancing the therapeutic efficacy of cancer …*

Zuardi Brazilian Journal of Psychiatry 2008 Sep;30(3):271–280 *Cannabidiol: From an inactive cannabinoid to a drug …*

https://www.goodjobsfirst.org/ *Tracking subsidies, promoting accountability in economic development.*

https://www.theguardian.com/sustainable-business/2016/oct/22/recreational-marijuana-legalization-big-business *Inside big pharma's fight to block recreational …*

https://www.cancer.gov/about-cancer/treatment/cam/hp/cannabis-pdq *Cannabis and cannabinoids (PDQ …*

https://www.philly.com/Health *Medical marijuana reduces prescription …*

https://www.npr.org/sections/health-shots/2016/07/06/484977159/after-medical-marijuana-legalized-medicare-prescriptions-drop-for-many-drugs *After medical marijuana …*

https://www.epilepsy.com/learn/treating-seizures-and-epilepsy/other-treatment-approaches/medical-marijuana-and-epilepsy *Medical marijuana and epilepsy …*

https://en.wikipedia.org/wiki/Hemp *Hemp.*

https://magazine.hempshopper.com/en/hemp-history-bce/171-scythians-smoked-cannabis *400 BCE – Scythians smoke marijuana from golden bongs.*

https://magazine.hempshopper.com/en/hemp-history-bce/126-2700-bce *2700 BCE: Shennong pen …*

https://www.cannabisculture.com/content/2018/05/02/new-archeological-evidence-of-cannabis-use-in-ancient-civilizations/ *New archaeological evidence …*

https://magazine.hempshopper.com/en/hemp-history/19th-century *19th century.*

https://www.ancient-origins.net/news-history-archaeology/first-hemp-weaved-fabric-world-found-wrapped-around-baby-9000-year-old *First hemp-weaved fabric in the …*

https://www.healthyhildegard.com/hemp-seeds-benefits/ *Hemp (seeds), a culinary and medicinal plant …*

https://britishsurnames.co.uk *British surnames.*

https://internationalhighlife.com/hempcrete/ *Everything you ever needed to know …*

https://www.consumerhealthdigest.com/cbd-hemp-center/hemp-seed-side effects.html *Top 10 side effects of hemp seed.*

https://www.nationalgeographic.com/adventure/article/marijuana-cannabis-pot-weed-burial-shroud-china-ancient-discovery-scythians-turpan-archaeology-botany *Ancient cannabis "burial rhroud" …*

http://hemp-ethanol.blogspot.com *Fuel of the future?*

https://www.palmoilextractionmachine.com/FAQ/oil_content_of_different_oil_seeds_506.html *What is the oil content of different seed oils?*

www.historyofplastic.com/lastic-history/plastic-timeline/ *Timeline of plastic history*

https://www.unodc.org/unodc/en/frontpage/2017/June/world-drug-report-2017_-29-5-million-people-globally-suffer-from-drug-use-disorders--opioids-the-most-harmful.html *World Drug Report 2017 …*

https://www.dailymail.co.uk/news/article-7123805/Insys-Therapeutics-files-Chapter-11-bankruptcy.html?src=ilaw *Drugmaker files for bankruptcy …*

https://www.bbc.co.uk/news/health-44197038 *Is the UK the world's biggest exporter of legal cannabis?*

https://returntonow.net/2019/04/14/americas-first-hemp-wood-factory-is-being-built/ *Hemp is the new oak …*

www.scholar.sun.ac.za/…/98872/coogan_hemp_2016.pdf?sequence=2 *The South African hemp story …*

http://www.japanhemp.org/uncleweed/history.htm *Hempen culture in Japan.*

https://www.australianindustrialhempconference.com.au *Australian Industrial Hemp Conference.*

https://internationalhighlife.com/hemp-ethanol-could-replace-gas-and-cost-4-times-less-at-the-pump-biofuel-expert-claims/ *Hemp ethanol could …*

https://www.theextract.co.uk/lifestyle/sustainability/lego-going-hemp/ *LEGO going Hemp?*

https://physicsworld.com/a/composite-chitin-film-could-replace-plastic-packaging/ *Composite chitin could replace plastic packaging.*

https://www.nbmcw.com/tech-articles/concrete/570-new-building-material-freshly-ground-lime-instead-of-cement.html *New building material …*

https://drhealthbenefits.com/herbal/leaves/health-benefits-of-cannabis-leaves *Health benefits of cannabis leaves.*

https://www.onlinemarijuanaguide.com/these-are-the-differences-between-flower-and-bud/ *These are the differences between flower and bud.*

https://openlibrary.org/books/OL15330635M/De_honesta_voluptate_et_valetudine.

https://www.druglibrary.org/schaffer/hemp/history/first12000/abel.htm *Marijuana – the first twelve thousand years.*

https://www.newcastle.edu.au/newsroom/featured/plastic-ingestion-by-people-could-be-equating-to-a-credit-card-a-week/how-much-microplastics-are-we-ingesting-estimation-of-the-mass-of-microplastics-ingested *How much microplastics are we …*

https://www.linkedin.com/pulse/hemp-future-week-day-1-graphene-from-fiber-richard-rose *Graphene from hemp.*

https://loudcloudhealth.com/resources/marijuana-tax-revenue-by-state-map/ *How much marijuana tax revenue can we expect?*

https://www.researchgate.net/publication/228603981_Physical_evidence_for_the_antiquity_of_Cannabis_sativa_L *Physical evidence for the antiquity of* Cannabis sativa.

https://www.cbdsloth.com/advice/cannabinoids/list-of-cannabinoids/ *List of cannabinoids.*

https://ourworldindata.org/drug-use Hannah Ritchie *2019 drug use.*

https://www.statista.com/statistics/1073451/global-market-value-pulp-and-paper/ *Global market value …*

https://loudcloudhealth.com/resources/marijuana-tax-revenue-by-state-map/ *Marijuana tax revenue by state.*

https://www.cannabisculture.com/content/2019/05/31/the-mother-plant-of-the-goddess-cannabis/ *The mother plant …*

https://www.ers.usda.gov/webdocs/publications/41740/15855_ages001ee_1_.pdf?v=42087 *Harvesting, retting and fiber separation.*

https://www.british-history.ac.uk/no-series/traded-goods-dictionary/1550-1820/hemp-herse *Hemp.*

https://financesonline.com/cannabis-industry-statistics/ *53 fascinating cannabis industry statistics.*

Boulac *HEMP Industrial Production and Uses* (2013 CABI Publishing).

Bowrey *Geographical Account of Countries Round the Bay of Bengal 1669 to 1679* (2015 Andesite Press).

Burton *The Anatomy of Melancholy* (original 1621 British Library reader).

Cohen *Marijuana: Its Effects an Mind & Body* (1985 Burke Publishing).

Erasmus *Fats that Heal, Fats that Kill* (1986 Alive Books).

Fine *Hemp Bound* (2014 Chelsea Green).

Herer *The Emperor Wears No Clothes* (1985 Independent).

Preston *Pot Planet* (2002 Atlantic Books).

Robinson *The Great Book of Hemp* (1995 Park Street Press).

Robinson *The Hemp Manifesto* (1997 Park Street Press).

Schofield *The Strange Case of Pot* (1971 Penguin Books).

Sparrow & Stanwix *The Hempcrete Book* (2014 Green Books).

Hawthorn

Aierken et al. Journal of the Science of Food and Agriculture 2017 Oct;97(13):4557–4561 *Hypoglycemic effect of hawthorn in type II …*

Alp et al. Anatolian Journal of Cardiology 2015;15(12):970–975 *Protective effects of Hawthorn* (Crataegus oxyacantha) …

Benabderrahmane et al. Natural Product Research 2021 Feb;35(4):627–632 *Polyphenolic content and bioactivities of* Crataegus …

Bernatoniene et al. Acta Poloniae Pharmaceutica 2009 Jul–Aug;66(4):415–421 *The comparison of anti-oxidative kinetics in vitro of the …*

Bleske et al. Medical Sciences Monitor 2007 Dec;13(12):BR255–258 *Evaluation of hawthorn extract on …*

Chang et al. The Journal of Clinical Pharmacology 2002 Jun;42(6):605–612 *Hawthorn.*

Cuevas-Duran et al. International Journal of Molecular Sciences 2017 Nov;18(11): 2412 *Extracts of* Crataegus oxyacantha *and* Rosmarinus …

Cui et al. Pharmaceutical Biology 2018 Dec;56(1):528–534 *Green synthesis of selenium nanoparticles with extract …*

Dahmer & Scott American Family Physician 2010 Feb 15;81(4):465–468 *Health effects of hawthorn.*

Daniele et al. Drug Safety 2006 Feb;29(6):523–535 *Adverse-event profile of* Crataegus *Spp.*

Dong et al. Journal of Ethnopharmacology 2017 Feb 23;198:479–488 *Hawthorn (*Crataegus pinnatifida *Bunge) leave …*

Edwards et al. Phytochemistry 2012;79:5–26 Review of the chemistry of the genus *Crataegus.*

Froelicher et al. Food Chemistry 2009 Aug;115(3):897–903 *Phenolic profiles and antioxidative effects …*

Fu et al. Chinese Journal of Integral Medicine 2013 Aug;19(8):582–588 *Hawthorn leaves flavonoids decreases inflammation related …*

Gundogdu et al. Biological Research 2014;47(1):21 *Organic acids, sugars, vitamin C content and some …*

Han et al. Chemical Biological Interactions 2016 Sep 25;257:132–140 *Polyphenols from hawthorn peels and fleshes differently mitigate …*

Hatipoglu et al. PLOS One 2015 Jun 1;10(6):e0128134 *The effectiveness of* Crataegus orientalis *M. Bieber. (Hawthorn) …*

Hu et al. Experimental Biology and Medicine 2016 Mar;241(6):667–674 *Ethanol extract of Zhongtian hawthorn lowers serum …*

Huang et al. Fitoterapia 2018 Mar;125:6–12 *Seven new sesquineolignans isolated from* Crataegus …

Kallassy et al. Medical Science Monitor Basic Research 2017;23:270–284 *Chemical composition, antioxidant …*

Krajka-Kuzniak et al. Phytotherapy Research 2014 Apr;28(4):593–602 *Hawthorn* (Crataegus oxyacantha *L.) bark extract …*

Leuchtgens Fortschritte der Medizin 1993 Jul 20;111(20–21):35–4 (Crataegus *Special Extract WS 1442 in NYHA II …*

Liu et al. Food Chemistry 2011 Aug;127(1):1370–1377 *Quantitative analysis of phenolic compounds in …*

Mahmood et al. Pakistan Journal of Pharmaceutical Sciences 2010 Jan;23(1):119–124 *Herbal treatment for cardiovascular disease …*

Mustapha et al. Melanoma Research 2016 Jun;26(3):2–22 *Antitumoral, antioxidant, and anti-melanogenesis …*

Nabavi et al. Nutrients 2015 Sep;7(9):7708–7728 *Polyphenolic composition of* Crataegus monogyna …

Niu et al. African Journal of Traditional, Complementary and Alternative Medicines 2013 Apr 12;10(3):567–573 *Study on antibacterial effect of …*

Oliveira de Quadros et al. Journal of Toxicology and Environmental Health 2017;80(3): 161–170 *Fruit extract of the medicinal plant* Crataegus …

Orhan Current Medicinal Chemistry 2018;25(37):4854–4865 *Phytochemical and pharmacological activity …*

Pahlavan et al. Federation of American Societies for Experimental Biology Journal 2018 Mar;32(3):1440–1451 *Effects of hawthorn* (Crataegus pentagyna) *leaf …*

Papuc et al. Journal of the Science of Food and Agriculture 2018 Mar;98(4):1346–1361 *Comparative study of the influence of hawthorn …*

Pittler et al. American Journal of Medicine 2003 Jun 1;114(8):665–674 *Hawthorn extract for treating chronic heart …*

Pittler et al. Cochrane Database Systemic Review 2008 Jan 23;1:CD005312 *Hawthorn extract for treating chronic heart failure.*

Rastogi et al. Phytomedicine 2016 Oct 15;23(11):1082–1089 *Traditional herbs: A remedy for cardiovascular disorders.*

Rigelsky & Sweet American Journal of Health System Pharmacy 2002 Mar 1;59(5): 417–422 *Hawthorn: Pharmacology and therapeutic uses.*

Schlegelmilch & Heywood International Journal of Toxicology 1994 Apr 1 *Toxicity of* Crataegus *(hawthorn extract (WS 1442).*

Schmidt et al. Phytomedicine 1994 Jun;1(1):17–24 *Efficacy of the Hawthorn* (Crataegus) *preparation LI132 …*

Tankanow et al. The Journal of Clinical Pharmacology 2003 Jun;43(6):637–642 *Interaction study between digoxin and a preparation of …*

Walker et al. British Journal of General Practice 2006 Jun;56(527):437–443 *Hypotensive effects of hawthorn for patients …*

Wang et al. Evidence-Based Complementary Alternative Medicine 2013;2013:149363 *Effects of* Crataegus *usage in cardiovascular disease prevention …*

Wen et al. Food Chemistry 2015 Nov 1;186:54–62 *Phenolic contents and cellular antioxidant activity …*

Wen et al. Food and Chemical Toxicology 2017 Feb;100:149–160 *Major triterpenoids in Chinese hawthorn …*

Wu et al. BMC Complementary & Alternative Medicine 2017;17:151 *Phytochemical compositions of extract from peel of hawthorn fruit …*

Wyspianska et al. Journal of the Science of Food and Agriculture 2017 Jan;97(2): 669–678 *Physico-chemical, antioxidant, and anti-inflammatory …*

Yanar et al. Scientific Research and Essays 2011 Jan;6(1):35–38 *Morphological and chemical diversity among …*

Yang & Liu Journal of the Science of Food and Agriculture 2012 Jun;92(8):1578–1590 *Composition and health effects of phenolic compounds …*

Yao et al. Journal of Ethnopharmacology 2008 June;118(1):127–132 *A reproductive screening test of hawthorn.*

Yoo et al. Nutrients 2016 May;8(5):283 *Hawthorn fruit extract elevates expression of Nrf2/HO-1 and …*

Zhang et al. Journal of Atherosclerosis and Thrombosis 2014;21(2):119–128 *Hawthorn fruit attenuates atherosclerosis by …*

Zhu et al. Chemico-Biological Interactions 2015 Aug 5;238:42–47 *Comparative effects of hawthorn* (Crataegus …

Zhu & Xiao Journal of Agricultural Food Chemistry 2018 Nov 21;66(46): 12296–12305 *Characterization of odor-active volatiles in hawthorn …*

Zick et al. European Journal of Heart Failure 2009 Oct;11(10):990–999 *Hawthorn extract randomized blinded …*

Zorniak et al. Journal of Physiology and Pharmacology 2017 Aug;68(4):521–526 Crataegus *special extract WS 1442 …*

https://pfaf.org/user/cmspage.aspx?pageid=59 *Crataegus species – the hawthorns.*

https://www.cancertutor.com/laetrile/ *Vitamin B-17 vs. laetrile.*

https://whisperingearth.co.uk/category/herbs/hawthorn/ Archive for the *'hawthorn'* category.

http://www.glenniekindred.co.uk/articles/print/herbstrees7.html *The hawthorn tree.*

https://www.aliisaacstoryteller.com/post/the-curious-phenomenon-of-the-irish-fairy-tree *The curious phenomenon of the Irish fairy tree.*

https://en.wikipedia.org/wiki/List_of_people_named_in_the_Panama_Papers *List of people named in the …*

https://www.sunsigns.org/celtic-astrology-hawthorn-tree/ *Celtic zodiac: Hawthorn tree.*

https://druidry.org/resources/tree-lore-hawthorn *Tree lore: Hawthorn.*

http://www.secondactive.com/2009/08/mindfulness-exercise-read-marcel-proust.html *A mindfulness exercise …*

https://www.naturalhealthnews.uk/article/hawthorn-the-heart-healer/ *Hawthorn – the heart healer.*

https://www.thespiritualcentre.com/hawthorn-tree-huathe.html *Hawthorn tree – huathe – huath.*

https://ireland-calling.com/celtic-mythology-hawthorn-tree/ *Hawthorn – the fairy tree.*

https://witchipedia.com/book-of-shadows/herblore/hawthorn/ *Hawthorn.*
https://en.wikipedia.org/wiki/Crataegus_monogyna *Crataegus monogyna.*
https://www.eatweeds.co.uk/hawthorn-crataegus-monogyna *Hawthorn – a foraging guide …*
http://www.thegoddesstree.com/trees/Hawthorn.htm *Hawthorn.*
https://stoneartblog.blogspot.com/2010/05/blackthorn-shillelaghs-and-tree-of.html *Blackthorn, shillelaghs and the tree …*
https://threeacresandacow.co.uk/2013/10/a-peoples-history-of-england-by-a-l-morton/ *A people's history of England.*
http://wildfoodsandmedicines.com//?s=hawthorn&x=0&y=0 *Hawthorn.*
https://www.theguardian.com/environment/2020/jul/01/england-privatised-water-firms-dividends-shareholders *England's privatized water firms paid £57bn …*
http;//tlio.org.uk/john-clare-and-the-tragedy-of-the-enclosures/ *John Clare and the …*
https://en.wikipedia.org/wiki/The_May-Pole_of_Merry_Mount *The may-pole of …*
https://www.fatsecret.com/calories-nutrition/generic/chinese-hawthorn-berry-fruit *Chinese hawthorn berry fruit.*
https://archive.org/details/drwilliamkingcoo00king *Dr. William King and the Co-operator, 1828–1830.*
https://www.statista.com/statistics/263424/the-largest-energy-utility-companies-worldwide-based-on-market-value/ *Largest electricity utilities globally …*
http://www.historyofpencils.com/writing-instruments-history/history-of-ink-and-pen/ *History of pen and ink.*

Adams *The Wisdom of Trees* (2018 Head of Zeus).
Gifford *The Celtic Wisdom of Trees* (2000 Godsfield).
Graves *The White Goddess* (1999 Faber & Faber).
Proust *In Search of Lost Times* (2003 Penguin Classics).
Vaughn *Hawthorn* (2015 Yale University Press).

Blackberry

Afrin et al. Molecules 2016 Feb;21(2):169 *Chemopreventive and therapeutic effects of edible berries …*
Ali et al. BMC Complementary and Alternative Medicine 2017;17:59 *Pharmacological profile of the aerial parts of* Rubus …
Atchley et al. PLOS One 2012;7(12):e51474 *Creativity in the wild: Improving creative reasoning …*
Avila et al. Oxidative Medicine and Cellular Longevity 2017;2017:9808520 *Cytoprotective mechanisms mediated by polyphenols from Chilean …*
Azofeifa et al. Journal of Agricultural and Food Chemistry 2013 Jun 19;61(24): 5798–5804 *Antioxidant and anti-inflammatory in vitro activities of …*
Beata Frontiers in Pharmacology 2018;9:78 *Berry phenolic antioxidants – implications for human health?*
Biel & Jaroszewska Scientia Agricola 2017;74(5) *The nutritional value of leaves of selected berry species.*
Burton-Freeman et al. Advances in Nutrition 2016 Jan;7(1):44–65 *Red raspberries and their bioactive polyphenols …*
Choi et al. BMC Complementary and Alternative Medicine 2016;16:202 Rubus occidentalis *alleviates hyperalgesia induced by …*

Cuevas-Rodriguez et al. Journal of Agricultural and Food Chemistry 2010;58(17): 9542–9548 *Inhibition of pro-inflammatory responses and …*

Danaher et al. Oral Surgery, Oral Medicines, Oral Pathology, Oral Radiology and Endodontics 2011 Sep;112(3):e31–35 *Antiviral effects of blackberry extract against …*

Dayawansa et al. Autonomic Neuroscience 2003 Oct;108(1–2):79–86 *Autonomic responses during inhalation of natural fragrance …*

Feng et al. Nutrition and Cancer 2004;50(1):80–90 *Blackberry extracts inhibit activating protein 1 activation …*

Fidelis et al. Molecules 2019;24(21):3854 *Fruit seeds as sources of bioactive …*

Furuuchi et al. PLOS One 2018;13(8):e0202051 *Boysenberry polyphenol inhibits endothelial …*

George et al. BioMed Research International 2014;2014:204340 *Bioassay directed isolation and biological …*

George & Abrahamse Oxidative Medicine and Cellular Longevity 2019.2019:6797921 *Increased oxidative stress induced by* Rubus …

Giampieri et al. Molecules 2018 Aug;23(8):1918 *Phytochemical composition and cytotoxic effects on liver …*

Gonzalez et al. Journal of Periodontal Research 2013 Feb;48(1):80–86 *Antibacterial effects of blackberry extract …*

Gu et al. Journal of Agricultural and Food Chemistry 2014 May 7;62(18):3997–4006 *Characterization of black raspberry functional food …*

Hayes et al. PNAS 2010 Mar 9;107:4612–4617 *Atrazine induces complete feminization and chemical castration …*

Joung Advances in Horticultural Science 2014;28(2):90–94 *Physiological and psychological effects of olfactory …*

Jurikova et al. Molecules 2017 Jun;22(6):944 *Fruits of black chokeberry* Aronia melanocarpa *in …*

Keservani et al. Scientifica (Cairo) 2016;2016:3109254 *Medicinal effect of nutraceutical fruits for the …*

Klewicka et al. Molecules 2016 Jul;21(7):908 *Ellagitannins from raspberry (*Rubus idaeus *L.) …*

Kowalska et al. PLOS One 2017;12(11):e0188583 *Inhibitory activity of chokeberry, bilberry, raspberry and …*

Krauze-Baronowska et al. BMC Complementary and Alternative Medicine 2014; 14:480 *Chemical composition and biological activity of* Rubus …

Lee et al. Nutrition Research and Practice 2011 Oct;5(5):429–434 *Effects of a* Rubus coreanus *Miquel supplement on …*

Li et al. International Journal of Immunopathology and Pharmacology 2007 Apr–Jun; 20(2Suppl2):3–8 *Forest bathing enhances human natural killer activity …*

Li et al. International Journal of Molecular Sciences 2016 Aug;17(8):1258 *Bioactivities and health benefits of wild fruits.*

Ma et al. International Journal of Molecular Sciences 2018 Feb;19(2):461 *Evaluation of polyphenol anthocyanin-enriched …*

Mace et al. Cancer Immunology, Immunotherapy 2014 Sep;63(9):889–900 *Bioactive compounds or metabolites from black raspberries …*

Nesello et al. Gastroenterology Research and Practice 2017;2017:7089697 *Gastroprotective value of berries: Evidences from …*

Oh et al. Evidence-Based Complementary and Alternative Medicine 2016;2016:4357656 *Antiobesity effects of unripe* Rubus coreanus *Miquel …*

Oszmianski et al. Molecules 2015 Mar;20(3):4951–4966 *Determination of phenolic compounds and antioxidant …*

Pajari et al. Oncotarget 2016 Jul;7(28):43907–43923 *Ellagitannin-rich cloudberry inhibits hepatocyte …*
Park et al. Environmental Health and Preventive Medicine 2010 Jan;15(1):18–26 *The physiological effects of* Shinrin-yoku *(taking in the forest …*
Petreanu et al. Evidence-Based Complementary and Alternative Medicine 2015; 2015:341216 *Uncommon trimethoxylated flavonol obtained from* Rubus …
Purgar et al. Molecules 2012 Sep;17(9):10390–10398 *A comparison of fruit chemical characteristics of two wild grown …*
Sahin Journal of Health & Medical Informatics 2011;2(1):106 *Cyanide poisoning in a children caused by apricot seeds.*
Sangiovanni et al. PLOS One 2013;8(8):e71762 *Ellagitannins from* Rubus *berries for the control of gastric …*
Seo et al. BMC Complementary and Alternative Medicine 2019;19:89 *Differences in anti-inflammatory effect of immature and mature …*
Shukitt et al. Nutritional Neuroscience 2009;12(3):135–140 *Effects of blackberries on motor and cognitive …*
Skrovankova et al. International Journal of Molecular Sciences 2015 Oct;16(10): 24673–24706 *Bioactive compounds and antioxidant activity in different types of berries.*
Stoner et al. Seminars in Cancer Biology 2007 Oct;17(5):403–410 *Cancer prevention with freeze-dried berries and berry components.*
Stoner et al. Carcinogenesis 2008 Sep;29(9):1665–1674 *Laboratory and clinical studies of cancer …*
Stoner Cancer Prevention Research 2009 Mar;2(3):187–194 *Foodstuffs for preventing cancer: The preclinical …*
Stoner et al. Pharmacological Research 2010 Jun;27(6):1138–1145 *Multiple berry types prevent n-nitrosomethylbenzylamine …*
Suh et al. Frontiers in Plant Science 2018;9:1462 *Distinguishing six edible berries based on metabolic …*
Szymanowska et al. Molecules 2018 Jul;23(7):1812 *Antioxidant, anti-inflammatory, and postulated cytotoxic activity …*
Tatar et al. Nutrition and Cancer 2019;71(3):461–471 *Blackberry extract inhibits telomerase activity …*
Tavares et al. European Journal of Nutrition 2013 Feb;52(1):225–236 *Neuroprotective effects of digested polyphenols from wild …*
Verma et al. Pharmacognosy Review 2014 Jul–Dec;8(16):101–104 Rubus fruticosus *(blackberry) use as an herbal medicine.*
Zia-Ul-Haq et al. Molecules 2014 Aug;19(8):10998–11029 Rubus fruticosus *L.: Constituents …*

https://www.madaboutberries.com/types-of-berries.html *Most common types of berries.*
Pesticide Action Network, North America, PAN Pesticide Database (Berkeley, CA, 2019), http:www.pesticideinfo.org
https://nutritiondata.self.com/facts/fruits-and-fruit-juices/1848/2 *Blackberries, raw …*
https://www.nutritionvalue.org/Blackberries%2C_raw_nutritional_value.html *Blackberries, raw.*
http://www.pesticideinfo.org/DS.jsp?sk=1002 *Pesticide usage wild blackberries California.*
https://androgenhacker.com/atrazine-effects-on-humans *Atrazine: 7 harmful …*
https://www.nytimes.com/2007/07/25/dining/25blac.html *At last, sweet blackberries stay the course.*
https://www.yesmagazine.org/environment/2018/07/18/a-brief-history-of-the-feral-blackberry/ *A brief history of the feral blackberry.*

https://www.lifeextension.com/magazine/2010/ss/cutting-edge-berry-extract-for-improved-night-vision *Berry extract improves night vision.*
https://en.wikipedia.org/wiki/Blackberry *Blackberry.*
https://en.wikipedia.org/wiki/Rubus *Rubus.*
https://en.wikipedia.org/wiki/List_of_Lepidoptera_that_feed_on_Rubus *List of …*
https://www.prweb.com/releases/herpes_simplex_virus/hsv_market/prweb9431137.htm *Increasing number of HSV cases to drive …*

Heaney *Opened Ground: Selected Poems 1966–1996* (1999 Farrar, Straus & Giroux).
Lai *Determination of Phenolic Hydroxyl Groups* (1992 Springer).
Moss *Doctored Results* (2014 Equinox Press).

Solutional summations

Fidelis et al. Molecules 2019 Nov;24(21):3854 *Fruit seeds as sources of bioactive compounds: Sustainable …*
Ladha & Kirk Kosmos: Journal for Global Transformation 2016 Summer *Seeing wetiko: On capitalism, mind viruses …*
Rillig et al. Fruit Crops 2020 Glomalin (a compilation of a few studies)
Shin et al. Nutrients 2016 Feb 16;8(2):92 *Phlorizin supplementation attenuates obesity …*

https://www.opendemocracy.net/en/opendemocracyuk/new-report-reveals-englands-farmland-tax-haven/ *The tax haven hidden in plain sight …*
https://www.theguardian.com/commentisfree/2019/jul/03/land-reform-brute-power-billionaire-press-attacks *After urging land reform …*
https://www.euractiv.com/section/agriculture-food/news/new-uk-farming-bill-guarantees-subsidies-for-2020/ *New UK farming bill …*
https://publications.parliament.uk/pa/cm199798/cmselect/cmenvtra/560-iii/560iii02.htm *The future for allotments.*
https://en.wikipedia.org/wiki/Global_warming_potential *Global warming potential.*
https://www.theguardian.com/us-news/2020/oct/07/purdue-pharma-sackler-family-guilty-agreement *Purdue Pharma nears guilty plea …*
https://www.bbc.co.uk/news/uk-england-48880939 *Brexit: EU subsidy loss 'could …*
https://www.theguardian.com/science/2020/oct/27/sleeping-giant-arctic-methane-deposits-starting-to-release-scientists-find?ref=hvper.com *Arctic methane deposits …*
https://thehistoryofengland.co.uk/resource/the-forest-charter-1217/ *The Forest Charter, 1217.*
https://returntonow.net/2019/06/27/atlanta-turns-7-acres-vacant-lot-into-largest-free-food-forest-in-the-country/ *Atlanta turns 7 acres …*
https://orionmagazine.org/article/medicine-after-oil/ *Medicine after oil.*
https://www.theguardian.com/environment/2016/sep/29/the-queen-aristocrats-and-saudi-prince-among-recipients-of-eu-farm-subsidies *The Queen, aristocrats and …*
https://ourworldindata.org/diet-compositions *Diet compositions.*
https://www.businessinsider.com/no-deal-brexit-percentage-british-food-imported-shortages-2019-1?r=US&IR=T *Say goodbye to tea and carrots: 80% of British …*
https://foodprint.org/issues/how-industrial-agriculture-affects-our-soil/ *How industrial agriculture …*

https://www.sciencealert.com/planned-self-sustainable-off-the-grid-towns-dubbed-tesla-of-eco-villages *The 'Tesla of eco-villages' …*

https://returntonow.net/2017/11/01/study-names-tyson-foods-1-culprit-largest-dead-zone-u-s-history/ *Investigation reveals Tyson Foods …*

https://www.drugwatch.com/manufacturers/glaxosmithkline/ *GlaxoSmithKline (GSK).*

https://blogs.lse.ac.uk/brexit/2017/05/26/britains-farmers-get-3bn-a-year-from-the-inefficient-cap-that-has-to-change/ *Britain's farmers get £3bn a year …*

https://en.wikipedia.org/wiki/Agriculture_in_China *Agriculture in China.*

https://en.wikipedia.org/wiki/List_of_industrial_disasters *List of …*

https://en.wikipedia.org/wiki/List_of_environmental_disasters *List of …*

https://en.wikipedia.org/wiki/Garden_Organic *Garden organic.*

https://theecologist.org/2014/nov/22/want-double-world-food-production-return-land-small-farmers *Want to double world food production? …*

https://www.alliedmarketresearch.com/press-release/anti-inflammatory-therapeutics.html *Anti-inflammatory therapeutics market analysis.*

https://www.goodshomedesign.com/learn-how-to-create-your-own-1-acre-self-sustaining-homestead/ *Learn how to create your …*

https://www.goodshomedesign.com/silent-rooftop-wind-turbines-could-generate-half-of-a-households-energy-needs/ *Silent rooftop wind turbines …*

https://oilprice.com/Energy/Crude-Oil/How-Much-Crude-Oil-Do-You-Unknowingly-Eat.html *How much crude oil …*

https://www.nissan-global.com/EN/HERITAGE/short_story/en_p32-01.html *Tama Electric vehicles.*

https://www.euractiv.com/section/agriculture-food/news/ecological-farming-can-be-more-profitable-than-conventional-says-new-report/ *Ecological farming can be more …*

https://www.vitacost.com/blog/how-is-soil-depletion-affecting-your-food/ *How Is soil depletion affecting your food?*

https://www.telegraph.co.uk/news/2020/03/27/allotments-should-brought-back-help-feed-nation-study-says/ *Allotments should be brought back …*

https://caat.org.uk/news/2000-06-09-2/ *The Arabian connection: The UK arms trade to Saudi Arabia.*

https://sistersofmercy.ie/2020/10/this-is-my-message-to-the-western-world-your-civilisation-is-killing-life-on-earth/ *This Is my message to the …*

https://www.greencarfuture.com/electric/making-of-ev-batteries *How electric car batteries are made …*

https://www.countryfile.com/news/can-the-uk-feed-itself-after-brexit/ *Can the UK feed itself after Brexit?*

https://www.thelandmagazine.org.uk/articles/ask-not-what-soils-can-do-us *Ask not what soils …*

https://www.bbc.co.uk/news/stories-53640382 *How the oil industry made us …*

https://www.hinrichfoundation.com/research/article/protectionism/agricultural-subsidies/ *Agricultural subsidies: Everyone's doing it.*

https://www-cdn.oxfam.org/s3fs-public/file_attachments/dp-extreme-wealth-is-not-merited-241115-en.pdf *Extreme wealth Is not merited.*

https://core.ac.uk/download/pdf/288287899.pdf *Two centuries of farmland prices.*

https://measuringworth.com/datasets/ukgdp/ *What was the U.K. GDP then?*

https://www.nytimes.com/2019/02/06/opinion/abolish-billionaires-tax.html *Abolish billionaires.*

https://www.ukpublicspending.co.uk/spending_chart_1900_2020UKp_17c1li111tcn_G0t_UK_National_Debt_since_1900 *National debt since 1900.*

https://www.resolutionfoundation.org/comment/who-owns-britains-13tn-wealth/ *Who owns Britain's £13tn wealth?*

https://www.bbc.co.uk/news/science-environment54238571 *Why do farmers get paid by taxpayers?*

https://www.shh.mpg.de/1638262/salmonella *Oldest reconstructed bacterial genomes …*

https://dmarron.com/2009/08/17/the-50-most-important-economic-theories/ *The 50 most important economic theories.*

https://tribunemag.co.uk/2019/04/the-kinder-scout-mass-trespass *The Kinder Scout mass trespass.*

https://www.theguardian.com/environment/2019/jan/27/revealed-the-mps-and-peers-receiving-millions-in-eu-farm-subsidies-cap *Peers and MPs receiving millions …*

https://listverse.com/2019/09/26/edward-bernays-freud-tricked-us/ *10 ways that Freud's nephew duped us all (and still does).*

https://www.healthline.com/health/laws-of-karma *What are the 12 laws of karma?*

https://www.opendemocracy.net/en/opendemocracyuk/why-youve-never-heard-of-charter-thats-as-important-as-magna-carta/ *Why you've never heard of a Charter …*

https://www.businessinsider.com/bitcoin-uses-as-much-energy-as-sweden-and-is-on-course-to-use-even-more-2021-3?r=US&IR=T *Bitcoin uses as much energy as …*

https://www.linkedin.com/pulse/global-pharmaceutical-market-report-2022-dr-richard-raven *Global pharmaceutical market report 2022.*

https://www.britannica.com/money/topic/money *Money.*

https://www.britannica.com/money/topic/money/After-Bretton-Woods *Money.*

Balfour *The Living Soil* (1943 Faber & Faber).

Dawson *Ecovillages: New Frontiers for Sustainability* (2006 Green Books).

Douglas & Hart *Forest Farming* (1993 Practical Action Publishing).

Easwaran (trans.) *The Upanishads* (1987 Nilgiri Press).

Evans *Soil* (2021 Murdoch Books).

Howard *An Agricultural Testament* (1943 Oxford University Press).

Howard *Farming and Gardening for Health or Disease* (1945 Faber & Faber).

Huang et al. *Small-scale Farmers in China in the Face of Modernisation and Globalisation* (2012 Chinese Academy of Science).

King *Farmers of Forty Centuries* (2004 Dover Publications).

Krishnamurti *The First and Last Freedom* (1958 Victor Gollancz).

Lawrence *Apocalypse* (1931 Heinemann).

Levy *Dispelling Wetiko: Breaking the Curse of Evil* (2013 North Atlantic Books).

Martin *Money: The Unauthorised Biography* (2014 Vintage).

Mendoza *Austerity: The Demolition of the Welfare State and the Rise of the Zombie Economy* (2015 New Internationalist Publications).

Monbiot et al. *Land for the Many* 2019.

Parkin *Fuck It* (2008 Hay House).

Piketty *Capital in the Twenty-First Century* (2017 Harvard University Press).

Quinn *Beyond Civilization* (1999 Harmony Books).

Russell *Mortals and Others: American Essays 1931–35 Volumes I & II* (2009 Routledge Classics).

Simard *Finding the Mother Tree* (2021 Penguin).

Smith *Tree Crops: A Permanent Agriculture* (1988 Island Press).

Stamets *Mycelium Running* (2005 Ten Speed Press).
Steiner *Agriculture Course* (2004 Rudolf Steiner Press).
Tanner *Nietzsche: A Very Short Introduction* (2000 Oxford University Press).
Tawney *Religion and the Rise of Capitalism* (1938 Pelican Books).
Tesla *My Inventions* (2011 Penguin Books).
Thomas *The Truth Agenda* (2009 Vital Signs Publishing).
Toffler *The Third Wave* (1981 Pan Books).
Tolle *Oneness With All Life* (2008 Penguin Books).
Tolle *A New Earth* (2016 Penguin Books).
Wells *First and Last Things* (1929 The Thinker's Library).
Wilson *The Illuminati Papers* (1980 Ronin Publishing).

Appendix

Alavanja Review of Environmental Health 2009 Oct–Dec;24(4):303–309 *Pesticide use and exposure extensive worldwide.*

https://thegreenfarmacygarden.com/jim-dukes-legacy/
https://worldpopulationreview.com/world-cities/london-population *London …*
https://www.wanttoknow.nl/gezondheid/elektro-magnetische-straling-ems/5g-in-actie-mussen-vallen-dood-van-t-dak/ *5G in actie: mussen …*
https://en.wikipedia.org/wiki/Pesticide *Pesticide.*
https://en.wikipedia.org/wiki/List_of_countries_by_GDP_(nominal) *List of countries by GDP (nominal).*
https://www.census.gov/data/tables/timeseries/demo/internationalprograms/historical -est-worldpop.html *Historical estimates of world population.*
https://www.forbes.com/sites/jonathanponciano/2018/06/06/worlds-largest-telecom-companies-2018/?sh=4bd517eb7d39 *The world's largest telecom companies …*
https://www.statista.com/statistics/664868/global-top-energy-companies-by-revenue/ *Leading energy companies worldwide in FY …*
https://edis.ifas.ufl.edu/pdffiles/PI/PI00800.pdf *Toxicity of pesticides.*

Schultz et al. Food Research International 2019 Aug;122:627–634 *Blackberry* (Rubus ulmifolius *Schott): Chemical composition …*

The following were pertinent to more than one of the chapters, had inspired further investigation, helped to set the foundations or are just worthy of a shout-out—

Gibbs *Planet of Humans* 2019 Documentary.

Bergfjord et al. Science 2010 Jun 25;328(5986):1634 *Comment on "30,000-year-old wild …*
Kvavadze et al. Science 2009 Sep11;325(5946):1359 *30,000-year-old wild …*
Santangelli et al. General and Comparative Endocrinology 2017;245:122–126 *Effects of BPA on female reproductive function …*

https://www.thecanary.co [The only newsfeed I'd recommend.]
https://www.organicconsumers.org [Information about food and agriculture.]
https://www.redpepper.org.uk/this-land-is-our-land/ *This land is our land.*

https://www.thelandmagazine.org.uk/articles/short-history-enclosure-britain *A short history of enclosure in Britain.*

https://www.newstatesman.com/life-and-society/2011/03/million-acres-land-ownership *The great property swindle …*

https://www.savills.co.uk/research_articles/229130/202022-0 *90 Years of farmland values & supply.*

https://www.newstatesman.com/politics/staggers/2018/08/british-wealth-concentrated-not-property-land-underneath *British wealth is concentrated not …*

https://en.wikipedia.org/wiki/List_of_countries_by_carbon_dioxide_emissions *List of countries by carbon dioxide emissions.*

http://www.naturallifemagazine.com/0904/bamboo_textiles.htm *Bamboo textiles …*

https://www.connexionfrance.com/index.php/French-news/New-French-study-finds-32-toxic-pesticides-in-the-air *New French study …*

https://www.biodynamic.org.uk/wp-content/uploads/2018/03/128-Feed-your-soil-the-biochar-revolution.pdf *Feed your soil – the biochar revolution.*

https://www.youtube.com/@kirstendirksen tune in for some sustainable positivity.

https://context.capp.ca/articles/2019/feature_petroleum-in-real-life_pills *Petroleum in real life: Pills.*

https://www.globalpetrolprices.com/articles/39/ *Breakdown of oil consumption by sector.*

https://www.newsweek.com/glyphosate-now-most-used-agricultural-chemical-ever-422419 *Glyphosate now most used …*

https://www.theguardian.com/environment/2020/may/16/the-end-of-plastic-new-plant-based-bottles-will-degrade-in-a-year *The end of plastic? …*

https://www.scientificamerican.com/article/soil-depletion-and-nutrition-loss/ *Dirt poor: Have fruits and …*

https://www.darrinqualman.com/global-plastics-production/ *Global plastics …*

https://www.bba.bristol.sch.uk/_site/data/files/users/11/D61CB183E9104F-B10E35A62B70338589.pdf *Crime and punishment through time …*

https://www.collective-evolution.com/2017/03/17/two-new-bills-quietly-slipping-through-congress-that-will-give-big-pharma-unlimited-power-zero-accountability/ *Two new bills quietly slipping through …*

https://www.gardeningknowhow.com/garden-how-to/soil-fertilizers/soil-types-and-weeds.htm *Weed identification control: Weeds as indicators of soil conditions.*

https://www.consumerlab.com/rdas/ *Recommended daily intakes and upper …*

http://wildplantguide.com/2017/04/23/medicinal-plant-constituents-and-actions/ *Medicinal plant constituents and actions.*

https://www.newsweek.com/2015/03/13/bpa-fine-if-you-ignore-most-studies-about-it-311203.html *BPA Is fine, if you ignore most studies about it.*

https://www.academia.edu/42295029/Rockefeller_Vakfının_Mayıs_2010_Raporu_Scenarios_for_the_Future_of_Technology_and_International_Development_ *Scenarios for the …*

https://www.forbes.com/consent/?toURL=https://www.forbes.com/global2000/ *Forbes Global 2000.*

https://www.pan-uk.org/25-year-environment-plan/ *Using pesticide policies as a tool for implementing …*

http://www.adsb.co.uk/date_and_time/calendar_reform_1752/ *Calendar reform in England, 1752.*

https://www.hempmarketreport.com/an-acre-of-hemp-could-be-worth-40000/ *An acre of hemp could be worth $40,000 …*

https://cosmosmagazine.com/biology/weeds-growing-in-poor-city-areas-more-nutritious-than-store-bought-produce/ *Weeds in poor city areas have nutritional value.*
https://www.thornber.net/medicine/html/medgloss.html *Glossary of medical terms used in the 18th and 19th centuries.*
http://www.food-info.net/uk/e/e200-300.htm *E200–E300 preservatives.*
https://diggersanddreamers.org.uk *Intentional community In Britain.*
https://www.righttoroam.org.uk/ [please join as it is our human right to roam.]
https://www.drugs.com [FDA-approved independent site used for all side effects information.]
https://www.forbes.com/lists/global2000/ *Forbes top 2000 companies.*
https://en.wikipedia.org/wiki/List_of_countries_by_GDP_(nominal) *GDPs.*
http://irishhedgerows.weebly.com/flora.html *Folklore of the hedgerow.*

Acemoglu & Robinson *Why Nations Fail* (2013 Profile Books).
Allen *A Dictionary of Sussex Folk Medicine* (1995 Countryside Books).
Bakan *The Corporation* (2004 Simon & Schuster Inc.).
Baker *Discovering the Folklore of Plants* (1969 Shire Publications).
Beresford-Kroeger *The Global Forest: 40 Ways Trees Can Save Us* (2010 Particular Books).
Blamey et al. *Wild Flowers of Britain & Ireland* (2003 A & C Publishers).
Brady *The Nature and Properties of Soils* (1990 Macmillan Publishing).
Breverton *Complete Herbal* (2011 Quercus).
Brickell (Ed.) *Gardeners' Encyclopedia of Plants and Flowers* (1989 Dorling Kindersley).
Brown (Ed.) *The New Shorter Oxford English Dictionary Vols. I & II* (1993 Oxford University Press).
Bruton-Seal & Seal *Hedgerow Medicine* (2008 Merlin Unwin Books).
Bruton-Seal & Seal *Wayside Medicine* (2017 Merlin Unwin Books).
Carson *Silent Spring* (1962 Penguin Books).
Castleman *The Healing Herbs* (1995 Bantam Books).
Chomsky *Media Control* (2002 Seven Stories Press).
Conway *The Magic of Herbs* (1975 Panther Books).
Culpeper *Complete Herbal and English Physician* (1828 Thomas Kelly).
Culpeper *Culpeper's Complete Herbal* (Foulsham 1950).
Davies et al. *Soil Management* (1972 Farming Press).
Davies *The Drugs Don't Work* (2013 Penguin Specials).
Deakin *Wildwood: A Journey through Trees* (2007 Hamish Hamilton).
Deitch et al. *Keith Haring* (2008 Rizzoli).
Diamond *The Rise and Fall of the Third Chimpanzee* (1991 Radius).
Dorland's Illustrated Medical Dictionary (1974 Saunders).
Druker *Altered Genes, Twisted Truth* (2015 Clear River Press).
Elliot-Gough *The Seven Deadly Whites* (2016 Earth Books).
Field-Lewis *Anthology of Flowers* (2016 Quadrille).
Forsyth *The Etymologican* (2011 Icon Books).
Fowler *The Thrifty Forager* (2011 Kyle Books).
Froud *Faeries* (1978 Rufus Publications).
Fukuoka *The One-Straw Revolution* (2009 New York Review Books Classic).
Gawande *Being Mortal* (2014 Profile Books).
Gibbons & Brough *The Hamlyn Photographic Guide to the Wild Flowers of Britain and Northern Europe* (1992 Hamlyn).

Goldacre *Bad Pharma* (2012 Fourth Estate).
Goodall *Ten Technologies to Fix Energy and Climate* (2008 Profile Books).
Grill *Nutrition and the Inner Sense of Giving* (2007 Lammers-Koll-Verlag).
Hamaker & Weaver *The Survival of Civilization* (1982 Hamaker-Weaver Publishers).
Hartley-Hennessy *Healing by Water or Drinking Sunlight and Oxygen* (1950 Essence of Health Publishing Company).
Harvey *The Killing of the Countryside* (1998 Vintage).
Hawken et al. *Natural Capitalism: The Next Industrial Revolution* (1999 Earthscan Publications).
Hayes *The Book of Trespass* (2020 Bloomsbury).
Hills *Fertility without Fertilisers* (1975 Henry Doubleday Research Association).
Holden & Abbott *RSPB Handbook of Garden Wildlife* (2008 Bloomsbury).
Holtom & Hyiton *The Complete Guide to Herbs* (1979 Rodale).
Iburg *Herbal Remedies* (2004 Rebo).
Imhoff *Food Fight* (2012 A Watershed Media Book).
Inglis *The Diseases of Civilization* (1983 Granada Publishing).
Jensen & Anderson *Empty Harvest* (1990 Avery Publishing).
Jones *The Establishment and How They Got Away with it* (2014 Penguin Books).
Jordan *Spirit of the East* (2003 Carlton Books).
Keen *The Cult of the Amateur* (2007 Nicholas Brealey Publishing).
Khan & Abourashed *Leung's Encyclopedia of Common Natural Ingredients* (2010 Wiley).
Kindred *A Hedgerow Cookbook* (2002 Wooden Books).
Kingsnorth *Real England: The Battle against the Bland* (2009 Portobello Books).
Klein *The Shock Doctrine* (2007 Penguin Books).
Lappe & Collins *Food First* (1977 Houghton Mifflin).
Le Fanu *The Rise and Fall of Modern Medicine* (1999 Little, Brown and Company).
Lee & Daly (eds.) *Hunters and Gatherers* (1999 Cambridge University Press).
Levitt & Dubner *Superfreakonomics* (2009 Penguin Books).
Lloyd *The QI Book of General Ignorance* (2006 Faber & Faber).
Lovelock *Gaia* (1979 Oxford University Press).
Lucas *Honourable Friends?* (2015 Portobello Books).
Luzel *Celtic Folk Tales from Armorica* (1985 Llanerch).
Mabey *Food for Free* (1972 William Collins Sons & Co.).
Mabey *The Common Ground* (1980 Arrow Books).
Mabey *Flora Britannica* (1997 Chatto & Windus).
Maddocks *Hildegard of Bingen: The Woman of Her Age* (2001 Headline Books).
Markowitz & Rosner *Deceit and Denial: The Deadly Politics of Industrial Pollution* (2002 University of California Press).
McIntosh *Soil and Soul: People versus Corporate Power* (2001 Aurum Press).
McKenna *True Hallucinations* (1993 Harper Collins).
Mears *Wild Food* (2007 Hodder & Stoughton).
Meek *Private Island: Why Britain Now Belongs to Someone Else* (2014 Verso).
Merrington *Agricultural Pollution* (2002 Spon Press).
Milham *Dirty Electricity* (2010 iUniverse Star).
Mollison *Permaculture: A Designer's Manual* (2002 Tagari Publications).
Murray et al. *The Encyclopedia of Healing Foods* (2007 Index Books).
Myss *The Creation of Health* (1999 Bantam).
Onstad *Whole Foods Companion* (1996 Chelsea Green).
Peattie *Flowering Earth* (1951 The Country Book Club).

Petulengo *The Roots of Health* (1968 Souvenir Press).
Phillips *Mushrooms and Other Fungi of Great Britain & Europe* (1981 Pan Books).
Phillips *Wild Flowers of Britain* (1977 Pan Books).
Phillips *Wild Foods* (1983 Pan Books).
Pilger *The New Rulers of the World* (2002 Verso).
Pilger (Ed.) *Tell Me No Lies* (2005 Vintage).
Postgate *Karl Marx* (1933 Hamish Hamilton Publisher).
Price *Nutrition and Physical Degeneration* (2010 Benediction Classics).
Roberts *The End of Food* (2008 Bloomsbury).
Rodale *Organic Manifesto* (2010 Rodale).
Sax *Industrial Pollution* (1974 Van Nostrand Reinhold Company).
Schneidau *Botanical Folk Tales of Britain and Ireland* (2018 The History Press).
Schultes *Plants of the Gods: Their Sacred, Healing, and Hallucinogenic Powers* (2001 Healing Arts Press).
Seymour *The Ultimate Heresy* (1989 Green Books).
Sheldrake *Entangled Life* (2020 The Bodley Head).
Sherman *Indian Tribes of North America* (1996 Tiger Books).
Shiva *Tomorrow's Biodiversity* (2000 Thames & Hudson).
Shiva *Stolen Harvest* (2000 South End Press).
Shrubsole *Who Owns England* (2019 William Collins).
Simmonds et al. *The Gardener's Companion to Medicinal Plants* (2016 Frances Lincoln).
Skeat *English Etymology* (1993 Wordsworth Editions).
Snyder & Furst *Mushrooms: Psychedelic Fungi* (1988 Burke Publishing).
Stamets *Growing Gourmet and Medicinal Mushrooms* (2000 Ten Speed Press).
Taylor et al. (Eds.) *Soils, Sustainability and the Natural Heritage* (1996 Scottish Natural Heritage).
Tierra *Planetary Herbology* (1988 Lotus Press).
Vogel *The Nature Doctor* (2003 Mainstream Publishing).
Waldman *Encyclopedia of Native American Tribes* (1999 Checkmark Books).
Wells *100 Flowers and How They Got Their Names* (1997 Workman Publishing).
White & Foster *The Herbal Drugstore* (1998 Rodale).
Whitefield *How to Make a Forest Garden* (1997 Permanent Publications).
Whitman *The Psychic Power of Plants* (1975 A Star Book).
Wild Flowers of Britain (1981 Reader's Digest).
Wohlleben *The Hidden Life of Trees* (2017 William Collins).
Wood *The Earthwise Herbal Vol. 1* (2008 North Atlantic Books).
Woolley *The Herbalist* (2005 Harper Perennial).
Wright *The River Cottage Hedgerow Handbook* (2010 Bloomsbury).

INDEX

Abnaki, 81, 117, 233
Abraham, 230
Abrams, 178
Acetaria: A Discourse of Sallets (Evelyn), 6
Acetoria, grande salad (Evelyn)
acetylcholine (ACh), 4, 11, 18, 20, 30, 31, 39, 41, 76, 96, 112, 207, 210, 211, 261, 272, 322, 325, 326
  inhibition, 19
acetylcholinesterase, 18, 19
acetylsalicylic acid, 264
achilleine, 86
Achilles, 78, 273
Achilles' Heel, 78, 94
acne, 12, 62, 84, 173
acrylonitrile butadiene styrene (ABS), 164
acute myocardial ischaemia, 207
adipogenesis, 243
Aeschylus, 141
agamospermy, 194
AgLogic 15G, 30
Agricultural Bill, 310–311
agricultural chemicals, 238–239
Agricultural Revolution, 283
agricultural system, 283, 318
  Agricultural Bill, 310–311
  Agricultural Revolution, 283
  Agroecology, 297
  allotment revolution, 306–307
  Amanita muscaria, 289
Basic Payment Scheme, 311
Biodynamic farming, 296–297
County Farms, 298
Crown Estate, 304
'Ecological Farming', 296
food security, 302–305
Forest Charter, 305–306
Forest Farming, 297
genetically modified organisms, 298
geosmin, 290
great agricultural myth, 283–288
Green Revolution, 285, 286, 287–288
Haughley Experiment, 297
land prices and subsidies, 308–311
land reform, 301–305
modern agriculture, 294
monoculture, 283–288, 294–295, 298
Mycobacterium vaccae, 290
mycoremediation, 293
mycorrhizal fungi, 289–296
PCCs, 307
permaculture, 288–289, 295, 296–298
PermaLife, 288–289, 303
petrichor, 290
pharmas to farmers, 311–312
phytonutrients, 293
regenerative agriculture, 296–298
self-sufficiency and sustainable smallholdings, 298–301
smallholding farming, 294

soil-enriching crop rotations, 299
soil health, 289–296
sustainable food systems, 296–298
10-acre smallholdings, 298–301, 303
Urban Farming, 307–308
An Agricultural Testament (Howard), 290–291
Agroecology, 297
ague, 46
Aguirre-Velazquez, 176
Akihito, Emperor, 150
Albertus Magnus, 80
aldicarb, 30
Aleut, 50, 81, 116
alewives, 108
Algonquin, 50, 81, 117
allopurinol, 123–124
allotment revolution, 306–307
alpha-linolenic acid (ALA), 170
alpha pinene, 85
Alzheimer's, 18, 19, 98, 242, 263, 275
amygdalin, 247
analgesic, 91, 177, 178
Anatomy of Melancholy (Burton), 145
Andersen, Hans Christian, 8
Anslinger, Harry, 149
Anthony, St., 72
anti-inflammatory plants, 23
Apache, 50
apigenin, 86, 87, 89, 122, 244, 247, 322, 323, 324, 325
apocalypse, 2
apoptosis, 22, 23, 54, 55, 63, 211, 248
arachidonic acid (AA), 171
Aristotle, 265
Arnemunde, Queen, 142
Arnulf, St., 72
Ashurbanipal, King, 140
Asian Tiger mosquito (*Aedes albopictus*), 92–95, 124
aspartame, 177
Aspdin, Joseph, 155
aspirin, 264
Assassins, 135
AstraZeneca, 123
Atharvaveda (*Science of Charms*), 139
atherogenesis, 210
atherosclerosis, 22
Atkins, Julia, 175
Atrazine, 239
Attenborough, Saint, 44
Attlee, Clement, 268–269
Australian gympiegympie (*Dendrocnide moroides*), 3
Avesta, Zend, 139
Avicenna (Ibn Sina), 49, 142
Avocados from Mexico, 37
Ayurvedic Medicine, 84, 120, 139, 205
Aztecs, 192
azulene, 87, 103

Bakelite, 164
Balfour, Eve, 31, 297
Bank of England, 266, 267
Barker, Cicely Mary, xiii
Basic Payment Scheme, 311
bast fibre, 28
batologist, 228
Bavarian Reinheitsgebot, 107
'bawming the thorn', 196
Baxter, John, 159
Bayer, 30
Bella Coola, 50, 81
Beltane, 193, 219
Benhaim, Paul, 175
Benk u Bode (Foruli), 145
Benn, Tony, 218
Bernays, Edward, 91–92
berry, 229
-based therapeutics, 240–245
beta-sitosterol, 5, 11
bhang, 139, 146
Big Pharma, 128–130
Binance coin, 270
biodynamic farming, 296–297
bioplastic, 166
Bitcoin, 270
bitters, 70–71
blackberry (*Rubus fruticosus agg.*), xiv, 227, 263
agricultural chemicals, 238–239
amygdalin, 247
anti-carcinogenic, 235, 240, 246, 247, 248, 249
anti-inflammatory, 235, 242, 243, 244, 246, 247, 256
anti-tumour, 246
anti-viral, 241, 247, 256
berry-based therapeutics, 240–245
against cancer, 247–250
'caneberries', 229
caution, 236–237
cultivation, 254–255
Cutleaf, 254
ethnobotanical legacy, 233
etymology, 229–230
in folklore and culture, 230–235
foraging, 250–254
history, 232–233

multifaceted uses of, 235–236
nutraceuticals, 243, 247
nutritional profiles, 325–326
oxymel, 234
plantality, 229
polyphenols, 245–247
primocane, 228–229
recipes, 255–257
roots, 228
Rubus armeniacus, 254
Rubus genus, 227–228, 233
Rubus laciniatus, 254
Rubus ulmifolius, 234, 246
Shinrin-yoku, 252
vitamin B17, 250, 263
woodland resonance, 251–252
blackcurrants (*Ribes nigrum*), 227
Blackfoot, 81
black raspberries (*Rubus occidentalis*), 240
Blackwell, Elizabeth, 6–7
blood stasis elimination, 203
The Book of Changes (Emperor Fuxi), 101
borek, 40
borneol, 85
Botanologia. See *The English Herbal*
Bowrey, Thomas, 139
BPA (Bisphenol A), 165, 166, 239
Bram, 230
bramble. *See* blackberry
Bredemann, Gustuv, 37
Brigid (Brigit), St., 49
Bristol-Myers Squibb, 207, 248
The British Pharmacopeia, 70
Brown, Allan, 28
Buddha, 43, 170
building industry, 277
Bupropion, 243
Burbank, Luther, 254
Burns, Robert, 193
Burrell, Lloyd, 99
Burton, Robert, 145
butylated hydroxyanisole (BHA), 212
butylated hydroxytoluene (BHT), 212

Caesarion, 140
caffeic acid, 123
caffeoylquinic acid, 11
Cahulla, 50
Calvin Klein, 159
Cambric, 159
Cambri Formosa, 141
camphene, 85
camphor, 85
cancer
drugs, 56, 245, 248
industry, 56, 126, 248, 249, 272
'caneberries', 229
cannabichromene (CBC), 173
cannabichromevarin (CBCV), 173
cannabidiol (CBD), 173–176, 178, 262
cannabidiolic acid (CBDA), 173
cannabidivarin (CBDV), 173
cannabigerol (CBG), 173–174
cannabigerovarin (CBGV), 174
cannabinoids, 172, 173, 174, 179, 184, 262, 324
cannabinol (CBN), 174
cannabis (*Cannabis sativa*), xiv, 133, 274–275, 319
ancient legacy, 138–142
anti-inflammatory, 171, 172, 173, 177
being 'stoned', 180
bhang, 139, 146
cannabidiol, 173–176, 178
Cannabis indica, 134
Cannabis ruderalis, 134
and civilization, 142–148
decorticator, 162–164
endocannabinoid system, 173
epilepsy benefits, 140, 147, 176, 177, 179, 184
etymology, 135
evolution of consumption, 135–136
fabric, 157–160
hash, 134, 135, 136, 141, 142, 146, 147, 151, 152
hashish, 135–136
hemp, 133, 136–137, 168–169
hempcrete, 153–157
hempire, 168–169
hemp seed, 167, 169–173
history, 134–135, 137
industrial use to global prohibition, 148–153
marijuana, 133, 140, 152, 180
Mary Jane, 134, 145
names for, 134
nutritional profiles, 324–325
opioid crisis and, 181–183
paper production, 160–162
politics of, 148–153
research and pharmaceutical interests, 176–180
rituals to remedies, 184–185
semieniatka, 141
side effects, 180–181
skunk, 136
sustainable plastics and energy, 164–168
weed, 136
Canon of Medicine (Avicenna), 49
Canute, King, 197

Capital in the Twenty-First Century (Piketty), 269
Captain Pouch. *See* Reynolds, John
captopril, 207
cardiovascular problems, 203
carotene, 59, 60
Carrier, 81, 117
Carson, Rachel, 31, 326
Carthan, Hattie, 308
catalpol, 120
cell apoptosis. *See* apoptosis
Cellophane, 164
Celluloid, 164
Celsus, 75
Cesamet, 178
chakras, 101–102
chamazulene, 85, 103
Chamberlain, Joseph, 298, 306
Charaka Samhita, 139
Charles III, King, 281, 301
Chehalis, 81
Cherokee, 7, 50, 81, 116, 192
Cheyenne, 81
China Grass. *See* ramie
Chinese hawthorn (*Crataegus pinnatifida*), 202, 206, 207, 210, 211, 214
The Chinese Herbal Materia Medica, 68
Chinese Plantain (*Plantago asiatica*), 114
Chinese shamanism, 141
Chippewa, 81, 117, 192
Chiron, 78
chitin, 167
chlorhexidine, 125
chlorogenic acid, 64, 123, 261
cholesterol dysfunctionality, 205–206
choline acetyltransferase (ChAT), 18, 19, 20
chrism, 135
chrismation, 135
'Christ', 135
Christ, Jesus, 48, 194, 225
Christopher, St., 72
Cialiam, 82
cichoric acid, 62
cineol, 85
City of London, 269
civilization, xv
Clare, John, 222
Cleopatra, 140
Clingfilm, 164
Cochrane Collaboration, 208
Codex Alimentarius, 33
collective consciousness, 316
Columbus, Christopher, 51, 143–144
Comanche, 192
communism, 318
The Complete Herbal (Culpeper), 6
composite companies, 168
Compulsory Purchase Orders (CPOs), 304, 306
concrete, 155, 156, 275, 277
Confessions of an English Opium-Eater (De Quincey), 147
Continental Tyres, 66
The Co-operator, 314–315
Costanoan, 117
cotton, 28
  consumerism and sustainability, 32
  hidden costs of, 30–31
  vs. nettles, 28–33
  pesticides, 30–32
  sustainable shift from cotton cultivation, 32–33
  toxic legacy of, 29–30
coumarins, 60, 105
County Farms, 298
Cowitz, 82
Cree, 7, 82, 117, 315
Crow, 82
Crown Estate, 304
cryptocurrencies, 270–271
The Culinary Caveman, xv, 73, 79
Culpeper, Nicholas, 6, 49, 80, 118, 119, 145, 201, 232
'cultural amnesia', 79
cultural change, 259–260
cultural system, 312
  capital, 317
  collective consciousness, 316
  co-operative vision, 314–315
  The Co-operator, 314–315
  money, 316–317
  obscenity law, 312–314
  redistribution and corporate accountability, 312–314
  seven plants and path to harmony, 318–320
  socialism, 317–318
  wetiko, 315–316
A Curious Herbal (Blackwell), 7
Cutleaf, 254
cystoliths, 16
cytoprotection, 88–89

Dale, Henry, 18
dandelion (*Taraxacum officinale*), xiv, 43–45, 72–73, 261, 272–273
  anaemia, 50, 56, 62, 71
  antibacterial, 64
  antibiotic, 59, 64

anti-carcinogenic, 54, 59, 72
anti-inflammatory, 62, 63
benefits of, 70–72
in cancer research, 53–57
cholagogue, 61, 62
choleretic, 61
coffee, 72
colorectal disease, 54, 61
contraindications, 58–59
culinary dandelions, 70–72
as diuretics, 57–58
ethnobotanical uses, 50
etymologicum, 45–46
etymology, 47–48
First Nations' medicine to European colonization, 50–51
folklores, 51–52
foraging/harvesting, 70
for health and digestion, 70–72
heart disease, 61, 65
historical uses, 49–51
IBD, 64
liver protection, 49, 57, 58, 59, 61, 62, 63, 64, 65, 68
medicinal benefits, 61–65, 68–69
myths, 49
nutrients, 59–60, 322–323
from paradise gardens to chemical monocultures, 52–53
vs. pharmaceuticals, 57–59
phytonutrients, 59–61
regional knowledge, 46–49
skin cleanser, 50, 55, 59, 62, 68
Taraxacum laurentainum, 51
Taraxacum mongolicum, 59, 63
Taraxacum palustre, 51
Taraxagum, 65–66
tyres, 65–67
dandelion root extract (DRE), 54
1925 Dangerous Drugs Act, 148
Dark Waters, 238
Das Kapital, 266
Davies, Dame Sally, 17
decarboxylation, 170, 173
decorticator, 162–164
de Courtive, Edmond, 147
DEET (dimethyl toluamide), 103
DeFelice, Stephen L., 243
De Honesta Voluptate et Valetudine (*Of Honest Voluptuousness and Health*), 170
Delaware, 50, 82, 117
De Materia Medica (Dioscorides), 5
De Quincey, Thomas, 147
de Rivaz, Francois, 167
detox, 17
de Verrazano, Giovanni, 144
diabetes, xv, 12, 22, 24, 56, 61, 65, 84, 87, 94, 96, 124, 139, 166, 171, 178, 204, 209, 210, 252, 261, 263, 273, 275, 295
diabetic foot ulcer (DFU), 124
diazepam, 88, 273
dichloromethane, 22, 23
Diclofenac, 12, 13, 14, 15, 36, 56, 90, 111, 271, 273, 327
Diesel, Rudolf, 280
Dioscorides, Pedanius, 5, 79, 142, 201, 232
Directive 2004/24/EC, 44, 130
Dirksen, Kirsten, 308
disease, xv
diseasemongery, 177
diseases of civilization, xv–xvi, 128, 209, 261
diuretics, 57–58, 59, 120
Docosahexaenoic Acid (DHA), 171
Doctrine of Signatures, 75, 76
Dodoens, Rembert, 80
dragonflies, 95
The Drugs Don't Work (Davies), 17
Du Chesne, Joseph, 203
Duke, Jim, 321, 323
DuPont, 238, 280
Dylan, Bob, 151
dysmenorrhea, 86, 273

E140, 33, 67
Eclectics, 201
ecocide, 31, 182, 313
'Ecological Farming', 296
economic system, 265
cryptocurrencies, 270–271
fault lines of modern economics, 269–270
fragile foundations of, 266–267
Green Currency, 270–275
green economy revolution, 276
monetary system, 266
money, 265–270
sociopolitical roots of economic inequality, 267–269
economic theories, 314
Ecotricity, 278
Edison, 67
Edison's goldenrod (*Solidago edisoniana*), 67
Edward the Confessor, 196, 197
eicosapentaenoic acid (EPA), 171
Einstein, Albert, 269
electric cars, 278
electromagnetic pollution, 95–99

electro-magnetic radiation (EMR), 95–97
Eliade, Mercia, 140
Elizabeth II, Queen, 200
ellagic acid, 240
ellagitannins, 240, 241
Elon Musk, 154
emmenagogue effect, 89
The Emperor Wears No Clothes (Herer), 133, 153, 185
enclosure, 218–223
Enclosure Act of 1773, 221, 223
endocannabinoid system, 173
endothelial cells, 241
'engineering consent', 91
English hawthorn (*Crataegus monogyna*), 191, 325
The English Herbal (Salmon), 119
The English Physician (Culpeper), 6, 201
epidemiological study, xv, xvii, 21–24, 112, 178, 179, 207, 208, 210, 212, 242, 245, 246
Epidiolex, 177
Erasmus, Udo, 170, 171
essential fatty acids (EFAs), 170–171, 172
Ethereum, 270
Eurasian blackberry (*Rubus laciniatus*), 254
Europa, Goddess, 48
European dewberry (*Rubus caesius*), 228
Evelyn, John, 6, 49
Extremely High Frequency (EHF), 95, 96
Extremely Low Frequency (ELF), 95, 96, 99

fabric, 157–160, 277–278
Fairfax, Thomas, 220
fake news, 98, 99
Farben, I. G., 165
Farmers of Forty Centuries (King), 296
Farming and Gardening for Health and Disease (Howard), 291
Fen. *See* Stingless Nettle
fentanyl, 181, 182
fibromyalgia, 112
First Nation People, 7, 8, 50, 81, 116, 144, 192, 202, 315
fish oils, 170
5G, 98–99
Flathead, 82
flavonoids, 11, 60, 65, 86, 87, 89, 121, 122, 125, 206, 207, 208, 211, 212, 244, 246, 247, 322
flavonol, 240
flaxseed (linseeds), 21
Flora Britannica (Mabey), 232
Flower Fairies (Barker), xiii
fly agaric/amanita (*Amanita muscaria*), 289
folic acid (Vitamin B9), 256
Folk-Liv 1942 (Hald), 7
Food For Free (Mabey), 232
food security, 300, 302–305
foraging, 9, 38, 39, 70, 105, 214, 222, 228, 245, 250, 251, 253, 255, 263, 272
Ford, Henry, 280
forest bathing, 252, 263
Forest Charter, 222, 305–306
Forest Day mosquito. *See* Asian Tiger mosquito
Forest Farming (Hart), 297
Forest Farming, 297
Foruli, Mohammed Ebn Soleiman, 145
Fouquet, Nicholas, 53
Frammento di un libro di Cucina (*Fragment of a Cookbook*), 170
Freud, Sigmund, 91
fruit leather, 215
Fuchs, Leonhart, 135
Fukuoka, Masanobu, 295
furosemide, 57
Futomono, 158
Fuxi, Emperor, 101

Gaia, 316
galactagogue, 17
Galen, 49, 142
gamma-linoleic acid (GLA), 171
Gaoni, Yechiel, 176
garden, 53
Gargantua and Pantagruel, 145
The General Theory of Employment, Interest and Money (Keynes), 270
genetically modified organisms (GMOs), 298
A Geographical Account of Countries Round the Bay of Bengal (Bowrey), 139
geosmin, 290
Gerard, John, 6, 80
German Commission E, 57, 84, 120, 202
Giambattista della Porta, 76
Giles, St., 72
Glastonbury Thorn, 195–196
GlaxoSmithKline, 250
glioblastoma, 178
global economy, 329–332
global fertiliser industry, 272
Global Soil Partnership, 286
glomalin, 293
glucantime, 88
glucoside scopoletin, 16
glutathione, 23
gluten intolerance, 32
glycoprotein, 71
glycosides, 60, 120, 247

glyphosate, 239, 292, 327
Gobekli Tepe, 287
Goebbels, Joseph, 92
government, 151
Grauballe Man, 118
Graves, Robert, 194
Green Currency, 270–275
green economy revolution, 276
Green Man, 198
Green, Philip, 305
Green Revolution, 285, 286, 287–288
Green, Thomas, 26
gross domestic product (GDP), xvi, 310, 329–332
Gulliver's Travels (Swift), 145
Guterres, António, 298
Guzmán, Manuel, 178
GW Pharmaceuticals, 175, 177, 262

Haber, Fritz, 165–166
Haisla, 233
Hald, Margrethe, 7
hamp. *See* hemp
Hanaksiala, 233
Haraldskaer Woman, 232
Hart, Robert, 297
Hasan-i Sabbah, 135
hash, 135
The Hasheesh Eater (Ludlow), 147
hashish, 135–136
Hashish and Mental Illness (Moreau), 147
Hassassin, 135
Haughley Experiment, 297
hawkbit (*Leontodon autumnalis*), 47
hawthorn (*Crataegus monogyna*), xiv, 187, 188, 262–263
- agamospermy, 194
- ancient reverence, 194–197
- angina, 202, 203, 204, 205, 207, 208
- antibacterial, 210, 211, 213
- anti-carcinogenic, 210, 211, 213
- anti-inflammatory, 205, 209, 210, 211, 212, 225
- antioxidant, 205, 207, 209, 211, 213
- arrhythmia, 203, 205, 206, 208
- atherosclerosis, 203, 206, 208
- 'bawming the thorn', 196
- *Crataegus aestivalis*, 215
- *Crataegus arnoldiana*, 189
- *Crataegus aronia*, 190
- *Crataegus azarolus*, 189
- *Crataegus baroussana*, 189
- *Crataegus calpodendron*, 192
- *Crataegus laevigata*, 201, 212
- *Crataegus macrosperma*, 189
- *Crataegus mexicana*, 192, 214
- *Crataegus missouriensis*, 189
- *Crataegus monogyna* spp. *biflora*, 195, 196
- *Crataegus orientalis*, 190, 210
- *Crataegus pinnatifida*, 202, 210, 211, 214
- *Crataegus pontica*, 190
- *Crataegus schraderiana*, 189
- *Crataegus spathulata*, 192
- *Crataegus submollis*, 192
- *Crataegus succulenta*, 189
- culinary uses, 214–215
- cultural significance, 190–194, 195
- enclosure, 218–223
- etymology, 188
- family, 189
- folk stories and traditions, 190–193
- Glastonbury Thorn, 195–196
- global science, 209–213
- Green Man, 198
- heart disease, 203–209, 210
- hedges, 215–218
- high-blood pressure, 202, 204, 205, 208
- in literature, 194
- multifaceted nature of, 188–190
- names, 193
- nutritional profiles, 325
- parthenogenesis, 194–195
- vs. pharmaceutical interests, 213
- pleaching, 217
- procyanidins, 211–212
- review of therapeutic potential, 209–213
- sacred hawthorn, 225
- 'Sale of Britain', 223–225
- *sceach gheal*, 199
- Sidhe, 199
- supernatural, 197–201
- superstitions, 197–201
- Thorney Island, 196–197
- traditional uses, 201–203
- *zalzalak*, 215

Hawthorne, Nathaniel, 193, 248
healing plants, botanical names, 45–46
health, xv–xvi
health system, 261
- blackberry, 263
- CBD, 262
- dandelions, 261
- hawthorn, 262–263
- hemp, 262
- marijuana, 262
- medicine, 263–265
- nature's cure for civilization's diseases, 261–263

nettles, 261
pharmaceuticals and petroleum, 264
plantain, 262
yarrow, 261–262
Hearst, William Randolph, 148, 149
heart disease, 203–209, 210
hedgerow, 215, 216, 217, 218, 221, 228, 231, 251, 312
Hedgerows Regulations 1997, 217
hedges, 215–218
Helicobacter pylori, 263
heliotropism, 52
hemp, 133, 136–137, 168–169, 262
hempcrete, 153–157, 277
hempire, 168–169
hemp seed, 167, 169–173
Henry Doubleday, 296
Henry Doubleday Research Association (HDRA), 296
Henry VIII (King), 144
herbalism, 118
Herball (Gerard), 6, 8
Herer, Jack, 133, 153, 173, 185
Herodotus, 141
Hesquiat, 8, 50, 82, 117
Higgins, Polly, 313
Hildegard of Bingen, 6, 79–80, 143, 193
Hill, Octavia, 220
Hills, Lawrence, 296
Himalayan blackberry (*Rubus armeniacus*), 254
Hippocrates, xvi, 49
Hirohito, Emperor, 150
historical pandemics, 333–335
HMS *Victory*, 144
Hoh, 50
Hokey Pokey, 9
Holmgren, David, 296
homeostasis, xvi, 30, 44, 54, 62, 72, 173, 241, 293
Homer, 78
Homo denisova, 25, 158
Homo neanderthalensis, 78, 79, 157
Homo sapiens, 8, 78, 283
horta, 71
hortopita, 40
Howard, Albert, 31, 290
Hua-T'o, 142
humanness, 20
Human Plant, xiv, 96
hunter-gatherer, xvi, xvii, 79, 128, 139, 157, 205, 320
hurd. *See* shiv
Huxley, Aldous, 112
hydrogen fuel cells, 279
Hymen, 192
Hymenaios, 192

iatrogenic death, 61
iatrogeny, xv, 61, 261, 335
Ibn al-Baytar, 143
I Ching. See *The Book of Changes*
'incense bowls', 138
indomethacin, 90
Industrial Revolution, 33, 168, 216, 259
industrial system, 276
building, 277
fabrics, 277–278
petroleum, 280–283
transport, 278–279
Innocent VIII, Pope, 143
insulin, 22, 58, 61, 65, 71, 87, 173, 210, 243, 247, 252, 254, 256
Insys, 181–182
intermittent claudication, 205
International Food Policy Research Institute, 163, 284
inulin, 61
insoluble, 71
Iroquois, 8, 50, 82, 117, 192
Irving, Miles, 39
Ispaghul (*Plantago ovate*), 114

James I (King), 65
Jefferson, Thomas, 146
Jenks, Jorian, 297
Johnson, Thomas, 80
Joseph, St., 72
Josselyn, John, 6
Julius Caesar, 140

kaempferol, 87, 244, 322, 325, 326
Kagawa, Toyohiko, 297
Kapoor, John, 181–182
karma, 315, 316
Kawalisu, 117
Kenward, Paul, 174–175
Kett's Rebellion, 219
Keynes, John Maynard, 270
kief, 143
kif. *See* kief
1932 Kinder Scout trespass, 220
Kindleberger, Charles, 268
King, Franklin Hiram, 296
King, Mervyn, 267
King, William, 314
Kiowa, 50
Kneipp, Sebastian, 80

Krebs, Ernst T., 250
kudzu, 158
Kwakiutl, 8, 192

laetrile. *See* amygdalin
Lakota, 8
Lancelot 'Capability' Brown, 53
land reform, 301–305
Land Reform Act, 304
Language of Flowers, 51, 76
The Last Judgement (Martin), 77
Lauren, Ralph, 159
The Law Against Demons (Avesta), 139
Lawrence, D. H., 318
Lawson, Nigel, 223
Lawton, Geoff, 296
laxative, 62
leishmaniasis, 23, 88, 262, 273
Leo, 48
Leuchtgens, H., 207
leukaemia, 23, 54, 59, 96, 123, 240, 241, 243, 247, 249, 263
levulose, 61
lignans, 21
Lilly, Eli, 152
limonene, 85
Linnaeus, Carl, 45, 47, 80, 107, 135
linseeds. *See* flaxseed
lion, 48
Lipitor, 131, 274
lipogenesis, 243
Loewi, Otto, 18
Longfellow, Henry Wadsworth, 116
Louis XIV, 53
low-density lipoprotein (LDL), 21
Ludlow, Fitz Hugh, 147
lutein, 60
luteolin, 61–62, 125

Mabey, Richard, 232
Madame Pierre, 154
Makah, 8, 233
Malcolm X, 149
Malecite, 50
Marigold (*Calendula officinalis*), 45
marijuana, 133, 140, 152, 180, 262
1937 Marijuana Tax Act, 148, 149
Marinol, 178
Mark the Evangelist, 72
Marley, Bob, 318
Martin, Felix, 268
Martin V, Pope, 143
Marx, Karl, 29, 266, 318
Mary Jane, 134, 145. *See also* cannabis
Materia Medicas, xv
May Day, 190, 193, 196, 200, 219
The Mayflower, 50, 51, 82, 191
May-Honey, 72
May, Philip, 175
The May-Pole of Merry Mount (Hawthorne), 193
May, Theresa, 175
McCloud, Kevin, 154
Mears, Ray, 80
Mechoulam, Raphael, 176
medical marijuana, 133, 140, 152, 180
medicine, 263–265
melatonin, 97
Mendocino, 117
Menominee, 50
Mesopotamia, 140
Meswaki, 50
Methicillin Resistant *Straphyloccus aureus* (MRSA), 21
metsulfuron, 238
Micmac, 50
Midland Revolt, 219
Milarepa, 5
Miwok, 82
modern agriculture, 294
Mohegan, 50, 82
Mollison, Bill, 296
Monbiot, George, 302
monetary system, 266
Money: The Unauthorised Biography (Martin), 268
money, 265–270
Monier, Joseph, 155
monoculture, 283–288, 294–295, 298
monoterpenes, 122, 126
Monsanto, 292
Moore, Michael, 238
Moreau, Jacques-Joseph, 147
moroidin, 3
Moss, Ralph, 249
Mother Nature, 73, 130, 173
mucilage, 130
multiple sclerosis (MS), 87
  drugs, 273
mycelium, 104, 163, 289, 291, 292, 293, 297
Mycobacterium vaccae, 290
mycoremediation, 293
mycorrhiza, 104
mycorrhizal fungi, 289–296

Naltrexone, 243
Napoleon Bonaparte, 146–147
Naproxen, 13

natural dyes, 28
nature's cure for civilization's diseases, 261–263
Navajo, 50, 82
Nemonte Nenquimo, 320
neonicotinoids, 326
Nervii of Flanders, 216
nettle (*Urtica dioica*), xiv, 1, 3, 41–42, 72, 261, 327
  anti-arthritic, 12
  antibacterial properties of, 21–22
  anti-inflammatory, 10, 11, 12, 23, 24, 41
  antiviral properties, 22–23
  balancing benefits and risks, 16–17
  in beauty industry, 35
  cancer prevention, 22–23
  choline acetyltransferase, 20
  commercial potential, 37–38
  cotton vs., 28–33
  environmental disruptions and need, 19–20
  epidemiological study, 21–24
  ethnobotanical uses, 7–8
  etymology of, 25–26
  as fertiliser, 34
  foraging and harvesting, 38–39
  healing powers of, 11–12
  hidden potential, 1–2, 10–11, 33
  history, 5–7
  industrial revolution, 33–38
  limitations of standardized nettle supplements, 15–16
  magical and folkloric connections of, 8–9
  natural remedies vs. profitable pharmaceuticals, 13–14
  nettlecrete, 36
  nettlesfortextiles.org.uk, 28
  in neurological health, 18
  neurological power of, 18–19
  nutritional and medicinal benefits, 9–10
  nutritional profiles, 321–322
  paper manufacture, 35–36
  phenols, 21
  phytonutrients, 10–11, 21
  as phytoremediation, 24
  recipes, 39–41
  research and corporate interests, 9–10, 21–22
  revitalizing agriculture, 41
  SARS, Corona, 23
  side effects and risks of common drugs, 12–13
  sustainable future, 33–37
  for textiles, 24–28
  as therapeutic herbal drug, 23–24
  toxic shock of pharmaceuticals, 15
  Urticaceae family, 2–5
  vulture crisis, 14–15
The Nettle Dress, 28
Nettle Fibre Company, 26
neurological-type diseases, 18
New-Englands Rarities (Josselyn), 6
Newton Rebellion, 219
Nitinaht, 8, 82
nonsteroidal anti-inflammatory drugs (NSAIDs), 13, 90, 110
nuclear pollution, 95–99
nuclear radiation, 95, 99
nutraceutical, 22, 28, 89, 111, 112, 120, 125, 206, 210, 233, 237, 240, 243, 244, 245, 246, 247, 263, 264, 271, 272, 274, 275, 294, 300, 311, 319

Oak Apple Day, 9
obscaenus, 314
oikonomia, xvi, 239, 265–266, 267
oil industry, 282
Ojibwa, 50
Okanagan-Colville, 8, 50, 82, 192
Okanagon, 82
Olwen, Goddess, 198
Omega3, 10, 170, 171, 172, 173, 256, 323, 324, 325
Omega6, 171, 172, 256, 323, 324, 325
Omega6s, 171
omega-oils, 184
omeprazole, 104
The One-Straw Revolution (Fukuoka), 295
1,001 Nights, 143
On the True Doctrine (Celsus), 75
opioid crisis, 181–183
opobalsam. *See* opopanax
opopanax, 103
Organic Manifesto (Rodale), 312
organochlorines, 326
Orwell, George, 112
Osbert, William Fitz, 219
O'Shaughnessy, William Brooke, 147
over-the-counter (OTC), 91–92, 274
oxalic acid, 4, 16, 322, 323, 325
OxyContin, 182
oxymel, 234

Paine, Thomas, 146
Palute, 8, 82
panacea, 103
pandemics, historical, 333–335
pangamic acid (vitamin B15), 250
paper production, 160–162
papyrus (*Cyperus papyrus*), 160
Paracelsus, 75–76
paradise, 53
Parkes, Alexander, 164

Parkesine, 164
Parkinson, John, 80
Parkinson's, 18, 19, 80, 87, 96, 120, 140, 166, 178, 184, 261, 262
Parsi, 14
parthenogenesis, 194–195
Patrick, St., 49
peanuts, 31
Pepys, Samuel, 6
Peralta, Marina Galvez, 125
pericarp, 229
permaculture, 288–289, 295, 296–298
PermaLife, 288–289, 303
pesticide, 326
 arsenic, 326
 carbamates, 19, 326
 glyphosate, 327
 lethal dose, 327–328
 neonicotinoids, 326
 nicotine sulphate, 326
 organochlorines, 326
 organophosphates, 19, 326
 parathion, 29, 327
 sulphur compounds, 326
Peter the Great, 141
petrichor, 290
petroleum, 280–283
 plastic, 165, 166
pharmaceutical(s), 58–59
 corporations, 128–130
 diuretics, 58
 OTC market for, 91–92
 and petroleum, 264
 toxic shock of, 15
 yarrow as alternative to, 88–91
phenolic acids, 246
phenolic compounds, 21, 211, 246, 325
phenols, 21, 245
Phillips, Roger, 119
phototoxic, 104
Physica, 6
phytochemicals. *See* phytonutrients
phytonutrients, xiv–xv, 10–11, 59–61, 85–86, 293
phytoremediation, 24, 127, 162
phyto, xiv
Piketty, Thomas, 269, 313, 314, 317
'Pinch-bum-day', 9
'pixilated', 200
plantain (*Plantago major*), xiv, 109, 262, 274
 Alisma plantagoaquatica, 113
 antibacterial, 117, 119, 125, 126
 anti-carcinogenic, 123, 130
 anti-inflammatory, 112, 114, 121, 122, 123, 124
 anti-viral, 121, 123
 cytotoxic, 122, 125
 ethnobotanical uses, 116–117
 forbidden knowledge of herbalism, 127–130
 healing power, 120–121
 history, 117–120
 Musa paradisiaca, 113
 names, 115–116
 nutritional profiles, 324
 phytoremediation, 126–127
 Plantago amplexicaulis, 114
 Plantago asiatica, 114
 Plantago coronopus, 113
 Plantago elongata, 114
 Plantago lanceolata, 113
 Plantago leiopetala, 114
 Plantago major, 113
 Plantago maritima, 113
 Plantago media, 113
 Plantago ovata, 114
 Plantago psyllium, 114, 130–131
 post-surgery, 109–111
 resilience of, 131–132
 revolution in healthcare, 111–113
 scientific insights, 121–127
 species specific, 113–114
 traditional uses, 120–121
 varieties, 113–114
 wound healing, 111, 114, 117, 119, 126
plantality, 229, 232, 243, 253, 319
Plantamajoside, 123
plastic, 77, 156, 157, 158, 164, 165, 166, 167, 169, 182, 235, 251, 280, 281, 282
 bioplastic, 164, 166, 167, 168, 255, 319
 plant plastic, 167, 280
 producers, 165
 sustainable plastics and energy, 164–168
pleaching, 217, 218, 221, 222
Pliny the Elder, 142
pollution, industrial, 328–329
polyphenolic, 121, 210, 211, 240, 242, 243, 246
polyphenols, 11, 21, 87, 214, 241, 244, 245–247, 250, 254, 256, 261, 263
polyvinyl chloride (PVC), 164
population dynamics, 332–333
Portland cement, 155
Potawatomi, 8, 50, 82, 117, 192
Povlsen, Anders Holch, 305
prescription drug revenues, 129
Price, Weston A., 83, 121
primocane, 228–229
probiotics, 59
procyanidins, 208, 209, 211, 212
Propaganda (Bernays), 92
Proust, Marcel, 194

psyllium husk (*Plantago psyllium*), 130–131
'psyllium plug', 131
Public Cultivation Centres (PCCs), 307
Puddlicott, Richard, 143
Purdue, 182
Purin d'ortie, 34

Qingsongling, 214
quercetin, 11, 60, 122, 210, 244, 247, 256, 322, 323, 325, 326
Quilleute, 8, 50
Quinault, 82

r > g equation, 313–314, 317
Rabelais, François, 145
radiation, 95–99
Rameses II, 140
ramie (*Boehmeria nivea*), 27, 158
Rappahannock, 50
raspberry (*Rubus idaeus*), 228, 234, 241, 253, 326
reductionism, 85
reductionist, xvi, 85, 211, 242
Reefer Madness, 149
refractory, 176
refractory epilepsy, 176
regenerative agriculture, 296–298
retting, 27–28
Reynolds, John, 219, 220
Reynolds, Russell, 147
Rhino tank, 216
Ribena, 250
rielitos, 214
Rig-Veda, 139
River Nettle (*Laportea peduncularis*), 3
Rockefeller, John D., 264
Rodale, Maria, 312
Roddick, Anita, 168
Romeh, Ahmed, 126–127
Rose family, 247
Rothman, Benny, 220
Rousseau, Jean-Jacques, 137
rubber, 66
Rubus genus, 227–228, 233
Runic scripture, 193
rutin, 247

sabinene, 85
Sackler, Arthur, 182
'Sale of Britain', 223–225
salicylic acid (aspirin), 264
Salmonella, 22, 89, 241, 262, 263, 273, 334
Salmon, William, 119
Sambucus ebulus, 21
Sams, Craig, 287
Sam the Skunkman, 177
Sarin, 19
SARS. *See* Severe Acute Respiratory Syndrome
Sassoon, Siegfried, 194
Sativex, 177, 178
sceach gheal, 199
Schopenhauer, Arthur, 317
Schultes, Richard Evans, 184
Scopoletin, 16, 60, 322, 323
scopoletin, 60
Scythians, 141
Second Law of Thermodynamics, 96
Semedo, Maria Helena, 291
semieniatka, 141
sesquiterpenes, 103, 126
The Seven Deadly Whites (Elliot-Gough), 129, 205, 289, 329
seven plants, xiv
Severe Acute Respiratory Syndrome (SARS), 23, 83, 272, 273
Shakespeare, 145
Shanidar Cave, 78
Shan-Zha. *See* hawthorn
Sheikhoumi, Ottoman Emir Soudan, 143
Sheldrake, Merlin, 293
Shen Neng, Emperor, 138
Shepherd's Clock, 52
Shigella, 89
Shinrin-yoku, 252
Shinto Buddhist, 150, 252
shiv, 28
Shoshoni, 8, 82, 117
Shrubsole, Guy, 302
Shuswap, 8
side-effects, 12, 13, 15, 17, 22, 24, 36, 39, 54, 56, 58, 59, 61, 65, 87, 88, 90, 105, 111, 112, 122, 125, 128, 177, 178, 180, 203, 204, 207, 208, 210, 213, 242, 243, 273, 274, 275, 286, 308
Sidhe, 199
Silent Spring (Carson), 31, 326
skunk, 136
Sloan Kettering Center, 248
smallholdings, 298–301, 319
  farming, 294
  Small Holdings Act of 1892, 298
  Small Holdings and Allotments Acts of 1908 and 1925, 298
  10-acre smallholdings, 298–301, 303
Small Nettle (*Urtica urens*), 4
smeuse, 218
Smith, Adam, 268
Smith, J. Russell, 297

socialism, 269, 317–318
Socrates, 49
Soil Association, 288, 296, 297
soil-enriching crop rotations, 299
soil health, 289–296
solution, 259
Spanish Inquisition, 108
spicule, 4
spotted wing drosophilia (*Drosophilia suzukii*), 236
stablecoin, 270
Stamets, Paul, 291, 293
Staphylococcus, 22, 89, 211, 241, 262, 263, 273
statins, 122, 208, 210, 275
Steiner, Rudolf, 296
Stingless Nettle (*Urtica dioica* ssp. *galeopsifolia*), 4
St. Michael's Day/Michaelmas, 231
Stoner, Gary, 240
Straw, Jack, 219
stray electricity, 96
Streptococci, 125
Streptococcus, 126
Sugiura, Kanematsu, 248, 249
The Suicide Plant. *See* Australian gympiegympie
Sumerians, 140, 326
Summerlin, William, 248
surnames, 137
survival roots, xvii
Sushrata Samhita, 139
Sushruta, 139
sustainable food systems, 296–298
Swift, Jonathan, 145
Sykes, Friend, 297
systemic reformation, 261
  agricultural system, 283–312
  cultural system, 312–320
  economic system, 265–276
  health system, 261–265
  industrial system, 276–283
tannins, 120, 236
Taraxacum mongolicum, 59, 63, 68
Taraxagum, 65–66
taraxasterol, 63, 64
taraxinic acid, 59
taxol, 55, 249
TCM. *See* Traditional Chinese Medicine
Teflon, 238
10-acre smallholdings, 298–301, 303
terpenes, 85
  aromatic, 253
Tesla, Nikola, 153, 167
Tether, 270
Teton Dakota People, 82
tetrahydrocannabinol (THC), 174
tetrahydrocanna-binolic acid (THCA), 174
tetrahydrocannabivarin (THCV), 174
Tewa, 50
textile, 26
The Theater of Plants (Parkinson), 80
The Beatles, 151
The Theory of Moral Sentiments (Smith), 268
Theseus, 49
Thetis, 78
Thomas, Andy, 316
Thomas Aquinas, St., 49
Thompson, 8, 82, 192
Thorney Island, 196–197
Tiberius, Emperor, 267
tinctures, 234
Tollund Man, 117
Tower of Silence, 14
toxic flush, 17
Traditional Chinese Medicine (TCM), 59, 63, 85, 120, 123 202, 206, 211, 262
transport, 278–279
Tree Crops: A Permanent Agriculture (Smith), 297
Tresham, Thomas, 219
trichome, 4
triethylamine, 200
Trolox, 246
The Truth Agenda (Thomas), 316
tryfydin, 2
tryptophan, 97
Ts'ai Lun, 160
Turner, J. M. W., 43–44
Twain, Mark, 92
Tyler, Wat, 219
Tyson, Mike, 176

Universal Herbal (Green), 26
universal value, 265, 271, 316
University of Padova, 89
Urban Farming, 307–308
urbanization, 332–333
urolithiasis, 123
Urticaceae family, 2. *See also* nettle
  Britain's nettles, 4–5
  nutritional and medicinal wonders, 3
  perils and peculiarities, 2–3
  Urtica aestuans, 3
  Urtica chamaedryoides, 3
  Urtica ferox, 2
  Urtica incisa, 3
  Urtica lalibertadensis, 3
  Urtica massaica, 3
  Urtica pilulifera, 4

Urtica thunbergiana, 3
Urtica urentissima, 2
USDA's National Nutrient Database, 10
USD Coin, 270

valium, 182
varicose veins, 80, 84, 120, 205
Vaseline, 264
vellum, 160
Vitamin B12, 71, 290
Vitamin B15, 250
Vitamin B17, 250, 263
Vitamin C, 61, 253–254
Vogel, Alfred, 119, 234
vulture crisis, 14–15

Warburg, Otto, 56
warfarin, 17, 104
War on Drugs, 152
Washington, George, 146
water plantain (*Alisma plantagoaquatica*), 113
Wealth of Nations (Smith), 268
Wearyall Hill, 195, 196
weed, 136
Weiss, Frederick, 81
Weiss, Rudolf, 201
Wellbeing, 52, 128, 285, 295, 328
Wells, H. G., 318
Westcott, Brooke Foss, 194
Western medicine, 264
Westminster Abbey, 143, 196, 197
wetiko, 315–316
Whiteman's Footprint, 115, 117
White-Rumped Vulture, 14
Whitney, Eli, 164
Who Owns England? (Shrubsole), 302
Wickham, Henry, 66
Wild Swans (Andersen), 8
Winnebago, 8, 82, 192
Winnie the Pooh, 75
Winstanley, Gerrard, 219
woodland resonance, 251–252
Wood, Matthew, 84, 119–120
Wootton's report, 151
World Nettle Eating Championship, 41
world population, 332–333
historical plagues and impact on, 333–335
wu-wei, 315

Yaoxing Lun, 49
yarrow (*Achillea millefolium*), xiv, 75, 108, 261–262, 273–274
Aedes albopictus, 92–95
agriculture, 104
as alternative to pharmaceutical drugs, 88–91
antibacterial, 86, 89
anti-carcinogenic, 85, 89
anti-inflammatory, 85, 86, 90, 102, 103
and blood, 78, 80, 81, 93, 84, 87, 93, 98, 105, 107
The Book of Changes, 101
brewing traditions, 107–108
cautions, 104–105
chakras, 101–102
across cultures and conditions, 83–85, 100
ecological and historical significance, 105–106
electromagnetic and nuclear pollution, 95–99
emmenagogue, 89
ethnobotanical, 81–83
growing, 106–107
healing potential, 83–86, 102–104
history and names, 78–81
indigenous uses of, 81–83
insect repellent, 78, 82, 85, 94, 103, 104, 106
mystical powers of, 99–101
nutritional profiles, 323
OTC market for pharmaceuticals, 91–92
phytonutrients, 85–86
protection from ionising radiation, 87
for radiation, 77, 85, 87, 92, 95–99, 102
reduces anxiety, 88
scientific research on, 86–88
skin care, 103–104
Yarrow Environmental Solution (YES), 99
The Yarrow Oracle. See *The Book of Changes*
yew medicine, 55
yew tree, 55, 249
yin and yang, 101
Young, Coleman A., 307

zalzalak, 215
zeaxanthin, 60
zoopharmacognosy, 69
Zoroaster, 140
Zoroastrians, 14, 140
Zuni, 8, 82, 116, 117, 233
Zyklon B, 165

"Nothing is an end because it always can be the basis for something new and different."

—Keith Haring

www.ingramcontent.com/pod-product-compliance
Ingram Content Group UK Ltd.
Pitfield, Milton Keynes, MK11 3LW, UK
UKHW050359040626
471660UK00020B/19